HISTORY

OF THE

12th

(THE SUFFOLK) REGIMENT

1685–1913

J. B. GILPIN,
Ensign to Captain,
1721–38.

LIEUT.-COLONEL
W. H. FORSSTEEN,
1815
(*in the uniform of an*
aide-de-camp).

LIEUT.-COLONEL
W. BELL,
1847
(*in the uniform of the*
Grenadier Company).

LIEUTENANT W. G. SHAFTO,
1818.

COLONEL R. BAYLY,
1828.

Plate 7

1796
LIGHT COMPANY OFFICER.

1796
PRIVATE, BATTALION COMPANY.

Plate 12

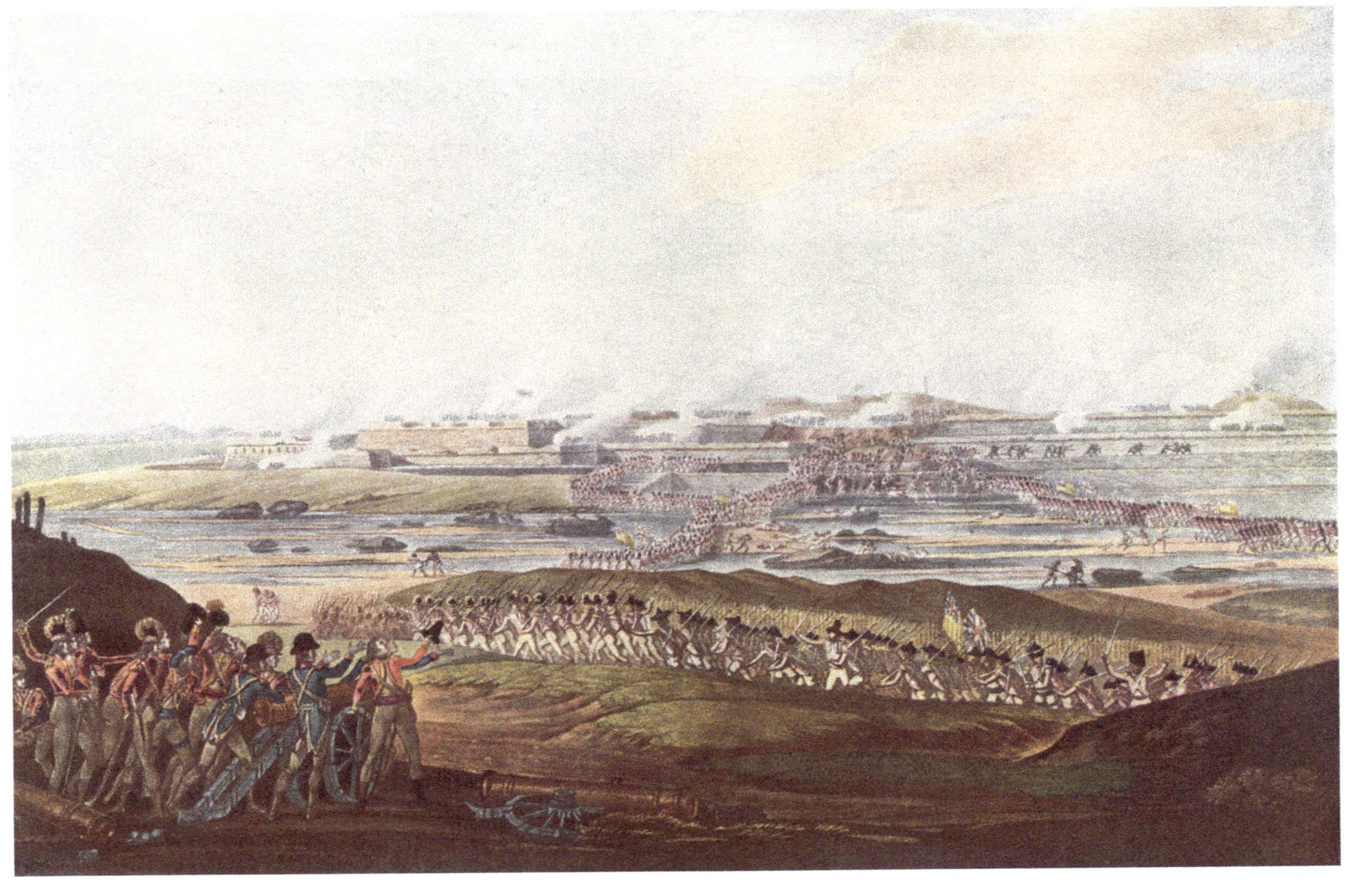

CROSSING THE RIVER CAVERY.—SERINGAPATAM.
(The 12th leading the Left Assaulting Column.)

Plate 13

THE STORMING OF SERINGAPATAM.

By the courtesy of Lieut.-Colonel A. E. Mainwaring, Royal Dublin Fusiliers.

THE SERINGAPATAM GROUP OF TROPHIES.

TAKEN AT THE STORMING OF SERINGAPATAM IN MAY, 1799.

PRESERVED IN THE CHAPEL OF THE ROYAL HOSPITAL AT CHELSEA.

Nos. 31 and 32. Two standards which were borne in state before the Sultans Hyder Ali and Tippoo Sahib.

Nos. 29 and 33. Two pendants belonging to the standards (Nos. 31 and 32) which were fixed in front and rear of the howdah of Hyder Ali and Tippoo Sahib's state elephants.

No. 20. Flag of a French Corps in the service of Tippoo (Republican type).

No. 28. Flag of a French Corps in the service of Tippoo (Monarchical type).

1812–16

OFFICERS,
1825–30.

OFFICER AND MEN, BATTALION COMPANY, SERGEANT, LIGHT COMPANY.
(SUMMER DRESS.)
1834

1846–55

1846–55

1857–61

THE BAND.
1861-71.

1876

1913

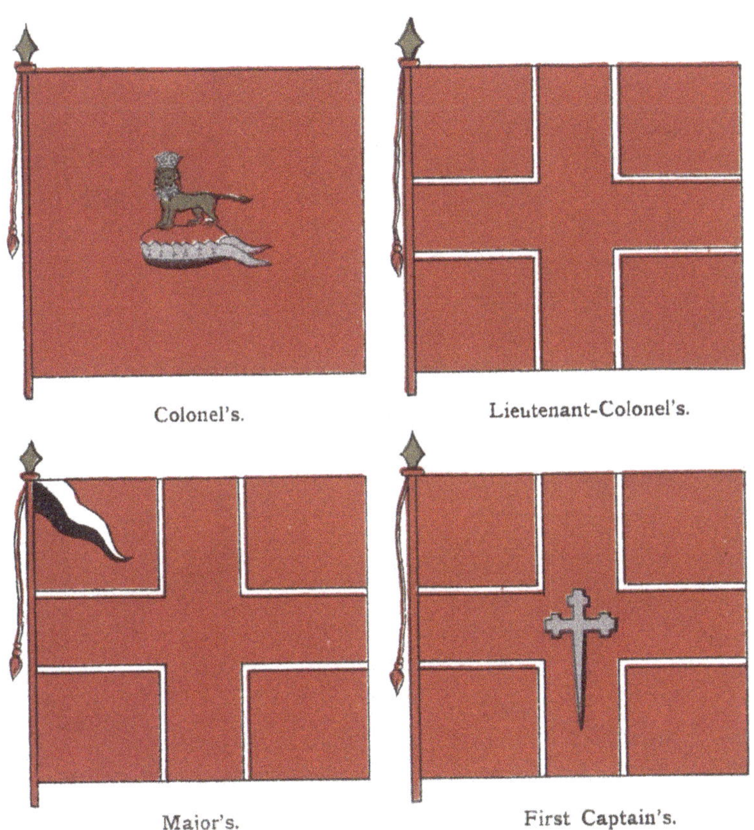

Colours of the Duke of Norfolk's Regiment, 1686.

THE 1ST BATTALION COLOURS

Presented in 1849.

HIS GRACE HENRY, 7TH DUKE OF NORFOLK, K.G.

Raised the Regiment, 20*th June*, 1685.

(*Colonel until June*, 1686.)

(*By the courtesy of His Grace Henry*, 15*th Duke of Norfolk*, K.G.)

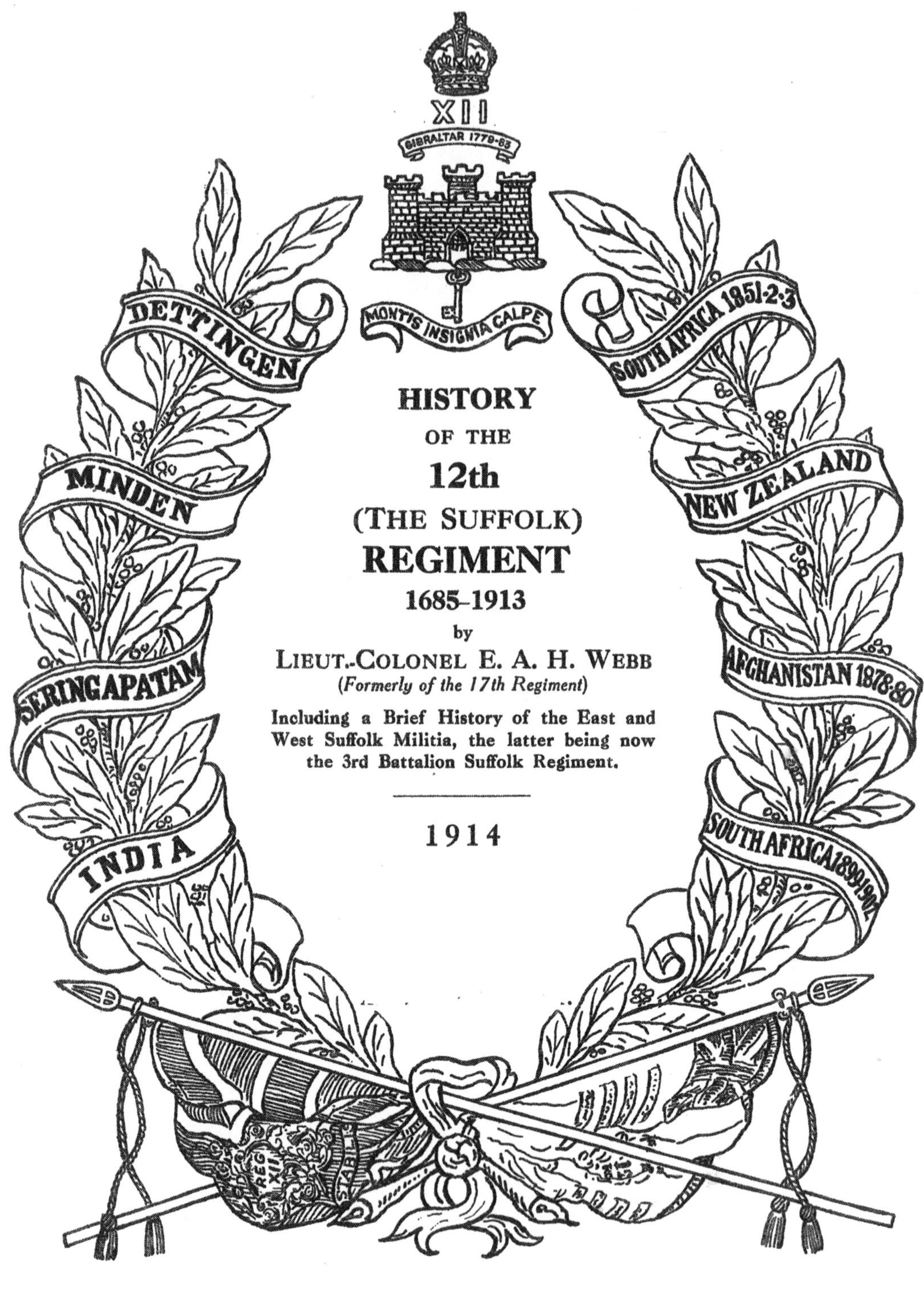

HISTORY
OF THE
12th
(THE SUFFOLK)
REGIMENT
1685–1913

by

LIEUT.-COLONEL E. A. H. WEBB
(Formerly of the 17th Regiment)

Including a Brief History of the East and West Suffolk Militia, the latter being now the 3rd Battalion Suffolk Regiment.

1914

DEDICATED

TO

THE OFFICERS,

NON-COMMISSIONED OFFICERS

AND MEN

NOW SERVING, OR WHO HAVE SERVED

IN THE

SUFFOLK REGIMENT

(OLD 12TH).

THE

12TH

(THE SUFFOLK)

REGIMENT OF FOOT

BEARS ON ITS REGIMENTAL COLOUR

THE CASTLE AND KEY SUPERSCRIBED "GIBRALTAR, 1779–83"
With the Motto, *Montis Insignia Calpe*,

To commemorate the Glorious Defence of Gibraltar
during the Great Siege;

THE WORDS

"DETTINGEN," "MINDEN,"

"SERINGAPATAM" AND "INDIA,"

IN COMMEMORATION OF ITS DISTINGUISHED SERVICES

AT THE BATTLES OF DETTINGEN AND MINDEN.

JUNE 27TH, 1743, AND
AUGUST 1ST, 1759;

THE STORMING AND CAPTURE OF SERINGAPATAM,
MAY 4TH, 1799;

And its Gallant Conduct on many arduous Duties in INDIA,
1798–1807.

ALSO

"SOUTH AFRICA, 1851–2–3," "NEW ZEALAND,"

"AFGHANISTAN, 1878–80," "SOUTH AFRICA, 1899–1902."

PREFACE

THE inception of a History of this distinguished old regiment dates from the autumn of 1910, when Colonel E. Montagu, commanding the 1st Battalion, represented to his brother officers of both battalions the necessity for such a work. The Official History, by Richard Cannon, ending at 1848, not only gives a most meagre account of the regiment's great services, but is undoubtedly untrustworthy in many respects, especially as to dates.

Colonel Montagu's suggestion having met with general approval, a committee of regimental officers was formed, and I, having just then completed the Regimental History of the 17th Foot, had the honour of being invited to write one for the 12th (Suffolk) Regiment, a task which has been considerably lightened by the assistance I have had from the officers. It is doubtful, indeed, whether there is another regiment in the army so rich in personal diaries, journals, and varied reminiscences as the old 12th, all of which have been very kindly placed at my disposal, thereby not only lessening my labours, but helping to impart human touches at intervals, the personal reminiscences of officers being invariably interesting.

With a view to gaining the greatest accuracy, recourse has, in every instance, been had to original despatches, and to the *London Gazette*, whilst the different moves and changes of quarters have been principally obtained from the old official War Office books at the Public Record Offices in London and Dublin, and, for the purpose of completing the Irish service of the regiment, a special visit to Ireland has been made. While no literary merit has been aspired to, it is hoped that the wish of the regiment has now been fulfilled, in the publication of a work containing sufficiently full descriptions of the many sieges and other memorable actions in which it has been engaged, without troubling the reader with too much general history.

The spelling of names throughout, of towns, rivers, etc., is in accordance with the contemporary books or documents from which the information has been obtained.

The officers of the regiment are particularly fortunate in having met with Major E. D. Pickard Cambridge (late Bedfordshire Regiment), a direct descendant of Colonel Scipio Duroure, by whose courtesy, the origin of the "Stabilis" Motto, which had so long puzzled all interested in the history of the 12th, is now made clear (See Chapter XIV, "Notes on Colours"). Colonel Duroure's letters to his sons, and his "Review of the Battle of Dettingen," also very kindly supplied, which, with one exception, have never been published until now, are most valuable assets to military history. Apart from other incidents of the battle, Colonel Duroure's observations (in his constant attendance, as Adjutant-General, on King George II), reveal to the utmost the extreme composure, and the great bravery, of His Majesty, in the heat of action. Details, moreover, of the terrible privations undergone by the British troops for some days after the battle appear to be now brought to light for the first time.

Another piece of luck, just prior to closing the history, must be ascribed to the help of Major C. H. Wylly, Librarian of the Royal United Service Institution, who courteously placed at my disposal the recently discovered valuable General Orders of Prince Ferdinand of Brunswick, issued during the war in Germany (1760–62), thereby giving the regiment possession of a very important one, referring to the bravery of Captain William Picton, at Zierenberg, in 1760.

With reference to the discovery at the Institution, shortly before this, of a number of valuable old documents, which include the above, it may be here stated that the thanks of all army officers interested in research, are due to the veteran Colonel Sir Lonsdale Hale, late R.E. (Chairman of the Council), for his skill and indomitable energy, in so quickly compiling them in book form.

The regiment has been designated in these pages by the names of its Colonels, up to 1751, in accordance with the custom which prevailed until regimental numbers became officially recognised in that year. As a tribute to the soundness and simplicity of the old numbers (which, it is to be hoped, may yet be restored at no distant date) it may be as well here to recall the recommendation of Lord Esher's Committee, in 1903, for their restoration, the Committee's plea being that "the connection between the regiments and the finest pages in the history of the British Army would be re-established, and, at the same time, the great convenience of the numbers—in war time especially—would be regained."

My very special thanks are due to Mr. H. Tregoning, of Ashford, Middlesex, who has given most valuable help, and, had it not been for his unstinted co-operation, it is certain that this book could not have passed through the printer's hands as quickly as it has. His photographs (as an amateur) of regimental ornaments, are, of themselves, works of art.

There are few officers in the army who could ever have worked harder in the interests of their regiments than has Colonel Montagu, on behalf of the old 12th. His marvellous capacity for literary work is well known to all his friends. I have to acknowledge, with grateful thanks, most valuable notes he has sent me, and the officers and men of his old regiment will always greatly appreciate the indefatigable way he has worked for them.

The regiment is fortunate in its happy selection of a most zealous and painstaking Honorary Secretary to the Committee, Lieut.-Colonel Massy Lloyd, now commanding the 3rd Battalion, from whom I have always received the utmost kindness.

All coloured drawings in these pages which bear the initials "P. W. R." will be easily recognised by readers of military histories, as the handiwork of the celebrated military artist, Mr. P. W. Reynolds. The complete collection of notes and sketches he possesses of army uniform, and the close study he has given to the subject, entitle him to be considered, in this respect, one of the leading experts of the day, and I most gratefully acknowledge his kind help.

I am much indebted to the authorities at the Public Record Offices in London and Dublin, to Colonel Sir Arthur Leetham, Secretary Royal United Service Institution; also to Mr. A. D. Cary, Parliamentary Librarian at the War Office, in himself a mine of information, which is often imparted with genuine Irish humour.

Amongst officers of the old 12th, my best thanks, for their ready and generous assistance, are due to Brigadier-General Townley; Colonels Mair, Gardiner, Boyes, Cutbill, Harris, Williams, Cubitt, Routh, Montague and van Straubenzee; Lieut.-Colonels Giles, Wallace, Massy Lloyd, and Brett; Majors Clifford, Crooke and Doughty; Captains Davies, Reid, Walker and Cutbill; Lieutenant and Quarter-Master A. Wills, and others.

The following celebrated old London firms have, with their usual courtesy, given whatever information was in their power, namely, Messrs. Hawkes & Co., 1 Savile Row, W.; Messrs. Jennens & Co., 56 Conduit Street, W.; and Messrs. Herbert & Co., Bedford Street, Covent Garden.

And lastly my thanks are due to Mr. K. R. Wilson, representative of the firm of Messrs. Spottiswoode & Co. Ltd., who has always been most courteous and obliging.

The officers are grateful for many valuable contributions from a large circle of friends of the regiment, including ladies, who have very kindly assisted with information, or by giving access to relics, documents, and portraits. It would be a pleasure to name them all, but the list would be too long.

E. A. H. Webb,
Lieut.-Colonel, Retired.
(Formerly of the 17th Regiment.)

The Following Abbreviations occur as References:—

"W.O." War Office. "F.O." Foreign Office.
"H.O." Home Office. "P.R.O." Public Record Office.
"R.U.S.I." Royal United Service Institution.

CONTENTS

CHAPTER I. 1685–1696.

HOME SERVICE, WARS IN IRELAND, FLANDERS.

CHAPTER II. 1697–1742.

IRELAND, WEST INDIES, ENGLAND, FLANDERS, ENGLAND, SPAIN, MINORCA, AND GREAT BRITAIN.

CHAPTER III. 1742–1751.

FLANDERS, GERMANY, BATTLE OF DETTINGEN, FLANDERS, BATTLE OF FONTENOY, GREAT BRITAIN, HOLLAND, ENGLAND, AND MINORCA.

CHAPTER IV. 1751–1762.

GREAT BRITAIN, GERMANY, BATTLES OF MINDEN, WARBURG, FELLINGHAUSEN, AND WILHELMSTAHL.

PAGE

CHAPTER V. 1763-1783.

GREAT BRITAIN, GIBRALTAR, "THE GREAT SIEGE."

CHAPTER VI. 1783-1798.

GREAT BRITAIN, IRELAND, WEST INDIES, FLANDERS, HOLLAND, GERMANY, ENGLAND, FRANCE, EAST INDIES.

CHAPTER VII. 1799-1817.

SIEGE OF SERINGAPATAM, EAST INDIES, ISLE OF BOURBON, MAURITIUS, IRELAND, ENGLAND, FRANCE.

CHAPTER X. 1867–1878.

EAST INDIES, ENGLAND, IRELAND, EAST INDIES.

CHAPTER XI. 1879–1898.

ENGLAND, EAST INDIES, AFGHANISTAN, EAST INDIES, EGYPT, EAST INDIES, ENGLAND.

CHAPTER XII. 1899–1902.

ENGLAND, EAST INDIES, MALTA, SOUTH AFRICAN WAR, ENGLAND.

CHAPTER XIII. 1903–1913.

ENGLAND, EAST INDIES, MALTA, EGYPT, ENGLAND.

CHAPTER XIV. 1685–1913.

CHAPTER XV. 1660–1913.

APPENDICES

Errata.

Page 1, line 12. *For* "considerably to" *read* "to considerably."
Page 504, Index, 2nd column. *After* line 8, *insert* "Second Battalion reduced, 92, 243."

ILLUSTRATIONS

(The asterisks * denote coloured plates.)

ILLUSTRATIONS IN THE TEXT.

"*Esprit de corps—an attachment to everything connected with their Regiment—ought to exist in the breasts of all soldiers who have served, and are serving, in the Army.*"

HISTORY

OF

THE 12TH (THE SUFFOLK) REGIMENT

CHAPTER I

HOME SERVICE, WARS IN IRELAND, FLANDERS.
1685–1696

1685

AT the death, this year, of King Charles II, his brother James, Duke of York, succeeded to the throne, but his possession of it was soon disturbed by the landing in May and June of two expeditions, which sailed from Holland. The first of these was under the outlawed Duke of Argyll, who arrived off the Scottish coast, with three ships, about the middle of May, his object being to raise Scotland against the King in order to restore its independence. The second expedition consisted of a small force under the Duke of Monmouth, who landed at Lyme in Dorset, in June, with a view to enforce his claim to the throne, on the ground of his mother's supposed marriage with Charles II; the Duke of Monmouth (born in 1649) being the natural son of King Charles by Lucy Waters (Mrs. Barlow).[1]

King James II therefore found it necessary considerably to increase his regular army, with the result that, between June and August this year, twelve regiments of cavalry and nine of infantry were raised, the latter consisting of those in numerical order from the 7th Fusiliers to the 15th Foot. The regiment which eventually became the 12th (in 1751) was raised at Norwich, some of its companies, on formation, being also quartered at Yarmouth and Lynn, and the colonelcy was conferred on His Grace, Henry (7th) Duke of Norfolk (See Frontispiece), dated 23rd June, 1685,[2] though this date seems to have been since officially recognised as the 20th June, 1685.

[1] *Evelyn's Diary.*

[2] *W. O.* 5, Book 1, p. 81.

The following is a copy of the order for raising Henry, Duke of Norfolk's company of Foot:—

"To Our Right Trusty and Right Entirely Beloved Cousin and Councillor, Henry, Duke of Norfolk, Captain of a Company of Foot in the Regiment, whereof he is Colonell." (The order for raising the same verbatim with Capt. Fowler's company.)

These are to authorise you, by beat of Drumm, or otherwise, to raise Voluntiers to serve for Soldiers in your own Company of Our Regiment, whereof you are Colonell, which is to consist of one hundred private Soldiers,[1] Three Sargeants, Three Corporalls, and two Drummers in each Company. And, as the said Soldiers shall be respectively raised in the said Company, they are to be produced to muster, to the intent that they may enter into Our Pay and entertainment, and when that Number shall be fully or near compleated, They are to march to the then Generall Rendezvous of their Regiment, where they are also to be mustered. And you are to appoint such Person or Persons as you shall think fit, to receive Armes for the said Soldiers, and Halberts for the said Sargeants out of the Store of Our Ordnance.

And Wee do hereby require all Magistrates, Justices of the Peace, Constables, and other Our Officers whom it may concern, at the places where you shall raise, march, or rendezvous Our said Company to be assisting therein as there shall be occasion."

By His Mats. Comand,
WILLIAM BLATHWAYTE."

"Given at Our Court, &c.
dated, 23rd June, 1685."

Between the 25th June and the 2nd July, similar orders for raising companies in the Duke of Norfolk's Regiment were issued to Major Trapps, and Captains Mottett, Brathwaite, Trant, Macartney, Howard, Wharton, and Paston.

The following is the first list of officers of

THE DUKE OF NORFOLK'S REGIMENT OF FOOT.

(Taken from Dalton's Army List, Vol. ii, in which all commissions are shown, as dated, 20th June, 1685.)

Colonel Henry, Duke of Norfolk.
Lieut.-Colonel Thomas Salusbury.
Major George Trapps.

Captains: Sir Alphonso Mottett, Francis Brathwaite, Dominick Trant, Chas. Macartney, Chas. Howard, Henry Wharton, Jasper Paston.

[1] The origin of "Private" dates from the time when each borough or smaller subdivision found troops for general service. Individuals (usually land-owners) of over a certain valuation were also taxed, and in certain cases these land-owners contracted to supply men fully equipped and armed, who were distinguished from the other levies as Private-men.

Lieutenants: John Berners (Capt.-Lieutenant), Chas. Houston, Edw. Rupert, Wm. Foster, Alexr. Waugh, Robt. Doughty, Chas. Bill, Robt. Moxam,[1] Robt. Seppins, John Broder.

Ensigns: Jas. Stourton, Thos. Halworthy, John Beverley, Isaac Foxley, Jas. Carlisle, Ferdo. Paris, Valentine Saunders, Miles Burke, Wm. Howard, Terence O'Bryan.

Adjutant, Chas. Houston; Quarter-Master, Jas. Hayley; Chaplain, Wm. Denny; Chirurgeon (surgeon), John Rosse.

A Royal Warrant, dated 25th June, directed Mr Mangridge, Drum-Major-General, "to Raise or Impress 20 Drummers for the Duke of Norfolk's Regiment of Foot."

All Army Orders, including those for marching, were now (and up to a much later period) issued in the form of Royal Warrants, and signed "By the King's Command." The regiment's first marching order, dated 11th July, directed that 4 companies were to remain at Yarmouth, and 1 to proceed to Landguard Fort, and concluded: "And so Wee bid you heartily farewell."[2]

On the 16th July, a reduction was ordered in Foot regiments, from 100 to 60 private soldiers per company, except in the Guards.

Immediately prior to receiving his commission as Colonel, His Grace the Duke of Norfolk had been in command, at Windsor, of one of the numerous Independent Companies which had been embodied for garrisoning our fortified towns, and was shown as follows:—

AN INDEPENDENT COMPANY IN THE GARRISON OF WINDSOR, 1685.[3]

Garrison	Governor	Lt.-Govnr.	Captains	Lieutenants	Ensigns	Gunners
Windsor.	Henry, Duke of Norfolk, Constable.	Chas. Potts.	Thos. Cheek, Henry, Duke of Norfolk.	Alexr. Shenton. Chas. Potts.	Benjamin Bloor. Thos. Dousett.	5

A Royal Warrant, dated 18th July, directed the Duke of Norfolk to disband a company in his regiment, and "that you Incorporate, on or before the 22nd instant, the Independent Company whereof you are Captain into the said Regiment, as also, that the officers deliver in the Armes, Partizans, Halberts and Drumms, belonging to the said Company, to the officers of Our Ordnance, And, for so doing, this shall be your Warrant."

The formation and arming of the regiment were in rapid progress, but were not completed in time for it to take part in the decisive action

[1] Killed in a duel with Captain Henry Wharton of the same regiment, in February, 1686. Dalton, Vol. ii. p. 33.

[2] *W. O.* 5, Book 1, p. 154.

[3] Dalton, Vol. ii. p. 38.

at Sedgmoor,[1] which resulted soon after in the Duke of Monmouth being captured and beheaded.

On the 27th July, a further reduction was made to 2 sergeants, 3 corporals, 1 drummer, and 50 privates in each company.

On the 31st July, the Duke of Norfolk's (12th) Regiment was directed to march from their present quarters to "The Hamlets of Our Tower of London," the company at Windsor excepted, and, on the 3rd August, the company at Landguard Fort rejoined headquarters.

By Royal Warrant of King James II, dated 3rd August, 1685, it was ordained that the following regiments were to have precedence as here ranked :—

1. Our Royall Regiment.
2. Our Dearest Sister, the Queen Dowager's Regiment
3. The Regiment of Prince George, Hereditary Prince of Denmark.
4. Our Holland Regiment.
5. Our Dearest Consort, the Queen's Regiment.
6. Our Royall Regiment of Fusiliers.
7. Our Most Dear and Most Entirely Beloved Daughter, the Princess Anne of Denmark's Regiment.
8. Our Regiment under the Command of Our Trusty and Welbeloved Henry Cornwall, Esqre.
9. Our Regiment under the Command of Our Right Trusty and Right Welbeloved Cousin and Counsellor, John Earl of Bath.
10. Our Regiment under the Command of Our Right Trusty and Right Entirely Beloved Cousin and Counsellor Henry Duke of Beaufort.
11. Our Regiment under the Command of Our Right Trusty and Right Entirely Beloved Cousin and Counsellor Henry Duke of Norfolk.
12. Our Regiment under the Command of Our Right Trusty and Right Welbeloved Cousin and Counsellor, Theophilus Earl of Huntingdon.
13. Our Regiment under the Command of Our Trusty and Welbeloved Sir Edward Hales, Barrt.
14. Our Regiment under the Command of Our Trusty and Welbeloved Sir William Clifton, Barrt.

Troops marching at this period were mostly billeted, every marching order being accompanied by the caution "to take care that the soldiers behave themselves civilly, and pay their landlords," and, to this, in Ireland, was added: "to pay only for their provisions at the accustomed rates, and that the men march orderly and entire, and do not desert their colours."

[1] The 1st Foot Guards were on duty during the Monmouth Rebellion, and the Royal Fusiliers served on the outbreak of it.

The following troops fought at Sedgmoor: Royal Horse Guards, 1st and 6th Dragoon Guards, and a number of Volunteer Troops of Horse. Infantry: A detachment of the 1st Foot Guards, the Royals, the Queen Dowager's (2nd) Regiment, the Queen Consort's (4th) Regiment (5 companies)—Dalton, Vol. ii.

On the 13th August, the regiment, including the company at Windsor, was ordered to march to Hounslow Heath, and whilst encamped there was reviewed by the King.

A Route, dated 25th August, directed it to march to Portsmouth, where it was due on the 29th.

In December, the army had fresh quarters assigned to it, and the regiment marched to Banbury.

1686

On the 1st January, the establishment was fixed at ten companies, with the following numbers and rates of daily pay:—

	£	*s.*	*d.*
Colonel: as Colonel, 12*s.*; as Captain, 8*s.*	1	0	0
Lieut.-Colonel: as Lieut.-Colonel, 7*s.*; as Captain, 8*s.*		15	0
Major: as Major, 5*s.*; as Captain, 8*s.*		13	0
Captain		8	0
Lieutenant		4	0
Ensign		3	0
Adjutant		4	0
Quarter-Master		4	0
Chirurgeon (surgeon)		4	0
Surgeon's Mate		2	6
Chaplain		6	8

Pay of one Company.

2 Sergeants at 1*s.* 6*d.* each; 3 Corporals, and 1 Drummer at 1*s.*; 50 Privates, at 8*d.* each.

All field officers at this period, had, in addition to their field duties, to look after a company.

A Route, dated 21st April, directed the Duke of Norfolk's (12th) Regiment to march to Uxbridge, and, on the 23rd May, to Hounslow Heath. It proceeded on each occasion with a company of grenadiers, which was attached to it until the 11th June.

To overawe the people, King James now assembled at Hounslow Heath a force consisting of 32 squadrons of Horse, 14 battalions of Foot, and 28 guns,[1] a body such as had not been seen massed in England for a long time.

On the 14th June, the Duke of Norfolk resigned, and the colonelcy of the regiment was conferred on Edward, 2nd Earl of Litchfield,[2] by commission of the same date. (See Plate 2.[3])

[1] *Buffs' History*, Capt. Knight.

[2] The name spelt thus at this period.

[3] His portrait (signed by Sir Godfrey Kneller) shows him wearing a blue coat, gold laced, red breeches, with armour on body and arms, and a gold hilted sword.

On the 30th, a grand review was held by the King of the whole of the troops. In King James's letters to the Prince of Orange, he had made much mention of the Hounslow Camp. Referring to this review, he wrote: "They are all good men, and the Horse and Dragoons very well mounted, and all very orderly."[1]

A Return of the Troops in camp, before it was broken up, shows the Earl of Litchfield's (12th) Regiment in the 1st Encampment, and mustering 500 men besides officers.[2]

On the 10th August, the regiment marched to Southwark, whence 4 companies proceeded to Southampton, to embark for Jersey and Guernsey, 2 other companies marching, on the 22nd, to Windsor.

1687

Under a new Act of Parliament, the authorised punishment now for desertion was death, and "Luttrell's Diary" (of an early date) reports a few instances this year of soldiers in different regiments being "shott to death" for "running from their colours," two of these executions having taken place at Covent Garden and Tower Hill respectively.

Luttrell also relates (in the case of a soldier convicted at the Berkshire Assizes of "running from his colours," and sentenced to be executed at Plymouth), that, on the morning of April 29th, "several maids, in white, went and mett the King in St. James's Park, and presented a petition in the soldier's behalf." The sentence, however, was in due course inflicted.[3]

On the 7th August, 2 companies of the regiment, which had proceeded to Sheerness, were ordered to Southwark, and, on the 14th, the 4 companies arriving from Jersey and Guernsey were to join headquarters at Hounslow Heath, where a grenadier company was added to the regiment this summer, and the officers posted were Captain Lord Jermin and Lieutenants George Raleigh and Elie Le Mountais.

The force now encamped included, amongst other troops, 17 infantry battalions, 5 of which were on the right and 12 (including the Earl of Litchfield's) in the centre.[4] On the breaking up of the camp, the regiment moved to Southwark.

1688

On the 16th April, 2 companies marched to Windsor, and, on the 12th May, a detachment of 80 men, with officers in proportion, was also ordered there until the 25th, in temporary relief of 5 companies of the Queen Dowager's (2nd) Regiment.

On the 22nd June, the Earl of Litchfield's (12th) was directed to march to Hounslow Heath.

[1] King William's Chest (formerly a sealed bag), 1674–86.
[2] *W. O.* 26, Book 6, p. 203.
[3] *Luttrell's Diary*, pp. 400–5.
[4] Dalton.

EDWARD HENRY, EARL OF LITCHFIELD.
Colonel, June, 1686, *to November*, 1688.

(By the courtesy of the Viscount Dillon.)

King James II had a bigoted and passionate affection for Popery, and had now by bad policy alienated the hearts of his subjects, by violating their laws, confiscating their property, and persecuting their religion, the result of which was general discontent, even extending to the army. The King dreaded this, and being anxious to ascertain how far he could depend on his troops for support in the abrogation of the religious tests, he selected the Earl of Litchfield's Regiment as one most likely to support him, by reason of there being so many Roman Catholics in the ranks. The men were accordingly drawn up in the King's presence, and the major (George Trapps), a Roman Catholic, explained to them that His Majesty wished them to sign an agreement, binding them to assist him in carrying out his intentions, and all who did not wish to comply were directed to lay down their arms.

To the King's surprise and disappointment, the whole regiment grounded arms, with the exception of two officers and a few soldiers who were Roman Catholics. After a few moments, he bade the men take up their arms, remarking, "Another time, I shall not do you the honour to consult you."

At last, the English nobility and gentry, disgusted with the conduct of the King, invited the Prince of Orange from Holland to dethrone him.

On the 7th August, the regiment returned from the camp to Southwark, having been directed to leave a detachment at Hounslow Heath "as a guard on His Majesty's quarters and the chapel," and, on the 21st and 22nd September, detachments of 50 men in each were ordered to Windsor and Sheerness respectively; the headquarters, on the 27th, marching to Dartford, and supplying a detachment to Erith, whilst those at Windsor and Hounslow Heath rejoined.

A Royal Warrant, dated 23rd September, directed that the Earl of Litchfield's (12th) Regiment was to be increased by ten men per company, to be raised by Beat of Drum; and each company to consist of 3 sergeants, 3 corporals, 2 drummers, and 60 private soldiers.[1]

It was directed on the 3rd October, that three men per company of the Earl of Litchfield's (12th) Regiment (the grenadier company excepted), were to be sent "on board such Vessels as shall be appointed, to carry them to Our ffleet, where they are to be employed in Our Service, their vacancies being forthwith filled by recruits."

On the 16th October, the regiment was ordered to Rotherhithe and Southwark, on the 7th November to Egham, and, on the 16th, to Winchester, to receive orders from the Earl of Faversham. On the 20th, two additional companies, which had been raised, were to join the headquarters at Amesbury, and, on the 22nd, the regiment was ordered to march to Portsmouth, which destination, on the following day, was altered to Putney and places adjacent.

The Prince and Princess of Orange landed in England in November,

[1] *W. O.* 26, Book 6, p. 225.

and, soon after, the Earl of Litchfield, on transfer to the 1st Foot Guards, was succeeded in the colonelcy by Robert, Lord Hunsdon, whose commission was dated 30th November, 1688.

On the 29th December, the regiment was directed to march to Leicester, there to join Sir John Guise's Regiment, and to detach single companies to York, Tynemouth, Scarborough Castle, and Hull respectively.

Lord Hunsdon having refused to take the oath of allegiance to the Prince of Orange,[1] His Highness, on the 31st December, conferred the colonelcy of the regiment on Henry Wharton (who had raised a company on its formation), described as "a gallant officer, possessing the confidence and affections of officers and men of the regiment." At the same time, Captain Richard Brewer, from Sir E. Hale's (14th) Regiment, was promoted to the lieut.-colonelcy.

The clothing and equipment for the army, except the Ordnance Corps, were, from the earliest period up to 1855, supplied by the colonel, who received a fixed sum for every man on the establishment, which was termed "off reckonings,"[2] and any balance remaining, after paying the cost of the articles, was a source of emolument. The first regulations on this subject were issued this year by King James II.[3]

1689

On the 3rd January, the regiment was ordered to relieve Lord Montgomery's at Hull, and, on the 9th February, the officer commanding the company at Scarborough was directed "to take the castle into his possession, and to quarter there."

King James fled to France early in the year, and, on the 13th February, the Prince and Princess of Orange were elevated to the throne by the titles of King William and Queen Mary, with great pomp. The sum voted for the expenses of the regiment this year, including clothing, was £16,145 3*s.* 4*d.*, and the rates of daily pay for all ranks were the same as in 1686.

The regiment was to consist of 13 companies (including one grenadier company) of 60 men in each, total 780, besides officers; and the payment of levy money for each recruit raised was 40 shillings. The following was the weekly subsistence allowed to a foot regiment:—

	s.	*d.*
To a Lieutenant	14	0
,, an Ensign	10	6
,, a Sergeant	6	0
,, a Corporal or Drummer	4	6
,, a Private	3	6

[1] The colonel, the lieut.-colonel, the major, and four senior captains adhered to James II at the Revolution, and Captain Sir A. Mottett was, in 1689, appointed major of Colonel Dan O'Donovan's Regiment of Foot, in King James's Irish Army.

[2] "Off Reckonings," a specific account, so called, which existed between the Government and the colonels for the clothing of their men.

[3] Old Pimlico Regulations.

On the 12th March, James landed at Kinsale from France, with about 1800 men; on the 18th April, he started for Ulster, and, after a short interval, went to Dublin, where he was received with every demonstration of joy by Lord Tyrconnell, and all the Popish party. The Earl of Tyrconnell had, during King James's reign, been forming and modelling an Irish Army that might be ready to serve the Popish interest, and with such success that he was able to attach to his cause 4 or 5 regiments of cavalry and 19 of infantry, with 20,000 stand of arms and several guns.[1]

On the 27th May, the company of Colonel Wharton's (12th) Regiment at Scarborough was moved to Pontefract, and, on the 1st June, the remainder of the regiment was ordered to Wakefield, Leeds, and Halifax, further orders, on the 5th, directing a move to Doncaster, Rotherham, and Sheffield. Two Routes were issued, on the 29th, for the regiment to march from Doncaster and Sheffield to Manchester.

James's first act, on the offensive, had been to besiege Londonderry, which, having refused to surrender, made a gallant defence until the 4th August, (the siege lasting 105 days,)[2] when the besiegers, finding that, the town being relieved, there was no hope of its capture, raised the siege and retreated.

King William now resolved to make strenuous efforts to end the war in Ireland, and placed the Duke of Schomberg in command of the troops he was sending there.

The *London Gazette* of May 16th, 1689, shows that, on that date, Frederick, Count de Schomberg, General of His Majesty's Forces, was created by the King, "a Baron, Earl, Marquis and Duke of this Kingdom, by the names and titles of Baron Teys, Earl of Brentford, Marquis of Harwich and Duke of Schomberg." By birth, a German and of good family, he was now 81 years of age, but still hale, active, and soldierlike, and his seat on horseback was reported to have been the envy of every cavalry officer.[3]

At about 6 A.M. on the 12th August, the Duke of Schomberg, accompanied by a large staff, embarked in sixty or seventy ships at Hoylake, with a force of 10,000 Horse and Foot, which included Colonel Wharton's (12th) Regiment, and, arriving in Bangor Bay on the 13th, the troops disembarked and encamped near the shore.

On the 14th, the garrison of Carrickfergus, apprehending a siege, burnt their suburbs, and, on the following day, the Duke sent Sir Charles Fielding with a party of 250 men to inspect the position of the Irish at Belfast, when it was found that the enemy had retired to Lisburn, whereupon the Duke sent Colonel Wharton's (12th) to take the place.[4] On the 17th, the General marched with his army to Belfast, and, on the 20th, he sent 5 Foot regiments to Carrickfergus (the advance guard consisting principally of Colonel Wharton's

[1] Lieut.-Colonel Macartney-Filgate, p. 7.
[2] *Ibid.*, p. 8.
[3] Clifford Walton, Vol. i.
[4] Story.

(12th) Regiment),[1] followed, the next day, by seven more regiments, which nearly surrounded the place.

The Duke having refused the terms of a parley which were submitted to him, the siege of Carrickfergus began, and, on the 22nd, a fleet of 50 ships brought him a further reinforcement of 4 Foot regiments and 1 of Horse. Another parley ended in the continuance of the siege; on the 27th, the garrison surrendered, and, escorted by a party of Horse, were allowed to march on the 28th, with their arms and some baggage, to the nearest Irish garrison, which was Newry. A breach had been made in the wall of Carrickfergus, and, whilst the Irish were discussing the terms of surrender, Colonel Wharton (12th) lay before the breach with his regiment, and was ready to enter, when the Duke sent to command his men to forbear firing, from which they were restrained with some difficulty, having been anxious to enter by force.[2] The siege had been aided by ships in the harbour, and the losses were about 150 killed and wounded on each side.

On the 28th and 29th August, our army marched to Belfast and encamped, and, on the 31st, it was mustered as follows:—

HORSE.

Earl of Devonshire's Regiment
Lord Delamere's ,,
Colonel Coy's Regiment (5th Dragoon Guards)
Duke of Schomberg's Regiment (7th Dragoon Guards)
Colonel Leveson's Dragoons (3rd Hussars)

Total of Horse, five regiments.

FOOT.

Princess Anne's Regiment		(8th)
Colonel Wharton's	,,	(12th)
Earl of Meath's	,,	(18th)
Colonel Erle's	,,	(19th)
Colonel Herbert's	,,	(23rd)
Sir Edward Deering's	,,	(24th)
Duke of Norfolk's[3]	,,	(22nd)
Earl of Drogheda's	,,	
Earl of Kingston's	,,	
Lord Lisburn's	,,	
Lord Roscommon's	,,	
Lord Lovelace's	,,	
Sir Thos. Gower's	,,	
La Melonière's	,,	The three French Huguenot Regiments
Du Cambon's	,,	
La Callimot's	,,	

One Battalion of the Blue Dutch Regiment.
One Battalion Col. Carlesan's White Dutch Regiment.

Total, eighteen regiments.

The Duke now gave orders for the English artillery train to be sent by sea to Carlingford. On September 2nd our army marched beyond Lisburn,

[1] Lieut.-Colonel Macartney-Filgate. [2] Story. [3] See page 452.

and, on the 5th, to Newry, which had been burnt down. The General, not knowing this, nor that the enemy had deserted the pass to it, gave orders for 170 men from each regiment of Foot, with a party of Horse and 4 guns (in all 1200), under Colonel Wharton (12th), to march at 3 A.M. on the 5th, followed by the remainder of the army, and a halt having been made on the 6th, the Duke issued a warning to the Irish that, if they burnt down any more towns, he would give no quarter.

On the 7th, the army marched to Dundalk and encamped about a mile from the town on low moist ground. Here, the force was fortunate in securing about 2000 sheep, there having been hitherto a great scarcity of provisions in general, and, even now, bread was so scarce as to necessitate orders being issued that whatever bread could be procured was to be exclusively for the men.

News was now received that part of James's army was at Drogheda. Having then advanced from Ardee, his outposts were constantly scouting up to our camp, and, on the 21st September, his whole army appeared, displaying the Royal Standard. The British troops stood to their arms, and some regiments were ordered to the trenches, but, beyond that, the Duke declined to entertain the idea of fighting, and merely watched the enemy's movements until they withdrew.[1]

On the 1st October, a move was made to higher ground, and later huts were built. Hitherto Schomberg had marched through a country full of bogs and mountains, where the enemy's cavalry could not annoy him. The fleet on which he depended for artillery and provisions had not yet reached Carlingford. His men, undisciplined and ignorant, had already experienced the hardships of their present service, wasted by a fatiguing march in rain and tempest, in cold and hunger, through a country dispiriting in its aspect, and rendered still more dreary and depressing by the inclemency of the weather. Several had sunk under these difficulties; a contagious fever made havoc amongst them. The sick lay languishing on the roads, and gave full employment to parties detached from every regiment in the army to collect and convey them to the camp. In such circumstances, Schomberg deemed it imprudent to advance. The enemy were elated at the intelligence that he had halted. Marshal Rosen of James's army immediately drew his forces nearer to Dundalk, while the Duke fortified his camp so as to make it impossible for the enemy to force him to action.

On the 12th October an order was issued appointing a "reserve guard" to every regiment, consisting of a captain, lieutenant, ensign and 50 men, to be always ready to turn out at short notice. About 27 ships having now come to Dundalk, it was ordered, on the 13th, that all the sick should be sent on board of them; but the contagion had spread too widely and raged too violently to be subdued. The doctors, moreover, had been supplied with medicines and appliances intended only for the cure of wounds.

[1] Story, p. 23.

Every regiment in those days had a surgeon, who received the somewhat moderate fee of 4*s.* a day, and he had two assistants, who received 2*s.* 6*d.* a day, their business being chiefly dressing wounds. They had apparently no knowledge of anything referring to disease or general health, and the care of the troops was confided to apothecaries who received the magnificent sum of 1*s.* a day.[1] The English troops, unaccustomed to severities, confined to a low and moist station, drenched with perpetual showers, and without sanitation, or the relief necessary in sickness, died in great numbers, and several of their most distinguished officers caught the infection, languished, and died.

Colonel Wharton's (12th) Regiment sustained a very serious loss in non-commissioned officers and men, and, on the 28th October, their commanding officer, the gallant Colonel Henry Wharton, died. Story states: "Colonel Wharton was a brisk bold man, and had a regiment that would have followed him anywhere, for the officers and men loved him, and this made him ready to push on upon all occasions." Colonel Sir Thomas Gower died on the preceding day, and the remains of these two officers were interred on the 30th October, in Dundalk Church, their regiments attending and firing three volleys.

In succession to Colonel Wharton, King William promoted Lieut.-Colonel Richard Brewer to the colonelcy, by commission dated 1st November, 1689. On the 16th October the greater part of our Horse marched towards Carlingford, for the convenience of getting forage.[2] The Duke's inactivity, however, had caused great discontent amongst his troops, who accused their General of an intention to prolong the war, and of indifference to their unhappy state; they also took exception to his great age.[3]

While the enemy prepared to retire into winter quarters in November, Schomberg was reinforced by some regiments from England and Scotland. To prevent these troops from catching the infection, he resolved to abandon his fatal station, and temporarily pitched a new camp beyond the town.

The sick had been ordered on board the ships, which, however, could not contain their numbers. Waggons were provided to carry them to Belfast, and the officers were employed in attendance on them, some of the men dying on their first attempt to move, whilst the General, deeply afflicted at this scene of wretchedness, stood for hours at the bridge of Dundalk, directing every means for alleviating the miseries of his men.

Among the private papers of King William the following particulars are found, of the losses of Colonel Wharton's (12th) Regiment during its stay at Dundalk. The Return shows Colonel Wharton "dead," and the following establishment of officers, viz:—2 majors, 9 captains, 12 lieutenants, and 12 ensigns, but there are no remarks as to whether all were effective.

[1] Lieut.-Colonel Macartney-Filgate. [2] Story. [3] Leland.

The total of "effective men" is shown as 136 only, also 281 sick and 363 dead.[1]

Some regiments had not above 60 men left effective. The Continental regiments, on the other hand, lost but few; one regiment of Dutch infantry lost only 11. To make matters worse, a formidable conspiracy broke out in the Duke's camp, which he dealt with by shooting six of the conspirators and deporting all the rest.

While Schomberg's men had been dying at Dundalk, James's soldiers were being carried off equally rapidly at Ardee and Drogheda, where, out of 40,000 men, nearly 15,000 died.

In King William's Chest there is an interesting paper, giving the names of all regiments placed in brigades at Dundalk, entitled "Liste de l'Infanterie passe en Revue a Camp de Dundalk, le $\frac{28}{18}$ d'Octobre, 1689." In this list, Colonel Wharton's (12th) Regiment is placed in the 2nd Brigade, the other corps in the brigade having been Stuart's (9th), Herbert's (23rd), and Kingston's and Gower's Regiments.

The following is a copy of the document relating to Colonel Wharton's Regiment, the report in French, at the end, being also a true rendering of the original.

CAMP DE DUNDALK, LE $\frac{28}{18}$ D'OCTOBRE[2]
2 Brigade.

Wharton.	Lieuts.	Anseig.	Sargant.	Tambour.	Soldats. Au filde Rangs.	Soldats. Malades.	Soldats. Morts.
1. Grenadier. Captain Beverley.	Malade	1	2.1.s.	2	39	17	1
2. Colonel	Malade	1	3	2	30	19	9
3. Lieut. Col. Barnes	Malade	1	3	1.1.s	36	18	0
4. Major Dowsett	Malade	1	3	1.1.s	34	18	3
5. Colt	1	1	2.1.s.	2	39	11	1
S.6. Smith	1	1	3	1.1.s	30	11	2
7. Bayrons	Malade	1	2.1.s	1.1+	39	15	3
8. Livesay	1	Malade	2.1.s	2	35	12	2
9. Baron	1	1	3	1.1+	30	29	3
S.10. Purefoy	1	1	3	1	35	19	5
11. Carlisle	Malade	Malade	2.1.s	2	37	13	2
12. Kempton	Mort	1	3	2	37	14	2
13. Seppens	Malade	0	2.1.s	1.1.s	25	23	9
2.S. Summa[3]	s.7. Six	10.2.s	33.6.s.	19.4.s2+	441	209	32

Ce Regiman a beaucoup des bons Hommes ausis for bien vestus. Le Colonel a ausis anvoyé au Escosser pour facher d'avoire des surtouts. Il prent ausis ases de soin des Regiman, mais la mechante compagnie qui frequante, le gate, et fest que for souvant, il est an deboche, et a trop boire. Les officiers sont entre deux. La compagnie des Grenadiers a esté trez belle, et le feras ancore, quand seulement les Malades seront Geris.

[1] King William's Chest, No. 13.
[2] *Ibid.*, No. 6.
[3] This peculiar rendering of the totals, many of which are wrong, no doubt refers to duties with other companies.

The list of our own army, dated November, 1689, was as follows :—

HORSE AND DRAGOONS.	Troops
Lord Devonshire	6
,, Delamere	6
,, Hewett	6
Colonel Coy (5th Dgn. Gds.)	6
,, Langston (4th Dn. Gds.)	6
,, Villiers	6
,, John Lanier	6
Duke of Schomberg's French	9
Colonel Wolseley's Inniskillings	12
Mr. Harbord's Troop	1
Captain Matthew White's Troop	1
Provost-Martial's Troop	1
Captain Hifford's Dragoons	9
Colonel Levison's ,, (3rd Huss.)	6
Sir A. Cunningham's Dragoons	6
Colonel Guronne's ,,	6

FOOT.

Major-General Kirk
Colonel Sir John Hanmer (11th)
Brigadier Stuart
Colonel Beaumont (8th)
,, Brewer (12th)
,, Lord Meath (18th)
,, ,, Kingston
,, ,, Drogheda
,, Sir Henry Bellasis
,, ,, Henry Ingoldsby
,, Zancby
,, Lord Roscommon
,, ,, Lisburn
* ,, Hamilton
* ,, Hastings
,, Deering (24th)
,, Herbert (23rd)
,, Sir Thos. Gower (late)
,, Erle (19th)
,, La Melonière
,, Du Cambon
,, La Callimot
A battalion of Blue Dutch
Carleson's White Dutch.

INNISKILLING AND DERRY FOOT.

*Colonel Gustavus Hamilton
* ,, Lloyd
* ,, White
Colonel Mitchelbourne
* ,, St. John
,, Tiffin

* Those marked with an asterisk were not at Dundalk, but in garrison.

Early in November both camps were struck, and on the 8th and 10th, our wretched troops, on leaving Dundalk, prepared to march into winter quarters.

On the 7th, four regiments were ordered towards Newry, and Colonel Brewer's (12th), with the regiments of Beaumont (8th), Hanmer (11th), Deering (24th), Drogheda, Bellasis, and Roscommon, towards Armagh. Story relates that several of the men, unable to march, were left to die. On the 9th, the rear of our army marched from Dundalk.

The English frontier, while the troops were in winter quarters, was formed by Lough Erne, and by garrisoned posts, stretching from it to Newry and Belfast.

On the 12th December, Colonel Wolseley, with a detachment of Inniskilling troops, took Belturbet, from whence, next year, he was able to inflict a disastrous defeat on the Irish troops at Cavan.

1690

Colonel Brewer's (12th) having, with the "Queen's," marched to reinforce Colonel Wolseley at Belturbet, the first operation in which these regiments were engaged this year was to put the place in safety by covering its approaches with entrenchments, and strengthening the garrison.

Information having been received by Colonel Wolseley that the Irish had resolved to attack him there, and were assembling troops at Cavan, eight miles off, he determined at once to attack them, and, on the night of the 10th February, left Belturbet with a mixed force, in the hope of engaging, if possible, the Irish Commander (O'Reilly) at Cavan, before the Duke of Berwick could arrive there. Wolseley's force only mustered 300 Horse and 700 Foot, from detachments of the Inniskillings, the "Queen's," and Brewer's (12th), whereas the Cavan garrison mustered 2000 before Berwick's arrival with 1700. On the night that Wolseley started from Belturbet, Berwick marched into Cavan.

Wolseley was marching with his Inniskillings as an advanced guard, and as he neared the place, the Irish cavalry charged his front, whilst their infantry opened a cross fire on his flanks. The Inniskilling Horse were thrown into confusion, the Irishmen pursuing, and driving them amongst our own infantry. So furious were the latter at the idea of running from the Irish, that some of the "Queen's" and Brewer's (12th) fired on the fugitives, and killed seven or eight of them. Wolseley now brought forward his Foot, the Irish retiring on their main body, and when the English emerged on more open ground, they deployed into lines before moving up the hill. A volley from the Irish proved so ineffective that, before they had time to re-load, the English poured into them a heavy fire at a short distance, and they fled forthwith. The English pursued, and following them into the town, began to loot. From this, their officers could not restrain them, until the idea was conceived of firing the town, and thus compelling them to quit the burning houses. Fortunately there was a reserve of about 300 men who managed to keep the enemy at bay until the plunderers rejoined, when the Irish infantry were driven back like sheep into the fort, while their cavalry disappeared altogether. The enemy's magazines were then blown up, their stores destroyed, and Wolseley returned to Belturbet unmolested, his men having been on the move since 4 P.M. the previous dav. The enemy's loss was

10 officers and 300 men, besides 12 officers and about 60 men taken prisoners, our loss having been 3 officers and about 30 men killed, and Captain Blood (engineer) and a French officer wounded.

Throughout the winter constant incursions were made by both sides on each other's camps. On the Irish side, these predatory measures were mostly carried on by Rapparees, of whom the Rev. G. W. Story, (the most reliable historian of the war,) gives an account.

They were country people, not of the army, who had constituted themselves thieves and vagabonds, not caring to work or take up any employment, a mixture of Irish with other nations, who, herding together, took every opportunity, especially when in any numbers, to plunder, burn, and murder, their hands being against all, and all hands against them, to destroy as beasts of prey. Consequently, in time of war, not being in concert with their avowed friends, they were a distinct annoyance to both sides. Some were armed with half pikes, and some with scythes or muskets. Story relates that, "as early as 1684, the priests would not allow an Irishman to go to mass unless he brought his Rapparee with him, which signified a half stick or broken beam resembling a half pike." It was from that the men got their name, whilst some called them "Creaughts" from the little huts they lived in, built so conveniently with hurdles and long turf, removable in summer to the hills, and in winter to the valleys. Their principal stronghold was the Bog of Allan, though they overran the country wherever damage could be done to the allies. Their skulking places were bogs, woods, or mountains, and their apparel, unless they rigged themselves by plunder, was so miserable, that they went, in a manner, naked.

When they feared detection, their cunning was so great that they would sink down into long grass or some other cover, and, dismounting the locks of their pieces, would conceal them in some dry spot or in their clothes; or, stopping up the muzzles with corks and the touch-holes with small quills, would throw away the firelocks into a pond or some other equally secure place.

One might see a hundred of them without arms, resembling the poorest, humblest slaves in the world, and they might have been searched without a musket being found; yet, when they had a mind to do mischief, they could all be ready at an hour's warning.[1]

At the beginning of March, the English received a reinforcement, by the arrival at Carrickfergus under the Duke of Wurtemburg, of 6000 Danes,[2] and, on the 14th March, James's army also received a reinforcement of 5000 French Foot, under Count Lauzanne and the Marquis de Levy, five Irish regiments having been sent to France in exchange.[3]

[1] Story's Wars. [2] *Ibid.*, pp. 96–7. [3] Colonel Davis's History, *The "Queen's."*

A LIST OF JAMES'S ARMY, TAKEN APRIL 19TH, 1690.

REGIMENTS OF HORSE.

Regiments		
Duke of Tyrconnell's Regiment Lord Galway's ,, Colonel Sarsfield's ,,	Six troops in each and 53 men in a troop	954
,, Sutherland's ,, Lord Abercorn's ,, Colonel Henry Luttrell's ,, ,, John Parker's ,, ,, Nicholas Purcell's ,,	As above	1,590
Lord Dungan's Dragoons Sir Neal O'Neal's Dragoons Colonel Simon Luttrell's Dragoons	Eight troops in each and 60 men in a troop	1,440
Colonel Robert Clifford's Horse Sir James Cotton's ,, Colonel Thomas Maxwell's ,, Lord Clare's ,,	Six troops in each and 60 men in a troop	1,440
	Total of Horse	5,424

FOOT.

The Royal Regiment (22 companies and 90 men in each). . . . 1,980.
The Earl of Clancarty's Regiment
Colonel Henry Fitzjames' ,,
,, John Hamilton's ,,
The Earl of Clanrickard's ,,
,, Antrim's ,,
,, Tyrone's ,,
Lord Germanstown's ,,
,, Slane's ,,
,, Galloway's ,,
,, Tonth's (?) ,,
,, Duleek's ,,
,, Kilmallock's ,,
,, Kinmare's ,,
Sir John Fitzgerald's ,,
,, Maurice Eustace's ,,
Colonel Nugent's ,,
,, Henry Dillon's ,,
,, John Grace's ,,
,, Edward Butler's ,,

Colonel Thomas Butler's Regiment
Lord Bophins' ,,
Colonel Charles Moor's ,,
,, Cormack Neal's ,,
,, A. Macmahon's ,,
Earl of Westmeath's ,,
Colonel Cavanagh's ,,
,, Uxbrugh's ,,
,, Macarty Moore's ,,
,, Gordon O'Neal's ,,
,, John Barrett's ,,
,, Charles O'Bryan's ,,
,, O'Donovan's ,,
,, Nicholas Brown's ,,
,, O'Gara's ,,
Sir Michael Creagh's ,,
Colonel Dom Brown's ,,
,, Bagnal's ,,
,, Mackellicut's ,,
Lord Inniskillin's ,,
Colonel Hugh Macmahon's ,,
,, Walter Bank's ,,
,, Felix O'Neal's ,,
,, Lord Iveagh's ,,
,, O'Kelly's ,,

44 regiments of Foot of 13 companies in each, and 63 men per company . . 36,036

Total of James's Army. . 43,440

In the spring, a numerous body of recruits from England replaced the losses of the regiment, and, in June, it brought 500 musketeers, 160 pikemen, and 60 grenadiers into the field, to serve under King William, who, on the 14th, arrived at Carrickfergus, where he was met by the Duke of Schomberg, Major-General Kirk, and other officers. At Loughbrickland, his army assembled, on the 21st, from their different quarters, and were joined by the King and his suite. On the 25th, he marched towards Dundalk, where, on the 27th, the whole army joined—English, Dutch, Danes, Germans and French—in all, not above 36,000, and encamped about a mile beyond the town. The Irish Camp was at Ardee, and James crossed the Boyne on the 28th. On the 29th, King William reviewed the Danish troops, and, at daybreak on the 30th, marched towards Drogheda, where he found James's army encamped along the River Boyne, but, it being late before the artillery and the Foot arrived, he could do no more that day than inspect the enemy's position and the fords around, which he found very difficult to pass. Whilst thus occupied, he received a flesh wound on the right shoulder, large but not deep, only raising the skin, and, as soon as it was dressed, he mounted his horse and remained on horseback for four hours.[1]

At 9 P.M. a council of war was held, when the King ordered his right wing, under command of Count Schomberg (one of the Duke's sons), to march to the bridge of Slane and cross there, by way of turning the left flank of the Irish army, whilst he, at the head of the left wing, which was composed exclusively of cavalry, was to pass the river not far above Drogheda, and the main or centre attack,[2] which consisted of a large body of Foot, under the Duke of Schomberg, was to force the position at Oldbridge.

The King, in spite of his wound, and his delicate state of health, after all was arranged, rode through the camp at 12 at night, with torches, for a final inspection, and then retired to his tent, impatient for the eventful day.[3]

Orders had been given that every soldier was to be provided with a good stock of ammunition, and also a green bough or spray in his hat to distinguish him from the enemy.

BATTLE OF THE BOYNE.

The 1st July dawned clear and bright, and King William's army was on the move at daybreak.

The following account of the battle is taken from the *London Gazette* dated July 3rd, 1690.

> "Count Schomberg, upon his arrival (at Slane) found eight squadrons of the enemy ready to receive him, but in a little time, and without much resistance, he beat them off and passed the ford, driving them before him, and drew up his men on the other side, ready to march, on receipt of the King's orders, after making his report. Upon this, the enemy detached a great number of their troops, who formed up for action. His Majesty, understanding that the right wing had passed the ford, commanded three attacks to be made. The first, at a good ford, before a small village, which the enemy had advantageously possessed; at the second ford, the Foot waded up to their armpits; at the third, the horses were fain to swim. The Dutch Regiment of Foot Guards passed over first, with the water up to their waists, sustaining well the enemy's fire. Thereupon, all who were in the village and the ditches fell back. Before, however, the 3rd Battalion of the same corps had passed the ford, five of the enemy's regiments had advanced to within a pike's length, but our fire was so severe, that they were forced to retire with the loss of a great many men and one of their Colours. After this, our men, advancing beyond the village, were twice vigorously attacked by the enemy's Horse, but all in vain.
>
> In the meantime, the Danes reinforced our left, and the brigades of

[1] *London Gazette*, 3rd July, 1690.

[2] The following British regiments constituted the main, or centre, attack, viz.: Columbine's (6th), Beaumont's (8th), Hanmer's (11th), Brewer's (12th), Meath's (18th), G. Hamilton's (20th), and Lord George Hamilton's (22nd).

[3] Harris's *Life of William III*, Vol. iii. p. 71.

The Battle of The Boyne 1690.*

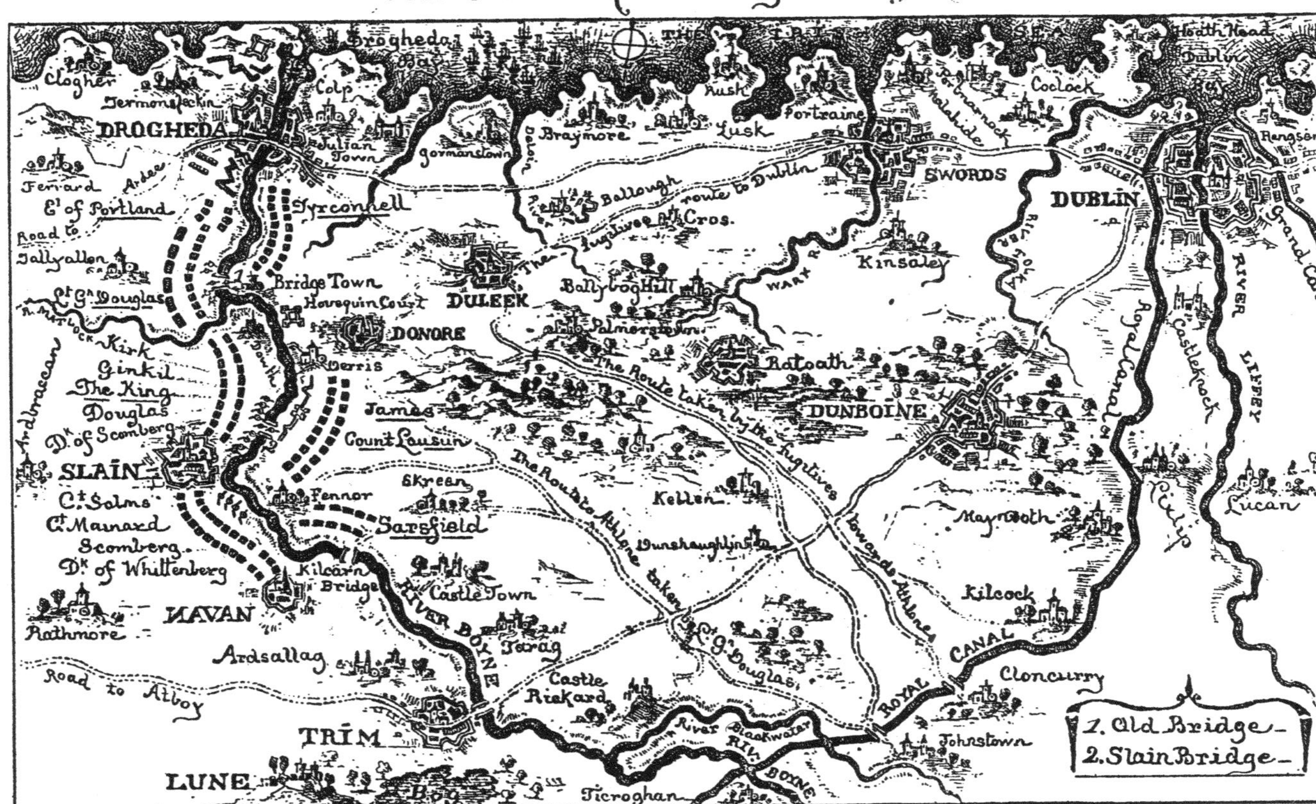

Compiled from The Battle of ye Boyne. London, 1693. 4° Fol. 22, B.M.
Battle of the Boyne, English Hist. Prints, 1690, B.M.
Two Views of the Boyne Battle, 12905 (1), B.M.

Maps Consulted { Boyne River, Map Catalogue, King's Library, 54–39, B.M.
The Battle of ye Boyne, Story's History of the Wars in Ireland.
and Author's MSS.

* *History of the 2nd "Queen's,"* by Colonel Davis. By kind permission.

> Hanmer and Melonière our right. The former were attacked by the enemy's Dragoons, and the latter by their Horse, and, in this interval, 30 officers and others, of the enemy's Life Guards, advancing, were all killed but five, who, endeavouring to escape through the village, unfortunately met the Duke of Schomberg, who was there killed with a pistol shot.
>
> The King then marched with the few Horse he had with him and 17 battalions of Foot, to assist those of our troops who were engaged elsewhere (under Lieut.-General Ginkel), whom His Majesty reinforced with 12 battalions of Foot, and 9 squadrons of Horse, and, on advancing, the enemy retired in great haste and confusion. Our Horse endeavoured to overtake them with five small guns, which did some execution. After this, our Dragoons, supported by the Horse, attacked the enemy's rearguard, when, upon the King appearing with his troops upon the mountain, they all fled. The country being full of bogs and narrow passages, and the Irishmen being more used to such ground, they easily escaped.
>
> Our Foot pursued them to Duleek, and here the King stayed, our Horse still pursuing four miles further until night intervened. He then commanded the Horse to return and encamp where the Foot were, and sent to the camp, at Drogheda, for the tents and baggage, intending to encamp there and rest, for a day, his troops, who were much fatigued, and then continue his march in pursuit of the enemy. We have taken upwards of 6000 weapons, 70 or 80 waggons, besides great quantities of tents and baggage."

The Inniskilling men were particularly fortunate in seizing all the plate, money, and jewels of Tyrconnell and Lauzanne, valued at £10,000.[1] On some of the bayonets of James's force found on the field was inscribed :— " Le Roi de France fera couper la tête du Prince d'Orange " ; on the reverse were the words " St. Louis."[2] There were great demonstrations of joy in London on receipt of the news of James's defeat. James, during part of the action at the Boyne, stood at a little old church on a hill, but when he saw how things were going, he marched off to Duleek, and thence to Dublin, with a few attendants, where he immediately called a council and acquainted the magistrates of the city with the misfortune that had befallen him, declaring he would never trust himself again at the head of an Irish army, who could not stand a single charge. He then told those about him they must shift for themselves, as he intended to do. The next morning he became so alarmed at hearing that King William was on his way to Dublin, that, accompanied by the Duke of Berwick and some others, he scarcely gave himself time for a meal, but left for Waterford, where a ship was ready to convey him to France,[3] landing him eventually at Brest.

On the 2nd July, our army marched about a mile to encamp on more suitable ground ; on the 3rd, to Ballybriggan, and, on the 5th, to Finglas, where the King held a review of his troops on the 7th and 8th July, of which the following was a muster :—

[1] *Luttrell's Diary*, Vol. ii. p. 71. [2] King William's Chest, 1691. [3] *Wars in Ireland*, pp. 72–3.

KING WILLIAM'S ARMY.[1]

English, French, and Irish Horse.

Two troops, 1st and 3rd Life Guards	273
Grenadiers attached	95
Earl of Oxford's Regiment ("The Blues")	368
Sir John Lanier's	360
Colonel Villiers'	245
,, Russell's	242
,, Coy's (5th Dragoon Guards)	236
,, Beverley's	244
,, Langston's (4th Dragoon Guards)	225
Count Schomberg's (7th Dragoon Guards)	242
The Duke of Schomberg's (French) Regiment	395
Colonel Wolseley's Regiment (Inniskilling)	423
Captain Harbord's (one troop)	38
Colonel Mathew's Royal Regiment of Dragoons (1st Royals)	406
,, Leveson's Dragoons (3rd Dragoons)	246
,, Gwinn's Dragoons	260
Sir Albert Cuningham's Dragoons (Inniskillings)	358

Dutch Horse.

Troop of Dutch Guards	145
Lord Portland's Regiment	357
Minopovillan's	171
Lieut.-General Ginkel's Regiment	152
Colonel Scholk's Regiment	167
Van Oyen's	164
Reideffel's	174
Rancour's	178
Nyenberg's	175
Colonel Eppinger's Dragoons	621

Danish Horse.

Colonel Jemel's Regiment	268
,, Desnop's ,,	263
,, Schescad's ,,	281
Total Cavalry	7,772

English, Irish, and French Foot.

Major-General Kirk's Regiment	(2nd)	666
Brigadier Trelawney's ,,	(4th)	553
Colonel Beaumont's ,,	(8th)	526
Brigadier Stuart's ,,	(9th)	660

[1] Story, Vol. ii. pp. 95–8.

Sir John Hanmer's Regiment	(11th)	593
Colonel Brewer's ,,	(12th)	571
,, Hasting's ,,	(13th)	606
Earl of Meath's ,,	(18th)	678
Colonel Foulke's ,,		439
,, Gustavus Hamilton's Regiment	(20th)	560
Sir Henry Bellasis' Regiment		628
Lord Lisburn's ,,		611
Lieut.-General Douglas's Regiment		648
Earl of Drogheda's ,,		660
Colonel Erle's ,,	(19th)	693
Brigadier La Melonière's (French Huguenots)		529
Colonel Cambon's ,, ,,		640
,, Callimot's ,, ,,		562
,, Mitchelbourne's (Londonderry Foot)		664
,, Tiffin's Regiment	(27th)	625
,, St. John's ,,		589
Lord George Hamilton's Regiment	(22nd)	583

DUTCH FOOT.

Count Solm's three battalions	1850
Two Companies of Cadets	81
Count Nassau's Regiment	652
The Brandenburg ,,	631
Colonel Babington's ,,	416
,, Cutt's ,,	543
,, Groben's ,,	490

DANISH FOOT.

Danish Guards	698
The Queen's Danish Regiment	634
Prince Frederick's ,,	555
,, Christian's ,,	547
,, George's ,,	547
The Zealand ,,	527
The Jutland ,,	554
The Zuitland ,,	519
Total of Foot	22,528

The total forces were :

Horse	7,772
Foot	22,528
	30,300

No less than ten medals have been struck to commemorate the Battle of the Boyne. The following is a description of one of these, in bronze, in possession of the officers, 2nd Battalion Suffolk Regiment. (See Plate 3.)

Obverse.—Bust of William III., *r.*, laureate, hair long, in front of both shoulders, in armour and mantle. *Legend.* GVILH. III. D.G. MAG. BRI. FRAN. ET HIB. REX. On truncation, R.A.F. (R. Arondeaux fecit).

Reverse.—William on horseback, *r.*, crosses the river at the head of his cavalry; on the opposite bank, the enemy fleeing. *Legend.* ET VULNERA ET INVIA SPERNIT. (He despises both wounds and impassable places.) *Ex.* EIICIT JACOBUM RESTITUIT HIBERNIAM. (He expels James, and restores Ireland.) 1690. 1·9 inches.[1]

Two days after the battle, King William sent the Duke of Ormond with 9 troops of Horse to take possession of the city of Dublin, and, on the 9th July, divided his army into two portions, sending Lieut.-General Douglas with 5 Horse and 10 Foot regiments to Athlone to capture the place, and with the remainder, which included Brewer's (12th), he marched and encamped with his army at Cromlin, within two miles of Dublin.

James, whilst at Dublin in 1689, had established mints and debased the coinage by using copper from the guns, and stamping it as half-crowns, shillings, and sixpences, and the same at Limerick. King William, on arrival at Dublin, took over the mint and the specie called gun money, and issued a proclamation that all nominal value was to cease, and the coppers were to be taken at their intrinsic worth.[2]

On the 11th, the army marched to Kill Kullen Bridge. Story relates that:

> "the King this morning, passing by the Ness, saw a soldier robbing a poor woman, which so enraged him that he beat the man with his cane, and gave orders that he and several others who had disobeyed field orders were to be executed on the following Monday, two of the offenders being Inniskilling Dragoons."

By easy marches, the enemy arrived on the 21st at Carruck, within twelve miles of Waterford, whence, on the 22nd, the King sent General Kirk with the "Queen's," Colonel Brewer's (12th), and a party of Horse to summon the town to surrender, which, after some parleying, was complied with. On the 25th,[3] the governor of it marched out with 3 companies of Foot, and gave up possession of the town, leaving 14 guns mounted and 2 dismounted (but without ammunition), 350 barrels of oats, several other sorts of provisions, and 215 stone of wool, and was conducted to Mallow.

Early on the 8th August, a reconnaissance in force was made, by the King's order, towards Limerick, and, on the 9th, the whole army began its march to the town. The *London Gazette* of the 1st and 10th September states:—

[1] *Medallic Illustrations*, Vol. i. p. 176.
[2] Colonel Gardiner's *Centurions of a Century*, p. 9.
[3] Story, pp. 106, 109.

BOYNE MEDAL.
1690.

DETTINGEN MEDAL.
1743.

SERINGAPATAM MEDAL.
1799.

"On the night of the 17th, we opened the trenches with 7 battalions, and took two redoubts from the enemy. By the 19th, we had advanced towards another redoubt, which was captured with extraordinary bravery, driving out the enemy and killing about 40. After half an hour's possession of the fort, they made a great sally, with about 2000 Horse and Foot, but were driven back, our Horse pursuing them to the walls of the town. On the 21st, we finished our trenches, and on the 22nd raised a battery against the enemy's high towers. On the night of the 23rd, we fired the town in several places with red hot shot and bombs, and one, falling in their great magazine of hay, consumed it, and several houses were burnt, the fire lasting over six hours. On the 24th, we finished all our batteries, on which were posted 30 guns. On the 25th, we advanced our trenches within 30 yards of the ditch. A deserter, who came out of the town, reported that on the day we took the redoubt, the enemy lost over 300 men. On the 26th, we widened the breach we had made the day before in the wall of the town, and destroyed some of the enemy's palisades on the counterscarp. On the 27th, the King ordered an attack on the counterscarp, which was begun about 3 P.M. A detached party of grenadiers made the assault, supported by other detachments, who, having gained the counterscarp, and a fort the enemy had made under the walls, failed to obey orders to shelter there, and not to advance further, and, instead, they mounted the breach, following the Irish, who fled that way, and some were entering the town, but the enemy being entrenched, and provided with guns, the party was cut off. The fight lasted three hours, during which the enemy were reinforced, and sprang a mine in the ditch with but little effect; but, in conclusion, our men pressing too far, lost the opportunity of lodging themselves, so it was thought fit to retire to our trenches. What our loss was in these actions we cannot say precisely, but it is estimated at 400 or 500 killed and wounded since the beginning of the siege. Last night, we advanced our trenches about twenty yards, and to-day have enlarged the breach, and are preparing for another assault."

Three or four days previously, General Ginkel had captured a large castle two miles from here, in which were eighty men.

The King having thought fit, on account of continued bad weather, to withdraw his forces from Limerick, the guns and heavy baggage were sent away on the 30th August, and, the next day, the army marched off, without any opposition, towards Clonmel, whence a strong detachment was ordered towards Cork and Kinsale.

On the 17th September, news was received that the Duke of Tyrconnell had left for France, and, on the 28th, the town of Cork, after holding out for five days, surrendered to the Earl of Marlborough, which was followed by the surrender of Kinsale.

In October, the enemy gave permission to several Protestants to leave Limerick, who reported the scarcity of everything, wheat being generally £10 (brass money) a barrel; malt, £9; brandy, £3 a quart; ale, 2*s.* 6*d.*

do.; men's shoes, 30*s*. a pair; salt, 20*s*. a quart; and that the Irish army was in a very miserable condition in all respects.

Colonel Brewer's (12th) Regiment was employed in various services until the 31st December, when a force under his command marched to Lanesborough. To oppose its advance the enemy had cut several trenches across the causeways in the bog, and these trenches were disputed for some time; but, on losing some of their men, they retired into the town, of which our men eventually took possession, after defeating about 3000 of them.[1]

1691

From Lanesborough, Colonel Brewer's (12th) Regiment marched to Mullingar, where he was appointed Governor, our garrison consisting of 7 Horse and Foot regiments.

On the 2nd March, recruits were sent from Bristol and Chester to Colonel Brewer's Regiment.[2] From Mullingar, on the 28th April, Colonel Brewer (12th) and Lieut.-Colonel Hamilton, with 600 Foot and 20 Horse, marched into the enemy's quarters beyond the castle of Donore, between which place and Kilbeggan about 2000 Rapparees had for some time sheltered in huts. The rebels, at first, drew up on the hills in several bodies of horse and foot, but, on our approach, retired to the bogs and woods. Our men pursuing them, killed about 50, and, after burning a great many of their huts and cabins, returned to quarters on the 29th.[3]

On May 5th, a party of Colonel Brewer's (12th) was going towards Kinnegad, when a sergeant and four men, detached from the remainder, were set upon and murdered by the Rapparees, except one whom they left for dead, whose eyes they had most barbarously picked out with a skean. But the next day, three Rapparees were taken, and brought to Mullingar, one of whom giving evidence against the other two, the latter were thereupon hanged. On the 6th, Captain Poynes went out with 110 men from Mullingar, and encountering a party of Rapparees, near Monaghan, killed between 40 and 50 of them. Soon after routing various parties of Rapparees, Colonel Brewer and Major Board went with a detached party of 150 Horse and about 200 Foot from Mullingar to relieve Marescourt and Mayoon [*sic*] with provisions, and, returning via Ballymore, dislodged the enemy in the passes, arriving back without the loss of any men.

On the 30th May, General Ginkel left Dublin for Mullingar to commence operations.

The supplies expected from England had seasonably arrived, and everything that was thought necessary was in readiness except the orders to march, which were not delayed.

[1] Story, pp. 106, 147, 155.
[2] *State Papers Domestic, William and Mary*, Book 101, p. 288.
[3] *London Gazettes*, May 11th and 18th, 1691.

Our forces accordingly decamped from Mullingar, and, on the same day, were joined by Lieut.-General Douglas with the northern troops; on the 7th June, we came before Ballymore, seizing all the enemy's outposts, which they quitted on the approach of our men. Our guns played on the fort, and after having made two large breaches, the besieged hung out a white flag, and Colonel Burke (the Governor) and several officers came and submitted to the General, whereupon 400 of our men marched into the place, where there was a garrison of 780 men besides 259 Rapparees, well armed, and about 1000 women and children. In this action, we had 8 men killed, and the enemy about 150.

This important place being thus reduced, and the Prince of Wurtemburg having joined our army with about 12,000 Horse and Foot, the march was directed towards Athlone, which was reached on the 19th June, our troops finding that the enemy had erected French Colours in four parts of the town, to make believe that there were a great many French in the garrison.

Our advance party was under Major-General Mackay, with Brigadiers Stuart and Wittinghoff.

Part of the Great Bastion at Athlone having been destroyed by our guns, the General ordered the advance to be as follows :—

(1) 300 detached Grenadiers; (2) 50 Pioneers with fascines; (3) 200 Fusiliers supported by two regiments of Foot; (4) 200 Pioneers with more fascines, followed by 20 men with axes. At about 6 P.M. on the 20th, the signal was given for the attack, which consisted of a discharge from all our guns.

Cannon relates, with reference to this attack, that Major-General Mackay, commanding the troops, having observed that the advanced party had missed its way and halted, instantly hastened to Colonel Brewer's (12th) Regiment, and directed the way to the breach to the first captain he met, whereupon the regiment stormed it, and drove the French across the bridge to the Connaught side of the town. At first the enemy seemed resolved to defend the breach, but when they caught sight of the English grenades they fled in great confusion over the bridge, some, in their haste, falling into the Shannon and perishing in its waters.

Our men immediately entered that part of the town farthest from the Connaught side, where, becoming exposed to fire from the other side of the river, some few were killed and wounded, but the engineers soon erected shelter. The siege was then carried on with great resolution; and, at about 7 A.M. on the 28th, the storming was to have taken place in much the same order as the former attack, but with stronger detachments. It was, however, deferred until the 30th, our guns and bombs in the meantime doing terrible execution. At about 6 P.M. on the 30th, the signal for the attack was the ringing of the church bell, whereupon our men entered the ford below the bridge, and pressed with such vigour on the enemy that the latter were forced to quit the trenches, and within half an hour the English were masters of all the outworks and ruins of the castle, and had taken Major-General Maxwell and 200 officers and men prisoners. Two thousand of the Irish were computed to have been killed in this siege, and the place was taken within sight of the Irish Army, who were sending a reinforcement to its relief. Only six brass guns and two mortars were found, besides a small quantity of ammunition and provisions. St. Ruth and some of the other Irish commanders were reported to have assured themselves that it would have taken a great part of the summer to reduce the place, if the besiegers had not in the end been obliged to withdraw from it.[1]

Cannon wrote of the capture of Athlone :—" Never was a more desperate service, nor was ever exploit performed with more valour and intrepidity."

The Irish army, commanded by General St. Ruth, now crossed the Shannon and took up a position at Aughrim.

General Ginkel, with the Prince of Wurtemburg as chief of his staff, marched with our army on the 10th July to Ballinasloe, where he halted on the 11th, and encamped on the Roscommon side of the River Suck.

After crossing the river on the 12th, our army took up its position in two lines, which comprised the following regiments :—

First Line.—Kirk's (2nd), Brewer's (12th), Hamilton's (20th), Lord George Hamilton's (22nd), Herbert's (23rd), under Brigadier-General Bellasis—Major-General Mackay commanding.

Second Line.—Columbine's (6th), Stuart's (9th), Meath's (18th), Erle's (19th), Tiffin's (27th), under Brigadier-General Stuart—Major-General Talmash, commanding. The troops remained drawn up until noon for a heavy fog to lift.

The following account of the battle of Aughrim is mainly taken from the *London Gazette*, dated July 13 :—

After some skirmishing, we drove in their outposts and our left wing of Horse moved beyond the bog which covered the enemy's right, and enabled the Foot to advance between them and the unsafe ground. Our artillery had in the meantime dislodged the enemy's guard at the end of the defile that leads to Aughrim, and our Horse and Foot took post there, occupying a spot of firm land that lay between the two bogs and Aughrim Castle. The enemy opposed our left in great numbers, and were so strongly posted behind the banks that were above one another, that an engagement followed which lasted for two hours, when they were forced to give way, notwithstanding the reinforcements they had received.

An opportunity now availing for an attack on the Irish centre, Major-General Mackay ordered Colonels Brewer's (12th), Erle's (19th) and Herbert's (23rd) Regiments to plunge into a bog, and, after crossing a rivulet, to drive them from behind the hedges of the nearest enclosures. The troops waded through up to their waists in mud and water, and carried the attack right into the Irish position, but, from sheer force of numbers (owing to the Irish

[1] *Wars in Ireland*, pp. 131, 133.

having been reinforced) the English Brigade was driven back into the bog again, Colonels Erle (19th) and Herbert (23rd) being taken prisoners. The former was twice taken and twice rescued, and, though badly wounded, succeeded in joining his regiment; while the latter was never heard of again, and was supposed to have been murdered.[1]

At the same time, the advance of three of our battalions from another direction met with severe opposition from the enemy's troops, but we maintained our ground until reinforced, when they were compelled to retire. While this was going on the right wing of our Horse charged immediately up the hills where the enemy's squadrons were. By this time, our left had beaten the Irish from their ground, and our right pressing on at the same time, made an entire rout.

We pursued them four miles, the approach of night not permitting us to proceed farther, for it was 6 o'clock when the two armies engaged, and 8 before the enemy were put to flight.

The Irish were never known to fight with more resolution, especially their Foot. Their army, consisting of 20,000 Foot, and 8000 Horse and Dragoons, was superior to ours in numbers, and had every advantage in the way of ground. Their loss was about 5000 killed, including the French general, St. Ruth, and about 100 officers and 500 rank and file taken prisoners. We took all their baggage, tents, provisions, ammunition, and 9 guns, with a great many colours and standards, including that of their troop of Guards, and we have many weapons, which, to help their speed, they threw down in their flight. On our side we lost several officers, including Major Colt, of Brewer's (12th) and about 600 men. Colonel Brewer's (12th) Regiment lost 1 major, 1 captain, and 1 ensign killed, and 7 men wounded.

Never was greater bravery seen than among our troops, on this occasion, who, with surprising courage and undauntedness, pushed and forced the enemy from the most advantageous posts and entrenchments, and the vigorous conduct and gallantry of general officers contributed very much to this glorious success.

The English Army, resolute on the entire reduction of Ireland, marched, after some interval, towards Galway, and the Governor (Lord Dillon) having been offered advantageous terms of surrender refused, replying that "they were prepared to defend the place to the last," whereupon an attack was made on the fort, our grenadiers advancing with such caution that they were not discovered until they were at the foot of the glacis, where they delivered their grenades, and, mounting the palisades, entered the fort. The enemy, perceiving all was lost, cried for quarter, leaving our men in possession, so that they were able strongly to entrench themselves. The capture of the fort so alarmed the town that the Governor desired a parley, when, a treaty having been entered into, the garrison marched out on the 26th July, numbering (according to Story) 2300. Before 9 A.M., the guards of the town

[1] Lieut.-Colonel Macartney-Filgate, p. 55.

had been handed over to the newly appointed Governor, Major-General Sir Henry Bellasis, who, with his own regiment, occupied the place, together with Colonel Brewer's (12th) and Colonel Erle's (19th) Regiments.[1]

During the remainder of the campaign, Colonel Brewer's (12th) Regiment was quartered at Galway, and, on the 3rd October, the war in Ireland was concluded by the surrender of Limerick (after a second siege) and the castles of Clare and Ross, together with all other places that were at that time in the hands of the Irish.[2]

On the 23rd November, the regiment marched to Kinsale, and, embarking on the 20th December, sailed to Plymouth, whence, on arrival, it marched to Salisbury.

1692

A Route, dated 2nd January, directed it to march to London, where it was due on the 12th.

About the end of June, Colonel Brewer's (12th) Regiment was selected to form part of an expedition against the coast of France, under command of the Duke of Leinster, when 7000 troops were embarked at Southampton. The expedition menaced the French coast at several places, but the French had everywhere assembled in such force, that a council of war decided against landing. The transports accordingly returned, and joined the fleet at Portland on the 28th July, the troops again leaving England on the 17th August, with orders to sail to Ostend, where they landed on the 22nd, and were joined by a force from the confederate army under King William III.

On our taking possession of the towns of Furnes and Dixmude, and fortifying them, orders were issued to the Duke of Leinster (7th September) to bombard and destroy the town of Dunkirk, and, on the 30th, to remain on the side of Neuport nearest to Furnes, until Dixmude was put in a proper state of defence. On completion of this, the 2nd Troop of Guards and Grenadiers and 14 regiments of Foot, including Colonel Brewer's (12th), were to embark at Ostend for England. In 22 transports this force returned home to winter, part being landed in the Thames, some at Harwich, and some at Margate.

It is much to be regretted that, owing to the entire absence of any War Office "Marching Orders" from March 1692 to the end of the following year, it is not possible further to trace the movements of the regiment whilst on this tour of home service.

1693

The strength of regiments on the English establishment from 1693 to 1696 is shown as 44 commissioned officers, 104 non-commissioned officers, and 780 men, a total of 928.

[1] *State Papers, Domestic, William and Mary*, Book 101, p. 476. P.R.O. London.
[2] *Wars in Ireland*, p. 145.

On the 25th August, the King issued an order commanding the officers to use their utmost endeavours to suppress swearing in the army, "first by abstaining themselves from all oaths and execrations, and so giving a good example to their soldiers."

The loss this year of the battle of Landen (July 18th) by King William, rendered it necessary for the army in Flanders to be augmented.

The "State of Europe," for October, reports accordingly, orders being given for a fleet of 40 or 45 men-of-war, English as well as Dutch, to be ready by the end of November, with great quantities of ammunition and provisions, and for it to be reinforced with several vessels for the transportation of men and a train of artillery.

Colonel Brewer's (12th) Regiment remained in Great Britain, and joined, early in November, a large standing camp, for about 20,000 troops, which had been assembled between Petersfield and Portsmouth. Luttrell relates that His Majesty had resolved to send 12 Foot regiments of the above force to Flanders, and, on the 11th November, Lord Cutts proceeded to Hampshire to inspect them prior to embarking.[1]

1694

By the 6th February, the regiments of Colonels Collier, Hastings, and Brewer (12th) had been ordered to Flanders.[2] King William, who had left for England in October, 1693, arrived back in Holland about the 10th May this year, and, on the 17th, Sir Henry Bellasis received orders to march next day with the garrisons of Bruges, Ostend, and those quartered on the canal of Neuport, to form a camp by Ghent, numbering in all 19 battalions. Arriving at Ghent on the 19th, they encamped at Marykirk until the 21st, when they marched through the town of Ghent, and were joined by the forces that had been quartered there.

Our Artillery Train had, for want of horses, been left behind near Ghent, and the regiments of Brewer (12th), Leslie (15th), and Buchan (afterwards disbanded), were detailed to guard it. Early in June, it was sent by water to Malines, to take over the horses that had been ordered from Holland, and, together with the regiments escorting it, arrived at the King's camp near Tirlemont on the 6th June.

These, with some cavalry, were reviewed on the 7th, in the presence of the Electors of Bavaria and Cologne, by the King, who expressed his approval of the appearance and discipline of Colonel Brewer's (12th) Regiment.

The first line of our troops consisted of 33 battalions, and Colonel Brewer's was posted to the 2nd Brigade, under Brigadier-General Erle, commanded by Major-General Mirmont, the other five battalions in the brigade being the "Royals," "Queen's," Trelawney's (4th), Fitzpatrick's (7th Fusiliers), and Erle's (19th).

[1] *Luttrell,* Vol. iii. p. 223. [2] *Calendar of State Papers, Domestic Series, William and Mary.*

Colonel Brewer's (12th) Regiment took part in the operations of this year's campaign, which was uneventful, and ended without a general engagement, the numbers of our allied forces having been sufficiently augmented to enable us to arrest the progress of French conquest, whereby the enemy were forced to act on the defensive. In the autumn, the confederate army besieged and captured the fortress of Huy, Colonel Brewer's (12th) Regiment forming part of the covering army, after which it was stationed at Bruges for the winter.

Precedence of the first twelve old British Infantry Regiments, serving in the Low Countries, in 1694, from which it will be seen that the 12th Foot (Colonel Brewer's) then stood eighth in order of precedence:—

1. The Royal
2. Colonel Selwyn's
3. Major-General Churchill's
4. Colonel Trelawney's
5. Colonel E. Lloyd's
6. Royal Regiment of Fuziliers
7. Sir Bevil Granville's
8. Colonel Richard Brewer's
9. Colonel Tidcomb's
10. Sir James Lesley's
11. Colonel James Stanley's
12. Colonel Francis Collingwood's

The Royal Warrant settling the precedence of the above-named regiments was signed at Roosbeck on the 10th June, 1694.

While the crime of desertion was now met by the penalty of death, corporal punishment appears, from the following, to have been much resorted to for minor offences. An old War Office book shows a sentence passed on two soldiers, of different regiments, who, whilst serving in Flanders, were tried by court-martial "on suspicion of robbery at the Train, at 1 A.M. on the 29th June." No robbery was proved, but, for being absent at that hour, and found in another camp, they were sentenced "to be 'whipt' twice at the head of their regiments, and once at the Train."[1]

There are instances on record at this period of recruiting officers having been tried by court-martial, for not getting the full complement of recruits. The levy money the officers received was frequently insufficient, and no allowance was made for recruits lost through desertion, sickness, or other misfortunes over which they had no control.[2]

1695

King William's manœuvres in the previous year having held the French in check, he now resolved to undertake the siege of Namur.

On the 26th May, our garrisons in Flanders marched to take the field, and Colonel Brewer's (12th), with several other regiments, had orders to proceed to Dixmude.

On the 7th June, the Duke of Wurtemburg arrived for the purpose of making a diversion in favour of the main army, and, after reviewing the

[1] As the word "train," in a military sense, signifies "an aggregation of vehicles, men and animals accompanying a military body," it seems not unlikely under all the circumstances, that it implied in this instance the Commissariat.

[2] Fortescue, Vol. i. p. 574.

troops, and dividing them into four brigades, he encamped before the Kenoque, a fortress at the junction of the Loo and Dixmude Canals.

The regiment commanded by Colonel Brewer was in the 2nd Brigade, with Tidcomb's (14th), Sir James Leslie's (15th), and Courthorpe's (17th) Regiments, under Colonel Sir James Leslie, and was in the attack on the Kenoque, when the French were driven from their entrenchments and houses near the Loo Canal. The brigade lost 20 men killed and wounded. The total loss of the British engaged before the Kenoque was 587 men killed and wounded besides officers.[1] The French troops having taken post behind their lines, King William now found it a favourable opportunity to invest the town of Namur. The attack on Fort Kenoque was discontinued, and Colonel Brewer's (12th) Regiment marched into garrison at Dixmude, where one British Dragoon Regiment, and 3 British and 5 Dutch regiments of Foot were stationed under Major-General Ellenberg.

A powerful French army, commanded by Marshal Villeroy, marched towards Dixmude, and, on the 15th July, the place was invested by a strong division under General de Montal. Their trenches were opened on the same night, and next day, a battery of eight guns and three mortars commenced a heavy fire.

Major-General Ellenberg was a Dane, who, by his personal courage and merit, had risen from a private to Major-General in the Danish service, and was particularly recommended to the King by the Duke of Wurtemburg, commanding the Danish forces, as suitable for the command at Dixmude. As he now failed to make the spirited opposition to the enemy which the circumstances called for, his behaviour surprised all who had ever known him; he appeared to view the progress of the besieging army with apathy, and, the works now beginning to crumble under fire, he called a council of war, to whom he advanced several reasons as to why the town could not be defended, and proposed to capitulate, which was agreed to by the majority of the council, but opposed by other members, amongst whom was Colonel Brewer (12th), who recommended a resolute defence of the town to the last extremity.[2] When the rank and file were informed that they were to become prisoners of war, they became enraged, many destroying their arms, and tearing their regimental Colours from the staves, rather than such trophies should fall into the hands of the enemy.

D'Auvergne states: "The body of the garrison had the same heart and soul as their comrades who did such wonders before Namur; but the soldiers were delivered into the power of the enemy against their will."

The regiments in garrison were all made prisoners of war, and marched to Ypres, Arras, Bethune, and Bouchain, whilst the officers were placed in close confinement, and attempts made to induce the men to enter the French service, Louis XIV refusing to give them up on the conditions of the cartel agreed upon.

When the castle of Namur surrendered to the British, its garrison was

[1] D'Auvergne. [2] Cannon.

permitted to march out with the honours of war, but Marshal Boufflers was arrested and detained until the British and other soldiers of the Allied Army, detained as prisoners, were released. This produced the desired effect.

Colonel Brewer's (12th) Regiment and other corps in prison were released and rejoined the army, every effort being made, as speedily as possible, to enable the regiment to obtain the necessary arms, equipment, and clothing in order to resume its duties. It was afterwards placed in garrison at Malines, where it passed the winter.

A general court-martial assembled for the trial of the officers who delivered up Dixmude and its garrison to the enemy. Major-General Ellenberg was sentenced to be beheaded, and was executed at Ghent on the 20th November; Colonels Graham, Sir James Leslie, and a Dutch Colonel, Oare, were cashiered, whilst (according to Cannon) Colonel Brewer (12th) and the other commanding officers who remonstrated against the surrender of the town were acquitted.

The following is an extract from the "Calendar of State Papers, Domestic Series, 1695," with reference to the sentences of this court-martial:—

> "Letter from Welbeck, dated, 1st November, 1695.
>
> Last night, Lieut.-General Bellasis waited on the King, with the sentences of the court-martial on Major-General Ellenberg, and Brigadier O'Farrell, viz.:—
>
> That the said Major-General shall lose his head, and his goods be confiscated, and the said Brigadier be cashiered, and have his sword broken over his head by the common hangman. The like to be inflicted on Colonel Oare, a Dutch officer; Sir James Leslie, and Colonel Graham are to be broken, and Colonel Brewer suspended for three months and imprisoned during the said time, all of which the King approved of, and signed a warrant for it to be put into execution, and sent to Flanders."

The colonelcies which now became vacant in the regiments of Colonels Sir James Leslie and Graham were at once filled, but Colonel Brewer retained the colonelcy of his regiment until he retired from the army in September, 1702.

Advertisements in the *London Gazette* at this period, for the apprehension of deserters (of which the following is a specimen), show that upwards of £10 was offered as a reward for an individual capture:—

> "April 1st., 1695.—James Row, tall and well set, ruddy Complexion, wears a Perriwig, Soldier in Colonel Brewer's Regiment, is escaped from Tooley's, the Martial's in Holborn. Whoever secures him, and gives Notice thereof to Mr. Moyers in Freeman's Court, near the Exchange, in Cornhill, will have £10 reward."

1696

Owing to the superiority that had been obtained by the confederate troops, the French monarch had resolved on special efforts to increase his army, and had issued, at the end of last year's campaign, commissions for raising between forty and fifty additional regiments, with a view to invading England in favour of King James. Accordingly, in the spring of 1696, several of our corps were withdrawn from Flanders to seaport towns, when Colonel Brewer's (12th) Regiment marched from Malines to Ostend and Bruges, but the French did not venture to put to sea, and the regiment was not required to embark for England.

In the line of battle this year of His Majesty's Army in Flanders, commanded by Prince Vaudemont, Colonel Brewer's (12th) Regiment was placed in the right wing of the 2nd Line of Foot, in brigade with the 1st Battalion "Royals," Howe's (15th), and Collingwood's (afterwards disbanded), under Brigadier-General the Earl of Orkney.

On the 28th May, the King reviewed the infantry encamped at Marykirk, and, the same evening, Colonel Brewer's (12th) Regiment joined the camp from Bruges and Ostend.

According to D'Auvergne, the next position taken up by it was, in June, on the left of the Marykirk Bridge, in brigade with the Scots Guards, Bridges' (17th), Tiffin's (27th), and Saunderson's (afterwards disbanded) Regiments.

Our troops were encamped throughout the summer along the banks of the Bruges Canal, to cover Ghent, Bruges, and the maritime towns of Flanders; this year's campaign was passed in marching and manœuvring without any fighting, except a few slight skirmishes between detachments of ours and those of the enemy, and, in the autumn, Colonel Brewer's (12th) Regiment was ordered to occupy quarters at Bruges, where it passed the winter.

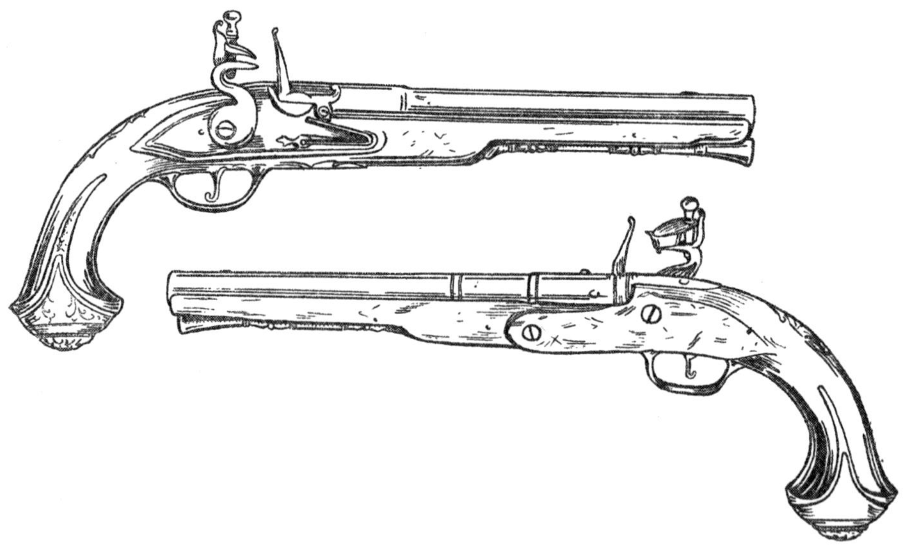

SILVER-MOUNTED PISTOLS, USED BY LIEUTENANT-GENERAL R. PHILIPPS (THEN A CAPTAIN) AT THE BATTLE OF THE BOYNE, 1690.

(By the courtesy of the Baron St. David's.)

CHAPTER II

Ireland, West Indies, England, Flanders, England, Spain, Minorca, and Great Britain
1697–1742

1697

Early in January, 1697, the regiment returned home to be quartered in Ireland, and, together with Colonel Tiffin's (27th) Regiment, entered Cork Harbour, in several vessels, on the evening of the 9th, when both corps at once landed.[1]

Colonel Brewer's (12th) remained at Cork until the 9th December, when orders were issued for the 13 companies to march as follows :—9 to Galway, 1 to Roscommon and Athlone, 1 to Loughrea, 1 to Ballinrobe, and 1 to Athenry and Tuam. The war in Flanders had in the meantime been terminated in September this year by the Treaty of Ryswick.

1698–1700

An Irish Return of Troops, dated 27th June, 1698, shows 12 companies of the regiment at Galway, and 1 at the Isle of Boffin, and, on the 4th August, 10 companies were ordered to be distributed as follows :—4 to Londonderry, 1 to St. Johnston, 1 to Moneymore, 1 to Magherafelt, 1 to Desertmartin, 1 to Ballymena, and 1 to Cashenden [*sic*].

In compliance with the Proclamation issued on the 1st May, 1699, revising the establishments of regiments in Ireland, Colonel Brewer's (12th) was now to consist of 11 companies, mustering 37 commissioned and staff officers, 22 sergeants, 22 corporals, 11 drummers, and 376 private men, a total of 488 of all ranks, and, on the 31st July, the regiment was directed to march :—3 companies to Belfast, 1 to Fall and Malone, 1 to Newtown and Bangor, 1 to Donaghadee, Ballywalter, and Ballyherbert, 1 to Carrickfergus, 1 to Larne and Isle Magee, 1 to Glenarm and Cashenden [*sic*], 1 to Killeleagh and Amber, and 1 to Strangford and Portersferry [*sic*].

The death, in November 1700, of Charles II, King of Spain, without

[1] *London Gazette*, January 9th, 1697.

issue, was followed by the accession to the throne of the Duke of Anjou, grandson of Louis XIV, to the prejudice of the Archduke Charles of Austria, and in violation of existing treaties, whereupon fresh hostilities were decided on.

1701

On the 7th April, the sheriffs of the city of Dublin were directed to provide convenient quarters for the officers and men of Colonel Brewer's (12th) Regiment, and, on the 20th May, the regiment was ordered to march there.

On the 13th July, orders were issued for Colonel Brewer's (12th) and Brigadier Tiffin's (27th) Regiments to proceed to the West Indies, the former to the Leeward Islands, and the latter to Jamaica, and each to be recruited to 80 private soldiers per company.

On the 31st July, Colonel Brewer's Regiment accordingly marched to Cork, and an order from Dublin Castle, of the same date, directed the sum of £1000 to be paid to him on embarking, which sum was not to be used, or disposed of, until after landing in the West Indies. The regiment was also to continue on the Irish establishment, and to be supplied with five barrels of gunpowder and a sufficient proportion of ball, before embarkation. On being settled with in Ireland to the 31st August inclusive, it sailed on the following day. Twelve infantry battalions were sent to Holland in June this year.

1702–1704

The sudden death of King William III took place on the 8th March, 1702, and war against France and Spain was declared by his successor, Queen Anne, on the 4th May.

As soon as hostilities commenced, Vice-Admiral Benbow, commanding the British naval force in the West Indies, proceeded to harass the commerce of the enemy with some success.

In the autumn of this year, Colonel Brewer retired from the service, and was succeeded in the colonelcy by Lieut.-Colonel John Livesay, by commission dated 28th September, 1702.

In the early part of March, 1703, an unsuccessful attack, under Colonel Codrington, was made on the island of Guadeloupe by troops in the West Indies; two regiments landed, and gained some advantages, but the expedition was not sufficiently strong to capture the island. Additional regiments were sent to the West Indies,[1] but nothing of importance took place, and Colonel Livesay's (12th) Regiment proceeded to Jamaica, where it was stationed throughout the year 1704.

[1] List of regiments in the West Indies in the summer of 1703:—

Columbine's, (6th).	Whetham's, (27th).
Livesay's, (12th).	Donegal's, (35th).
Erle's, (19th).	Charlemont's, (36th).
Handasyde's, (22nd).	Hamilton's, afterwards disbanded.

On the 27th April of that year, the regiment was authorised to raise, by beat of drum, another company "to consist of 2 sergeants, 2 corporals, 2 drummers, and 59 private men, servants included." [1]

1705–07

The regiment sustained very heavy losses from the effects of climate, and, in accordance with a Royal Warrant, dated 24th March, 1705, it transferred such of its non-commissioned officers and men as were fit for service to Colonel Handasyde's (22nd) Regiment, increasing it to a total of 70 per company. The remainder of Colonel Livesay's (12th) then returned to England to recruit its ranks to 48 privates per company.[2]

The War Office Marching Order Books show that, in 1704–05, two companies of the regiment, which had been left in England, had been busily employed marching through the country to recruit by beat of drum, so that by the 19th June the regiment (12 companies), on arrival home, was directed to march :—headquarters and 3 companies to Leicester, and 9 companies to out stations.

The moves now (and up to a much later period) of all regiments on home service followed each other in quick succession, and were frequently countermanded.

On the 11th May, 1706, the headquarters and 3 companies proceeded to Portsmouth, 2 other companies being distributed between Sheerness, Tilbury, Dover, and Southampton, and one marching to Landguard Fort.

On the 10th July, a draft of 100 men was ordered to be prepared, out of Colonel Livesay's (12th) and Colonel Whetham's (27th) Regiments, for the purpose of being employed in this summer's expedition.

A further interchange of quarters took place on the 29th October, the headquarters and 3 companies marching to Leicester, Harborough, and Stamford.

On the 3rd November, the regiment was ordered to York, and, on the 24th December, was directed to be split up into several detachments on election duty, the elections to take place on the 1st of the following month, after which it was concentrated at York.

On the 1st January, 1707, Colonel Livesay (12th) was promoted to the rank of Brigadier-General.

The first complete code of "Clothing Regulations" was issued during the reign of Queen Anne, dated 14th January, 1707, and states that :—

> "The sole responsibility for the pay and equipment of a regiment rests with the Colonel, who is held responsible in his fortune, and in his character, for the supplies of his regiment."

By Warrant of Queen Anne, in 1707, a Board of General Officers was

[1] *W. O.* 26, Book 13, pp. 4 and 5.

[2] *W. O.* 4, Book 3, p. 198.

instituted, and the duty of this Board was to select, seal, and issue patterns for the clothing of each regiment by the colonel. The supplies furnished by the colonel, when received at the regiment, were inspected by a Regimental Committee, and compared with the patterns sealed by the Board of General Officers. Any complaints made by the regiment were investigated by the Board, whose decision was final.[1]

An order, dated 12th February, 1707, directed headquarters and 6 companies to march from York to Hull, and, on the 8th March, the remaining 6 companies were ordered (2 each) to Leeds, Halifax, and Wakefield, and thence, on the 3rd April, to Dover, Tilbury, and Sheerness respectively.

By Routes, dated 12th August, the regiment was ordered to march to Salisbury and Winchester, and, on the 26th September, 4 companies moved from Winchester to Tilbury Fort and Sheerness respectively.

On the 30th October, 6 companies, at Salisbury, were moved in two company detachments to Shaftesbury, Blandford, and Sherbourne, and Routes, dated 26th December, directed these 6 companies to march to Plymouth Citadel to be quartered.

1708

The headquarters and 6 companies of the regiment at Plymouth were ordered to Portsmouth, the remaining companies joining them there early in July, when the whole regiment was to pass over to the Isle of Wight.[2] By a Royal Warrant of Queen Anne, dated 22nd June, Brigadier-General Livesay's (12th) was increased by an additional company from Lord Mountjoy's Regiment.[3]

In a Royal Warrant, dated Windsor, 12th July, Queen Anne decided to abolish all brevet rank in the army.

A Royal Warrant, dated August 23rd, laid down the following " annual bounties " to be paid to the several widows of officers who had been killed or died in Spain, Portugal, or the West Indies, viz. :—

The widow of a Colonel, £50 per annum ; a Lieut.-Colonel, £40 ; a Major, £30 ; a Captain, £26 ; a Lieutenant, £20 ; an Ensign, £16 ; a Cornet, £16 ; an Adjutant, £16 ; a Quarter-Master, £16 ; a Chirurgeon (Surgeon), £16 ; a Chaplain, £16.

It having been decided this summer to alarm the French coasts with a prospect of invasion, the regiment took part in an expedition for that purpose, the fleet being under the command of Admiral Sir George Byng. Brigadier-Generals Wynne and Livesay (12th) were appointed to command brigades under Lieut.-General Erle, and the troops to go on board the fleet consisted of 2 Dragoon regiments, 11 Foot regiments from England, and

[1] Old Pimlico Book. [2] *W. O.* 4, Book 7, p. 216. [3] *W. O.* 26, Book 12, p. 11.

5 from Ireland.[1] Lieut.-General Erle arrived on the 8th July to command the forces encamped in the Isle of Wight,[2] and, on the 19th, he inspected Brigadier-General Livesay's (12th) Regiment. The fleet sailed from Spithead on the 27th July, and, on the 1st August, anchored before Boulogne, where demonstrations were made the following day, as if to land troops. Similar demonstrations followed off other places, which possibly had the effect of inducing the French to withdraw some of their troops from Flanders for the defence of their own territory. Beyond the above, nothing of importance was accomplished.

In the meantime our allied forces in Flanders, commanded by the Duke of Marlborough and Prince Eugene of Savoy, were carrying on the siege of the celebrated city of Lille (the capital of French Flanders), which was being defended by 15,000 men under Marshal Boufflers. The French and Spaniards, in endeavouring to prevent the allied army receiving supplies from the coast, detached a body of troops under Count De la Motte towards Ostend, whereupon our troops, employed in alarming the French coast, were suddenly ordered to proceed to that port, where they arrived on the 21st September. Brigadier-General Livesay's (12th) and the other regiments of the expedition having landed, the French general retired, first cutting the dykes to swamp the country between Ostend and Neuport, in order to prevent General Erle's troops communicating with the Grand Army under the Duke of Marlborough.

A strong detachment from Livesay's (12th) and two other regiments now seized on Lessinghen, constructed some works, and established a post there.

Our army being at this time surrounded on all sides except that of Ostend, provisions began to be scarce, and especially ammunition, so, while the siege of Lille was proceeding, the Duke of Marlborough's time, on the 25th September, was taken up with measures for securing a great convoy that was expected from Ostend.

His Grace had ordered 12 battalions, 10 squadrons, and 800 Horse to guard it, but, on hearing that the English troops, under General Erle, had advanced to Lessinghen, and that the Count De la Motte had been reinforced by several French brigades to intercept him, he thereupon ordered Major-General Webb with 12 more battalions to join the force between Menin and Ostend, who accordingly marched, at daybreak on the 26th, direct for Ostend.

The troops employed to guard the convoy under General Webb left Ostend

[1] *Luttrell*, Vol. v, p. 221.
Cannon gives the following as some of the regiments employed in this expedition:—

Carpenter's Dragoons, now 3rd Hussars.	Essex's Dragoons, now 4th Hussars.
Livesay's, (12th).	Moore's, afterwards disbanded.
Farrington's, (29th).	Caulfield's ,, ,,
Hamilton's, afterwards disbanded.	Townshend's, ,, ,,
Johnson's, ,, ,,	Wynne's, ,, ,,

[2] *London Gazette*, 8th July, 1708.

on the 27th September, and, on his return march, he was met on Wynendale Plain by Count De la Motte, who, with 22,000 men from Ghent, vigorously attacked him on the march, as his force advanced on to the plain. Though General Webb's force was greatly inferior in numbers to that of the enemy, yet, by the skill with which he handled his troops, and the valour they displayed in repulsing the enemy's attacks, he firmly maintained his ground and convoy, so that De la Motte was forced to retire to Ghent, abandoning 9 guns, and leaving upwards of 4000 killed and wounded, a great many of his men having also deserted in the confusion of the retreat. That same night, the convoy arrived at Rosslaer unmolested, the following day at Menin, and from thence to the headquarters of the army, through this very creditable success, attributed to the masterly dispositions of General Webb, the number of the enemy being 4 to 1.[1]

The battle lasted two hours, our loss having been 912 officers and men killed and wounded, whilst that of the enemy was as stated above. We remained on the field of battle until 2 next morning, having first removed all our dead and several of the enemy's.

The military ability displayed by Major-General Webb in this eminent service was deservedly acknowledged by the unanimous thanks of the House of Commons, a reward of 1000 guineas from Queen Anne, and promotion to the rank of Lieut.-General, whilst the King of Prussia conferred on him the noble order of GENEROSITY.[2] He was later appointed Governor of the Isle of Wight. From December, 1695, to August, 1715, he had held the colonelcy of Princess Anne's (now the 8th "King's") Regiment of Foot, which political events caused him to vacate in the latter year.

Early in October, whilst the Duke of Marlborough was on the march, with 60 battalions and 130 squadrons, towards Rosslaer, he received advice that the French commander, the Duke of Vendome, prior to retiring on Bruges, had given orders for cutting the dykes in several places to swamp the country and hinder our communication with Ostend, whereupon His Grace ordered the army to encamp with its right at Torout and its left at Ichtegen, sending, at the same time, part of the troops back to Rosslaer to be nearer the siege, if required.[3]

Brigadier-General Livesay's (12th), with other regiments, was later employed in conveying necessaries, and also another supply of ammunition, to the besieging army, across the inundations in boats, which enabled our generals to continue the siege of Lille, and ensured the reduction of that fortress.

On General Erle being informed that the Duke of Vendome contemplated attacking Lessinghen on the 25th October, he reinforced the troops there with 50 men, whilst Colonel Caris sent another detachment of 109 men from Ostend. The attacking force, mustering 50 companies of grenadiers, supported by 10,000 Foot, landed, in the night, over 7000 men

[1] Millner's *Sieges*, p. 234. [2] Brodrick, p. 258. [3] *London Gazette*, 4th October, 1708.

behind the village, in a dry place, which it was impossible to swamp, and, at about 6 A.M., they attacked a churchyard wherein Colonel Caulfield had posted a guard of 150 men, who were overpowered and made prisoners of war, and after a short resistance by other troops Lessinghen was recaptured.

The enemy lost in that siege over 500 men, whilst some of the British officers distinguished themselves in a very marked manner, amongst them being Major Douglas of Livesay's (12th) Regiment, who, on the 21st October, had been ordered, with 100 English and Dutch grenadiers, to reinforce the troops at Lessinghen, and having taken post in a garden on the other side of the canal, with 50 grenadiers, kept the enemy at bay with such a heavy fire that they were forced to retire (with a loss, that night, of over 100 killed and wounded), and continued to hold his ground throughout the 22nd and 23rd, when he was relieved. After the capture of Lessinghen, the enemy menaced the camp at Raversein; Livesay's (12th) and other regiments under Major-General Erle being now quartered in the outworks of Ostend. The supplies furnished to the army had proved sufficient, and the citadel of Lille surrendered on the 9th December. The service for which the regiment proceeded to Flanders being now accomplished, it returned this month to England, and an order, dated 31st December, directed it to be quartered at Portsmouth.

1709–10

The regiment disembarked at Portsmouth from Ostend early in January, and later, 3 companies proceeded, at intervals, to the Isle of Wight.

A letter from the Secretary at War,[1] dated 16th April, represented that in consequence of Brigadier-General Livesay's (12th) Regiment then consisting of only 155 privates, every endeavour was to be made towards recruiting it.

This was followed, on the 18th July, by an order to transfer 50 men "without clothes or arms" to complete the strength of Colonel Dormer's Regiment ordered on immediate service abroad.[2]

On the 1st January, 1710, Brigadier-General Livesay was promoted Major-General.

Several British regiments were being raised at this period for service in Spain and Portugal, which were disbanded at the end of 1712.

Early in 1710, a "secret expedition" had been planned, the command of which was given to the Earl of Shannon, and Major-General Livesay's (12th) was one of the first corps ordered to embark, a letter from the Secretary at War, dated 10th May, directing him "to forthwith provide tents and all other camp necessaries for his regiment, which was to be held in readiness to embark for foreign service," and an order, dated

[1] His title at this period.

[2] *W. O.* 26, Book 13, p. 58.

30th May, directed the regiment "to be placed on board ship within six days, upon which the ships would proceed down the river."[1]

The intended expedition, however, never came off, and the troops who were to take part in it were massed in the Isle of Wight. Apparently, the above Embarkation Order for the regiment was cancelled, as the old War Office books show that, on the 25th June, 3 companies, detached at the Isle of Wight, were ordered to join headquarters at Portsmouth, and, on the 14th July, the whole regiment was ordered to the Isle of Wight, together with that of Lord Mark Kerr (on relief by the regiments of Colonel Windsor and the Marquis de Montandre), all four having, on the above date, been reviewed in brigade at Portsmouth by Lieut.-General Erle.

On the 7th August, 1710, all the troops encamped at the Isle of Wight were inspected by the Duke of Bolton, and, on the 26th October (N.S.), they embarked on transports to join four regiments at Cork, and thence to sail for Portugal.

The regiment, on arrival at Lisbon, proceeded with several others to Catalonia.

1711

The following Royal Warrant was issued in May this year, with reference to officers' commissions in the army :—

IT IS OUR WILL AND PLEASURE :

That, no Person shall have a Com^{n} in the Army, who is under the age of 16, unless in some extra^{dy} Cases, as We shall think fitt.

That, in future, no person be taken into any of Our Troops of Horse Guards, Grend^{r} Guards, or Reg^{ts} of Horse and Dragoons in Great Britain, but such as are Our Natural Born Subjects, and, that all ye Private Gent: who now are or shall be Entertained in Our Troops of Horse, and Gren^{dr} Guards, do take the Oaths and Test required by Law.

That, no commissions in Our Army be sold but by Our Approbation and Our Royal Sign Manual.

That, no Officer have Leave to Sel his Com^{n} who has not served 20 Years, or been disabled in the Service, unless upon some Extra^{dy} Occurrence

Where We shall think fitt for ye good of Our Service to allow thereof, and that in cases where We shall Consent to ye disposal of any Com^{n,} 12d. in ye pound be paid both by ye Buyer and Seller, which it is Our Intention should be apply'd to encrease ye fund of Our Royal Hospitall at Chelsea.

Given at Our Court at St. James's, this 1st day of May, 1711,
In the Tenth year of Our Reign,

By Her Majes^{ty} Comand

Granville.[2]

[1] *W. O.* 4, Book 11, p. 46.

[2] *W. O.* 26, Book 13, p. 58.

The following was the Regulation and Tariff of rate for settling the Quarters (otherwise billeting in the absence of barracks) of Regimental Officers and Troops in the Pay of Her Majesty the Queen of Great Britain, and their High Mightinesses the States General of the United Provinces, during the Winter, 1711.

General Officers having been otherwise provided for, it was ruled that:—

> "A Colonel, whether of Cavalry or Infantry, and a Lieut.-Colonel, who commands a Corps in which the Colonel is never present, shall be accommodated with two rooms, a kitchen, a stable, with the use of a Cellar, three Beds or Mattresses, Sheets and Coverlets for his servants; and he shall also be supplied daily with three fires, for each of which, a great Mande shall be provided in place of Coals and Faggots, and also six Candles, weighing a small Pound.
>
> A Lieut.-Colonel or Major of Brigade shall have a Chamber, Kitchen, Stable, a Bed for himself, two Beds or Mattresses with Sheets and Coverlets for his Domesticks, the use of a Cellar, two Fires in the same manner as a Colonel, and four Candles.
>
> A Major of Infantry or Captain of Cavalry shall have a Chamber, Kitchen, Stable, a Bed and Bedding for himself, a Bed or Mattress, with Sheet and Coverlet for his Footman, and Fire as in the preceding Article.
>
> A Captain of Infantry shall have a Chamber, Kitchen, Stable, a furnished Bed for himself, a Bed or Mattress with two Sheets and Coverlet for his Servant, and two Fires and four Candles per day.
>
> A Lieutenant, Adjutant, Second Lieutenant, Cornet, Ensign, Quarter Master, Chaplain and Surgeon shall have furnished Beds for themselves, Fire and two Candles each, but the same Chamber and the same Fire shall serve two, provided they are not married; and two of the same Company and Regiment shall be put together as nearly as can be done.
>
> The Chaplain, Adjutant, Quarter Master, Surgeon and Lieutenants and Cornets of Cavalry shall, moreover, each have a Stable for his Horse, if he has one.
>
> A Serjeant will have a Bed with Sheets and Coverlet, and two must lay in the same Room, and in the same Bed, if neither of them be married; and they may make use of the Fire of the Landlord.
>
> The rest of the Troops must content themselves with a Single Coverlet, and their Landlord's Fire; but the Master of the House must always have convenient Room for his Family, and Places to put up his Cattle, and deposit his Grain. Soldiers must be satisfied with their Bedding, and must live upon their pay without demanding anything from their Landlord for a Welcome on their Arrival, or at their Departure, or on any other Pretence."

Plate 8 shows a portrait of Field-Marshal, John, Earl Ligonier, K.B., who entered the army as a volunteer under Marlborough, and, from 1702 to 1709, was engaged with distinction in nearly every important battle and siege of the War of the Spanish Succession. One of the first to mount the breach at the siege of Liége, he commanded a company at Schellenberg and Blenheim, and was present at Menin, Ramillies, Oudenarde, and

Malplaquet, where he received twenty-three bullets through his clothing and remained unhurt. He continued to take a prominent part in most of the great sieges and other affairs in the Low Countries up to 1710.[1]

In November 1711, he obtained the rank of Colonel in the army,[2] and, during the years 1711–12, was posted as Lieut.-Colonel to Major-General John Livesay's (12th) Regiment,[3] being appointed, in the latter year, Governor of Fort St. Philip, Minorca.

In 1716, he became Colonel of the 4th Horse (now 3rd Dragoon Guards), and, in 1720, Colonel of the 7th Dragoon Guards, Brigadier-General in 1735, Major-General in 1739, and accompanied Lord Stair in the Rhine Campaign, 1742–43. Was made a Knight of the Bath by King George II on the battle-field at Dettingen, and commanded the British Foot at Fontenoy. On being raised to the Peerage in 1757, he was, in the same year, appointed Commander-in-Chief, and, in 1766, was created an Earl and a Field-Marshal. He died in 1770.

On the 23rd December, 1711, his younger brother Francis was gazetted Ensign in Major-General Livesay's (12th) Regiment.'[4]

The War of the Spanish Succession, which had been in progress since 1702, was now drawing to a close, and as regards the main operations in Catalonia, the severity of the recent campaign prevented the French Commander (the Duke of Vendome) from assembling his forces till September, and the army with which he then took the field, at Cervera, numbered about 19,000 men. On the side of the Allies, General Staremberg had been indefatigable in recruiting his troops, and, in August, when he took up a fortified position to cover Tarragona and Barcelona, his total strength was about 15,000. During this year, the French Marshal met with one defeat from the allied troops.

1712

Vendome and Staremberg again took the field, and though the operations were of no consequence, the Bourbons received a heavy blow in the death, from natural causes, of their Marshal.

A letter from the Secretary at War, dated 1st March, was sent to Major-General Livesay (12th) and five other colonels of regiments in Catalonia, calling on them for an explanation as to why they all required entire sets of arms and clothes, stating when they received their last issues of each, and asking how it came to pass they were then in so bad a condition.

Major-General Livesay retired from the command of the regiment early this year, and was succeeded by Lieut.-Colonel Richard Philipps, whose

[1] *Dictionary of National Biography*, Vol. xxxiii, p. 240.
[2] Cannon's History, *7th Dragoon Guards*, p. 83.
[3] *King George the First's Army*, C. Dalton, Vol. i, p. 150.
[4] *Army List*, C. Dalton, Vol. vi, p. 82.

commission was dated 16th March, and who later became a Lieut.-General. (See Plate 4.)

So far as Spain was concerned, the War of Succession practically terminated in April, 1712. The *London Gazette*, under the heading "Lisbon, April 12th," states:—"Our troops are marching from their quarters to their rendezvous in the province of Alemtejo."

The order for the departure of the British troops excited great and general indignation among the inhabitants of Catalonia, who loudly complained of England's selfish policy, first to kindle and blow the flame of civil war, and then coolly leave them to be devoured by it.[1]

Colonel Philipps' (12th) Regiment, after encamping for a short time, embarked in an English squadron, and sailed for Port Mahon, Minorca.

In August, a suspension of hostilities took place between England, Holland, France, and Spain.[2]

A letter from the Secretary at War to the Agent of Colonel Philipps' (12th), dated 7th November, states:—"H.M.S. 'Greenwich' being now in the river, under orders to sail the beginning of next week, you must take care that the clothing of Colonel Philipps' Regiment is immediately put on board, or there will be no opportunity of its being conveyed by her to Spain, as the ship does not touch at Portsmouth."

The regiment had, however, left some months previously.

Minorca, where the regiment was quartered, is the second of the Balearic Islands in the Mediterranean (near the coast of Spain), which had been captured by the British in 1708, and was ceded to Great Britain at the Peace of Utrecht in 1713. It is about thirty miles in length and twelve in breadth, and is chiefly valuable for the excellent harbour of Port Mahon, with deep water, and capable of sheltering in former days all the fleets of Europe. At the entrance to it is the castle of San Felipe.

The people of the island were well housed in solid stone buildings, the farmhouses being generally of two stories, with the granary under the roof. The farmers have to contend against frequent and violent gales, a very stony and shallow soil, and scarcity of water. They are very laborious, and work under a system of partnership.[3]

1713–14

On April 11th, 1713, the Peace of Utrecht was signed, the Archduke Charles, who had become Emperor in 1711, refusing to be a party to it. The Duke of Savoy was to have Sicily; Gibraltar and Minorca were ceded to England; the Netherlands, Naples, Milan and Sardinia to the Emperor.

By a Royal Warrant of Queen Anne, dated 23rd April, 1713, the following order of precedence was laid down for the first 12 old British infantry

[1] Mahon, p. 362. [2] Parnell, p. 301.
[3] Sir Clements Markham's *Story of Majorca and Minorca*, pp. 260–69.

LIEUTENANT-GENERAL RICHARD PHILIPPS.

(*From an Oil Painting.*)

Colonel of the Regiment, March, 1712, *to August,* 1717.

(*Appointed Governor of Nova Scotia,* 1719.)

(*By the courtesy of Major Raymond Smythies, late 40th Regiment.*)

regiments, when the 12th came in the order which has since been allotted to it :—

1. Royal.
2. Colonel Kirk's.
3. Colonel Selwyn's.
4. Our Own Regiment (Lieut.-General Seymour).
5. Major-General Pearce's.
6. Colonel Hamilton's.
7. Royal Fusiliers.
8. Our Own Regiment (Lieut.-General Webb).
9. Lieut.-General Stuart's.
10. Lord North and Grey's.
11. Major-General Hill's.
12. Colonel Philipps'.

Queen Anne died in 1714, and was succeeded by King George I, of the House of Hanover.

1715–19

In August, 1717, Colonel Philipps was transferred from his command to that of a regiment which later became the 40th, then newly formed from independent companies in different parts of America, and was succeeded in the colonelcy by Colonel Thomas Stanwix (from a corps which afterwards became the 30th), whose appointment was dated the 25th August, 1717, with the rank of Brigadier-General.

An order, dated 20th May, 1718, directed that, upon Colonel Cosby's (18th) Regiment disembarking at Minorca, Brigadier-General Stanwix (12th) was to transfer to it a draft of 2 companies, consisting of the 2 junior captains, lieutenants and ensigns, and each company to muster 4 sergeants, 4 corporals, 4 drummers, and 74 private men, also, in addition, 10 drummers and 18 private men, grenadiers, with all their arms and accoutrements. Also, of the men so transferred, each was to be allowed £4, and the draft to be made up of such men only who were willing to remain on the island, and were capable of doing duty.[1]

1720–22

On being relieved in 1720 from duty at Minorca, the regiment returned to England, and arrived in three ships in the months of August and November.

Routes issued in August directed 3 companies which had disembarked at Portsmouth, in advance of the headquarters, to march to Taunton, and, in October, to Bristol, where the headquarters and 7 remaining companies joined them in November, it having been the lot of the latter, on arrival, to be placed in quarantine for a few weeks.

[1] *W. O.* 26, Book 15, and *W. O.* 4, Book 22.

Early in April, 1721, the regiment marched to Exeter, where it remained for a year.

In April, 1722, a move was made to Bristol, and, in May, to Devizes, furnishing detachments to be quartered in the neighbouring villages. Early in August, it was encamped at Hungerford, moving later in the month to Salisbury Plain, where, on the 30th, it was reviewed by King George I, accompanied by His Royal Highness the Prince of Wales.

In September, the regiment, encamping near Hungerford, was directed to march to Gloucester, in October to Worcester, and in November to Manchester.

1723–24

In January, 1723, 4 companies were distributed:—1 to Knutsford, 1 to Stockport, and 2 to Macclesfield, and, on the 7th May, the headquarters and 6 companies moved to Shrewsbury, the detached companies joining them there.

A Route, dated 1st October, directed the headquarters and 6 companies to march to Leeds, detaching 2 companies to Wakefield and 2 to Halifax.

On the 14th April, 1724, the companies at Leeds and Wakefield were ordered to march to Berwick.

1725–26

The regiment proceeded to Edinburgh in 1725. On the 14th March, Brigadier-General Thomas Stanwix died, and King George I conferred the colonelcy of the regiment on Major-General Thomas Whetham, from a regiment which later became the 27th of the British Line, General Whetham's commission having been dated 22nd March.

The troops at this time in Scotland were chiefly employed in making roads, guarding the coast from smugglers, and in preventing and quelling riots, to which the country was then not a little addicted.

The regiment moved to Inverness in 1726. Major-General Whetham (12th) having represented that his regiment (lately commanded by Brigadier-General Stanwix, deceased) was short of 380 firelocks, the Master-General of Ordnance, on the 30th December, was directed to issue the above number to it, with a corresponding number of steel rammers and bayonets, upon the widow of the said Brigadier-General Stanwix paying for the same, the value of each, with steel rammer and bayonet complete, being 32 shillings a firelock, the sum total amounting to £608.[1]

1727–29

King George I died on the 18th June, 1727, and was succeeded by King George II. In 1727, the regiment moved to Berwick, and Major-General

[1] *W. O.* 26, Book 17, p. 138.

Whetham was promoted to the rank of Lieut.-General. An order, dated 31st January, directed an increase of ten men per company, and also two additional companies to Lieut.-General Whetham's (12th) Regiment, each of the latter to consist of 3 sergeants, 3 corporals, 2 drummers, and 60 effective privates,[1] to be quartered at Leeds until the 13th June, when they moved to Bradford, and, on the 22nd July, were directed to join headquarters at Berwick, where the regiment remained throughout the years 1728 and 1729.

In the latter year, the two additional companies were disbanded, and the increase of ten men per company was reduced.

1730

A Route, dated 11th April, directed Lieut.-General Whetham's (12th) Regiment to march from Berwick in three divisions, as follows :—

3 companies to Colchester, 2 to Ipswich, 2 to Chelmsford, and 1 each to Woodbridge, Braintree and Witham, and, on the 14th May, the companies at Colchester were ordered to Coggeshall, Halstead, and Sudbury respectively, until the 30th June, when the regiment marched to encamp at Windsor Forest. On the 8th July, it was directed to march from Kingston-on-Thames to Wynfield Plain, and encamp, in order to be reviewed by His Majesty, and thence to move in two divisions to Sunninghill, and encamp near the Seven Conduits, in order to mend the roads and other avenues in Windsor Forest.

An extract from notes of the regimental career of Lieutenant John Bernard Gilpin, 12th Regiment, at this period (See Plate 5), states :—" The regiment was employed in cutting roads through the woods. These roads or openings were constructed without any taste or design but that of making easy communications between the several parts of the forest, chiefly indeed for the Queen, and other ladies of the Court, to enjoy the diversion of stag-hunting in their carriages."

A letter from Army Headquarters to the Governor-General of Ireland, dated 23rd July, directed the appointment of Lieut.-General Whetham as a " Major-General of Our Forces " on the Irish establishment, at a salary of £485 per annum.

All men of the regiment falling sick, whilst at Windsor Forest, were, up to the 11th August, to be removed to Bracknell and Ockingham. An order, dated 22nd September, directed the regiment, at Windsor Forest, to decamp and march :—3 companies to Ipswich, 4 to Colchester, and 1 each to Woodbridge, Braintree, and Coggeshall, the two latter supplying detachments to Bocking and Kelvedon respectively, whilst the company at Woodbridge supplied detachments to Martlesham and Melsham, and, in December, the 4 companies at Colchester moved to Witham.

[1] *W. O.* 26, Book 17, p. 153.

1731–34

Several changes of companies to places in Essex and Suffolk took place until July, 1731, when the regiment was concentrated at Chelmsford for the Assizes, and immediately after a fresh redistribution of the companies found them at Romford, Colchester, Brentwood, Ingatestone, Barking, Braintree, Maldon, Witham, Halstead, and Chipping Ongar, in some instances supplying small detachments to adjacent villages, and, in October, the detachments were quartered at Arundel, Shoreham, and Emsworth.

The regiment again assembled at Chelmsford for the Assizes early in July, 1732, and, on the 7th, was directed to march :—headquarters and 4 companies to Lincoln, 2 to Norwich, and 1 each to Mansfield, Gainsborough, Worksop, and Horncastle, its quarters being later extended to Newark, Sleaford, and Doncaster.

On the 10th April, 1733, the several companies of Lieut.-General Whetham's (12th) Regiment were to assemble at Lincoln by the 18th, to be reviewed (inspected) by himself, and then to return to their former quarters, the detachment at Arundel rejoining headquarters at Lincoln, and, on the 16th June, the regiment received orders to march to Berwick.

A Royal Warrant, dated 21st February, 1734, directed Lieut.-General Whetham's (12th) to be augmented by 3 sergeants, 3 corporals, 2 drummers, and 70 men per company, under commissioned officers, to be raised by beat of drum.[1]

1735–38

A Royal Warrant, dated 14th January, 1735, directed a reduction in the regiment of 11 private men per company.[2]

Though the old War Office "Marching Orders" give no information relative to moves of troops, either in Scotland or Ireland (and there is a blank in them concerning the corps up to 1739), there seems reason to conclude that a move was made from Berwick to Glasgow, as, on the 19th September, 1735, the freedom of that city was conferred on Lieut.-Colonel Scipio Duroure, of the regiment.

Letters which have come to hand, from him to his two sons, dated from Berwick in November, 1738, appear to show that the regiment was again quartered there in that year.

1739

Lieut.-General Whetham was this year promoted to the rank of General.

A Route, dated 15th May, directed the headquarters and 3 companies

[1] *W. O.* 26, Book 18.

[2] *W. O.* 26, Book 19, p. 17.

to march from Berwick to St. Albans, detaching 2 companies to Dunstable, 2 to Barnet, and 3 to Watford, Hatfield, and Hoddenden respectively.

In June, the headquarters were transferred to Barnet, and, on a further redistribution of companies taking place, the regiment furnished a great number of small detachments to places around London, until July, when orders were issued to march from their present quarters in three divisions :—headquarters and 3 companies to Canterbury, 5 single companies to Arundel, Cranbrook, Lewes, Dorking, and Maidstone, 1 company between Battle, Hastings, and Rye, and 1 between Dover and Ashford (Kent).

In accordance with a Royal Warrant, dated 12th June, the reduction in the establishment of the regiment, ordered in January, 1735, was cancelled, and the increase was now confirmed, the establishment being augmented to a total of 900 officers and men.

Towards the completion of six new corps of Marines now being raised, Lieut.-General Whetham's (12th) was ordered to transfer 100 men, including 10 eligible to be corporals, and 4 drummers (one of the latter being qualified as drum-major), to Colonel Wolfe's Regiment.[1]

1740

On the 1st January, the regiment furnished a subaltern's guard at Upnor Castle, and, on the 1st April, the headquarter companies were transferred from Canterbury to Sittingbourne, and places adjacent.

By Royal Warrant, dated 21st February, General Whetham's (12th) Regiment was directed to transfer 50 men, qualified for sergeants, corporals, or drummers to Colonel Cholmondeley's Regiment at Canterbury.

On the 7th May, the following scheme was drawn out for soldiers serving as marines on board His Majesty's ships :—

On board a ship of 90 to 100 guns, a complement of 1 captain, 1 lieutenant, and 100 men ; in a ship of 80 guns, 2 lieutenants and 80 men ; in one of 70 guns, the same ; in one of 50 to 60 guns, 1 lieutenant and 60 men ; in one of 40 guns, 1 lieutenant and 50 men ; in one of 20 guns, 1 lieutenant and 30 men. In sloops, 15 men supernumerary to their complement of 150 men.[2]

In the summer, the regiment pitched its tents near Newbury, where an encampment was formed of two regiments of Horse, three of Dragoons, and four of Foot, under Lieut.-General Wade.

On the 12th August, orders were received to march to Portsmouth, and remain "until such ships were available to embark them" (the regiment), presumably as marines on board the fleet. This order was, however, cancelled, and, on arrival at Basingstoke, the following distribution took place—4 companies to Reading, 2 to Abingdon, 2 to Thame, and 2 to Henley.

[1] Royal Warrant, 21st December, 1739.

[2] *W. O.* 26, Book 19.

Under instructions, dated 11th November, the headquarters and 3 companies marched to Lincoln, detaching 2 other companies to Newark, and the remaining 5 to Mansfield, Grantham, Worksop, Gainsborough, and Horncastle respectively.

1741

On the 8th January, a further distribution of companies took place, the quarters of the regiment being extended to Louth, Retford, Histon, and Brigg.

General Whetham died on the 28th April, whereupon the Lieut.-Colonel of the regiment (Scipio Duroure) sent the following petition to the King, undated, but apparently written at the end of April, or the beginning of May :—

The petitioner represented that :—

" 1. He had served the Crown for 37 years, 19 of which were of field rank; he had 'carried arms' two years, served four as Ensign and Lieutenant, and twelve each in the ranks of Captain and Major, and had been seven years Lieut.-Colonel.

2. That he was present at the battles of Hochstet (Blenheim), Ramillies, Oudenarde, and Malplaquet, and at the sieges of Southlewen, Ostend, Menin, Lille, Tournay, Mons, Douay, St. Venant and Bouchain, and had been wounded at Malplaquet and Southlewen.

3. That he had served as Brigade Major for fourteen years in Scotland.

He therefore prays that in consideration of the above, His Majesty would be graciously pleased to grant him the command of the regiment, now vacant by the death of General Whetham, or of such other regiment as His Majesty should think fit."

To this petition, the King graciously assented by appointing Lieut.-Colonel Duroure, on the 12th August, 1741, to the colonelcy of his own corps.

Whilst encamped this summer on Lexden Heath, Colchester, the regiment had a most serious affray with another corps, in which one man was killed and several on both sides wounded. It was ended by the prudence of the commanding officer, assisted by the other regiment.[1]

An order, dated 27th September, directed Colonel Duroure's (12th) Regiment to march as follows :—4 companies to Colchester, 2 to Ipswich, 1 each to Sudbury, Witham, and Woodbridge, and 1 between Halstead and Coggeshall, a further distribution taking place later to Milton, Martlesham, Kelvedon, Dedham, Strafford, and Barfield.

1742

England was now to shortly participate in a war which had been threatening for some time.

[1] MSS. Records, R.U.S.I.

DISTINGUISHED OFFICERS WHO SERVED IN THE 12TH.

JAMES WOLFE, AS A SUBALTERN.*

Ensign, 27th March, 1742; *Adjutant, 13th July,* 1743; *Lieutenant, 25th July; promoted Captain, 3rd June,* 1744, *in Colonel Barrell's (4th) Regiment.*

GENERAL SIR THOMAS PICTON, G.C.B.

Ensign and Lieutenant, 1771-78.

(**By the courtesy of the Editor of " The Connoisseur."*)

Charles VI, Emperor of Germany, having died in 1740, the succession of his daughter, Maria Theresa, as Queen of Hungary and Bohemia, was disputed by the Elector of Bavaria, who was aided by a French army. King George II resolved to support the House of Austria, and took steps to put his army on a war footing.

On the 27th March, James Wolfe, who later became the hero of Quebec, was gazetted Ensign in Colonel Duroure's (12th) Regiment. Two years previously, his father had been appointed Adjutant-General of the unsuccessful expedition against Cartagena, and it was intended that James, then thirteen years of age, should accompany him, but an opportune illness prevented. In the following year, James was appointed a 2nd Lieutenant in his father's regiment of Marines, his commission having been dated November 3rd, 1741, and, on his desiring a transfer to the Line, the King nominated him as above. (See Plate 6.)

It was now decided to send a force of 17,000 men to Flanders, as auxiliaries to the Austrian forces, and Colonel Duroure's (12th) was one of the corps selected to proceed on foreign service.

On the 5th April, the headquarters of the regiment were transferred from Colchester to Ipswich, and, eight days later, a move was ordered to Woolwich, Dartford, Eltham, and Erith. On the 22nd April, instructions were given that the regiment was to be concentrated at Romford, until required for embarkation for Flanders, which took place shortly after, the recruits being ordered to march to Gravesend.

A letter from the Secretary at War directed Colonel Duroure not to give up any deserters now serving in the regiment, who might be claimed by other corps; but they were to continue to do duty, on which condition they were to have His Majesty's pardon.[1]

The command of the force under orders for Flanders was given to Lord Stair, who was made a Field-Marshal, and he had with him 5 generals and 8 brigadier-generals.

On the 27th April, the force was reviewed on Blackheath by King George II, who was accompanied by his two sons, the Prince of Wales and the Duke of Cumberland, and it consisted of 3 troops of the Horse Guards (Blue), 5 regiments of dragoons, and 13 of infantry, the occasion attracting a vast number of spectators, many years having elapsed since such a military display had been seen in England.

At this parade, one of the Colours of Colonel Duroure's (12th) Regiment was carried by Ensign James Wolfe, whose stature (though his age was now only 15 years and 4 months) was already that of a tall man, and he eventually attained the height of 6 feet 3 inches.[2]

On the 11th May, Colonel Duroure's Regiment was ordered to be completed forthwith to its full establishment, the recruits to join it, and a report to be made on its completion.

[1] *W. O.* 4, Book 37. [2] Beckles Willson, *Life and Letters of James Wolfe*, p. 18.

It was now the custom (and for some years later) for a proportion of the women of a regiment to accompany it when proceeding either on field or foreign service, and the limit now allowed on board the transports was 5 per company.[1]

The regiment embarked at Dover for Ostend at 10 P.M. on the 18th May (N.S.).[2]

At this juncture, France and Great Britain, with their armies, were in a singularly embarrassing state, since both countries professed to act merely as auxiliaries to their respective allies, and no declaration of war had been made by either.

[1] Beckles Willson, *Life and Letters of James Wolfe*, p. 19.
[2] Letter from Colonel Doroure to his son Frank, of the above date.

CHAPTER III

FLANDERS, GERMANY, BATTLE OF DETTINGEN, FLANDERS, BATTLE OF FONTENOY, GREAT BRITAIN, HOLLAND, ENGLAND, AND MINORCA. 1742–1751

1742

ON arrival at Ostend, on the 19th May, the troops of Lord Stair's Army marched to Bruges and thence to Ghent, and had, at every opportunity, shown them, *en route*, a not too friendly feeling on the part of the Belgians.

There was some difficulty at first in finding accommodation for our troops, until they gradually became quartered on the unwilling inhabitants, whilst affrays with the populace were of frequent occurrence.

Referring to one of these, Beckles Willson relates how a British soldier, when testing the quality of some meat in the market place at Ghent, was slashed across the face with a butcher's knife, which resulted in the butcher being run through the body by the soldier's comrade. Thereupon followed a general tumult between an infuriated soldiery and the burghers, with several casualties on both sides, the burgomaster finally issuing an edict that "whoever should offer the least affront to the subjects of the King of Great Britain should be whipped, burnt in the back, and turned out of the town."[1] The troops of Lord Stair's army passed the winter at Ghent.

1743

James Wolfe was joined in February by his brother Edward, who had been gazetted on the 19th to an ensigncy in Colonel Duroure's (12th) Regiment, and, on the departure of the army from Ghent towards the Rhine, the discomforts of a strenuous march, combining bad weather, roughness of the country, bad water, and scarcity of food, were a severe trial for the two delicate lads.[2]

[1] Beckles Willson, *Life and Letters of James Wolfe*, p. 20. [2] *Ibid.*

In a letter to his mother, dated St. Tron, February 12th, Edward Wolfe writes :—

> "Dear Madam,—I got your letter of the 23rd January. . . . This is our fifth day's march, and we have had very bad weather all the way. I have found out by experience that my strength is not so great as I imagined, but I have held out pretty well as yet. . . . We have lived pretty well all the way, but I have already been glad to take a little water out of a soldier's flask and eat some ammunition bread."

The following extract is from a letter from Colonel Duroure (12th) to his son Frank, dated Aix-la-Chapelle, 25th March (N.S.).

> "Upon some rumour that we were soon to march, I came here yesterday, and return to Hambach tomorrow. All the grenadiers with my regiment join the 1st Division under Lieut.-General Ligonier and Brigadier Hughes, which moves this day.
>
> *1st April.* Brigadier Frampton leads the 2nd Division, the three battalions of Guards, and Lord Rothes the 3rd Division, consisting of Howard's (19th), Pulteney's (13th), and Handasyde's (31st); a proportionate number of Austrians and Hanoverians are to pass the Rhine at the same time, and the rest of the Foot, with the Horse and Dragoons, are to follow soon after."

In a letter, dated, Bonn, April 7th, Edward Wolfe writes to his father as follows :—

> "Dearest Sir,—I am sent here with another gentleman, to buy provisions, for we can get none upon our march, but eggs and bacon and sour bread; but I have lived on a soldier's ammunition bread, which is far preferable to what we get on the road. We are within two leagues of the Rhine, which, in most peoples' opinion, we shall pass on the 14th, and then encamp. . . . We had a sad march last Monday morning. I was obliged to walk up to my knees in snow, though my brother and I have a horse between us, and, at the same time, I had it with me. . . . This is the first opportunity I have had since I wrote to you from Aix-la-Chapelle, which letter I hope you have received."[1]

In the middle of May, the allied forces of Queen Maria, 40,000 strong, mustered at Höchst, between Mayence and Frankfort.[2]

On the march from Flanders, Lord Stair had been reinforced by some Austrian regiments, under the Duke D'Aremberg, and by 16,000 Hanoverians who had been wintering in the neighbourhood of Liége.

Marshal Noailles, the French commander, with 60,000 men, had planned to engage Lord Stair's forces, while 50,000, under M. de Broglie, were to guard Alsace, and prevent the Austrians from crossing the Rhine.

On May 14th, Marshal Noailles crossed the Rhine six miles below

[1] Beckles Willson, *Life and Letters of James Wolfe*, pp. 25–6. [2] Skrine's *Fontenoy*, pp. 75–6.

Worms, which induced Lord Stair to leave Höchst for Aschaffenburg, where he established his headquarters, and wrote to D'Aremberg to join him, which the latter failed to do. Marshal Noailles then occupied the Maine on either side of Aschaffenburg, cutting off supplies.

Letter from Colonel Scipio Duroure (12th) to his son, Frank, dated, Höchst, near Frankfort, 11th June (N.S.) 1743.

"After being encamped for some time on this side of the Maine, it was judged proper to form a Camp on the other side, to preserve communication with Mayence, for the supply of our Magazines of Forage up the Rhine. Ten days ago, we crossed over all the British Foot and Dragoons with the Hanoverian Foot, and three regiments of their Horse, leaving our Horse encamped on this side, between this and Sillingen, the Duke D'Aremberg's Quarters, whose Austrian Foot, and two regiments of Dragoons remained likewise on this side.

The Camp on the other side of the River extended to the Right to Keltersbach, and the left to Schwanheim, the River in our Rear, and the Wood of Franckfort, about half a mile in our Front.

The English Generals were in the village of Keltersbach, and the Hanover Generals in that of Schwanheim. The inconvenience of this camp was that it was commanded on the Right by a rising ground above the village of Keltersbach, where the River makes an Elbow, and turns to the Right, with high Grounds on the other side, which flanked the whole plain on which we were encamped, but as, at that time, the French had not yet crossed the Rhine or the Neckar with any considerable Force, it was of no great matter. But as soon as Intelligence was had, that they were come over with their whole Army, the Brigade of Guards and that of Onslow's were ordered to take possession of the Hill and encamp there, and the Grand Guard we had all along kept upon the Hill was now posted further, and a field of Battle was traced with the Camp Colours, so that every Corps should know its ground. It is a little Plain that extends from the high Ground above the village of Keltersbach, and it was here that the Duke D'Aremberg, with Marshal Nieperg, met Lord Stair the day before, when all the General Officers of the three Nations were also desired to be present.

They all approved the Disposition Lord Stair had made, of the Infantry in two Lines within half Muskett shot of the Wood, with the Horse and Dragoons in a third Line, to support the Foot, a Battallion to face the Gorge or opening between the Wood and the River, with a Regiment of Dragoons to support them. While we were on the Spot, the three Marshalls received fresh Intelligence that the French Army was come that day within three Leagues of us, on this side of Darmstadt, much Superior to ours, there still remaining 17 Squadrons of Hanoverians on the march; notwithstanding which, they came to an immediate Resolution to draw up on the plain above, not knowing but what the Enemy might force a March in the Night, and take possession of it by break of Day, the Consequence of which might have proved fatal to us. Therefore orders were immediately given for the Troops and the Artillery to march up directly to the Ground marked out for them in the Morning, and to leave the Tents standing with a small Guard over them and the Baggage. As we had three Bridges between this and Sillingen, it facilitated bringing over our Horse in a Trice, and all the British and Hanover Horse, Foot and Dragoons were all upon their Ground before One in the Morning. The Austrian Foot had further to come, and by using only one of the three Bridges, when they might have used all after our Horse had gone over, they did not come to their ground before Light. When the whole were got up, it made up 44 Battallions and 43 Squadrons, besides 10 Squadrons of Austrian Dragoons who were ready to come over, when the Resolution was taken to retire.

Our 13 Battallions with the two Grenadier Comps. of Pulteney and Bligh (whose Regiments were still behind to guard the Bridge at Newit over the Rhine), with 9 Battallions of Austrians on their Left formed the 1st Line; the 13 Battallions of Hanoverians, with the other 9 Battallions on their Left, formed the 2nd Line, and our 29 Squadrons, with 14 Squadrons of Hanoverians on their Left, formed the 3rd Line. Our 20 Pieces of Cannon and 6 of the Hanoverians, besides the Train of the Austrians, were to be placed where they should be most wanted. In this Situation we continued till 3 in the Afternoon without any Tidings of any movement of the Enemy towards us. They had not stirred from the Ground they came to the Day before, which made our Generals judge, they had no mind to attack us in that Stronghold, but would still draw nearer to us, to oblige us to keep in a Spot, proper enough to fight in, but not to encamp in and Subsist. Therefore the Resolution was taken, to march back to our Ground below, instantly to strike Tents, and send off the Baggage and the Troops to follow, and get over on this side, each Nation having a Bridge allotted to it, by which means, we were all got over before 12 that Night, a Battallion and 200 Horse being left to make up the Rear of all. Both in marching up the plain, and coming back on this side, altho' most of it Night Work, and very sudden, was managed with the greatest Order, without the least Confusion or Accident, both officers and men observing the profoundest Silence, and showing as true a spirit for Action without Bombast, as ever I have seen in the Actions of last War, and I would not on any account have wished that we had crossed the River without drawing up yesterday as we did. Had we not done so, it would have damped the men's Spirits and confirmed them in the Opinion that their Fatigues were only to turn out into Spithead Expeditions. They are now convinced to the contrary and acquainted with the Nature of what it is to prepare for Action, and were they now to return home without striking a Stroke, Yesterday's appearance and Disposition, at such short warning, has done them more Service, and taught them more than ten Summers' Encampments could have in England. As we can no longer subsist here, from the difficulty of getting our Magazines from the Rhine up the Maine, the next thing will, I believe, be to remove our Camp between Frankfort and Hanau, and to secure the upper part of the Maine beyond Aschaffenburg, there to await the arrival of the Hessians, and additional Hanoverians, now on their march, as well as the remaining 17 Squadrons of the latter, the last of which will be with us in six Days. We shall then be in a condition to undertake something, and provided the Enemy gives us fair play, and does not chicane us with Rivers, Woods, and Entrenchments, we shall very little mind the odds of a Superiority of 8 or 10 thousand against us, with, we may safely say, the Flower of all the Armies in Europe, on our side."

King George II, accompanied by the Duke of Cumberland and Lord Cartaret, arrived at the camp of the Allies on the 19th June, and found to his disgust that the Earl of Stair and the Duke D'Aremberg, the two commanders, were not only divided in their counsels, but at actual enmity with each other.

The army he found drawn up in a valley, altogether in a faulty position; the men were on half rations, and the horses dying for want of forage. Nothing but a choice of difficulties remained, either to surrender to the French, or attempt a retreat. The King determined on the latter alternative.

Two days later, James Wolfe wrote to his father, when encamped before Dettingen, as follows:—

"Camp near Aschaffenburg.
June 21st, 1743, (N.S.)

"Dear Sir,—Captain Rainsford joined the regiment yesterday; he brought us your letter and made us both very happy. . . . My brother is at present very much fatigued with the hard duty he has had for some days past. He was with a party last night and saw shot fired in earnest, but was in no great danger, because separated from the enemy by the River Maine. The French are on the other side of the river, about a mile from us. We have now and then small skirmishes with them. They attacked the other night a party of our men, but were repulsed with the loss of an officer and four or five men killed, and some made prisoners. They desert prodigiously; there were yesterday no less than forty deserters in the camp that came over in the middle of the day, and brought with them great numbers of horses, for the river is fordable. 'Tis said there are 2000 Austrian Hussars come to us; I fancy they will harass the French a little. The Hessians, Pulteney's (13th), and Bligh's (20th) Regiments have not yet joined us, as likewise some Hanoverian Horse. . . . We shall soon know what we are to do now that our King is come. His Majesty came two days ago. The Duke of Cumberland is declared Major-General. . . . Colonel Duroure (12th) who acts as Adjutant-General, was thrown from his horse yesterday by a Hanoverian discharging his pistol just by him, and was much bruised. We are all sorry for it. He has been very good to his ensigns this march; we have had the use of his canteens whenever he thought we had occasion for them. We are now nearly forty miles from Frankfort, which we marched in two days and two nights, with about nine or ten hours' halt, in order to gain a pass that is here, and now in our possession. The men were almost starved on that march. Both they and the officers had little more than bread and water to live on, and that very scarce because they had not the ammunition bread the day it was due. But I believe it could not be helped.

We have left a very fine country to come to the worst I ever saw. I believe it is in the Prince of Hesse's dominions. The King is in a little palace in such a town as, I believe, he never lived in before. It was ruined by the Hanoverians, and everything almost that was in it carried off by them, some time before we came. They and our men now live by marauding. . . . The French are burning all the villages on the other side of the Maine, and we ravaging the country on this side.

I am now doing, and have done ever since we encamped, the duty of an adjutant. I was afraid when I first undertook it that the fatigue would be too much for me, but I am now used to it, and hope it will agree very well with me. . . . We both join in love and duty to you and my mother, and I am, dear Sir

Your dutiful and affectionate Son,

J. Wolfe."

After Tattoo, on June 26th (N.S.), Lord Stair, at Aschaffenburg, issued orders for the army to march at daybreak, every precaution being taken to observe silence when striking tents, and in moving off; and on arrival on their new ground, the troops to remain under arms, still preserving silence and no fires allowed.

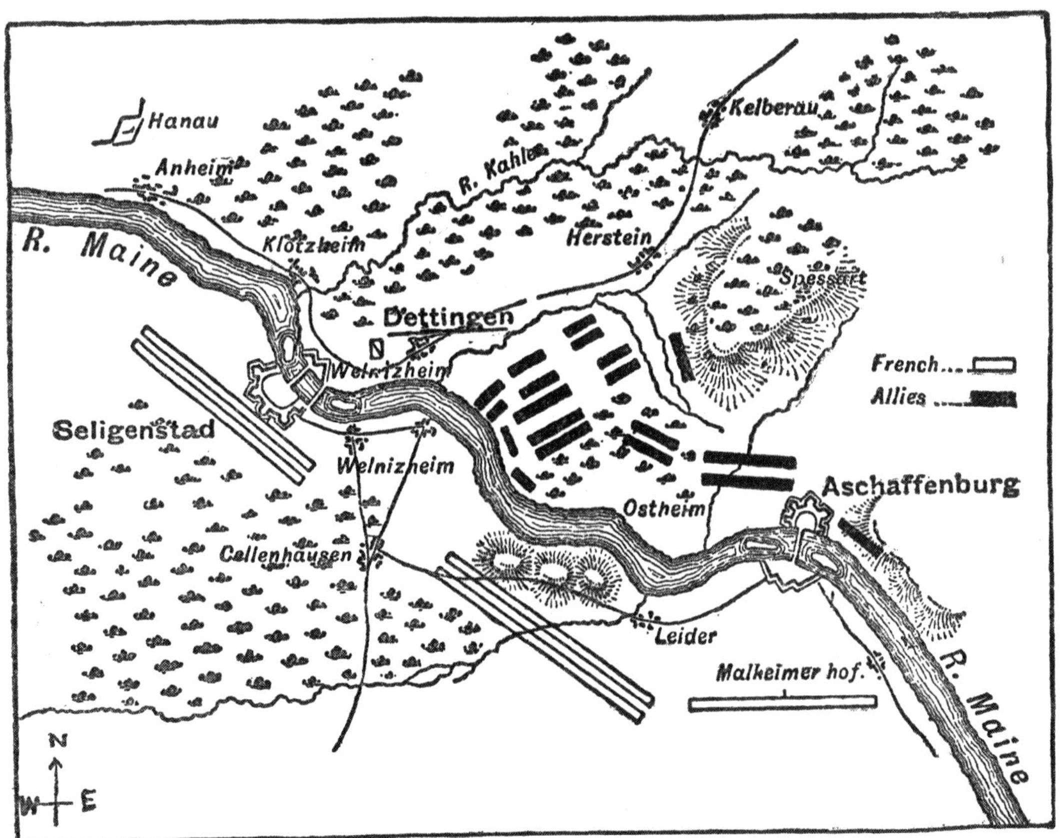

Battle Plan of Dettingen.

On Noailles becoming acquainted with our move, he sent his nephew, the Duc de Gramont, with 3000 men, to cross the river at Selingenstadt, three miles in advance of the allies, thereby occupying the defiles of Dettingen, and blocking the British retreat.

At 4 A.M. on June 27th, Duroure's (12th) and the other regiments began their march to Dettingen. When the King of England saw the French pouring over the bridge, he halted his columns, and drew up his army in order of battle.

The following was the formation of troops in the first line :—

To the left of all, within a furlong of the river, were the 3rd Dragoons, Johnson's (33rd), Campbell's (21st), Royal Welsh Fusiliers (23rd), Duroure's (12th), Sowle's (11th), Onslow's (8th), and Pulteney's (13th). On the right of Pulteney's, stood an Austrian Brigade, and then, in succession, the Blues, Life Guards, Inniskilling and Royal Dragoons.

To the left of the second line were :—Bligh's (20th), Huske's (32nd), Ponsonby's (37th), Handasyde's (31st), T. Howard's ("Buffs"), and, in rear of the cavalry on the right, the 7th and 1st Dragoon Guards, 4th and 7th Dragoons, and Scots Greys.

Opposite to them were the French in two lines, with a reserve in a third line, the infantry being in the centre, and cavalry on both flanks.[1]

Battle of Dettingen

The following account is an extract from the *London Gazette*, dated, Whitehall, June 29th, 1743.

Despatch from the Right Hon. the Lord Cartaret, dated at Hanau $\frac{\text{20 June}}{\text{1 July}}$ 1743.

"The King, having received certain intelligence that the Marshal de Noailles was endeavouring to prevent the junction of the Hessian Troops under command of Prince George of Hesse, and the eight Hanoverian battalions under General Druchleben, with the main body of the Army, sent orders to Prince George and the said general, on their march, to halt at Hanau, and determined to join them with the whole army ; accordingly on the $\frac{15}{26,}$ in the Evening, His Majesty gave orders that the army should hold itself in readiness to march the next morning early, and at about 4, the troops began to file off in two columns, when the Duke de Noailles, perceiving this motion, immediately ordered a detachment of his army to march along the Maine towards Selingenstadt, where the French infantry passed that river over two bridges, and the cavalry forded it a little above the said village, in order to prevent the junction of our armies, their artillery forming the rear guard. As soon as it was within reach of our army, it began at 10.30 A.M. to play on our flank.

Notwithstanding, our army continued its march, and, by perseverance, arrived in a little wood, behind which the French army was ranged in order of battle ; their right wing was covered by the Maine, and supported by a battery erected near Maynsling on the other side of the river. The left extended itself towards the hills, and had behind it a little rivulet and the village of Dettingen.

[1] Fortescue, Vol. ii. p. 95.

The French army, amounting to nearly 30,000 men, was drawn up in two lines and a rear guard, commanded by the Duke de Noailles, and the Duke de Chartres and several other Princes of the Blood were present; the Household Troops were in the centre, supported by the infantry.

The King, having given his orders to the respective generals of the army, with the greatest calmness and resolution placed himself on the left wing, at the head of the British Infantry, on foot, sword in hand.

Our army drew up in order of battle as well as it could in the wood, and extended itself as far to the front of the enemy as the ground would allow.

On the right of our army, at the entrance to the wood, the Hanoverians erected a battery which flanked the enemy, and did great service in the heat of the action; another was erected by the English on the left, and a third by the Austrians in the centre.

Such were the dispositions of the two Armies till about 12 o'clock, when ours advanced to charge the enemy.

The troops of the French King's Household attacked with great vigour our centre, which gave way a little, but soon rallied, and repulsed and drove them off the field.

After this defeat of their Household Troops, the French Army, perceiving itself attacked on all sides, quitted the field, passed the rivulet behind it, and posted itself in order of battle, upon an eminence commanding the plain. This they abandoned on our troops marching towards them, and, retiring to Kleinevelsheim, they at last retreated in great disorder towards the village of Selingenstadt, where they re-passed the Maine with precipitation and confusion; several were drowned, and a great many died of their wounds in the pursuit. The great number of killed who were found dispersed on all sides, shows that their loss must be considerable, and it is computed at about 4000 killed, wounded, drowned, and taken prisoners.

On our side, our loss is computed to be nearly 1500 killed and wounded. The British Army, and all the forces of the Allied Army who were engaged in this action, behaved with the utmost resolution, bravery and intrepidity."

Another account in the *London Gazette* says:—"His Britannic Majesty was in the heat of the action throughout, sometimes mounted, but more often on foot, and, after he dismounted, a cannon-ball broke the bough of a tree immediately over his head. The Duke of Cumberland was extremely active—had a horse shot under him, and a wound through his leg."

Letter from Colonel Duroure (12th) to his son Charles, dated "Field of Battle at Dettingen," between Aschaffenburg and Hanau.

"27th June, 1743, (N.S.)

"We have this day had a pretty warm action against 25 or 30 thousand of the French Army, who began at 1 A.M. to cross the Maine, about half a league below Dettingen.

We came up with them in time to make them re-cross it faster than they came, by obtaining a complete Victory over them, owing (next to God) to the Presence of His Majesty, who charged himself on foot, and to the gallant Behaviour of His Royal Highness the Duke of Cumberland, who was Major-General of the Day. . . . The Action lasted till 3, with a continued Cannonade from 9 o'clock. The Brigade of Foot Guards did not engage, so that all our Friends there are well. . . . The Morning of the Action, I was lolling in Lord Stair's Coach, not dreaming I could have put on boots for several days, but, when Necessity drove, I forced them on and find myself this Evening none the worse, although I have fairly tired three Horses; one of them indeed was shot under me, as likewise one I had lent to Ensign Wolfe, who acted as Adjutant."

The defeat of the French army opened a passage for the Allies to Hanau, where they marched the morning after the battle.

Another letter from Colonel Scipio Duroure (12th) to his son Charles, describing more fully the Battle of Dettingen dated, Hanau, 27th June, 1743.

"I had the pleasure of writing to your Brother three days before the Battle, and glad I am to have the pleasure of writing to you three days after it.

Orders were sent to Camp to march on the night of the 15th. Accordingly, about 8 o'clock, we struck our Tents, and lay upon our Arms till 1 A.M., when we began our March to Hanau, arriving at about 9 at our halting ground; at about 9.45., the French began to play their Cannon on the Hanoverian Horse on our Left. A Detachment of a few battalions, including Colonel Duroure's, was at first sent to the Right, to occupy a Wood towards which the French were moving, and we soon after saw their whole Army assembling towards that direction, upon which, the Queen's whole Armies were ordered towards the Right; but, in the meantime we had raised a Battery and were replying to the Enemy's Cannon, which, some time after, they shifted against our Flanks, upon our Cavalry and Infantry, the former of which suffered pretty much by it. The French then raised another Battery to our Front, which pointed directly towards our Regiment, and killed a few men, but, being by good fortune situated under a little rising Ground, which was very sandy, we could see their shots strike upon the Earth, and fly about over us, so that we were able to avoid most of them. They kept up that fire and the one on our flanks very briskly, and we stood both above an hour without stirring. It was thought proper at last to advance, which, in spite of its being in Face of their Cannon, our Soldiers did very boldly. . . . The two Armies drawing nearer each other, the French Cavalry, composed of their Gendarmerie and Mousquetaires (the prime of the Nobility and Gentry of France, and consequently the flower of their Army) marched up to charge the Foot, facing our Regiment. We let them come almost to within pistol Shot, when letting go all our Fire upon them, and loading and firing again with all the Quickness imaginable, we made some Slaughter among them and forced their horses to incline to the Left of the Line, where they received a fresh fire, which did them considerable damage, but, at last, they broke through our first Line, cutting at one or two Regiments pretty severely; When advancing on our Cavalry in the Second Line, they forced them down to the Austrian Infantry, who pouring in all their shot on them, destroyed almost every one of the French cavalry that had passed the first Line, so that, in short, those Troops were entirely demolished.

This Business done, the French Infantry advanced upon us, and our Soldiers let them come on with all the patience and Resolution in the world, till at last, the British begged of their Officers, to let them fire, which (when the Enemy were about 60 Yards from us), they did, not by Platoons (for there was no governing them) but with perpetual Volleys from Right to Left, loading almost as fast as they fired without ceasing, so that the French were forced to retreat. We let them rally again without advancing upon them, and they came on us a second time, when they were received exactly as before, our men actually making a Joke of them, and firing as if it were at a parcel of Sparrows, in such a manner as forced them to run away. While we were making our Disposition for this second Onset, we had the pleasure of seeing our Horse repulsing theirs, till, at last, both their Horse and Foot being routed, we remained Masters of the Field of Battle, which they left with great precipitation and Confusion, our Cannon advancing upon them as they fled both from Right and Left, and making great Havoc among them in their crowding through Dettingen, where there was a narrow Pass which they could not avoid, and which was afterwards seen covered with dead Bodies. Amidst their Hurry and Confusion, their Bridge over the Maine broke down, and many were drowned.

They threw some of their Cannon into the River ('tis said 16 Pieces) and we took 12 more, also some of their Colours and Standards, and have made prisoners of some of their Gendarmerie and Mousquetaires, among whom are several of their top Gentry and Nobility, whom we have sent back on their Paroles. We have lost Lieut.-General Clayton, and the Duke D'Aremberg is wounded, but 'tis thought he will recover; the Duke of Cumberland, who behaved with the greatest gallantry, is shot through the Calf of his Leg, but 'tis hoped he is doing well. *Il se comporta comme un Ange.* The British Foot have had 7 officers and 143 men killed, and 12 officers and 309 men wounded, which, considering they constituted the first Line, and stood the Brunt of the whole Engagement, is a very trifling Loss. Your Father's Regiment has lost Captain Phillips, whose leg was taken off by a Cannon Ball, but he lived till night, and Lieut. Munro, who was killed on the Spot by a Musquet Shot, 26 men killed, 68 wounded, a great many of whom will recover. . . . It was 11 at night before we came to the Ground where we were to encamp after the Battle, and the men having left their Knapsacks, Haversacks and Tent Poles behind (so as not to be encumbered with them in the Fight), could not pitch their Tents, any more than the Officers (whose Baggage had not yet joined them), till late next morning, and the Rain poured down as if a second Deluge had been coming. At last, the men who had been for the Tent Poles came back with most of them (but all the Knapsacks and Haversacks had been carried off), and the Men pitched their Tents for Shelter only, for the Ground was so wet, there was no lying down, and not a bit of Straw to be found. This was the second Night we had been without Sleep, besides the fatigue of our Marching and Engaging; for our comfort, about five hours after we had got the

Tents up, there came an Order to march, and it fell to the Lott of Colonel Duroure's (12th) Regiment among others, to form the Rear Guard for the Baggage and Artillery of the whole Army, so that we marched all the next Night again, and until 12 Noon the day following. Thus we were three Days and three Nights without Sleep, and the two last days without a bit of Bread, which fatigued us most terribly. (*Nous n'avons pas cuillé nos Lauriers à trop bon Marché.*)

At last we came to our Encampment half way between Hanau and Frankfort. I have forgotten to tell you that the 6000 Hessians and the additional Hanoverians, which the King furnishes as Elector, lay encamped near Hanau, and were not present at this Action; and our Foot Guards were on Outpost Duty, and not in the Field of Battle.

The whole ended between 3 and 4 o'clock in the Afternoon, the Engagement having begun about one. Many who had seen actions before, say they never knew or heard of a hotter one than this was for the time it lasted, and this Action was not expected by any of our Generals, though, as it turned out, it was of the utmost Consequence to the Queen's affairs; for, had not the Day been ours, we must either have been all cut to pieces, or starved, in a part of the World from whence we could not get, except by cutting our Way through the Enemy.

We have lost some of our Baggage by the pusillanimity of Servants, whom a Cannon Ball made run away, but they pretend that the French Hussars plundered it all; some of the Officers of our Regiment have suffered that way, but I have had the good fortune to escape as well as in my Person."

James Wolfe, busily employed with an adjutant's duties in the field, was in the very thick of the fight, as was his brother. The latter wrote three days afterwards to his mother the following account:—

"June 30th, 1743.

"Dearest Madam.—I take the very first opportunity I can to acquaint you that my brother and self escaped in the engagement we had with the French, the 16th of June last (O. S.), and, thank God, are as well as ever we were in our lives, after not only being cannonaded two hours and three-quarters, and fighting with small arms two hours and one quarter, but lay the two following nights on our arms, whilst it rained for about twenty hours in the same time; yet are ready and as capable to do the same again. We lost one Captain and a lieutenant. . . . Our Colonel had a horse shot under him, but escaped himself. The King was present in the field, The Duke of Cumberland behaved charmingly. . . . Duke d'Aremberg is dangerously wounded. We took two or three general officers, and two princes of the blood, and wounded Marshal Noailles.

Our regiment (12th) has got a great deal of honour, for we were in the middle of the first line, and in the greatest danger. . . . My brother has written to my father, and, I believe, has given him a small account of the battle, so I hope you will excuse it me. The Emperor is come to Frankfort, and we are encamped about two leagues from it; it is said that the King is to meet him there, and that there's a peace to be made between the Queen of Hungary and the Emperor.

I hope I shall see you some time or another, and then tell you more; but think now that I have given you joy and concern enough. Pray my duty to my dearest father, who I hope is well.

I am dearest Madam,

Your dutiful and Affectionate Son

E. Wolfe." [1]

Colonel Duroure's (12th) Regiment had Captain Phillips, Lieutenant Munro, and 27 rank and file killed; Captain Campbell, Lieutenant Williams, Ensign Townshend, 3 sergeants, 2 drummers, and 60 rank and file wounded on this occasion, more than any other regiment.

The French loss is variously estimated between 4000 and 6000 killed and wounded, whilst the Allies had over 2300 *hors de combat*.

A week after the battle, James Wolfe was able to send an account of it to his father.

"Hochst, July 4th (N.S.), 1743.

"Dear Sir,—This is the first time that I have been able, or have had the least time to write, otherwise I should have done it when my brother did. The fatigue I had the day we fought, and the day after, made me very much out of order, and I was obliged to keep my tent for two days. Bleeding was of great service to me, and now I am as well as ever.

The army was drawn out this day s'en-night between a wood and the river Maine, near a village called Dettingen, in five lines—two of foot and three of horse. The cannon on both sides began to play about nine o'clock in the morning, and we were exposed to the fire of theirs (said to be above fifty pieces) for near three hours, a great part of which flanked us terribly from the other side of the water. The French were all the while drawn up in sight of us on this side. About twelve o'clock we marched towards them; they advanced likewise, and, as near as I can guess, the fight began about one. The Gens d'Armes, or Mousquetaires Gris, attacked the first line, composed of nine regiments of English Foot, and four or five of Austrians, and some Hanoverians. They broke through the Scotch Fusiliers, whom they began the attack upon; but before they got to the second line, out of two hundred there were not forty living, so they wheeled and came between the first and second lines (except an officer with a standard, and four or five men, who broke through the second line and were taken by some of Hawley's Regiment of Dragoons), and about twenty of them escaped to their army, riding through an interval that was made for our horse to advance. These unhappy men were of the first families in France. Nothing, I believe, could be more rash than their undertaking.

The second attack was made on the left by their Horse against ours, which advanced for the first time. Neither side did much, for they both

[1] Beckles Willson, *Life and Letters of James Wolfe*, p. 35.

retreated; and our Horse had like to have broke our first line in the confusion. The Horse fired their pistols, which, if they had let alone, and attacked the French with their swords, being so much stronger and heavier, they would certainly have beat them. Their excuse for retreating—they could not make their horses stand the fire!

The third and last attack was made by the Foot on both sides. We advanced towards one another; our men in high spirits, and very impatient for fighting, being elated with beating the French Horse, part of which advanced towards us, while the rest attacked our Horse, but were driven back by the great fire we gave them. The Major and I (for we had neither Colonel nor Lieutenant-Colonel), before they came near, were employed in begging and ordering the men not to fire at too great a distance, but to keep it till the enemy should come near us; but to little purpose. The whole fired when they thought they could reach them, which had like to have ruined us. We did very little execution with it. As soon as the French saw we presented they all fell down, and when we had fired they got up and marched close to us in tolerable good order, and gave us a brisk fire, which put us into some disorder and made us give way a little, particularly ours and two or three more regiments, who were in the hottest of it. However, we soon rallied again, and attacked them with great fury, which gained us a complete victory, and forced the enemy to retire in great haste. 'Twas luck that we did give way a little, for our men were loading all the while, and it gave room for an Austrian regiment to move into an interval, rather too little before, who charged the enemy with great bravery and resolution. So soon as the French retreated, the line halted, and we got the sad news of the death of as good and as brave a man as any amongst us, General Clayton, who was killed by a musket ball in the last attack. His death gave us all sorrow, so great was the opinion we had of him, and was the hindrance of anything further being done that day. He had, 'tis said, orders for pursuing the enemy, and if we had followed them, as was expected, it is the opinion of most people, that of 27,000 men they brought over the Maine, they would not have repassed with half that number. When they retreated, several pieces of our artillery played upon them, and made terrible havoc; at last we followed them, but too late; they had almost all passed the river. One of the bridges broke, and in the hurry abundance were drowned. A great number of their officers and men were taken prisoners. Their loss is computed to be between six and seven thousand men, and ours three thousand.

His Majesty was in the midst of the fight; and the Duke behaved as bravely as a man could do. He had a musket shot through the calf of his leg. I had several times the honour of speaking with him just as the battle began, and was often afraid of his being dashed to pieces by the cannon-balls. He gave his orders with a great deal of calmness, and seemed quite unconcerned. The soldiers were in high delight to have him so near them. Captain Rainsford behaved with the greatest conduct and bravery in the world. I sometimes thought I had lost poor Ned when I saw arms, legs, and heads beat off close by him. He is called 'The Old Soldier,' and very deservedly. A horse I rid of the Colonel's at the first attack was shot in one of his hind legs, and threw me; so I was obliged to do the duty of an Adjutant all that and the next day on foot, in a pair of heavy boots. I lost with the horse, furniture and pistols which cost me ten ducats ; but three days after the battle got the horse again, with the ball in him, and he is now almost well again, but without furniture and pistols.

A brigade of English and another of Hanoverians are in garrison in this town, which we are fortifying daily. We are detached from the grand army, which is encamped between Frankfort and Hanau, about twelve miles off.

They talk of a second battle soon. Count Khevenhuller and Marshal Broglie are expected to join the two armies in a few days. We are very well situated at present, and in a plentiful country.

Had we stayed a few days longer at Aschaffenburg we had been all starved, for the French would have cut off our communication with Frankfort.

Poor Captain Merrydon is killed. Pray mine and my brother's duty to my mother. We hope you are both perfectly well,"

I am, dear Sir,
Your dutiful and affectionate son,
J. Wolfe."

James Wolfe was now 16½, and the ability he had displayed at Dettingen obtained for him a fortnight later, July 13th (N.S.), the Adjutantcy of his regiment, which was followed, on July 25th, by promotion to the rank of lieutenant.

It was at Dettingen that Lord Townshend, who was with Wolfe at Quebec and present at his death, made his first essay at arms ; and a drummer boy, who was near him, being struck on the head by a cannon ball, which scattered his brains all over him, an old soldier, who was standing by, told him not to be afraid. "Oh ! " said the young officer, "I am not afraid, I am only astonished that a fellow with such a quantity of brains should be here."

The following is a Review of the Battle of Dettingen, written by Colonel Duroure, Colonel of the Regiment, who was Acting Adjutant-General throughout the battle, dated, "Hanau, 19 July, 1743."

"When we marched from Aschaffenburg, 3 battalions of Guards, with two battalions and 14 squadrons of Hanoverians, formed the Rear Guard, which was the reason they had no share in the Action, although it was expected otherwise, and that most likely an attempt would be made upon our Rear.

This body was under the command of Lieut.-General Ilsen, and, by Marshal Nieperg's Advice, was posted in the Wood, close to the Hill, on our Right. . . . Had the Brigade of Guards, however, been with us after the first, or second engagement, and marched down the Hill on the enemy's Left Flank, it would have raised such a New Spirit in our two Lines of Foot that must infallibly have made a second Hochstet (Blenheim) of that Day, with the help of our Cavalry, for they would certainly have been over the Fords (which lay much nearer than the two Bridges at Selingstadt), before the Foot could have reached those Bridges ; and, attacked by both Horse and Foot, without any manner of resource, the enemy would have had no option but to quietly lay down their Arms, with the alternative (in case of retreat) of crossing a plain from Dettingen—a long mile to the village of Klein Weltsgheim—and another mile thence to Selingstadt, and between these two places were the Fords for the Cavalry.

On the other hand, had the Enemy made their Stand at the Village of Dettingen, upon the rising Ground with the Rivulet in their Front, and the Wood on their Left, with a Battery in the Village, and a Brigade of Foot in the Wood, to Flank us both Ways (as we did at Wynendal), it would most likely have turned the Day into what Malplaquet was for the Dutch ; for, penetrate we must, there was no Retreat for us, 'Dye or Conquer' must have been the Word, and had we actually been worsted, the turn it would have given to the affairs in Europe is readily conceived, and we saw it once very near taking that turn, but it is only within the past three days I have become clear on a point which gave us all great uneasiness at the beginning of the action. I should have observed, before this, that if our Friends had been more tractable, and had not laid Impediments in the way (which can't be further explained in a Letter) we should, in our March to Aschaffenburg, have shortened the work much by laying a Bridge or two at Selingstadt, and taking the very ground across the Wood the enemy took, with Aschaffenburg in the Rear ; but the Maine could not on any account be crossed at that time, altho' the rest of the Hanover Cavalry had joined us, which we wanted at Keltersbach.

His Majesty was yesterday upon the field of Battle, and he ordered the Commanding Officers of regiments to meet, and give an Account in Writing of the manner their regiments were drawn up during the Action, what Regiments were on the Right and Left of them, their further movements, and what orders they received. After the Foot had been drawn up in two Lines, with the Horse in the Rear, there remained a vacant space between our Foot and the River, which was ordered to be occupied by our Cavalry, there having been room for 8 or 10 Squadrons, but very much exposed to the Enemy's Cannon. Bland's (3rd Dragoons) three Squadrons had already taken up their Ground, and the rest were following, when Colonel Sowle (who with his Regiment closed the Front Line on the Left) perceived an Officer, riding at a slow pace, with a paper in his hand, towards the Dragoons, and took him for an Austrian, or Hanoverian ; Major Honeywood did the same, when, all of a sudden, he turned about, clapped spurs, and rode away. Everybody cried out to fire at him, but he was out of reach before anyone was ready with his piece, and, immediately after, the Squadron of Mousquetaires Noirs and the Chevaux Legers advanced at a trot, and, whether designedly, or for want of good Order, they appeared to be eight or ten deep. I am inclined to think it was the Chevaux Legers, close at the heels of the Mousquetaires, to reinforce them against our Squadrons.

General Clayton, who was then at the head of Sowle's (11th) Regiment, called out to Major Honeywood :—'Honey, charge them,' and, at the same time, ordered Colonel Sowle to fire some of his Left-hand Platoons on the Enemy's Squadrons, which had so good an Effect that it stopped the Gendarmes and Carabiniers from supporting the Mousquetaires and Chevaux Legers, as some of them told me that night. So, after they had made great havoc with Bland's Squadrons, finding themselves peppered by the fire of both Lines of Foot of our Left, and not at all supported, there was nothing left for those who did not drop by our shot, than to wheel about at a full

Gallop, and make their way through the openings they perceived in Sowle's, and the next Regiment to him, whom they had broken in some places at their first onset, but who soon recovered, and gave them another fire on their return, which prevented more than 40 out of the 150 returning.

When Sowle's Regiment gave the first fire, which certainly was very *àpropos*, and did great service, Ill Luck would have it that it ran along to the Right like a running fire.

Lord Stair was then giving some orders to the Left of the second Line, and judged that the whole fire had been given without Orders, against the Directions to preserve ours, and first to receive the Enemy's, then giving ours, and charging with Bayonets. Under this Impression, he rode up to the Front Line to stop it, while, at the same time, a Battallion of the Austrian new Regiment, in the Second Line, fell to firing thro' great eagerness, because they saw the Front Line fire. This was just when his lordship was riding up, and, how any of us, eight or ten with him, came to escape it, I can't conceive. I was sent to stop them, but to no purpose, until every man had fired his piece. The only Excuse the Commanding Officer could make was that his Men were very young, and did not know what they were about; this truth might, I believe, have been said of a good many of us in an Army in which, perhaps out of 40 thousand, 500 had ever seen a shot fired.

Luckily, when I was speaking to the Commanding Officer, the Mousquetaires and Chevaux Legers had turned about, and were galloping like mad to the Front, to the Left of the Battallion. I took them, at first sight, to be some of our own people, but soon found out my Mistake by distinguishing the Mousquetaires' *soubreveste*,[1] a part of their dress, so called, that has no Sleeves and comes over their Coat, which is red, trimmed with silver; the *soubreveste* blue, with the *Croix de St. Louis* before and behind. Their appearance happily drew the fire of my Battallion (12th) on them, and saved our Friends in the Front from any more of it, but, at the same time, it is true that, whilst the firing of the Left of both Lines did them great harm, it also did great Mischief to Bland's men, who were intermixed with them.

It is plain, from what has been said, that the Officer with the paper in his hand, had come to take a view of our Position, and, finding that we had occupied the spare plot of Ground with only three Squadrons of Bland's, was the reason for their beginning the Attacks sooner than they intended, which, had they supported as they might have done, would have given odds in their favour, for they would have found our Lines of Horse in the Rear pretty much *les uns les autres*, which would not have happened if the Ground I have mentioned had been occupied in time, as had been ordered, when we might have met them with an equal Front, and, I am sure, with more weight, supported in the Rear with much more Cavalry than they would have.

The following, we are informed, was the Formation of their troops (all in one line) before they marched down the Hill over the Rivulet.

Brigade des Gardes Francais	Brigade de Noailles	Brigade de Turenne	Brigade de Orleans	Brigade de Auvergne	Brigade de Navarre
	27 Esc.				
6 Battns.	5 Battns.	5 Battns.	4 Battns.	5 Battns.	6 Battns.

[1] Super-vest.

When they came on to narrower Ground, they formed into two or three Lines, and, when retreating, the three Brigades in the Front Line formed into three hollow Squares, and kept so until they had past the brow of the hill, and had lost Sight of us ; then, we are told, they went off with great precipitation, for by the time we reached some of our Cavalry on the hill, they had all got beyond Klein Weltsgheim, where we stopped by the side of an open Wood that joins it, waiting for the Infantry, who were quite exhausted from fatigue and want of Bread, having laid on their Arms all the night before, which, I suppose, was the reason we proceeded no further, and lay on the plain all that night, some with their Tents and some without, with a continued downpour of Rain, which commenced two hours before darkness set in, and lasted until noon the next day.

The Army, in the morning, proceeded over the little river Kensig, to Hanau, where the Rear Guard did not arrive until late at night, resulting, for most part of the Army, in a third night's lying upon their Arms, the Consequence of which we daily felt by the increase of our Sick.

In the meantime, from the Accounts we have had of the Enemy, it took three days before their people had quite rallied to their Colours and Standards, having scattered themselves in the Woods on the other side of the Maine in a most unaccountable manner.

I should have observed before that their Brigade of Guards had orders to advance at the same time as the Mousquetaires and the Chevaux Legers, but their officers could not prevail on them, and they did not stir an Inch ; but, when retreating, they were the first to reach the Bridges, and a great many had not patience to cross them, but attempted to swim the River, in which great Numbers were lost. On that account, they have since got the Nickname in their Army of '*Les Canards et plongeurs du Maine*,' as Monsieur De Grammont has similarly that of '*Le Baton rompu*,' being charged with having exceeded his Orders in descending from the Hill over the Rivulet, with a view to signalise himself, and thereby obtain *un Baton de Maréshal*, which, I can only say, he once had a very fair chance of, if his Troops had only done their Duty in the most moderate manner, and without any extraordinary acts of valour.

I must add no more except to allude to His Majesty having given 'Life' to the Infantry, by heading it himself on foot,[1] Lord Stair doing the same on the Left, after having led on and *rafermi* (closed up) our Cavalry on the left of that again. The firmness and calmness of temper His Majesty showed during the whole Action, which I had often occasion to Witness, from the several messages I had the honour to deliver to him, and the orders he gave

The *London Gazette*, dated 1st July, 1743, states that "King George II placed himself at the head of the British infantry and led them on foot." Colonel Duroure and James and Edward Wolfe (in their accounts of the battle) do not mention that His Majesty led any particular corps.

In the Hon. J. W. Fortescue's *History of the British Army*, it is stated (on the British dispositions being complete) that the King had "with difficulty been prevented from stationing himself on the extreme left" (of the first line) which apparently was on the left flank of the 33rd Regiment.

It is recorded by more than one historian, that, as soon as the firing commenced, the King's horse taking fright, bolted with him to the rear, and on its career being checked, His Majesty dismounted, and rejoined the infantry on foot, but history fails to record the exact position he then took up. The honour, then, of having been led by the King at Dettingen (the last occasion on which a British monarch led his troops in action) may, it would appear, be equally shared by the whole of the British infantry present.

me to convey—that steady Countenance of His Majesty, I say, which was the same at the beginning, in the middle, and at the End of the Action, cheered us all to a pitch that I cannot express. . . . After missing the advantage we might have received from the Brigade of Guards, and the other four Battallions, had they joined us on the Left of the Enemy, before the latter could cross the Rivulet, it had been proposed to cross the Enemy's own Bridges that night or next morning, and to send to Prince George of Hesse, to cross at Hanau, with his six Battallions, and the 8 additional Battallions of Hanoverians then just arrived, the Consequence of which must have been an entire Rout of the only army left to France for dictating to the Powers of Europe.

I must not conclude this without giving you the history of one of our Guns the enemy carried off, as I heard it from a first cousin of mine. He said their Regiment did not claim much merit for it, as it was owing to the Horses taking fright, and running directly towards their Regiment, when we were placing a Battery about the centre of our Front Line, and the Drivers were wiser than to go up to the Mouth of their pieces to stop them."

The King fixed his headquarters at Worms, from which place Lieutenant and Adjutant Wolfe wrote to his father the following :—

"Camp near Worms, September 1st (N.S.), 1743.

"Dear Sir,—By a letter I received from you some days ago, I had the happiness and satisfaction to hear that you and my mother are well ; but it being my brother's turn to write (which we intend to do by turns every Saturday), I put off answering until to-day.

The army passed the Rhine the 23rd (N.S.), a little below Mentz, and came to this ground yesterday. It was possessed by the French before the action of Dettingen. The fortifications of the Swiss camp (who would not pass the Rhine) are just by, and those where the bridge was that the French went over upon is close to it. The boats that made our bridge below Mentz are expected here to-morrow for the Dutch troops to come over, who, we hear, will be with us in six or seven days. There are numbers of reports relating to Prince Charles's army, so that I won't pretend to send you any account of it, only that most people think he has not passed the Rhine. The French are now encamped between Landau and Wissemberg. Captain Rainford says, if they have any spirit they will attack us here before we are joined by the Dutch, and so I believe our Commanders think, for they have just given orders to have all encumbrances removed from before the front of each regiment, in order to turn out at a minute's warning, and a chain of sentries are to be immediately placed in front of the camp. Our camp is tolerably strong : we are open in front, with hills, from which cannon cannot do us much harm. At the bottom of these hills is a little rivulet ; in our rear is the Rhine. The left is secured by the town of Worms, the right is open, but neither the front nor right have greater openings than we have troops to fill them up ; so I think we are pretty safe. I am just now told that a party of our Hussars have taken a French grand guard : they have killed the captain and thirteen men, and have brought sixty-four to Worms. I'm convinced of the truth, because some gentlemen of our regiment saw them go along the line, and are going to buy some of the horses. I cannot tell if the Duke of Cumberland knows what you mentioned in your letter: I have never had any opportunity of enquiring. It is but a few days that he is come abroad: he has marched, since we crossed the Rhine, at the head of his second line of English, which is his post. He is very brisk, and quite cured of his wound. His presence encourages the troops, and makes them ready to undertake anything, having so brave a man at the head of them. I hope some day or other to have the honour of knowing him better than I do now; 'tis what I wished as much as anything in the world (except the pleasure which I hope to enjoy when it shall please God), that of seeing my dear friends at Greenwich. Poor Colonel Duroure is, I am afraid, in great danger: we left him on the other side of the Rhine very ill with a bloody flux. Our Major is at the same place likewise, very much out of order. Our Colonel was never more wanted to command us than now.

I shall say nothing now of the behaviour of the Blue Guards; I wish they may do better next time, and I don't doubt but they will. It would give me a great deal of sorrow if they did not.

We have a great deal of sickness amongst us, so I believe the sooner we engage (if it is to be) the better. I hope you, Sir, and my mother are perfectly well. I heartily wish it, and that you may continue so. My brother joins with me in duty and love to both.

I am, dear Sir,
Your dutiful and affectionate son,
J. Wolfe."

The excitement of the English populace in consequence of the Battle of Dettingen was excessive. In one of Walpole's letters we read:—" I expect to be drunk with hogsheads of Maine-water, and with odes to his Majesty and the Duke, and *Te Deums*. We are all mad—drums, trumpets, bumpers, bonfires. The mob are wild, and cry, 'Long live King George, and Duke of Cumberland, and Lord Stair, and Lord Carteret, and General Clayton, that's dead.' "

While such was the prevailing delirium at home, the camp at Worms was broken up, and the regiments repaired to their old stations for the winter.

Colonel Duroure's (12th) Regiment had been encamped several weeks on the banks of the Kinzig, and in August marched towards the Rhine, which it crossed above Maintz, and was employed in various services until October, when the Army marched in divisions back to Flanders.

The regiment formed part of the 5th Division under Major-General the Earl of Rothes, and arrived on the 22nd November at Brussels, whence they proceeded to Ostend for winter quarters.

Five medals have been struck to commemorate the Battle of Dettingen.

Two of these were made in gold and in silver, one in silver only, and two in bronze; of the different specimens, there is only one bearing the figure of "Justice" (on reverse), and, one of these, described as follows, is in possession of the Officers, 2nd Battalion Suffolk Regiment:—

Obverse: Bust of George II, *l*, laureate, hair short, in armour, and mantle fastened with brooch on the shoulder.

Leg. GEORGIUS SECUNDUS DEI GRATIÆ REX.

Reverse: Justice seated, facing, holding sword and scales, tramples upon Tyranny; around lie a yoke, broken chains, scourge &c.

Leg. PARCERE . SUBJECTIS . ET . DEBELLARE . SUPERBOS (To spare the humble, and subdue the proud.—Virg., *Æn.* vi, 854.)

Ex. OB GALLOS VISTOS APUD DETTINGEN. PER EXER: FŒD: SUB AUSPICIO GEO: II. Iun: 16. 1743. (On the defeat of the French at Dettingen by the Allied Army, under the auspices of George II, 16 June, 1743.) 1·45 inches. MB.Æ. bronze.

The device intimates that the French were justly punished by their defeat at Dettingen, for their attempt to oppress the Queen of Hungary.

This medal, executed in England, is dated according to the old style.[1] (See Plate 3.)

The following is an approximate list of Officers in Colonel Scipio Duroure's (12th) Regiment for the year 1743 (10 companies), with dates of commissions, prior to the Battle of Dettingen.[2]

Colonel Scipio Duroure	12 August, 1741	
Lieut.-Colonel Wm. Whitmore	30 March, 1742	
Major John Cosseley	30 March, 1742	
Captain Edward Phillips		Killed at Dettingen.
,, Charles Rainsford	2 October, 1731	
,, George Stanhope	5 January, 1739	
,, Matthew Wright	9 July, 1739	
,, Edward Harris	7 November, 1739	
,, Sampson Archer	14 August, 1740	
,, William Watson	30 March, 1742	
,, Joseph Phillips	8 April, 1742	
Captain-Lieutenant James Campbell	8 April, 1742	Wounded at Dettingen.
Lieutenant Martin Emmenes	18 February, 1729	
,, Henry Powell	26 August, 1731	
,, Maurice Gouldston	19 January, 1736	
,, Richard Field	9 July, 1739	
,, John Romer	23 January, 1741	
,, James Stevens	30 March, 1742	
,, John Salt	8 April, 1742	
,, Robert Munro	8 April, 1742	Killed at Dettingen.
,, Wm. Millington	8 April, 1742	
,, George Williams	11 May, 1742	Wounded at Dettingen.
Ensign John Whetham	17 July, 1739	
,, Corbet Parry	28 January, 1741	
,, John Scott	7 March, 1741	

[1] *Medallic Illustrations*, A. W. Franks, Vol. ii, pp. 578–9.

[2] No officers of the regimental staff are shown in the old manuscript Army Lists at the Public Record Office, London.

Ensign James Wolfe	27 March, 1742	
„ Ruvigny De Cosne	8 April, 1742	
„ Thomas Townshend	8 April, 1742	Wounded at Dettingen.
„ John Delgarno	8 April, 1742	
„ John Oswald	21 April, 1742	
„ Thomas Trenchard	11 May, 1742	
„ Edward Wolfe	19 February, 1743	(Brother to James Wolfe).

1744

On the 3rd June, James Wolfe was promoted to a captaincy in Colonel Barrell's (4th) Regiment, and his brother Edward was promoted Lieutenant, on the 18th, in Col. Duroure's (12th) Regiment.

In September, the latter was ailing, and, on the approach of the cold weather, became seriously ill, rapid consumption following, and a few weeks later he died in his 17th year, his military enthusiasm and sweetness of disposition having endeared him to all.[1]

Early in the spring, the Allied Army had begun to assemble at Brussels and Ghent, the total force available for field operations having been 14,370 sabres and 40,388 bayonets, and the command of the British contingent was conferred on Field-Marshal Wade, then 73 years of age. Much precious time was wasted in wrangling by the allied generals, their hesitation and discord having been due to D'Aremberg's persistent obstruction, and to the Dutch generals' timidity. By the 10th June, they had advanced southwards by the right bank of the Scheldt, and halted between Oudenarde and Bottelaer.[2]

A Royal Warrant, dated 27th June, directed an increase of 2 companies in Col. Duroure's (12th) Regiment, making 12 in all, each company consisting of 3 sergeants, 3 corporals, 2 drummers, and 70 private men, besides commissioned officers, to be raised by Beat of Drum.[3] When completed, they were quartered at Ipswich, and thence moved to Guernsey.

After traversing the Scheldt, the Allies moved slowly westwards, and took up a new position nine miles from Courtrai, and, on August 3rd, pursuing the left bank of the Scheldt, they halted between Espierre and Avelghem, where Wade fixed his headquarters.

These manœuvres did not escape the notice of the French Commander, Marshal Saxe, who was not dismayed by the numerical superiority of the Allies. October 2nd found the latter encamped on the left bank of the Scheldt, British and Hanoverians at Avelghem, and Dutch and Austrians at Helchin, which led Saxe to suspect that they at last intended to attack him.

General Wade shortly after concentrated his troops in winter quarters on the Deyntz, the Dutch and Austrians moving to St. Denis, near Ghent, and, obtaining permission to return home, he resigned the command in Flanders to Sir John Ligonier. Thus ended the year's ignominious campaign, in which no general engagement occurred.

[1] Beckles Willson, *Life and Letters of James Wolfe.* [2] F. H. Skrine's *Fontenoy*, pp. 95–100.
[3] *W. O.* 26, Book 20, p. 141.

1745

The *London Gazette* of March 12th announced the appointment of the Duke of Cumberland to a commission as " Captain-General " (Commander-in-Chief) of all His Majesty's Forces, in conjunction with the troops of His Majesty's Allies. His Austrian Commander was Marshal Count Königsegg, the Prince of Waldeck commanding the Dutch contingent. One of the six aides-de-camp, now appointed to the Duke's staff, was Captain Robert Napier, who, twelve years later (1757), was gazetted Colonel of the 12th Foot.

At the time of Cumberland's arrival, the French Marshal Saxe was busily besieging Tournay, one of the principal fortresses of Flanders. The object of the Allies being to relieve it, a number of our regiments were pushed forward from Brussels, and, when Saxe heard of the British advance, he massed four-fifths of his army at the village of Fontenoy, giving battle to Cumberland and the Allies on May 11th.[1]

In their march between Soignies and Moulbais, the allies were reinforced by 2 Hanoverian and 4 Dutch battalions, including a portion of the garrison of Ath.

The Army which fought at Fontenoy was now complete, and was thus composed :—[2]

	Battalions	Men	Squadrons	Men	Total
Right wing, English, Hanoverians and Austrians, under the Duke of Cumberland and Marshal Königsegg	25	16,500	45	6,750	23,250
Left wing, Dutch, commanded by the Prince of Waldeck	27	17,550	40	6,000	23,550
Grand Total . .	52	34,050	85	12,750	46,800

The following is a list of troops who fought at Fontenoy :—

Household Cavalry.

Horse Guards, 3rd and 4th (Scots) Troops, now 1st Life Guards.
Horse Grenadier Guards, 2nd (Scots) Troop, now 2nd Life Guards.
Blue Guards, now Royal Horse Guards (The Blues).

Regiments of Horse.

The King's Regiment of Horse (1st Dragoon Guards).
Ligonier's, or the Black Horse (7th Dragoon Guards).

[1] Beckles Willson, *Life and Letters of James Wolfe*, p. 49. [2] Skrine, p. 145.

Dragoons.

Hawley's, or the Royal (1st Royal).
Campbell's, or the Royal North British (Royal Scots Greys).
Bland's, or the King's Own (3rd Hussars).
Cope's, or the Queen's (7th Hussars).
Stair's (6th Inniskilling).

Infantry.

1st Foot Guards (Grenadier Guards).
2nd ,, (Coldstream Regiment).
3rd ,, (Scots Guards).
The Royals (1st Royal).
Lieut.-General T. Howard's (3rd, The Buffs).
Onslow's (8th, " King's ").
Sowle's (11th).
Duroure's (12th).
Pulteney's (13th).
Major-General the Hon. C. Howard's (19th).
Bligh's (20th).
Campbell's (21st).
Royal Welsh Fusiliers (23rd).
Earl of Rothes' (25th).
Bragg's (28th).
Late Handasyde's (31st).
Skelton's (32nd).
Johnson's (33rd).
Cholmondeley's (34th).
Lord John Murray's Royal Highlanders (42nd).

Saxe left 21,550 men to carry on the siege of Tournay, while he, with the remainder of his army, 66,000, took up a position between Tournay and the Allies, his right resting on the village of Antoing, his centre pushed forward to Fontenoy, and his left at Barri Wood; the whole of his position being carefully fortified. Fontenoy was strongly entrenched, three redoubts were erected between it and Antoing, and two on the edge of Barri Wood, facing Fontenoy, the nearest of which was the Redoubt d'Eu.

Over 100 guns were employed in the defence of these works, and the whole of the Irish Brigade, in the French service, amounting to 6 infantry regiments, and 1 cavalry (that of Fitz-James)[1] were posted on the French left.

[1] *Irish Brigades on the Continent*, by P. Higgins, p. 77.

BATTLE OF FONTENOY.

At 2 A.M. on the 11th May, the allied army advanced against the formidable position occupied by the enemy, and Colonel Duroure's (12th)

BATTLE OF FONTENOY
May 11th, 1745
IN WAR OF AUSTRIAN SUCCESSION 1741–1748

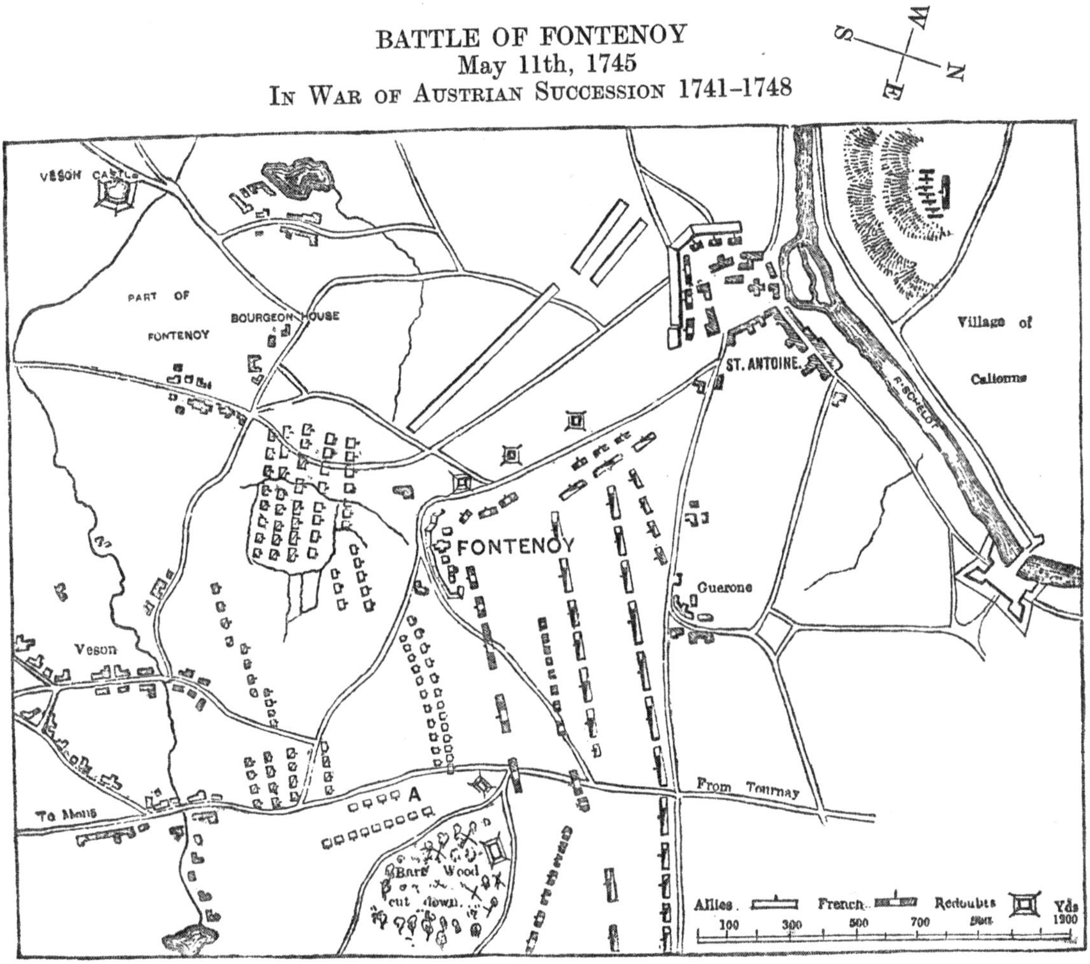

The French Army held the line St. Antoine, Fontenoy, Barri Wood, with the intervals defended by redoubts.
The British main attack was delivered between Fontenoy and Barri Wood.
The Dutch were ordered to attack Fontenoy and St. Antoine.
The 12th Regiment, with the 13th, 42nd, and Böschlanger's Hanoverian Regiment, at first formed a detached Brigade to attack Barri Wood on right flank.
A. 1st position of 12th Regiment.

Regiment formed a part of the right attack in the brigade commanded by Colonel James Ingoldsby, of the 1st Guards, with Pulteney's (13th), Lord John Murray's Royal Highlanders (42nd), and Böschlanger's Hanoverian Regiment.

The Brigadier's orders were to assault the Redoubt d'Eu, at the edge of Barri Wood, and carry it with the bayonet, whilst the Duke of Cumberland, at the head of 15,000 British infantry and 40 guns, attacked the French centre, and the Dutch advanced against Fontenoy and Antoing.[1]

When the right and left attacks had succeeded, the rest of the British infantry were to press forward through the gap between Fontenoy and Barri Wood.

Ingoldsby advanced with his brigade at 6 A.M., but halted before Vezon, and, whilst reconnoitring the ground in front, sent for guns, whereupon the Duke promptly complied by sending him (under command of Captain Mitchelson, R.A.) three 6-pounders, which he himself accompanied.

Mitchelson then advanced his guns to the front of Duroure's (12th), and opened fire. Meanwhile the Brigadier wheeled his regiments back by platoons. Duroure's (12th) and the Royal Highlanders were in front, supported by Pulteney's (13th) and the Hanoverians, and a slow advance began.

Brigadier-General Ingoldsby did not clearly understand his orders; his brigade again halted, and no further effort was made to carry the Redoubt d'Eu.

Cumberland and Königsegg were now confronted with the alternatives of beating a general retreat or pushing forward against the French position. They chose the latter. For the success of so daring a movement, it was necessary to storm the village of Fontenoy, on our left front, bristling with cannon, which would otherwise rake the advancing columns at point-blank range. The allied generals, therefore, resolved to make a second attempt on that improvised citadel, and, in order to stiffen the Dutch regiments operating against it, on our left, with a sterner element, Cumberland detached the Royal Highlanders (42nd) from Ingoldsby's command, and placed them, with another British regiment, under command of Prince Waldeck, with orders to attack Fontenoy.

There is some doubt as to the second regiment thus employed, and Captain Knight, in his history of "The Buffs," states that it was General T. Howard's, now "The Buffs." It is, however, of little importance, as this detached duty effected nothing. Moreover, the small number of casualties in these two regiments tends to prove that the operations on our left front were not of a serious nature.

Brigadier-General Ingoldsby, being now slightly wounded, was succeeded in his command by the Hanoverian General Zastrow, who ordered Pulteney's (13th) and Böschlanger's Regiments to fall in with Skelton's Brigade. Duroure's (12th) had been ordered to fall in on the right of the second line of the British main advance. The position of the regiment is said to have been on the right of a parallel, within 150 yards of a redoubt,[2] where it was exposed to a continual fire from the beginning of the action, and must have borne the full brunt of the Irish Brigade's onslaught, which two facts account for its heavy losses.

The main attack failed, the British regiments having been unsupported

[1] Shown on Plan as "St. Antoine." [2] *Gentleman's Magazine*, Vol. xv., p. 317.

by the Dutch, and, within seven or eight minutes, the prospect of a victory was converted into a defeat.

Thus deserted by their allies, and assailed by overwhelming numbers, our magnificent infantry were slowly pressed back, and it was with heavy hearts that Cumberland and Königsegg gave orders for a general retreat.

Our troops, still numbering 11,000, had been for twelve hours without food or rest. From 12 to 2 P.M. Duroure's (12th) took part in a general advance between Fontenoy and the Redoubt d'Eu, and at 2.15 P.M. the regiment retreated in perfect order on Vezon and Ath.

The battle in all lasted some eight hours, from 5 A.M. until 1 o'clock. The French losses amounted to 7139 killed and wounded, of whom 25 were Irish officers of rank.[1]

Colonel Duroure's (12th) Regiment had the following casualties :—

Killed ; Lieut.-Colonel Whitmore, Captain Campbell, Lieutenants Bockland and Lane, Ensigns Cannon and Clifton, 5 Sergeants, and 148 rank and file.

Wounded ; Colonel Duroure (mortally), Major Cosseley, Captains Rainsford and Robinson ; Lieutenants Murray, Townshend, Millington and Delgarno ; Ensigns Dagnia and Pearce, 7 Sergeants, and 142 rank and file.

Missing ; Captain de Cosne, Captain-Lieutenant Gouldston, and Lieutenant Salt.

The following official account is a copy of the second enclosure in the Duke of Cumberland's letter to the Earl of Harrington, Secretary of State, dated "Lessines, May 17th, 1745, N. S." It was therefore written 6 days after the battle. The original wording, punctuation, and spelling have been preserved.

A Relation of the Action between the Allied Army and that of France near Tournay, the 11th of May, 1745, N. S.

"The enemy opened their Trenches before Tournay the 30th of April at Night, and as they employed a very great and unusual Number of Workmen, the Siege advanced so fast that there was no Time to be lost ; but whatever was to be done towards obliging the Enemy to raise it was necessary to be put into Execution immediately.

The Generals of the Allied Army looked upon the raising this Siege as a Point of the highest Concern ; and H. R. H., the Marshal Königsegg, and the Prince of Waldeck resolved therefore to attempt it, though the Enemy was

[1] The Irish Brigade wore scarlet in the French service, and it is related that in their victorious charge against the British, they were met by the French Carabineers, who, mistaking them for English, fired a volley into them, killing and wounding several, until the shout of "Vive la France" made them sensible of their mistake. *Irish Brigades on the Continent*, P. Higgins, pp. 80, 81.

advantageously posted as well as superior in Number. With this View the Army marched on the 9th from Moulbay and encamped that Evening with the Right at Bougnies and the Left at Monbray, within a little more than Musket-shot of the advanced Posts of the Enemy.

The Generals went in the Evening to observe them, and could easily discover several of the Squadrons which were separated from our Army by a Country divided by a little Rivulet on our Left and by Underwood, Copses, and Hedges which they had filled with their Pandours and Grassins, and supported them by several little Squadrons drawn up in a Plain which rose by an easy Ascent to within a little Distance of their Camp, which was situated at the Top of that Rising beginning at Antoin leaving the village of Fontenoy in their Front, and extending itself towards their Left near a large Wood, which was beyond the Village of Veson towards the Center of our Right. This Village was also possessed by the Enemy, and covered by small Squadrons placed at little Distances from each other.

As we could not get into the Plain which was between their Camp and the Defiles on our Side, without first driving them from their little Posts; and as it was then late it was resolved to put off this Attempt till next Morning. Accordingly on the 10th, six Battalions and twelve Squadrons, with 500 Pioneers, six Pieces of Cannon and two Haubitzers were commanded from each Wing for this Service, which was performed with great Ease, the Enemy having been driven away everywhere to the very Top of the Rising near their Camp where they stood drawn up, as well to observe us, as to cover the Dispositions they were making behind that Line; H. R. H., the Marshal, and Prince Waldeck went upon the Plain, and having examined the Ground, we returned in the Evening to our Camp, after we had seen the Enemy burn a little Village somewhat short of Fontenoy, which they had fortified. We left the Detachments at the Posts they had taken, and the Order was given for attacking the Enemy in the Morning.

H. R. H. ordered that the Army should march at Two in the Morning; and as he had been informed that there was in front of the Village of Vezon, near the Wood, a Fort mounted with Cannon where 500 or 600 men might be lodged, he ordered Brigadier Ingoldsby with [1] *four good Battalions* and three Six-pounders to attack this Village Sword in hand, whilst the Prince of Waldeck should attack the Village of Fontenoy, which he had undertaken to do. Lieutenant-General Campbell was ordered to cover the Infantry of the Right Wing, which was commanded by Lieutenant-General Ligonier, whilst it should be forming, with 15 Squadrons by extending himself along the Plain from the Wood towards the Village of Fontenoy. But General Campbell having lost his Leg by a Cannon-shot, this Disposition, which had been trusted to him, did not take Effect. However General Ligonier formed the two Lines of Infantry, quite exposed, without any other Interruption from the Enemy than a brisk Cannonade, which did great Execution, till by order of H. R. H. he caused seven Pieces of Cannon to advance at the Head of the Brigade of Guards, which silenced the moving Batteries of the Enemy.

The Army was now in Order of Battle, and General Ligonier acquainted H. R. H. by an Aid de camp that he was ready, and if he approved it, would

[1] Battalions selected:—12th Foot, 13th Foot, 42nd Highlanders, and Böschlanger's Hanoverian Regiment.

march to attack the Enemy as soon as Prince Waldeck should march to the Village of Fontenoy as had been before agreed between them.

The Fort near the Wood should now have been attacked, and if that had been done, as H. R. H. ordered, it would, in all Probability have been carried, which would have greatly contributed to our further Success. But by some Fatality, Brigadier Ingoldsby did not attack the Fort, notwithstanding repeated orders sent to him by H. R. H. and General Ligonier.

When our two Lines were drawn up in very good Order with the Cavalry behind them, H. R. H. put himself at their Head, and gave Orders to march directly to the Enemy. Prince Waldeck moved at the same Time to attack Fontenoy, which the Left Wing did, but without Effect, and during this March there was a most terrible Fire of Cannon. We advanced nevertheless to the Enemy, and received their Discharge at the Distance of thirty Paces before we fired. Then Things had a very good Appearance, and there was a fair Prospect of a compleat Victory, for our Infantry bore down all before it, and the Enemy were driven three hundred Paces beyond the Fort and the Village, and we were Masters of the Field of Battle as far as to their Camp. But the Left Wing, though favoured by the Fire of our Batteries, and supported by two English Battalions which H. R. H. sent to favour the Attack of Fontenoy, not having succeeded in that Attack, and the Fort, as has been said before, not having been attacked at all, we found ourselves between cross Fires of Small Arms and Cannon, and were likewise exposed to that of their Front, so that we found it necessary to retire to the Height of Fontenoy and the Fort near the Wood, whence also there was a continued Fire which occasioned some Confusion. But by the Attention of H. R. H. and the Marshal it was soon stopt, and the Troops again put into Order.

It was then resolved to make a second Trial, and our Men, encouraged by the Generals, made the Enemy give Way once more, and they were driven to their Camp with great Loss ; but we also began to feel very sensibly the Diminution of our Numbers, and the Left Wing having remained where they were during this second Trial, we were again obliged to retire to the Ground between the Village and the Point of the Wood.

The Enemy's Cavalry attempted to break us as we retired, but, they were so well received by our Guards and Major-General Zastrow, of the Hanoverian Troops that the Regiment of Noailles was in good Measure destroyed, and the Carabiniers, by the Report of Deserters, had 32 Officers killed.

It was then resolved by H. R. H., the Marshal, and the Prince of Waldeck, that the whole Army should retire, and the Commanding Officers of Lieutenant-General Howard's Regiment[1] and of the Highlanders[2] were ordered to put themselves, the first in the Church yard of Vezon, and the others in the Hedges where they had been posted the Day before. The Cavalry was likewise drawn up to secure our Retreat, which was made in so good Order, the Battalions fronting the Enemy every 100 Paces, that there was not the least Attempt made by the Enemy to disturb us, which seems an argument that they had suffered very much.

The Baggage belonging to H. R. H. received Orders about Two to take

[1] The Buffs. [2] 42nd Highlanders.

the way to Ath. It remained during the Action at his Head Quarters at Brussoel, and marched about Three. The Marshal Königsegg had been hurt by a Fall from his Horse, and was a good deal fatigued, so after the Army was out of the Defiles, he went to Ath, where he arrived in the Evening; but H. R. H. kept constantly with the Right of the Army, and did not reach Ath till past Three in the morning.

The Infantry of the Right Wing has behaved very well, and suffered terribly upon this Occasion. The Hanover Troops as well Cavalry as Infantry, have had their Share with us in the Dangers, Fatigues and Loss. It is impossible to regret sufficiently the great Number of Officers, as well as private Men, who are missing. Most of them we know are dead. Lieut.-General Campbell had his Leg shot off, and is since dead. Major-General Ponsonby was killed on the Spot. Lord Albemarle and Major-General Howard, and the Brigadiers Churchill and Ingoldsby are wounded, General Howard in four Places.

Prince Waldeck on the Left, behaved with his usual bravery. Brigadier Salis and Colonel Boetslayer are killed.

The Behaviour of the Blue Guards is highly to be commended. The Lieut.-Colonel was wounded, and the Major distinguished himself particularly upon this Occasion by his Conduct and Care; The first Battalion of Guards remained the whole Day without once being put into Confusion, though they lost many brave Officers as well as Privates.

The Highlanders (42nd), late Handasyde's (31st), Duroure's (12th), and many others distinguished themselves.

The Honour gained by the Infantry is in a great Measure owing to the Conduct and Bravery of Lieut.-General Ligonier. Major-General Zastrow and Lord Albemarle did Everything that could be expected from brave and experienced Officers.

There are hardly any Prisoners but the wounded, and they were left at the Duke's Quarters at Brussoel, upon the Confidence of the Cartel, and the usual Behaviour upon such Occasions. We have not lost any Colours, Standards, or Kettledrums, but have taken one Standard. And the Cannon lost was left behind for want of Horses, the Contractors with the Artillery having run off with them so early that they reached Brussels that Day. The Army of the Allies was the next Day encampt in the Neighbourhood of Ath."

The following extract is from a letter, from Captain James Wolfe (late of Duroure's Regiment) to his father.

"Ghent, 4th May, 1745. O.S.

"Dear Sir.—I'm concerned I must send you so melancholy an account of a great but unsuccessful attempt to raise the siege of Tournay. I shall just tell what a letter before me from Captain Field, who commanded Colonel Duroure's (12th) Regiment, says of it.

We attacked a numerous army, entrenched with a multitude of batteries, well placed, both in front and flank. The action began at 5 o'clock in the morning, and lasted till 2 in the afternoon. There has been a great deal of slaughter, particularly amongst the infantry, officers more in portion than soldiers. The enemy's army was supposed to be 70,000, and ours about 50,000. The soldiers behaved with the utmost bravery and courage during the whole affair, but rather rash and impetuous. Notwithstanding the bravest attempts were made to conquer, it was not possible for us to surmount the difficulties we met with. . . . The old regiment (Duroure's) has suffered very much: eighteen officers and three hundred men killed and wounded: amongst the latter is Major Rainsford.

I believe this account will shock you not a little, but 'tis surprising the number of officers of lower rank that are gone. . . .

I am, dear Sir,

Your dutiful and affectionate Son.

J. Wolfe."

The number of subalterns killed was astonishing, so much that the French King, looking upon the English that were killed, said, 'Ma foy, ces gens meritoient de vivre,'[1] and Marshal Saxe said, 'Cette poignée de gens m'a fait plus de peine que tout le reste.' "[2]

The Allies' loss at Fontenoy was 7545 officers and men; the British regiments returned 1237 killed and 2425 wounded; the Hanoverians 432 and 798 respectively.

Our infantry of the right wing bore the burden and heat of that awful day, and owing to the bravery with which the troops fought, Fontenoy has been rightly called a glorious defeat.

The Duke of Cumberland, in his despatch, singled out the Brigade of Guards, Duroure's (12th), (late) Handasyde's (31st), and Lord John Murray's Royal Highlanders (42nd) Regiments for special praise, but every infantry regiment did its duty nobly.

To this Mr. Skrine adds:—"Fontenoy should be borne on the colours of all who shared in the glory of that day. May our country have defenders as staunch in the time of stress which is surely approaching."

Approximate list of Officers in Colonel Scipio Duroure's (12th) Regiment, for the year 1745 (12 Companies), with dates of commissions, prior to the Battle of Fontenoy.

Colonel Scipio Duroure	12 Aug.	1741	Wounded at Fontenoy, and died of his wounds.
Lieut.-Col. Wm. Whitmore	30 Mar.	1742	Killed at Fontenoy
Major John Cosseley	30 Mar.	1742	Wounded do.
Captain Chas. Rainsford	2 Oct.	1731	Wounded do.
,, Matthew Wright	9 July,	1739	
,, Sampson Archer	14 Aug.	1740	
,, Wm. Watson	30 Mar.	1742	
,, Henry Powell	14 July,	1743	
,, Richard Field	14 July,	1743	
,, James Campbell	14 July,	1743	Killed do.
,, Wm. Robinson	21 Sept.	1743	Wounded do.
,, Ruvigny De Cosne	11 Apr.	1744	Missing do.

[1] "My faith, these people deserved to live."

[2] "This handful of people has given me more trouble than all the rest."

Captain John Leader	25 June, 1744	
„ Edward Bate	26 June, 1744	
Capt.-Lieutenant Maurice Gouldston	14 July, 1743	Missing at Fontenoy
Lieutenant John Salt	8 Apr. 1742	Missing do.
„ Wm. Millington	8 Apr. 1742	Wounded do.
„ Geo. Williams	11 May, 1742	
„ Corbet Parry	14 July, 1743	
„ Thos. Townshend	14 July, 1743	Wounded do.
„ John Delgarno [1]	26 Mar. 1744	Wounded do.
„ Peter Chabbert	11 June, 1744	
„ Patrick Ogilvie	25 June, 1744	
„ George Bockland	2 Oct. 1744	Killed do.
„ Mathias Murray	29 Nov. 1744	Wounded do.
„ Samuel Lane	17 Apr. 1745	Killed do.
Ensign Hugh Adams	14 July, 1743	
„ Geo. Cockburn	17 Sep. 1743	
„ Chas. Cannon	26 Mar. 1744	Killed do.
„ Onesiphorus Dagnia [2]	11 Apr. 1744	Wounded do.
„ Robert Carter	23 June, 1744	
„ Hayward Stevens	24 June, 1744	
„ Peter Powrie	25 June, 1744	
„ Wm. Jenkins	25 June, 1744	
„ Gervas Clifton	25 Oct. 1744	Killed do.
„ Baker Pearce	22 Nov. 1744	Wounded do.

The 12th and 13th May were employed in examining the state of regiments, establishing hospitals, and despatching messengers to the French camp to enquire after the wounded and to complain of the treatment they received contrary to the rules of war. On the 15th, the Quarter-Masters were sent out to mark a camp, with Lessines on the right, and Rebay on the left, a small league from Ath.

On the 16th, the army marched, and took possession of the new camp, and a new Corps de Reserve was formed, consisting of the Highland Regiment, a squadron of the King's Dragoons, one of Stair's, 2 Austrian squadrons, 2 of Hop's Horse, and 2 of Marsau's Dutch Dragoons.

On the 19th, Colonel Ligonier marched with Duroure's (12th), Welsh Fusiliers (23rd), and Scots Fusiliers (21st) to relieve the King's Own (4th) Regiment at Ghent, late Ponsonby's (37th) at Bruges, and Ligonier's (48th) at Ostend, respectively.[3]

Colonel Duroure died of his wounds on the 21st May (N.S.), and was succeeded by Major-General Henry Skelton, from the 32nd Regiment, by commission dated 28th May. Major Cosseley recovered from his wounds, and was promoted to the lieut.-colonelcy, and Captain Rainsford, who had carried the standard at Fontenoy, was promoted Major.

A "State of His Majesty's British Forces in the Low Countries," dated 16th July, 1745, shows Major-General Skelton's (12th) Regiment in garrison at Ostend, with the Scots Fusiliers (21st Regiment).

[1] This name, spelt thus in the old manuscript Army Lists, is shown in Cannon's History as "Dalgaire."

[2] This name, spelt thus in the old manuscript Army Lists, is shown in Cannon's History as "Dagers."

[3] *London Gazette*, 21st to 25th May, 1745.

The regiment was afterwards encamped near Brussels, and the French, by their superior numbers, were enabled to capture several fortified towns.

James II, after his attempt in Ireland to regain the Crown of England, retired to France, pensioned by the French King. France, in order to trouble England, now offered Prince Charles, King James's grandson (the son of James Edward the Old Pretender), money and arms if he would go to Scotland and induce the Scots to rise in his favour.

Prince Charles Edward, encouraged by the result of the battle of Fontenoy, landed in Scotland, and was joined by several of the Highland clans, his design being to dethrone the King and restore the Stuart Line. Most of the King's troops being in Flanders, success attended his efforts for a short period. Major-General Skelton's (12th) Regiment was one of the corps ordered to England on this occasion, for the tide of invasion had already gone against the Royal Troops, they having met with a complete defeat at Preston Pans.

The regiment arrived at Gravesend on the 4th November, and was ordered to march:—headquarters and 5 companies to Ipswich, and the remaining 5 service companies were distributed in eleven detachments in Essex and Suffolk.

The headquarter companies at Ipswich were later transferred to Colchester. The rebels had entered England and penetrated as far as Derby, but, on the arrival of the reinforcements from Flanders, they had to retreat to Scotland; the Royal Troops met with yet another reverse at Falkirk, but recovered their lost honours by totally defeating the insurgents at Culloden, in 1746, which battle extinguished the rebellion.

A Route, dated 12th November, directed the regiment to march to Derby, and form part of the army assembled under the Duke of Cumberland, but, although Cannon states that the regiment pursued the Highlanders as far as Carlisle (and was before that town when the rebel garrison surrendered), the War Office Marching Order Books show that, on the 18th November, the regiment was ordered to Northampton, whence, on arrival, they were directed to march to Coventry.

In the following month they moved south, and an order, dated 22nd December, directed Major-General Skelton's (12th) Regiment, on arrival at Barnet, to march to Ewell, Dorking, Reigate, Westerham, and Brinstead, their quarters being further extended to Horsham, Grinstead, Epsom, and Leatherhead.

A dramatic occurrence had happened at Perth in September this year. A certain Captain Crosbie, of Prince Charles's French contingent, had been made prisoner. He was discovered to be a deserter from Skelton's (12th) Regiment, and it was determined to hang him. The Perth hangman, thinking it simple murder, refused to perform his office, and ran away. The Stirling hangman was then sent for, but, either through fright or from natural causes, he dropped dead. At last, a fellow-prisoner was induced by a pardon and a reward to do the deed, and performed it amid universal execration.[1]

1746–48

An order, dated 24th March, directed the ten companies of Major-General Skelton's (12th) Regiment to march from their present quarters to Rochester, Strood, and Chatham, and to remain until required to embark at Gravesend for Scotland, "with tents, camp necessaries, and whatever else may be necessary to take the field."[2]

A vexatious incident is reported to have occurred, which prevented the regiment from taking any active part in the Scottish campaign. While off the coast of Yorkshire, the transports, containing the 12th, 16th, and 18th Regiments, were warned that three French men-of-war were cruising in the neighbourhood; the ships went for safety to the Humber, and owing to the delay, the regiment did not arrive in Scotland until after the rebels had been finally crushed, on the 16th April, at the battle of Culloden. It was then quartered for several months at Perth.[3]

On the 26th December, its two additional companies, on their return from Guernsey to Portsmouth, were directed to march to Worcester and Ipswich respectively. The total establishment of the regiment this year was 977.

Throughout the summer of 1747 it was encamped in a rugged valley surrounded by gloomy precipices, near Fort Augustus in the Highlands of Scotland, until it was removed in the autumn to Berwick, and, on the 1st October, the two additional companies were ordered to march to Exeter. Major-General Skelton was this year promoted to the rank of Lieutenant-General.

An official order appears this year that "no officer under the degree of a brigadier shall ever appear, either in quarters or in camp, whether on duty or off, in any other coat than his regimentals or uniform, either old or new. No officer under the degree of a brigadier shall have either chariot or chaise."

On the 4th February, 1748, the additional companies were ordered to Croydon, and, on the 11th, the regiment, at Berwick, was to march to Newcastle, Gateshead, and North and South Shields, pending embarkation for Holland, the sick men left in Scotland being directed to join the additional companies in South Britain.

In the spring, the 12th took the field, and were engaged in several services in Holland until peace was concluded at Aix-la-Chapelle, on the termination, this year, of the war of the Austrian Succession, when the regiment received orders to return to England, and, on the 3rd November, was directed, on disembarking from Holland, to march to Exeter, excepting

[1] *History of Perthshire*, Marchioness of Tullibardine, Vol. i, p. 331. [2] *W. O.* 4, Book 41, p. 296.
[3] Colonel Gretton's *History of the 18th "Royal Irish."*

the sick, who disembarked at Whitby, and, on becoming convalescent, were ordered to march to Huntingdon, rejoining the regiment in the following year.

1749–51

On the 2nd May, 1749, the regiment was ordered to march to Plymouth and places adjacent, until required to embark for Minorca, where it was stationed two years.

It arrived there in relief of Colonel Wynyard's (17th) Regiment, which sailed for Ireland in July.

The garrison consisted of 5 foot regiments and a company of artillery, total, about 2400 effectives, of whom only a third could be put on duty at once, although there was a vast extent of works for 800 men to defend. Then, in case of attack, some allowance would have to be made for detachments, drawn from the above, for serving the platforms, repairing damage by the enemy, and other emergencies, apart from killed, wounded, sick, and deserters.

Minorca was divided into 4 Terminos,[1] the Termino of Mahon, the Termino of Alaior, the united Terminos of Mercadal and Fererias, and the Termino of Ciudadella.

There was generally a regiment quartered at Mahon. The officers had a house assigned to each of them, and the rank and file were dispersed in those of the lower burghers, which were converted into barracks for their use. The proprietors who were obliged to make room for these tenants received a small rent from the magistrates, and accommodated themselves elsewhere as they best could.

Alaior was considered good quarters for a regiment, though only nine companies were usually placed in garrison there. Ciudadella, was, however, allowed by everyone to be the best quarters in the island.

From June to the middle of September the weather is excessively hot.[2] The stay of the regiment at Minorca until June, 1751, was uneventful.

[1] The word Termino, from the Latin Terminus, signifying a boundary, limit, or border.

[2] John Armstrong, pp. 4 and 5.

CHAPTER IV.

GREAT BRITAIN, GERMANY, BATTLES OF MINDEN, WARBURG, FELLINGHAUSEN, AND WILHELMSTAHL.
1751–1762

1751

THE regiment was directed, on disembarking at Portsmouth from Minorca, to march to Godalming, with detachments at seven out-stations, but this was cancelled by a move of the headquarters and 8 companies to Winchester, 1 to Stockbridge, and 1 between Waltham and Alresford.

On the 1st July, a Royal Warrant was issued, assigning, for the first time, regimental numbers to cavalry and infantry regiments, and also regulating the clothing, standards and colours of the several regiments of the army, which (relating to the 12th) is referred to in Chapter XIV.

On the 9th August, the regiment was ordered to march in three divisions as follows :—3 companies to Newcastle and Gateshead, and 7 to Berwick ; and by the 31st the whole regiment was concentrated at Berwick.

1752–55

Until the spring of 1755, it was quartered in Scotland—in 1753–54 at Fort George—and, throughout the above period, was employed on the Highland roads.[1]

On being ordered, in March, 1755, to the south of England, the 12th, on arrival at Carlisle, were directed to march to St. Albans, and, on the 15th April, to continue the march, in three divisions, to Maidstone, where the regiment was reviewed (inspected), on the 4th June, by Lieut.-General Campbell.

The following is an extract from the Inspection Report :—

> " Officers ; properly armed, salute well, and are very alert on their duty.
>
> The non-commissioned officers do their duty well, and the sergeants wear sashes.

[1] MSS. Records, R.U.S.I.

Men ; of a good size. The Grenadier Company very tall, strong, well-made men.

Recruits ; well-made, strong young men.

Exercises ; the manual very well, and the evolutions extremely well.

Arms ; clean, but, in general, extremely bad. Swords remarkably good.

Accoutrements and Clothing clean and well put on."

The final remarks were :—

"It's an extremely fine regiment, perfectly well disciplined, and, when supplied with a new set of arms, will, in all respects, be fit for service."

A Royal Warrant, dated 15th October, directed that 2 companies were to be added to the 12th Regiment, each to consist of 4 sergeants, 4 corporals, 2 drummers, and 100 privates, besides commissioned officers; and a letter from the Secretary at War, dated 20th October, notified His Majesty's pleasure that Bat and Baggage Horses, with camp necessaries, were forthwith to be provided for the regiment;[1] also, on the 14th November, a lieutenant and 24 rank and file were ordered to march from Maidstone to Portsmouth and Stroud, to receive and escort two field-pieces for the use of the regiment.[2]

1756

FORMATION OF A SECOND BATTALION.

On the 13th May, the regiment furnished an escort of a captain and 50 non-commissioned officers and men, to escort the French prisoners to Cranbrooke, where the detachment was to do duty over them until further orders.

On the 17th, Routes were issued to march in two divisions to Trowbridge, Bradford, and Westbury, and, on the 22nd, the escort above mentioned was relieved by a detachment of Hanoverians, and rejoined headquarters at Trowbridge and Bradford.

On the 20th July, the regiment moved from Trowbridge and Bradford to Blandford.

The French having now made a descent on Minorca, which was lost to the British, a declaration of war against France had been issued on the 18th May, whereupon an augmentation of British troops followed, and, by an order, dated 7th August, the establishment of the 12th Regiment was increased by raising a Second Battalion, to consist of ten companies.

The 1st Battalion was inspected at Shroton Camp on the 11th October by Lieut.-General Sir J. Mordaunt, and reported on as "properly appointed and fit for service," and, in the same month, the headquarters were moved to

[1] *W. O.* 4, Book 50, p. 484. [2] *W. O.* 5, Book 42.

Reading, with detachments at Oakingham, Henley, and places adjacent, and, in November, Watlington was included.

On the 25th October, the 2nd Battalion was ordered to Lincoln.

1757

On the 25th January, a transfer was directed of 150 recruits from the 2nd to the 1st Battalion, at Reading, and an order of the same date decreed that :—" Several dragoons having been transferred, with a view to becoming sergeants, to the newly formed second battalions, the sum of £5 was authorised to be paid, for each of the dragoons so received, to the regiment from which they were taken."

On the 1st February, 2 of the headquarter companies of the 1st Battalion at Reading were ordered to march to Oakingham and places adjacent, and, on the 8th March, 3 more companies from Reading to Newbury and Speenhamland, where, on the 7th April, 5 companies of the 2nd Battalion were ordered to join them.

Lieut.-General Skelton died on the 9th April, and King George II conferred the colonelcy of the 12th Regiment on Major-General Robert Napier, from the 51st Regiment.

An order, dated 7th May, directed the 1st Battalion to march as follows :— 7 companies to Gosport, Fareham, and Wickham, and 3 to Cumberland Fort, and, on arrival, to furnish a detachment as a guard over the French prisoners at Portchester Castle, and, on the same date, the 2nd Battalion was ordered to march :—6 companies to Winchester, 3 to Southampton, and 1 to Rumsey.

Six days later (May 13th), the 1st Battalion was re-distributed as follows :— 5 companies to Gosport, 3 to Fareham, 1 to Titchfield, 1 to Wickham, the last named being further distributed in June between Cosham, Bedhampton, and Havant, and, on the 8th June, this company was ordered to Petworth, to suppress rioting.[1]

For the same purpose, 2 companies of the 2nd Battalion at Winchester were ordered to Alton and places adjacent, all three companies returning to their former quarters on the 25th June.

On the 9th August, the 12th Regiment, in conjunction with many others, was ordered to furnish a draft of 80 men towards reinforcing His Majesty's forces in North America.

The 2nd Battalion marched, on the 1st September, from its present quarters to the new barracks at Portsmouth, whilst the 1st Battalion was to encamp at Chatham, where it was inspected, on the 24th, by its own colonel, Major-General Robert Napier, and was reported on as :—" A tall, strong body of men, very well appointed, very ready in their exercise, and very fit for service." [2]

[1] *W. O.* 5, Book 44. [2] *W. O.* 27, Book 5.

An order, two days later, directed the 1st Battalion (now 9 companies) to march :—4 companies to Landguard Fort and 5 to Ipswich.

By a Route, dated 7th October, the 9 companies of the 2nd Battalion were to march from Portsmouth to Norwich, and, on the following day, 4 companies of the 1st Battalion at Landguard Fort were ordered to join them at Norwich.

An order of the 28th October stated that the King having been pleased to direct that a battalion of Foot, consisting of 9 companies, be immediately raised and formed, Major-General Napier (Colonel of the 12th) and eight other colonels of regiments were directed " to detach the youngest company of their 2nd Battalions (after having completed it to 4 sergeants, 4 corporals, 2 drummers, and 96 privates, under 1 captain, 1 lieutenant, and 1 ensign), and march them to Chelmsford to join the other companies of the newly formed battalion.

On the 7th November, the officer commanding the 2nd Battalion 12th at Norwich was ordered to detach 4 companies to Beccles, North Elmham, Wymondham, and Dereham.

1758

The 12th Regiment was directed to furnish a one-company detachment on the 21st January, to do duty over the French prisoners at Yarmouth, until an opportunity offered of bringing them southward.

The detachment was relieved regimentally on the 18th April, and this duty ended on the 30th June.

In April, the 2nd Battalion 12th Foot was constituted the 65th Regiment, under the command of Colonel Armiger (from captain and lieut.-colonel of the 1st Foot Guards),[1] and a letter from the Secretary at War, dated 4th October, directed Major-General Napier to pay the colonel of the 65th Regiment £5 for every man wanting to complete the said 65th Regiment to the establishment, at the time of its separation from the old battalion.

On the 13th May, the headquarters of the 12th (now a single battalion), which had moved to Ipswich, were ordered to send a sergeant's party to Southwold (Suffolk), to relieve a similar detachment of another regiment on smuggling duty.

[1] The second battalions of the undermentioned regiments were formed into distinct corps in April, 1758, and numbered from the 61st to the 75th Regiments, as shown in the following list, viz. :—

2nd Batt.	3rd	Foot, constituted the	61st	Regt.		2nd Batt.	24th	Foot constituted the	69th	Regt.
,,	4th	,, ,, ,,	62nd	,,		,,	31st	,, ,, ,,	70th	,,
,,	8th	,, ,, ,,	63rd	,,		,,	32nd	,, ,, ,,	71st	,,
,,	11th	,, ,, ,,	64th	,,		,,	33rd	,, ,, ,,	72nd	,,
,,	12th	,, ,, ,,	65th	,,		,,	34th	,, ,, ,,	73rd	,,
,,	19th	,, ,, ,,	66th	,,		,,	36th	,, ,, ,,	74th	,,
,,	20th	,, ,, ,,	67th	,,		,,	37th	,, ,, ,,	75th	,,
,,	23rd	,, ,, ,,	68th	,,						

The above 71st, 72nd, 73rd, 74th and 75th Regiments were disbanded in the year 1763, after the Peace of Fontainebleau.

The regiment was inspected at Ipswich on the 24th June by Brigadier-General A. Douglas, and was reported on as :—" A very good regiment, well disciplined, well appointed, and fit for service."

The Peace of Aix-la-Chapelle, in 1748, had conceded Silesia to Prussia, much to the annoyance of Maria Theresa, Queen of Austria, and, in 1756, she assembled large forces in Bohemia and Moravia ; England, being at war with France in North America, intended, if she could find an ally, to fight France also on the Continent. This ally she found in Frederick the Great, making the sides England and Prussia against France, Austria, and Russia.

Frederick selected Prince Ferdinand of Brunswick, one of his ablest generals, to conduct the war around Westphalia, the commencement of which had been a fiasco under the Duke of Cumberland. Ferdinand took with him as aides-de-camp Major Humboldt and Captain George Grey, the latter belonging to the 20th Regiment, recently commanded by James Wolfe, who had left a few months earlier for America, to command a brigade at the siege of Louisburg.

The 12th Regiment, having, on the 30th June, been directed to be held in readiness to march and embark for foreign service, moved early in July to Gravesend, and a notification was issued, on the 11th, that the Duke of Marlborough having been appointed to command the troops going to Germany, all orders for embarkation, &c., were to be taken from him.

After encamping for a short time in the Isle of Wight, the regiment proceeded to Germany, and, arriving at Embden on the 1st August, landed, on the 3rd, a few miles above the town, commencing, on the 5th, its march to join the army, which was accomplished in twelve days ; on the 20th, the regiment was reviewed by Prince Ferdinand.

During the remainder of the year, the regiment was actively employed, and performed many fatiguing services and long and toilsome marches, and towards the end of November moved into winter quarters at Minster, a city situated in a pleasant part of the country on the river Aa.

1759

The forces that were now to be matched against each other consisted of 60,000 of the Allies under Prince Ferdinand of Brunswick against 70,000 French under Marshal Contades.

Operations were commenced early in the spring, and the Allies gained some advantage ; the superior strength, however, of the French forces caused Prince Ferdinand to make a series of retrograde movements, which brought the allied army to the vicinity of Minden, where the French army, commanded by Marshal Contades, occupied a strong position, having Minden on the right, a steep hill on the left, and a morass in front.

In order to deceive his adversary, Prince Ferdinand detached one body of troops under his nephew, the Hereditary Prince of Brunswick, and so placed another under General Wangenheim as to induce Contades to direct Marshal de Broglie to attack it, under the impression that it was unsupported.

Immediately after sunset on the 31st July, the French army broke up

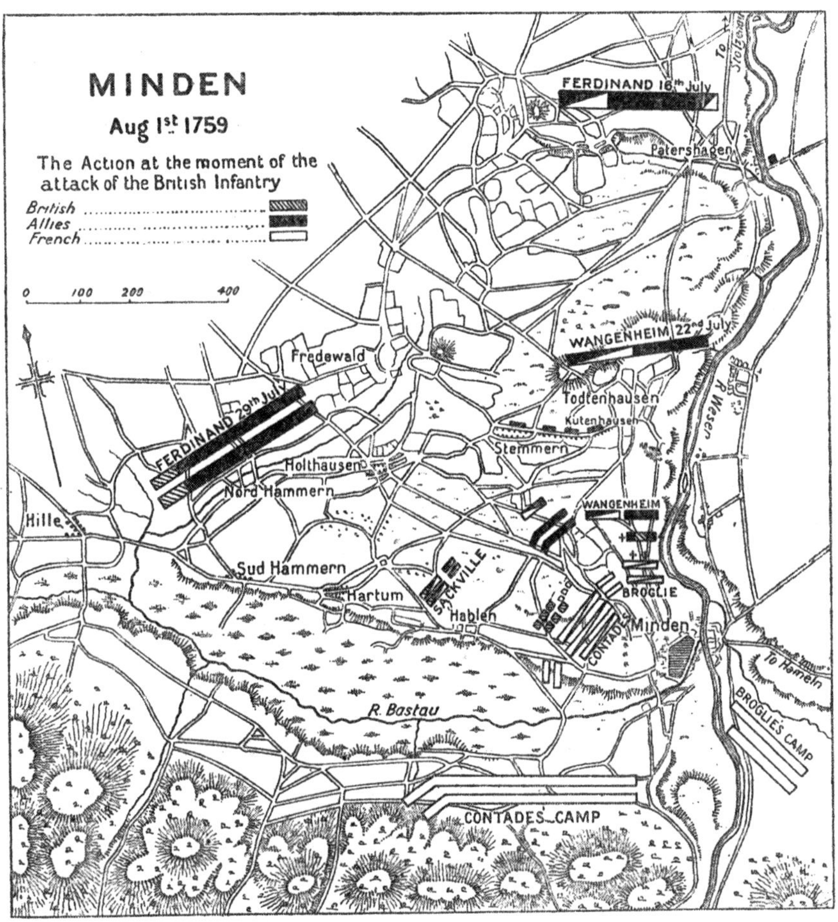

Battle Plan, by the courtesy of the Hon. J. W. Fortescue.

its camp on Minden Heath in nine columns, and crossed the rivulet of Barta by nineteen bridges, which also secured them a retreat.

Arriving at their destination about midnight, Contades formed up in order of battle. The two wings consisted of infantry, the centre of cavalry.

Battle of Minden.

On the 1st August, at 3 A.M., the French began to cannonade Prince Ferdinand's quarters at Hille, on his extreme right, from a battery of six guns which they had raised during the night. The Prince forthwith caused two guns to advance to Hille, with orders to the officer in command there to defend himself, while he caused General Greson to attack the enemy's post.

This service was successfully performed, and the Prince of Anhalt moved up his corps to the village of Halen, of which he took possession, and here established the right of the position clear of the battery.

Two deserters came into Prince Ferdinand's camp and revealed the enemy's movements; consequently, now apprehending their design, he sent advice of it to Wangenheim at 3 A.M.

De Broglie's corps was formed at 4 o'clock, and he, leading the attack, expected to surprise Wangenheim's corps, and thus penetrate between it and the Allied Army. He marched forward with great confidence; but as soon as he had gained an eminence, he was struck with the utmost surprise, when, instead of a few posts weakly guarded, he beheld the whole army of the Allies drawn up in complete order, and his corps exposed to a battery of thirty guns.

De Broglie immediately sent to Contades for reinforcements.

At about 5 A.M., the guns of both armies played briskly; ours, however, soon gained the ascendancy and maintained it, whilst our army advanced with rapidity, and shortly after the fire of the small arms had begun along the front, the heat of the action was turned to our right, where six English battalions and two of Hanoverian Guards performed prodigies of valour.[1]

The following was the order in which the Allies marched :—

Column 1. Cavalry of the right wing, Lord George Sackville :—3 squadrons of Bland's, 2 of Inniskillings, 3 of Blue Guards, 1 of Horse Grenadiers, 4 of Maxe Breisenbach, 1 of Body Guards, 2 of Howard's, 2 of Mordaunt's, 2 of Scots Greys, 2 of Bremer, and 2 of Veltheim.

Column 2. A brigade of heavy artillery, commanded by Major Hesse.

Column 3. General de Spörcke's :—a brigade of light artillery, and, in the 1st Line (counting from right to left), the 12th, 37th and 23rd Regiments under Brigadier Waldegrave; in the 2nd Line (which extended beyond the first on each flank) the 20th, 51st and 25th Regiments under Brigadier Kingsley, Hardenberg's Hanoverian Battalion, and two battalions of Hanoverian Guards.

Column 4. Prince of Anhalt :—1 battalion of Hardenberg, 1 of Keden, 1 of Scheele, 1 of Stoltsenberg, 1 of Brunswick, and 1 of the Hereditary Prince of Hesse.

Column 5. Centre brigade of heavy artillery under Colonel Braun.

Column 6. General Wurgnam :—1 battalion of Tolle, the brigade of light artillery of the centre of the first line, 1 battalion of

[1] Journal of a distinguished officer in the Allied Army.

Hessian Guards, 1 of Wangenheim, 1 of Mausback, 1 of Bischhausen, and 1 of the Prince Anhalt.

Column 7. General Imhoff :—1 battalion of Behes, 1 brigade of Brunswick's light artillery, 1 battalion of Hessian Grenadiers, 1 of Prince William's, 1 of Gilse, 2 of Imhoff, 1 of the Hessian Body Guards, and Brigade of heavy artillery of the left wing under Lieut.-Colonel Hutte.

Column 8. Cavalry of the left wing, Duke of Holstein :—5 squadrons of Holstein, 2 of Hessian Body Guards, 2 of Prince William's, 2 of Kamerstein, 4 of the Finkenstein, 2 of Militia, and 2 of Bruschenk.

The 12th, 23rd, and 37th British Regiments, followed by the 20th, 51st, and 25th, under the Prince of Bevern, advanced to attack the left of the French position.

On the left of the 25th were Hardenberg's two Hanoverian Regiments of the Guard, and, behind these again, were more Hanoverians, and a battalion of Hessian Guards. The advance was perfectly steady, and the 1st Brigade (12th, 23rd, and 37th), resuming the march after a short halt, encountered the enemy before the other regiments, which had been temporarily held back by the Staff. Sixty-six is given as the number of the enemy's guns which concentrated their fire on the British line.

The British cavalry, under Lord George Sackville, were on the right wing, and were advanced so as to threaten the enemy's centre. The French cavalry composed the entire centre of their line. They made no less than six charges on the advancing infantry, all being repulsed by the close-range volleys of the line.

The 12th, with the other five regiments of British infantry, supported by two battalions of Hanoverian Guards, with the aid of a powerful fire of cannon, not only bore the shock of the French Carabineers and Gendarmerie, but absolutely broke the whole body that advanced against them, who were utterly overthrown by the exertions of the British and Hanoverian Foot, whose behaviour on this day was valiant and courageous to a degree that was never perhaps exceeded. Contades, on seeing this, ordered a battery and some infantry to take this gallant body of Foot in the rear, but neither they nor a fresh body of French cavalry could shake the firmness of this advance. In vain did they attempt to rally : the continual fire, aided by that of the artillery, prevented them from again looking this splendid body of infantry in the face.

The 12th, with the corps before mentioned, be it said to their honour, had stood the reiterated charges of the many successive bodies of horse that were brought against them, with a resolution and steadiness which could not be surpassed. They endured six separate charges, and in the end cut to pieces and entirely routed two brigades of infantry who attempted to move up to assist the French cavalry attack.

At this period of the action, the Prince sent orders to Lord George Sackville, who commanded the British cavalry, to advance and charge the enemy as they retreated; but whether the orders were not clear, as Lord George said, or whether the personal courage of Sackville was in fault, as was the more general opinion, he would not at any rate understand the Duke's wishes, and the opportune moment passed away.

In the meanwhile the French cavalry formed again, and Colonel Fitzroy was sent once more to Sackville to desire him to advance. Lord George replied, "This cannot be; would he have me break the line?" The Aide-de-Camp answered, "My orders are positive, that the French are in disorder, and here is a glorious opportunity for the English cavalry to distinguish itself." Lord George said, "The orders given by the different messengers disagreed." "Perhaps they may," said Colonel Fitzroy, "but their destination is the same—to the left forward."

Lord George still hesitated, and it was not until he received the order from Prince Ferdinand's own mouth, that he proceeded to obey it, and then it was too late; so the British cavalry lost all share in the glory of the action.

After the battle Prince Ferdinand took his sword and sash from him, and his command was given to the Marquis of Granby, who had highly distinguished himself at this battle.

The weight of public opinion, both in the camp and at home, was generally against Lord George Sackville.

A court-martial was demanded, and Sackville returned to England to face and brave a disgrace which was more terrible in the eyes of most officers than all the risks and dangers of a battle.

He was found guilty of disobeying the orders of Prince Ferdinand, whom, by his commission and instructions, he was bound to obey, and declared unfit to serve his Majesty in any military capacity whatever.[1]

By 10 o'clock, the whole French army fled in disorder, totally routed, with the loss of 43 guns, 10 stand of Colours, and 7 Standards, with a considerable loss in killed, wounded, and prisoners, whilst that of the assailants did not exceed 500 men.

The 12th had Lieutenants William Falkingham, Henry Probyn, George Townshend, 4 sergeants, 1 drummer, and 77 rank and file killed; Lieut.-Colonel William Robinson, Captains Mathias Murray, William Cloudesley, Peter Campbell, Peter Dunbar, Lieutenants Thomas Fletcher, William Barlow, Thomas Lawless, Edward Freeman, John Campbell, and George Rose, Ensigns John Forbes, David Parkhill, John Kay, 11 sergeants, 4 drummers, and 175 rank and file wounded; Captains Peter Chabbert, Robert Ackland, and 11 rank and file missing.

The 12th Regiment was thanked in Orders, on the following day, together with the other British Regiments, and its distinguished conduct on this

[1] Lord George Sackville, however, was later employed; he became Lord George Germain, and Secretary of State for Colonies during the American Revolutionary War.

occasion was later honoured with authority to bear the word "Minden" on its Colours and appointments in commemoration of its gallantry.

The following was Prince Ferdinand's Order, dated at Headquarters, Suderhermain, 2nd August :—

> "The Army to be under arms at 6 o'clock this evening to fire a feu-de-joie. The brigades of heavy artillery are to take their respective posts on the wings, and in the front ; the three brigades of light artillery are to join those of the heavy brigade. When the order is given for firing, it will be continued to three rounds, beginning upon the right, first the artillery, and then the Army. The order of firing to be as follows :—1st, The Regiment of Gotha's volley ; 2nd, Captain Phillips' Brigade ; 3rd, Major Hesse's ; 4th, Captain Drummond's Brigade ; 5th, the Heavy and Light Artillery of the Centre Brigade ; 6th, Captain Foy's Brigade ; 7th, Lieut.-Colonel Hutte's Brigade ; 8th, Buckburgh's Regiment and Grenadiers. A volley, then the first line from right to left, followed by the second, from left to right ; the cannon taken from the enemy to be placed with the Centre Brigade of Heavy Artillery, and to fire with it.
>
> His Serene Highness orders his greatest thanks to be given to the whole army for their bravery and good behaviour yesterday, particularly to the British Infantry, the two battalions of Hanoverian Guards, and to all the cavalry of the left wing ; to General Wangenheim's corps particularly, the regiments of Holstein, the Hessian Cavalry, the Hanoverian Regiments, and Hamerstein's, and the same to all the Brigade of heavy artillery. His Serene Highness declares publicly that, next to God, he attributes the Glory of the day to the intrepidity and extraordinary good behaviour of these troops, which, he assures them, he will retain the strongest sense of, as long as he lives, and, if ever upon any occasion, he may be able to serve these brave troops, or any of them in particular, it will give him the utmost pleasure. . . ."

(The remainder of the order refers to the Prince's thanks to foreign troops and to his personal staff.)

The following is an extract from the Lords' address presented to His Majesty on the 14th November :—

> "The memorable victory gained over the French near Minden cannot but make a deep and lasting impression on every British mind. Whether we consider the great and able conduct of your Majesty's General, Prince Ferdinand of Brunswick, the valour of your Majesty's troops, the inequality of force, or the imminent perils of that important crisis ; the happy deliverance wrought by the action, and the glorious consequences of it, must ever be the subject of our praise and thankfulness."

Extract from the Commons' address presented to His Majesty, November 15th.

> "Nor can we ever forget that critical, signal, and memorable defeat of the French army near Minden, so justly the subject of lasting admiration and thankfulness, if we consider the superior numbers of the

enemy, the great and able conduct of His Serene Highness, Prince Ferdinand of Brunswick, or the unconquerable valour of your Majesty's troops."

Private Dorman, 12th (mentioned in Appendix X), was on the Continent throughout this war, and often engaged in different battles and skirmishes, the particulars of none of which (when relating his services) he remembered, except those of Minden, where he was severely wounded in the left hand, and carried out of the field on a waggon, with other wounded men, to a military hospital at Bremen, on the Weser.

The following is a copy of Prince Ferdinand's despatch, taken from the *London Gazette* (August 7th to August 11th, 1759) :—

> "Prince Ferdinand's headquarters at Minden, August 4th.
>
> The Prince having pushed forward a detachment as far as Vechte, in order to block the small position we had there, Prince Ferdinand determined to relieve it. This was executed by M. de Schleffen, his Serene Highness's aide-de-camp, with about 40 hussars, followed by 200 of Bretonbach's Dragoons. After some measures had been taken about provisions at Vechte, the Garrison of Bremen marched thither on the morning of the 28th July, under the command of General Drever, and thence to Osnabruck. M. de Schleffen forced one of the gates of the town, and made himself master of it. On the evening of the 27th the Hereditary Prince of Brunswick marched with 6000 men towards Lubeke, and next morning dislodged the enemy who occupied this pass.
>
> By easy stages, he advanced to Kirchlimieger, which was in the road of the Enemy's convoys coming from Paderhorn.
>
> Prince Ferdinand, with the Grand Army, made a short march on the 29th, towards Hille ; General Wangenheim remaining with a body of troops in the camp of Thornhausen.
>
> Some battalions of grenadiers, with the light troops, were left on the right side of the Weser, to observe the army under the Duke de Broglie. It was soon observed that the Enemy were not inattentive to their dispositions. In effect, Marshal Contades came to attack on the 1st August. This battle began at 5 A.M., and ended by the retreat of the enemy about noon. They attacked General Wangenheim briskly without making the least impression on him. Prince Ferdinand came up instantly with the main body of the army, and the heat of the action was then turned upon our right. The British infantry who were there, as well as the Hanoverian Guards, performed wonders.
>
> Every regiment that was engaged distinguished itself highly, and not a platoon in the whole army gave way one single step during the whole action. The Prince de Contades is among the slain. A considerable number of prisoners have been taken ; also 25 pieces of cannon, 10 pairs of Colours, and 7 Standards.
>
> M. de Contades passed the Weser at midnight on the 1st, and gave orders for burning the bridges. Prince Ferdinand entered the town of Minden at noon on the 2nd, the garrison having surrendered

themselves prisoners of war. The enemy took the route of Hesse, burning and pillaging all the towns and villages on quitting them. The same day an engagement took place between the Hereditary Prince and M. de Brissac, of which the following is an account.

Coveldt, August 2nd.—The body of troops under the command of the Duke de Brissac, consisting of 7000 or 8000 men, encamped on the night of the 31st July, with their left at the village of Coveldt, their front to the Werra, and their right towards the salt pits. Their position in front was impregnable, surrounding their left being the only other means of approach, for which purpose suitable dispositions were made.

The Hereditary Prince marched with the right, and we set out at 3 A.M. from our camp at Quernam. As soon as Count Kielmansegge came out of the defile of Beck, a cannonade began on both sides. The right was to pass the Werra, in order to turn the enemy's left at the village of Kirchlinger, upon a very narrow bridge. This difficulty, however, was in some measure removed by the spirit of the troops, the infantry fording the river partly behind the horsemen, and partly in peasants' waggons.

By the passage of the Werra, the enemy's position was entirely changed; the artillery fire was brisk on both sides and lasted two hours, though ours had always the superiority.

At last, upon our showing ourselves upon their rear, they immediately fled, and, in filing off, were received by M. de Bock with a discharge of artillery which was well supported. At last, finding themselves entirely surrounded, they had no other resource but in flight. Five pieces of the enemy's cannon, with their baggage, are in our hands."

The following is a copy of a letter written by an officer of the 12th Regiment to his mother, eight days after the battle of Minden :—

"Dear Madam.

The pursuit of the Enemy, who have retired with the greatest precipitation, prevents me from giving you so exact an account of the late most glorious victory over the French Army, as I would, had I almost any leisure; however, here goes as much as I can.

We marched from our Camp between 4 and 5 o'clock in the morning, about 7 drew up in a valley, from thence marched about 300 yards, when an 18 pound Ball came rowling gently up to us; now began the most disagreeable march that I ever had in my life, for we advanced more than a quarter of a mile thro' a most furious fire from a most infernal Battery of eighteen pounders, which was at first upon our front, but as we proceeded, bore upon our flank, and at last upon our rear. It might be imagined, that this cannonade would render the Regiments incapable of bearing the shock of unhurt Troops drawn up long before on ground of their own choosing, but firmness and resolution will surmount almost any difficulty. When we got within about 100 yards of the Enemy, a large body of French Cavalry galloped boldly down upon us; these, our men, by reserving their fire until they came within 30 yards, immediately ruined, but not without receiving some injury from them, for they rode down two Companies on the right of our Regiment, wounded 3 Officers, took one of them prisoner with our Artillery Lieut., and whipped off the Tumbrells. This cost them dear, for it forced many of them into our rear, on whom the men faced about, and 5 of them did not return. These visitants being thus dismissed, without giving us a moment's time to recover the inavoidable disorder, down came upon us like lightning the glory of France in the persons of the Gens d'Armes; these we almost immediately dispersed without receiving hardly any mischief from the harmless creatures. We now discovered a large body of Infantry consisting of 17 Regiments moving down directly on our flank in Column, a very ugly situation, but Stewart's Regiment and ours wheeled, and shewed them a Front, which is a thing not to be expected from Troops already twice attacked, but this must be placed to the credit of Genl. Waldgrave and his Aide-de-Camp. We engaged this Corps for about 10 minutes, killed them a good many, and as the song says, the rest they run away. The next who made their appearance, were some Regiments of the Granadiers of France, as fine and terrible looking fellows as I ever saw, they stood us a tug; notwithstanding, we beat them off to a distance, where they galded us much, they having rifled barrels, and our Musquets would not reach them. To remedy this we advanced, they took the hint, and run away. Now we were in hopes that we had done enough for one day's work, and that they would not disturb us more, but soon after a very large body of fresh Infantry, the last resource of Contades, made the final attempt on us; with them we had a long but not a very brisk engagement. At last we made them retire almost out of reach, when the 3 English Regiments of the rear line came up and gave them one fire, which sent them off for good and all, but what is wonderful to tell, we ourselves, after all this success, at the very same time also retired, but indeed we did not then know that the Victory was ours. However we rallied, but all that could now be mustered was about 13 files private, with our Colonel and four other Officers, one of which I was as fortunate as to be. With this remnant we returned again to the charge, but to our unspeakable joy no opponants could be found. It is astonishing that this Victory was gained by six English Regiments of Foot without their Granadiers, unsupported by Cavalry or Cannon, not even their own Battalion Guns, in the face of a dreadful Battery so near as to tear them with Grape shot, against 40 Battalions, and 36 Squadrons, which is directly the quantity of the Enemy which fell to their share. It is true that two Hanoverian Regiments were engaged on the left of the English, but so inconsiderably as to lose only 50 men between them. On the left of the Army, the Granadiers, who now form a separate body, withstood a furious Cannonade. Of the English there was only killed 1 Captain and 1 Sergeant. Some Prussian Dragoons were engaged, and did good service. Our Artillery, which was stationed in different places, also behaved well, but the grand attack, on which depended the fate of the day, fell to the lot of the 6 English Regiments of foot. From this account the Prince might be accused of misconduct for trusting the issue of so great an event to so small a body, but this affair you will have soon enough explained to the disadvantage of a great man whose easy part, had it been properly acted, must have occasioned to France one of the greatest overthrows it ever met with. The sufferings of our Regiment will give you the best notion of the smartness of the action. We actually fought that day, not more than 480 Private and 27 Officers; of the first 302 were killed and wounded, and of the latter 18; 3 Lieuts. were killed on the spot, the rest are only wounded, and all of them in a good way except two. Of the Officers who escaped, there are only 4 who cannot shew some marks of the Enemy's good intentions, and as perhaps you may be desirous to know any little risques that I might have run, I will mention those of which I was sensible. At the beginning of the action I was almost knocked off my legs by my three right hand men, who were killed and drove against me by a Cannon Ball, the same Ball also killed two men close to Ward, whose post was in the rear of my Platoon, and in this place I will assure you, that he behaved with the greatest bravery, which I suppose you will make known to his Father and Friends. Some time after I received a spent Ball, just such a rap on my Collar-bone as I have frequently had from that once most dreadful weapon, your crooked headed stick; it just swelled and grew red enough to convince the neighbours, that I was not fibbing when I mentioned it. I got another of these also on one of my legs, which gave me about as much pain as would a tap of Miss Mathew's Fan. The last and greatest misfortune of all fell to the share of my poor old coat, for a musquet ball entered into the right skert of it, and made three holes. I had almost forgot to tell you, that my spontoon was shot thro' a little below my hand; this disabled it, but a French one now does duty in its room. The consequences of this affair are very great. We found by the papers that the world began to give us up, and the French had swallowed us up in their imaginations. We have now pursued them about 100 miles with the advanced armies of the Hereditary Prince, Wanganheim, and Urff in our front, of whose success in taking Prisoners and baggage, and receiving Deserters, Francis Joy will give you a better account than I can at Present. They are now intrenching themselves at Cassel, and you may depend upon it, that they will not show us their faces again during this Campaign.

I have the pleasure of being able to tell you, that Captain Rainey is well, he is at present in advance with the Granadiers plundering French baggage and taking Prisoners. I would venture to give him forty Ducates for his share of prize money.

I have now, contrary to my expectations and in spite of many interruptions, wrote you a long letter. This paper I have carried this week past in my pocket for the purpose, but could not attempt it before. We marched into this Camp yesterday evening, and shall quit it early in the morning. I wrote you a note, just informing you that I was well, the day after the Battle. I hope you will receive it in due time. Be pleased to give my most affectionate duty to my Uncles and Aunts, my love to Bob Maxwell, Mathews, Nancy, Kitty, &c., and believe me to be

Dear Madam

with the greatest Affection
Your very dutiful Son

Camp at Paderborn
9th August, 1759.

H[h] Montgomery.

As a list as long as that of a Pedlar would not contain the names of all my Friends, and as you know them, I shall not at present make it out, but beg of you to remember me to them every one.

The noise of the Battle frightened our Sutler's Wife into labour; the next morning she was brought to bed of a Son, and we have had him christened by the name of Ferdinand."

The following is an extract from a letter written to a relative by Lieutenant Thomas Thomson, 20th Regiment, wounded at Minden:—

"Hospital at Minden, August 18th, 1759.

(At least 70 officers wounded here—several men dead, others dying daily.)

"Dear Sir,

Agreeably to promise and inclination I take the earliest opportunity of paying my respects to yourself and the good family at Pethem, this being the first day my health would admit of so long a letter. My wounds, I thank God, are by report of my Surgeons now in a good way, but by no means free from great pain at times, both night and day; insomuch that my Surgeons cannot prevent frequent attacks of fever, which consequently raise the inflammation and keep the wounds back by causing great torture, and not a proper discharge of matter ; hope when I get rid of splintered bones, which they say are endeavouring to find their way out, I shall then be more at ease, but this must be a work of time.

When I reflect on the miraculous escape (that my friends and myself had by coming out of the field alive) it fills one with a just sense of the power and goodness of the great God above. I shall attempt to give you a sketch of the proceedings of that day, notwithstanding I am certain a much abler pen than mine must fall greatly short, as no word can sufficiently paint the horrors and shocking sights which were every moment presented to the eyes of the living.

At one of the morn of the glorious 1st of August we received orders to turn out accoutred with all speed. The regiment was under arms in less than eight minutes. Our tents left standing, and baggage unpacked ; notwithstanding all this very few thought of an action, as we have often been alarmed in the same way ; exactness is always highly necessary. We marched towards Minden about two miles, found many Regiments preparing, and all the English on their march ; at length the scene appeared, a Battery of six guns began to play on the camp we had left very smartly.

We proceeded about a mile further, joined the Brunswick troops of Infantry, and our own, and got into a regular line of Battle march. Everything was still quiet before us until we got about a mile further, when we discovered the enemy with the greatest advantage over us, being already formed in Battle array ready to receive us. On the immediate sight of us they opened a battery of eighteen heavy cannon, which, from the nature of the ground (which was a plain), flanked this regiment in particular every foot we marched. Their cannon was ill served at first, but they soon felt us, and their shot took effect so fast, that every officer imagined the battalion would be taken off before we could get up to give a fire, notwithstanding we were then within a quarter of a mile of their right wing, and absolutely running up to the mouth of their cannon in front.

I saw heads, legs and arms taken off every moment, my right-hand file of men not more than a foot from me were all by one ball dashed to pieces, and their blood flying all over me ; this, I must confess, staggered me not a little, but, on my receiving a contusion in the bend of my right arm by a spent musket shot, it steadied me, immediately all apprehensions of hurt vanished ; revenge, and the care of the company I commanded took place, and I was *then* much more at ease than now.

By this time we were within two hundred yards or less of them, and plainly perceived the Fusiliers, Stewart's, and Napier's (12th) Regiments engaging an amazing number of their troops ; all the time their right wing was pelting us both with small arms, cannon, and grape shot, and we were not suffered to fire, but stood tamely looking on, whilst they, at their leisure, picked us off as you would small birds at a barn door. I cannot compare it to anything else, as their shot came full and thick ; had one quarter of them taken effect there could not have been a man left.

Everyone, I believe you will allow with reason, cursed our leader Beckwith, who was more confused than was consistent with his character, and I not only hope, but imagine, an enquiry will take place, notwithstanding his being made Aidecamp to the Prince. The French charged them, the three Regiments mentioned above, with at least 20 Squadrons, but by their steadiness and bravery, keeping their firing until the enemy were close up to them, they gave them such a terrible fire that not *even lions* could have come on; such a number of them fell, both horses and men, that it made it difficult for those not touched to retire. This charge over, a second and a third came on, and were repulsed in the same manner. Now was the time the English Cavalry should have come up; every eye was looking with impatience. Just at this time I got my wound, after having been hit three times before by spent balls, but this seared me like a red hot iron, I found myself fainting and quitted the Regt., after having called for a fresh officer, but found no one to supply my place, several having gone off wounded, or being already dead. I had not got four rods in the rear, before I heard the battalion fire, which pleased me so much in my agony, that I stood stupefied, looking on; the many poor soldiers praying, begging me to come off; after a few moments, I recovered my senses, and finding I had no further business there, made the best of my way, which was slow enough, over about a mile of common, where the balls came as thick as in front. By this time, a soldier of the regiment, slightly wounded in the leg, came up offering me his assistance; while supporting me, his left leg was carried away by a cannon ball, the wind of which fairly turned me round, but did not hurt me otherwise. The poor man is since dead. The common was strewed with dead and wounded men and horses; on the leeward side of those horses quite dead, lay wounded soldiers that could not get any further, to shelter them from the small shot. The action came on in such a hurry that we did not know where to look for the Surgeons. Captain—— and self walked three miles in this condition before we could get the blood stopped; at last we fortunately met with my Lord George Sackville's Coach, and a Surgeon he had in reserve for himself, which he need not have had, as there was no danger of his being hurt, as you will soon find by the *Vox populi*. I don't imagine, by what the surgeons say, I shall be able to take the field again sooner than the end of September, if then. I don't care who knows my sentiments, when I say my curiosity is satisfied, and that I never wish to see a second slaughter of my fellow creatures. We have little or no intelligence here, but find the enemy are entrenching by Hesse Capell, and the Duke alone at their heels; at least 10,000 of them have deserted. George is with me, to whom I owe my life from his great care before I came here, being two nights in a barn.

I hope to have the satisfaction to hear from you soon; the more letters from a family the greater comfort. To whom, as well to other friends, I beg my best respects.

From Sir, Your obedient servant.

(Sgd) THOS. THOMSON."

"P.S.—I hope the nation is now satisfied, as there was plenty of blood for their money."

While Lord Sackville had acted in this battle in open defiance of his Commander-in-Chief, the behaviour of the British infantry and Hanoverians was beyond all praise. The six battalions, together with two of the Hanoverian Guards, distinguished themselves in the highest degree. In resisting the attack of the absolutely intact French cavalry, they (as Smollett says) "not only bore the whole brunt of the French carabineers and gendarmerie, but absolutely broke every body of horse and foot that advanced to attack them on the left and in the centre."

The author of "Campaigns of Prince Ferdinand of Brunswick" observes :—

"Notwithstanding the loss they sustained before they could get up to the enemy; the repeated attacks of the enemy's cavalry, and a fire of musketry well kept up by the enemy's infantry; also their being exposed in front and flank; such was the unshaken firmness of those troops (12th, 20th, 23rd, 25th, 37th, 51st, and Brigade of Hanoverians), that nothing could stop them, and the whole body of French cavalry was totally routed."

To this may be added from the "Annual Register" :—

"The brunt of the battle was almost wholly sustained by the English infantry and some corps of Hanoverians, who stood the reiterated charges of so many bodies of horse, the strength and glory of the French armies, with a resolution, steadiness, and expertness in their manœuvres, which was never exceeded, perhaps never equalled: they cut to pieces, or entirely routed those bodies. Two brigades of Foot attempted to support the French Horse; but they also vanished before the English infantry."

Alluding to their cavalry, the French account says :—"They essayed in vain to break the British infantry; such a prodigious quantity of balls fell upon them, that, after a considerable loss, they saved themselves by flight, and in disorder," whilst the opinion expressed by the French Commander, Marshal Contades, is thus reported :—"I have seen what I never saw before, a single line of infantry break through three lines of cavalry, and throw them into confusion."

The splendid troops who conquered that day had avenged their glorious defeat at Fontenoy, and, by their success, had again raised the name of the British infantry as the first in the world.

	Killed.		Wounded.		Missing.		Totals.
Regt.	Officers.	Men.	Officers.	Men.	Officers.	Men.	
12th	3	82	14	190	2	11	302
20th	6	80	11	224			321
23rd		35	10	162		10	217
25th	1	19	6	119		9	154
37th			(No Details)				298
51st	1	10	9	78		4	102
							1394

Total 1394, or more than half of the loss of the whole army.

It would be hard to find a better example of local as compared with general incidence of losses in battle.

Of the 3198 British who commenced the action, there were 294 killed, 1037 wounded, 63 missing, and a loss of 500 horses. The enemy acknowledged their loss at Minden to amount to 7000 killed and wounded.

The following is reported with reference to James Campbell, an out-pensioner of Chelsea Hospital, who died at Dyke, on the 20th December, 1826, at the age of 86. For 26 years he had served in the 12th Foot, and for more than a year in Dumbarton Castle. On the 1st August, 1759, he fought at Minden ; when the battle commenced, the company to which he belonged had its full complement of men, but, on the following morning, seven only were able to do duty ; he had been slightly wounded, but nevertheless, was one of the seven who thus stood muster.

Minden was taken possession of the day after the battle, and the French army, suffering great losses in men, horses, baggage, &c., was pursued for upwards of two hundred miles by the Allied Troops.

On the 5th August, our army had marched to Hervorden, when Lieut.-General Urff (who had been detached with 7 battalions and 20 squadrons to Leinigon) took about 800 of the enemy prisoners at Detmold, together with the heavy baggage of the French army, amongst which was found part of Marshal Contades' papers and the military chest of the Saxons.

Every operation which followed resulted favourably to the Allies throughout the year. On the 11th September, Marburg surrendered, its garrison of 800 men being taken prisoners; and, on the 23rd October, another heavy loss befell the enemy, when Colonel Luckner, attacking a strong post, entirely defeated the French, with a loss to them of 4 officers and 50 men, and also taking 72 prisoners, with 99 horses and 112 waggons laden with forage.

About the 5th December, the main body of the French army having reached Frankfort, the Allies broke up their camp at Krossdorff, and went into cantonments at Osnaburg on the Lahn.

On the 13th September, this year, when in command of the British forces at Quebec, there had fallen gloriously in the hour of victory, Major-General James Wolfe, who had acted as adjutant to the 12th Regiment at the Battle of Dettingen, and later obtained the adjutantcy.

His death has been thus described :—

> " At the English right, though the attacking column was broken to pieces, a fire was still kept up, chiefly, it seems, by sharpshooters from the bushes and cornfields, where they had lain for an hour or more. Here Wolfe himself led the charge, at the head of the Louisburg Grenadiers. A shot shattered his wrist. He wrapped his handkerchief about it and kept on. Another shot struck him, and he still advanced, when a third lodged in his breast. He staggered, and sat on the ground. Lieutenant Brown, of the grenadiers, one Henderson, a volunteer in the same company, and a private soldier, aided by an officer of artillery who ran to join them, carried him in their arms to the rear. He begged them to lay him down. They did so, and asked if he would have a surgeon. 'There's no need,' he answered ; 'it's all over with me.' A moment after, one of them cried out, 'They run ; see how they run ! ' 'Who run ? ' Wolfe demanded, like a man roused from sleep.

'The enemy, sir. Egad, they give way everywhere!' 'Go, one of you, to Colonel Burton,' returned the dying man; 'tell him to march Webb's Regiment[1] down to Charles River, to cut off their retreat from the bridge.' Then, turning on his side, he murmured, 'Now God be praised, I will die in peace!' and in a few moments his gallant soul had fled."[2]

In Westminster Abbey we find a memorial to him with the following inscription:—

"To the memory of
JAMES WOLFE,
Major-General and Commander-in-Chief
of the British Land Forces,
on an expedition against Quebec,
who, after surmounting by ability and valour
all obstacles of art and nature,
was slain in the moment of Victory
on the XII of September MDCCLIX.
The King and Parliament of Great Britain
dedicate this monument."

EXTRACT FROM ARMY LIST 1759.

COLONEL ROBERT NAPIER'S (12TH) REGIMENT.

Showing casualties at Battle of Minden, August 1, 1759.

Rank	Name	Casualty
Lieutenant-Colonel	Wm. Robinson	(wounded)
Major	Corbet Parry	
Captains	Peter Chabbert	(missing)
	Robert Ackland	(missing)
	Thomas Brereton	
	Mathias Murray	(wounded)
	William Picton	
	William Cloudesley	(wounded)
	Peter Campbell	(wounded)
Captain-Lieutenant	Peter Dunbar	(wounded)
Lieutenants	William Falkingham	(killed)
	Edward Freeman	(wounded)
	William Armstrong	
	George T. Massey	
	Philip Stapleton	
	George Rose	(wounded)
	Thomas Lawless	(wounded)
	George Townshend	(killed)
	William Barlow	(wounded)
	Thomas Fletcher	(wounded)
	Hugh Montgomery	
	John Campbell	(wounded)
	Joseph Walworth	
	Robt. Geo. Bruce	
	Thomas Tutteridge	
	John Crozier	
	William Crompton	
	Henry Probyn	(killed)
Ensigns	John Forbes	(wounded)
	Thomas Trigge	

[1] 48th Regiment. [2] From *Montcalm and Wolfe*, by Francis Parkman.

Ensigns	David Parkhill	(wounded)
	John Kay	(wounded)
	John Walcott	
	Charles Blair	
	Charles Ward	
Adjutant	Thomas Lawless	(see previous page)
Quarter-Master	Wm. Falkingham	(see previous page)
Chaplain	T. M. Kay	
Surgeon	W. Hastie	

Five different designs of medals have been struck to commemorate the Battle of Minden. One of these was made in silver and in copper, two in copper only, and two in brass.

Two of the five specimens are in possession of the officers, 2nd Battalion Suffolk Regiment, and the following is a description of the silver medal, the only one showing the figure of "Victory":—

Obverse.—A British and a German soldier, in Roman costume, hold a globe, on which stands "Victory" crowning each with a laurel wreath. *Legend,* CONCORD OF THE ALLIES. *Ex.* AVG. MDCCLIX.

Reverse.—"Victory," seated, *r.*, upon a globe amidst captured French shields, holds a palm branch, and supports upon her knee a shield inscribed MINDEN. *Ex.* SOCIETY . PROM . ARTS . AND . COMMERCE. Size, 1·55 inches. (See Plate 10.)

1760

The Allied Army moved on the 4th January, Prince Ferdinand making Marburg his headquarters, his main body being quartered in the adjacent villages, with an advanced corps at Dillenburg and another towards the right of the French, who had retired to Friedburg. On hearing that Dillenburg was attacked and hard pressed, His Serene Highness set out at 1 A.M. on the 7th, to relieve it, when 700 of the French, with about 40 officers, were taken prisoners, also two guns and seven pairs of Colours.[1]

On the same evening, a similar success attended Keith's Highlanders, who, with Colonel Luckner's Hussars, attacked an advanced post of French Dragoons at Eybach, killing and dispersing a great part of them, and taking 80 prisoners and nearly 200 horses, with their baggage also. Another small affair on the following day resulted in 7 French officers and 50 men being made prisoners. Prince Ferdinand's headquarters were at Osnabruck from the 16th February to the 4th May.

On the 4th March, a Warrant was issued for raising, by beat of drum, an additional company to the 12th Regiment,[2] and, in the spring, a large body of recruits joined from England.

The troops of the Allied Army marched from their cantonments on the 5th May, the last division arriving at Paderborn and its adjacent villages

[1] *London Gazette*, January 15–19, 1760.

[2] *W. O.* 26, Book 24, p. 122.

on the 12th, whence they marched again on the 14th to encamp on the 20th at Fritzlar.

The year's campaign was opened two days later by the defeat, at Butzbach, of the French garrison, who, on fleeing from the place, were pursued and overtaken near a wood, where about 100 prisoners with 4 officers were captured.[1]

Thereupon, the whole army of Marshal Broglie was on the move from the 24th to the 27th May. Prince Ferdinand's headquarters had, by the 16th June, moved to Wavern, and, on the 18th, to Diederhausen; and Marshal Broglie's camp was at Neustadt.

The Allied Army being now vastly outnumbered by the French (who were being continually reinforced with fresh troops), Prince Ferdinand found it impossible to dislodge them from their posts, and was consequently obliged to act on the defensive.

On the 8th July, the French army, under Marshal Broglie, marched towards Franskenberg, whilst ours marched on the same date to Saxenhausen, the Hereditary Prince proceeding with an advanced corps to Corbach, where he found the French army already formed. A warm action ensued, when it was found necessary (from the enemy's superior numbers) for the Prince to effect a retreat, which was carried out with a loss on our side amounting to 500, and 15 guns, which could not be carried off for want of horses to replace those that were killed, and the Prince was wounded in the shoulder, but not dangerously. The only British infantry engaged in this action were the 5th, 24th, 25th, and Carr's Regiments.

This was followed by a reverse for the French at Emsdorff, on the 16th July, when the Hereditary Prince was detached with 6 battalions (2 Hanoverian and 4 Hessian) with Eliott's 15th Light Dragoons, which had just arrived, to oppose a body of the enemy, who had advanced on the left of an army towards Ziegenheim. Six guns were taken, and all their arms and baggage, and Eliott's Dragoons (the present 15th Hussars) distinguished themselves greatly on this occasion, their first appearance in the field.[2] Amongst trophies captured were nine pairs of Colours, and our loss was not very considerable. On the 24th July, orders were issued for the army to remain in order of battle until further orders, and, at night, to remain "lying on their arms."

From the vicinity of Saxenhausen, the Allied Army moved, at the end of July, towards Cassel, and encamped near Kalle. Thirty thousand French troops now crossed the River Dymel, and took post near Warburg, whereupon Prince Ferdinand quitted the camp at Kalle, and crossed the river to attack them, when a battle ensued on the following morning.

The British infantry who fought at Warburg were formed into one brigade of two battalions (under Lieut.-Colonel Beckwith, 20th Regiment)

[1] *London Gazette*, May 31–June 3, 1760. [2] *Ibid.*

which consisted exclusively of the grenadier companies of twelve regiments, made up as follows :—

1st Battalion, Grenadiers. Commanded by Major Daulbart :—the grenadier companies of the 5th, 8th, 11th, 24th, and Griffin's and Carr's Regiments.

2nd Battalion, Grenadiers. Commanded by Major Maxwell :—the grenadier companies of the 12th, 20th, 23rd, Fusiliers, 25th, 37th and 51st, also Keith's and Campbell's Highlanders.[1]

BATTLE OF WARBURG.

The following is an extract from the despatch of Prince Ferdinand of Brunswick, relative to the Battle of Warburg, July 31, 1760 :—

"It was agreed that the Hereditary Prince and M. Spörcke should turn the enemy's left, whilst I advanced with the army upon their front, which was done with all possible success, the enemy being attacked almost at the same instant by M. Spörcke and the Hereditary Prince in flank and in rear. As the infantry of the army could not march fast enough to charge at the same time, I ordered Lord Granby to advance with the cavalry of the right. The English artillery seconded the attack in a surprising manner. All the troops have done well, and particularly the English. The French cavalry, though very numerous, retreated as soon as ours advanced to charge them, except three squadrons who kept their ground, but were soon broken. . . . I ordered an attack to be made on the town of Warbourg by the Legion Britannique, and the Enemy, finding themselves thus attacked on their two flanks, in front and rear, retired with the utmost precipitation, and with the loss of many men, from the fire of our artillery and the attacks of our cavalry. Many were drowned in the Dymel, attempting to ford it.

The enemy's loss in men is very considerable ; I cannot exactly ascertain, but it is supposed that they left 1500 men on the field of battle, and the prisoners we have taken probably exceed that number. We have also ten guns, with some colours.

The loss on our side is very moderate, and falls chiefly on the brave battalion of Maxwell's English Grenadiers, which did wonders. Lieut.-Colonel Beckwith (20th), who commanded the brigade formed of English Grenadiers and Scotch Highlanders, distinguished himself greatly, and has been wounded in the head."

In conferring high praise on the troops who fought at Warburg, the following appears in Prince Ferdinand's General Orders, dated 31st July and 1st August :—

"His Serene Highness has infinite thanks to give to the corps of Grenadiers for the proof they gave of their Gallantry in the attack of this day. The corps of brave Grenadiers, who so much contributed to the Glorious Success of the Day, receives by this the greatest Praise

[1] *London Gazette*, August 19–23, 1760.

due to them. H.S.H. cannot enough acknowledge how much Esteem and Regard he has for them. He orders highest thanks to Lieut.-Colonel Beckwith, and Major Maxwell. . . ."

The loss sustained, in killed, wounded, and missing, amongst the Grenadiers and Highlanders was 415, and the total loss of the British troops, 590. The Casualty Returns, however, show that the number of killed alone (1 non-commissioned officer and 15 privates) of the grenadiers of the 12th (in Maxwell's Battalion, which received such high praise) exceeded that of any of the other grenadier companies engaged, and it also had Lieutenant Crozier, 3 non-commissioned officers, and 32 privates wounded, and Lieutenant Armstrong missing (apparently taken prisoner, as he rejoined later).[1]

The two grenadier battalions were next engaged in supporting the Greys and Inniskilling Dragoons, on the 22nd August, when the Hereditary Prince crossed the Dymel at the head of 12,000 men, in order to gain the left flank of the enemy. After the light troops on each side had been engaged, his timely arrival put an end to the affair in a quarter of an hour, by forcing the enemy to a precipitate flight with great loss.[2]

In General Orders, dated August 25th, H.S.H. published the following extract from a letter received from the King, with reference to the glorious action of the 31st July, at Warburg:—

"You will be pleased to testify to those brave troops, who so well executed your orders, that their good conduct gives me the most real pleasure, and that their bravery authorises me to hope that by God's assistance the Designs of my Enemy's will be frustrated."

Maxwell's Battalion of Grenadiers was next employed in a night attack on the French, at Zierenberg, on the 5th September, with a force of mixed troops, when H.S.H. the Hereditary Prince effected a complete surprise on them, whereby 37 officers and 380 men were made prisoners, and two guns taken.

The following appeared in General Orders, dated Camp Buna, September 9th, 1760:—

"H.S.H. the Duke has been informed more particularly since the Day before yesterday of what passed on the night of the 5th, at the Surprise of the Town of Zierenberg, and the fine Behaviour of many of the officers, in which they extremely signalised themselves. For instance, Major Maxwell, commanding a battalion of British Grenadiers, who forced the Guard and entered the Town on one side, while Captain Grey (20th) entered it at another, and made Monsieur Normand, a brigadier-general, prisoner of war with his troops. . . . Captain Wm. Picton, of the 12th Grenadiers, whose Bravery, presence of Mind, and the different Arrangements he made during the Dark,

[1] *London Gazette*, August 19–23, 1760.

[2] *Ibid*, 2nd to 6th September, 1760.

cannot be enough commended. . . . H.S.H., loving to do Justice to Merit, desires and orders that the fine Behaviour of these Brave Officers, and their names, be made public to the Army, and that his particular Thanks be given them, assuring them of his Friendship and perfect Esteem."

With reference to the above, on the decease of General William Picton, in October, 1811, a manuscript account of the following interview with King George III was found among his papers :—

"When appointed to the colonelcy of the 12th, Colonel Picton went to Court, and after kissing His Majesty's hand at the levée, he was admitted to an audience in the King's closet, when he acknowledged, in grateful terms, the honour conferred upon him ; and His Majesty replied, 'You are entirely obliged to Captain Picton, who commanded the grenadier company of the 12th Regiment, in the late war in Germany ;' at the same time alluding particularly to his gallantry at Zierenberg, for which he was thanked in general orders."

In an action which took place with the French on the 16th October, near the Convent of Campen,[1] the loss of the British was computed at about 1200, that of the enemy being much greater, with some hundreds of prisoners. We also took a pair of Colours and two guns, losing one of our own which burst.[2] Prince Ferdinand authorised the following rewards to the men who captured the guns and colours on this date :—2 guns at 100 dollars each, and the same for 2 of our own guns recaptured ; 2 pairs of colours at 50 dollars each.

Maxwell's Grenadiers fought in this engagement, and their casualties were :—2 non-commissioned officers and 11 rank and file killed, and 5 officers (including Lieutenant Armstrong of the 12th) and 40 rank and file wounded ; also 75 rank and file missing, but the different grenadier companies in which those losses occurred are not mentioned in the despatch.

King George II died on the 25th October, and was succeeded by King George III.

Various skirmishes and encounters continued to take place, and the Allies, by daring and rapid advances, kept the enemy in constant alarm. The 12th went into winter quarters in the Bishopric of Paderborn.

1761

An attack on the French at Duderstadt, on the 2nd January (when they were pursued as far as Witzenhausen), resulted in a loss to them, by their own account, of 600 men, 200 of whom were made prisoners, and amongst them were three complete companies of French Grenadiers, whilst the loss on our side was about 190 men.

The 12th continued to take part in several important captures of strong towns, and numerous stores of provisions, and by the 14th April, the allied

[1] See page 457. [2] Prince Ferdinand's Despatch, dated 23rd October, 1760.

DISTINGUISHED OFFICERS WHO SERVED IN THE 12TH.

EARL LIGONIER.
Lieut.-Colonel, 1711-12.

GENERAL SIR HENRY CLINTON, K.B.
Colonel, 1766-78.

GENERAL CHARLES, 1ST MARQUIS CORNWALLIS, K.G., K.P.*
Lieut.-Colonel, May, 1761, *to August*, 1765.

(*By the courtesy of the Editor of "The Connoisseur.")

troops had returned to their former cantonments with headquarters at Paderborn.

On the 21st April, the French met with a disaster at Wesel, when a large magazine of hay, which they had collected, was burnt, consuming 1,250,000 rations, computed at a loss of two million livres, and 33 soldiers of the Regiment of Normandy perished in the flames.

It was on the 1st May, 1761, that the distinguished General, Charles, 1st Marquis Cornwallis, K.G., K.P. (then Lord Chas. Brome), was appointed to the lieut.-colonelcy of the 12th Regiment (of which Lieut.-General Robt. Napier was Colonel), Lord Charles holding the rank until the 21st August, 1765, when he left the 12th, on appointment as A.D.C. to the King, whereby Major Wm. Picton obtained his promotion. On the 21st March, 1766, the Marquis was appointed Colonel of the 33rd Regiment. (See Plate 8.)

In June, the 12th, brigaded with the 5th, 24th, and 37th Regiments, again took the field, under the command of Brigadier-General Sandford, and was posted to the Marquis of Granby's Division.

On the 8th July, it was directed in General Orders that the sum of 2 dollars was to be awarded for every prisoner of war brought to Headquarters.

Battle of Fellinghausen.

The new French Marshal had, for some months, been Prince Soubise, and, on the 15th July, in conjunction with the Duke of Broglie, he attacked the light troops of the Allies, who were encamped in front of Lord Granby's Division, on the heights of Kirch Denkern, and near to Fellinghausen, when an engagement ensued, of which the following is an account from the despatch of Prince Ferdinand of Brunswick :—

> "The battle began at 3 A.M. on the 16th July, and the enemy redoubled their efforts against M. Wutgeman's Corps, who sustained them with the greatest firmness. The fire from our artillery and small arms continued for five hours, without the enemy gaining an inch of ground.
>
> It was near 9, when His Serene Highness ordered the troops nearest at hand to advance on the enemy (Lord Granby having previously ordered the posts of the light troops to be supported by the 24th, Keith's and Campbell's Highlanders, and Mansberg's Regiments) which, being carried out with the greatest intrepidity, soon obliged the French to give way, and retire with precipitation, abandoning their dead and wounded, and several guns, some of which were 16 pounders.
>
> Maxwell's Battalion of Grenadiers made prisoners of the Regiment of Rougé, consisting of four battalions, with its gun and Colours, and we have taken a great many prisoners besides.
>
> The victorious troops followed the enemy as far as Haltrup, and, the nature of the ground not admitting of the action of cavalry, His Serene Highness despatched some light troops in pursuit of them.

> A brisk cannonade continued on the side where the Hereditary Prince commanded, but the news of the defeat on their right probably induced them to give up attacks in that part also. The day ended with a general retreat of the enemy."

Other accounts mention that the French loss in killed, wounded, and prisoners was computed at about 5000 men, and that nine guns and six pairs of Colours were taken.

The loss of the 12th Regiment was 3 rank and file killed, and 9 rank and file wounded, whilst Maxwell's Battalion of British Grenadiers (including the grenadier company of the 12th) had 7 rank and file killed; 1 officer (Lieutenant Mercer), and 19 rank and file wounded; and 1 officer (Lieutenant Ferguson), and 32 rank and file prisoners. Neither of these officers belonged to the 12th, and, as the numbers of the rank and file are shown in one total, it is impossible to say in what regiments the casualties took place.

On the following day Prince Ferdinand published the following:—

> "The glorious Victory of yesterday furnished H.S.H. with a fresh opportunity to testify to the Troops he has the honour to command, the high Esteem, and perfect Consideration he has for them. . . . He hereby gives them his most Sincere and perfect Acknowledgments. . . . *The action of yesterday is to take the name of Fellinghausen*, which is to be declared to the Army."

The following appeared in General Orders, dated Camp, Hohenover, July 24th:—

> "At the intercession of the 12th Regiment, Lord Granby is pleased to pardon Corporal Wm. Ransin of the Said Regiment, condemned to Death for marauding, which Sentence nothing could have prevented His Lordship carrying into Execution, but for such an Application, and testimony of his former good Behaviour, but His Lordship now acquaints the Troops, they are not to presume upon this instance of his Lenity, as every Man detected hereafter in the same Crime must not expect Mercy, His Lordship being determined to put a Stop to such Shameful Excesses, which ruin our Army, and disgrace our Arms. Deserters will be treated with the same Severity."

A despatch, dated The Hague, July 31st, states that the Battle of Fellinghausen was followed by Colonel Freitag of the Allies destroying 50 boats laden with ammunition and corn, and burning the French magazines on the Fulde and the Werra without the loss of a single man of the expedition.

A letter from Colonel Clavering to the Earl of Bute, dated Brunswick, 11th August, reports that the enemy had established communication for the subsistence of their army between Gottingen and Hoxter, and how their first convoy, going towards the Weser, consisting of 250 waggons, had been captured by a detachment of the Allies' Chasseurs. The meal and bread were dispersed and given to the country people, and the horses and waggons sent back to their respective villages. Desertion also, from the French army, was now very great.

Further successes, on the 14th and 31st of August, attended the Allied Troops in the capture of a great number of prisoners, horses and waggons, &c., and the taking of Dorstein caused the Prince Soubise to retire towards Wesel.

By the 29th of September the two armies were opposed to each other in the neighbourhood of Cassel, and the French, committing great excesses in the course of their progress through the country, had provoked the peasants to rise against them, which caused one body of their troops to abandon the Principality of East Friesland.

The 12th Regiment was engaged in several skirmishes in the Electorate of Hanover in the early part of November, and by the middle of the month, the Prince of Soubise was putting his army into winter quarters. By the 5th of December, the allied troops were also marching off to their respective winter quarters, that of the 12th being at Osnaburg.

1762

A return of effective British troops, by Prince Ferdinand of Brunswick, dated 1st January, shows the 12th Regiment 552 "Under Arms," after deducting from its full strength (886) those "Sick," "on Command," and 172 "Wanting to complete."

On March 9th, the French garrison at Gottingen detached 4000 men to attack the chain of the allied army.

On April 16th, it was reported that our light troops had lately been very successful in several skirmishes with the enemy.

By May 4th, there was no appearance of the opening of the campaign, beyond Marshal Soubise, since his arrival at Cassel, having ordered a move of some of the Saxons who had crossed the Werra.

This caused a move on May 8th of the allied troops, whereupon 9 French battalions, at Cologne, commenced to march, "having (as the despatch says) 3 days' bread with them."

Accounts received from The Hague, dated June 11th, reported delay in commencing the campaign on the part of the Allies, on account of the scarcity of green forage.

By June 18th, 14 battalions had been ordered from the Lower Rhine, to reinforce our army in Hesse.

The 12th Regiment were engaged, on the 24th June, at Groebenstein, in a surprise attack on the French Army, who had encamped their troops, under Marshals D'Estrée and Soubise, on a very advantageous eminence between Groebenstein and Membusen, their left wing inaccessible by several deep ravines, and their right covered by Groebenstein, several little rivulets, and a body of troops posted at Carlsdorff.

BATTLE OF WILHELMSTAHL.

Prince Ferdinand having made his several dispositions for his attack on the French, some of his troops moved off at 3, and some at 4 A.M., on the 24th June, the British infantry consisting of 3 battalions of Foot Guards, the Welsh Grenadiers, Maxwell's Grenadiers, 5th Regiment, Keith's and Campbell's Highlanders, and Frasier's Chasseurs; also 11 battalions of Brunswickers, and 8 Hessian Regiments, together with the English and part of the German Cavalry, under the immediate command of the Prince himself, which passed the Dymel at 4 A.M., with the intention of forming up behind the ponds of Kalse.

The Marquis of Granby was to pass the Dymel, at Warburg, between 2 and 3 A.M., with the Reserve under his command, to march by Zierenberg and Zieberhausen, upon the eminence which is opposite to Furstenwald, in order to fall upon the left wing of the enemy. The whole plan was put into execution.

The attacking force was in presence of the enemy before they had the least apprehension of being attacked, and finding themselves opposed in front, flank, and rear, they were not long in striking their tents and retreating towards Cassel. Prince Ferdinand pursued and pressed them as closely as possible, and they would, no doubt, have been entirely routed, had not their commander (M. de Stainville) thrown himself, with the Grenadiers of France, the Royal Grenadiers, the Regiment of Aquitaine, and other corps (the flower of the French infantry), into the woods of Wilhelmstahl to cover their retreat.

That resolution cost him dearly, his whole infantry being taken, killed, or dispersed, after a very gallant defence, excepting two battalions which found means of escape. Some of these troops had before surrendered to Lord Granby's corps, and, upon the army coming up, the remainder, after firing once, surrendered to the 5th Regiment, the leading corps, in brigade, of the British troops.

All the troops behaved extremely well, and showed great zeal and willingness. . . .

The enemy's army retreated under the cannon of Cassel, and a great part of it passed very hastily over the Fulda. We took from their regiment of Fitz-James 300 horses and 2 standards. We also captured two or three thousand prisoners and several standards and colours. Our loss in killed was comparatively slight, amounting to between two and three hundred. The French infantry consisted of 100 battalions, and the allies had no more than 60.

Amongst their prisoners were upwards of 200 officers, including a brigadier and several colonels.[1]

Maxwell's Battalion of Grenadiers, which included the grenadier company of the 12th, took part in this action, and the casualties in the battalion

[1] *London Gazette Extraordinary*, July 1st, 1762.

were :—1 rank and file wounded ; 2 officers, and 58 rank and file missing, but to which regiments they belonged is not stated. Lieutenants Power and Irwin were missing, but they did not belong to the 12th.

Our total losses were 4 officers and 204 rank and file killed ; 2 officers and 271 rank and file wounded ; 4 officers and 311 rank and file missing.

On the 25th June, the following was published in General Orders :—

> "The happy Success of yesterday, under the Auspices of the Almighty, does so much credit to the Army under the command of the Duke, that H.S.H. finds a particular Pleasure, and looks on it as his Duty, to give public Thanks to the Army, which they so justly deserved. The Corps and Regiments who have had a particular Opportunity to distinguish themselves have shown so much Good Will, Courage and Eagerness to acquit themselves in doing their Duty in the Bravest Manner, that H.S.H. cannot testify to them his Satisfaction, or sufficiently express his Acknowledgments."

This was followed by an order, authorising a *feu-de-joie* to be fired by the troops at 7 o'clock that evening, a custom which had been strictly observed throughout the campaign, on the occasion of any success of the British arms.

The above General Order continued: "H.S.H. desires 1000 Loaves of Bread are to be borrowed from the British Troops for the use of the French Prisoners, who have no other Subsistence. This comes at the rate of 60 Loaves per battalion, and 12 per squadron." Its mode of delivery was then described, and the bread was to be returned to the regiments by their applying to Sir James Cockburn, at Kalle, on the 27th instant.

By the 9th July, four further small successes for the allied troops took place in Hesse.

On the 23rd July, the grenadier company of the 12th took part in driving the Saxons, under Prince Xavier, from Lutterburg. The despatch says :—"The allied troops marched through the Fulde up to their waists, and having clambered up the mountains, took four palisaded redoubts, one after the other, and drove the enemy from all their entrenchments."

In this action, a whole regiment of Saxon Horse was destroyed, and the Allies took over 1000 prisoners, 13 guns, and 3 standards, with a loss not exceeding 200 men.

During this attack, Prince Frederick of Brunswick advanced towards the Kratzburg (high ground which covered Cassel), and cannonaded the French lines, where Count Stainville was encamped with 10,000 men, who, hearing of the defeat of the Saxons, went to their assistance, whereupon Prince Frederick captured their camp without opposition.

The position then taken up by the Allies so embarrassed the French that, their communication with Frankfort being entirely cut off, they had only the choice either soon to retire or endeavour to disengage themselves by a general action.

After following the French for a considerable period without any decisive action, the army under Prince Ferdinand encamped at Homburg on the 19th August.[1]

On the 21st, the French evacuated Gottingen in such haste that they left behind three 12-pounder brass guns and a great quantity of ammunition of all kinds. This move was looked on as the prelude of the retreat of their Grand Army from the neighbourhood of Cassel. On the same date, an action was fought at Brucker-Muhl, near Podecker, on the Ohme, when, at 6 A.M., the enemy cannonaded and attacked the post with small arms. The firing, on both sides, which followed, was kept up with such unabated vigour, that it was only the approach of night which put an end to it. The whole affair lasted 14 hours without intermission, and the *London Gazette* says :—"Never did troops bear with more firmness and resolution so long and so severe a cannonade. Our loss is between 700 and 800 killed and wounded; that of the enemy by all accounts is much greater." Maxwell's battalion of grenadiers having been engaged, the grenadier company of the 12th took part in the action. The casualties of the grenadier battalion were :—1 captain, 5 rank and file killed; 1 lieutenant, 2 sergeants, 3 drummers, and 29 rank and file wounded. Neither of the officers belonged to the 12th.

Steps were now taken by the Allies for the siege of Cassel, which, however, on the 21st September, was temporarily suspended.

On the 25th September, Lord Granby approved of a sentence of 1000 lashes, awarded for desertion, by a general court-martial, to Private Thomas Kilborn, 12th Regiment.

On the 15th October, a surrender of Cassel was refused, and, on the following night, the trenches were opened by two attacks very near the town, when the allies had about 20 men killed and several wounded.

Cassel surrendered, and the capitulation was signed on the 1st November; the garrison was to march out on the 4th, but to leave in the town all the cannon and other effects belonging to the French King.

The 12th were employed in covering the siege, and, soon after the surrender of Cassel, a suspension of hostilities took place, which was followed by a treaty of peace, the preliminaries of which were signed at Fontainebleau on the 3rd November, but it was not until the 10th February, 1763, that the definite treaty was signed in Paris.

A resolution of thanks from the House of Commons to the Army in Germany, for their meritorious and eminent services to their King and country, was received in December.

The regiment went into winter quarters in the Bishopric of Münster, where it remained until required to embark for England early in the following year. For the march to the port of embarkation it was transferred to the 3rd Division.

[1] It would appear that, as the 12th had not arrived anywhere near Homburg until the above date, Richard Cannon's statement at page 37, para. 3, of his *Historical Record, 12th Foot*, is incorrect.

CHAPTER V

Great Britain, Gibraltar, "The Great Siege."
1763–1783

1763

In the beginning of this year, the thanks of Parliament were communicated to the army for its conduct during the war. In February, the regiment marched through Holland to Wilhelmstadt, and, by the 25th of the month, most of the troops returning home were embarked on board the transports.

The 12th Regiment had also embarked, and according to the Embarkation Return, its effective strength was 27 officers and 689 non-commissioned officers and men.

Arriving, on the 28th, at Harwich, Routes were issued on the 1st March, for the regiment (now 9 companies) to march in two divisions to Norwich, and, on the 20th, a Royal Warrant directed a reduction in the establishment to 1 captain, 1 lieutenant, 1 ensign, 3 sergeants, 3 corporals, 2 drummers, and 47 privates per company, except the Grenadier Company, which was to have two lieutenants instead of one, and 2 fifers.

On the 11th May, a move took place, in two divisions, to Berwick, and, in July, the headquarters were at Edinburgh Castle.

1764–67

Throughout the years 1764–65, the regiment was doing duty at Fort George, and marched in May, 1766, via Stirling, to Dumfries, arriving in June.

In November, 1766, Lieut.-General Robert Napier died, when King George III conferred the colonelcy of the regiment on Colonel Henry Clinton, from Captain and Lieut.-Colonel in the 1st Foot Guards, by commission dated 21st November, 1766.

On the removal of the regiment, in 1767, from North Britain, an order was issued, on the 5th February, to march to Carlisle, followed by one, on the 3rd March, to proceed from Dumfries in three divisions to Chatham, "taking care to avoid marching through any town where the assizes were being held."

Another move, on the 26th May, was to the Kensington Gravel Pits, and "such parts of Kensington and Knightsbridge as are not in the Liberties of Westminster, Brompton, and Great and Little Chelsea," and, on the 5th June, the 12th returned to Chatham.

On the 4th July, the regiment furnished a subaltern's detachment to Upnor Castle.

An increase to the establishment, in September, was authorised as follows :—9 sergeants, paid at the daily rate of 1*s*. 6*d*., and 9 corporals, 9 drummers, and 2 fifers at 1*s*. each, daily.

1768–69

A Commander-in-Chief's order, dated 23rd February, 1768, notified that His Majesty the King would review the 12th, 13th, and 25th Regiments of Foot on the same day, early in May, and, on the 30th April, the above regiments, then at Chatham, received the following order :—"The Ensigns who carry the Colours are not to take their Hats off, either when they are marching, or when Posted in Front of the Regiment, to receive His Majesty, or any General officer." [1]

The half-yearly inspection of the regiment took place on the 11th May at Chatham, by Major-General George Cary, when the inspecting officer reported on it as "an extremely fine corps, and fit for immediate service."

Following these inspections, orders were issued, on the 24th May, to march to Fulham, Putney, and Putney Bowling Green, and a detachment of the regiment, which had been doing duty at Dover Castle, was now ordered to Hampton Wick.

On the 7th June, the following moves were directed :—4 companies to Hounslow and Cranford Bridge, and 5 to be distributed between Egham Hythe, Staines, and Bedfont.

On the 21st June, 1768, the regiment marched to Kensington Gravel Pits, Knightsbridge, Marylebone, and Paddington, and, on the 26th, the headquarters and 6 companies were ordered to Winchester, 2 companies to Windsor, and 1 to Hampton and Hampton Court to do duty at the Palace, 3 companies moving from Winchester to Salisbury in November.

In February and March, 1769, further interchanges of companies took place, and 2 proceeded from Salisbury to Rumsey.

The regiment had, since January, been held in readiness to proceed on foreign service, and, on the 1st May, was directed to embark for Gibraltar, to relieve the 20th Regiment, at such time as the commanding officer might think proper, acquainting the War Office of the receipt of the order, and the date of embarkation.

The 12th arrived at Gibraltar on the 19th May.

[1] At this period, all the Commander-in-Chief's orders were issued from Cleveland Court, London, S.W.

1770

A letter from the Secretary at War, dated 22nd February, 1770, directed a reduction in the 12th Foot to 42 privates only per company, to take place from the 25th December, 1769.

On the 15th December, 1770, the addition of a light company was authorised for the regiment, the sum of five guineas being allowed to the Recruiting Officer from the Non-effective Fund, for each recruit approved of from the 25th inclusive, and the bounty money to any man on enlisting was not to exceed two and a half guineas.

1771

The regiment was reviewed at Gibraltar, on the 3rd May, 1771, by Lieut.-General the Honourable Edward Cornwallis, and reported on as " a fine body of men, and the marching incomparable. As good a regiment as, I believe, there is in our service, or in any in Europe. Performed their exercises and evolutions with great dexterity."

The 12th Regiment was the early school in which the "Illustrious Picton " developed that genius and strengthened that valour which shone so brightly in the Peninsula, and were at last consecrated by his life's blood on the field of Waterloo. (See Plate 6.)

On the 14th November this year he received his first commission as Ensign in the 12th Regiment, in which his uncle, William Picton, held, at this period, the rank of Lieut.-Colonel. In March, 1777, Thomas Picton was promoted Lieutenant in the regiment, and, after a sojourn of five years at Gibraltar, in time of peace, he requested his uncle to obtain for him a transfer to some regiment which had an earlier prospect of taking the field. In January, 1778, he was gazetted Captain in the 75th Regiment, to which his uncle William was, in the same year, posted as Colonel, and had much reason to regret his impatience in quitting the 12th, as, by so doing, he missed serving throughout the Siege of Gibraltar, and another five years of his life passed without active employment. The 75th Regiment was disbanded in 1783, and he was on half pay for twelve years.

Towards the end of 1794, tired of making applications, he embarked without employment for the West Indies, with no other inducement than a slight acquaintance with Sir John Vaughan (the newly appointed Commander-in-Chief in those parts), who got him transferred as a Captain and Brevet-Major from the 58th Regiment to a Captaincy in the 17th, part of which was then serving in the West Indies.[1]

Sir John, being much struck with Captain Picton's energy and ability, got him quickly promoted Major in the 68th Regiment, with the appoint-

[1] *London Gazette*, 25th April, 1795.

ment of Deputy-Quarter-Master-General, which entitled him to the brevet rank of Lieut.-Colonel. He then became Colonel of the 56th, and eventually a distinguished general in the Peninsula.[1]

On the 24th June, 1814, he received, for the seventh time, the unanimous thanks of the House of Commons. In 1815, having attained the rank of General, he was awarded the Grand Cross of the Bath, and was killed at Waterloo.

The sum of five guineas, which had been allowed from the 25th December, 1770, to recruiting officers (half of this being given to the recruit on enlistment), was changed in March, 1771, to three and a half guineas for officers, and one and a half for recruits.

In the early part of this year, a considerable correspondence took place respecting the poundage on the soldiers' pay. This poundage—a shilling in the pound—was to be taken out of a fund given by the King's bounty to the agents of the regiments, and was to be subjected to drafts by the regimental paymaster as might be required. The fund was also to be used to reimburse the rank and file for the deductions made upon them for the paymaster and surgeon. The saving to the soldier by this fund was calculated to be about 16*s*. 6*d*. per annum.

1772–78

Colonel Henry Clinton was in 1772 promoted Major-General.

The report of the Inspection, on the 18th April, at Gibraltar, showed that amongst 420 effective rank and file, 331 were English, 71 Scots, 15 Irish, and 3 foreigners, and the Inspecting Officer remarked on the battalion as "a very fine regiment, in its prime, and fit for any service whatever."

The year 1775 is memorable for the breaking out of the Civil War in America, and, on its commencement, Major-General Henry Clinton (Colonel of the regiment) was sent with reinforcements to Boston, with the local rank of Lieut.-General, and, throughout the war, distinguished himself greatly.

On the 25th February, 1776, an army notice was issued at home stations, offering pardon to all deserters who should give themselves up before the 10th April.

In 1777, Major-General Henry Clinton was promoted to Lieut.-General, and was rewarded with the dignity of Knight of the Bath.

A letter from the Secretary at War, dated 17th March, 1778, directed that a newly ordered increase to the 12th Foot was to be forthwith completed by drafts from the additional companies, and recalled all recruiting parties of the regiment, and recruits, to Chatham, in order to proceed to Gibraltar.

The regiment was inspected on the 24th March, at Gibraltar, by the Right Honourable Lieut.-General Eliott, the Governor, and was reported on as "fit for immediate service."

[1] *Memoirs of Sir Thomas Picton.*—H. B. Robinson.

The following regulations, dated 2nd June, relative to the supply of bread to troops in camp were to be observed :—

> "A well baked ammunition Loaf, made of Wheat, without anything taken from it, and weighing six pounds, will be delivered to each Soldier as the Allowance of Bread for four days, for which Loaf, five Pence is to be stopped in the Hands of the Officers paying Troops or Companies, to be collected from them by the Regimental Quarter-Master, for every fourth Delivery, &c." [1]

The cost of rations was fixed at 6*d*. a day, and the number allowed to each rank was as follows :—Colonel as Colonel and Captain, 11 ; Lieut.-Colonel and Captain, 9 ; Major and Captain, 7 ; Seven captains, each five, 35 ; Captain-lieutenant, 2 ; Nineteen Subalterns, each one, 19 ; Adjutant, Chaplain, Quarter-Master, and Surgeon's Mate, each one, 4 ; Surgeon and medicine chest, 2.

A regiment of 10 companies, 70 men in each, was to have 160 tents, 160 tin kettles and bags, 160 wood hatchets, 12 bell tents, 12 camp colours, 20 drum cases, 10 powder bags, 792 water bottles, haversacks, and knapsacks.

In December this year, Lieut.-General Sir Henry Clinton, K.B. (See Plate 8), was transferred from the 12th Foot to the colonelcy of the 84th Regiment, or "Royal Highland Emigrants" (then first embodied for service in North America) and later disbanded.

Lieut.-Colonel Wm. Picton was promoted Colonel, and transferred to the colonelcy of the 75th Regiment, newly raised, and disbanded five years later.

1779

On the 21st April, Colonel William Picton was appointed Colonel of the 12th Regiment.

The following regulations were issued at home stations, dated 5th July, relative to the issue of wood, straw and forage in camp :—

> "The officers of each company will be allowed eight rations of Wood for four days, in order, not only to supply the Common Sutler, but likewise to afford a sufficiency for such officers as may Chance to Mess by themselves.
>
> Straw will be allowed at the rate of 2 Trusses for the Bedding of every five Men for the first eight days ; one Truss for the Second period of eight days, and one for the Third, after which two will be again delivered, and so on for every succeeding period of 24 days."
>
> Forage allowance to all officers of a battalion of Foot was to commence on taking the field. The total number of horse rations per diem (at 6*d*. each) for a battalion of 46 officers (including two Surgeon's Mates) was computed at 109, and it was held that in this computation, the horses were supposed to be sufficient for carrying the tents and blankets of each company. The Sutler was allowed forage for his effective horses, not exceeding eight.

[1] *W. O.* 3, Book 26, p. 3.

The possession by the English of the fortress of Gibraltar, overlooking the Spanish provinces, had, ever since its capture, been regarded by the Spaniards with great jealousy, and, hostilities having been already carried on between Great Britain and France for nearly six months, Spain now adjudged this a favourable opportunity to offer her mediation, in terms, however, which could not be agreeable to the principal belligerent powers. No sooner had England refused her acquiescence than the Court of Madrid espoused the cause of France.

On the 16th June the Spanish Ambassador presented to the Court of London his hostile manifesto, and, on the 21st, all communication between Spain and Gibraltar was closed, by an order from Madrid.

Our guards were thereupon reinforced, the Grand Battery was made a Captain's Night Guard, and the picquets had their arms loaded, all other necessary precautions being taken in case of alarm, and most of the men off duty were employed in repairing the works, &c. The heaps of sand on the Isthmus were to be levelled by the Jews and Genoese of the town, and no one was to remain in it but those who had property or would assist in defending it.

The following was the strength of the Garrison at the commencement of the blockade, 21st June, 1779 :—

General G. A. Eliott, Governor.
Lieut.-General R. Boyd, Lieut.-Governor.
Major-General De la Motte, Commanding the Hanoverian Brigade.

British.		Officers.	Men.
Royal Artillery	Colonel Godwin	25	460
Engineers, with a company of Artificers	Colonel Green	8	114
12th Foot	Lieut.-Colonel Trigge	29	570
39th ,,	Major Kellett	29	557
56th ,,	Major Fancourt	27	560
58th ,,	Lieut.-Colonel Cochrane	28	577
72nd ,, (Royal Manchester Volunteers, disbanded in 1783)	Lieut.-Colonel Gledstanes	33	1013
Hanoverian.			
Hardenberg's Regiment	Lieut.-Colonel Hugo	29	423
Reden's ,,	Lieut.-Colonel Dachenhausen	27	417
De la Motte's ,,	Lieut.-Colonel Sclippergill	33	423
		268	5114

Total 5382.

On the 3rd July, four men from each battalion company of the 12th, 39th, 56th, and 58th Regiments, and six from the 72nd, were attached to the Artillery, to be taught great gun practice. The artillery in the garrison mustering only five companies, this reinforcement of about 180 men proved afterwards a very useful adjunct.

On the 5th, a Spanish squadron of eleven ships, lying to for some time off the garrison, necessitated special caution being given to the 12th and 72nd Regiments, in the South Barracks, to be alert.

The Spanish camp being daily reinforced with additional regiments of cavalry and infantry, and large parties still employed in landing ordnance and military stores at Port Mala, the Governor of Gibraltar considered it necessary to make several staff appointments amongst the officers of the garrison, and nominated Lieutenant Forch, 12th Regiment, a personal aide-de-camp to himself as Commander-in-Chief.

On the 21st July, an order was given for the guards to mount without powder in their hair.

In the beginning of August, the corps in garrison were ordered to give in returns of their best marksmen, and also of those who had been employed making fascines, and as affairs began to wear a more serious aspect, a general activity reigned throughout the garrison, promoted not a little by the example of the Governor, who was usually present at the daily parading, at dawn, of working parties.

Whilst our Engineers were busily employed, the Navy was not less diligent. A new battery for 22 guns was begun in the Navy Yard, in case it should be necessary to lay up the ships, and the stores were removed from the New Mole to the Naval Hospital.

Early on the 6th, a Portuguese schooner from Tangier brought 44 bullocks, 27 sheep, and a few fowls, and two days later another arrived with onions, fruit, and eggs ; the latter brought letters for the Governor, but no news from England. From this date, nothing material occurred till the 10th, when the enemy's cruisers captured a boat belonging to the garrison.

The enemy's blockade becoming more strict and severe, their army was in force before the place, and their present plan seemed to be to reduce Gibraltar by famine ; but for a considerable supply of provisions received in April, the troops would now have been reduced to the greatest distress. The inhabitants had been warned to provide against the impending calamities, and, as they could not expect to be supplied from the garrison stores, they were compelled, in general, to seek subsistence by quitting the place.

On September 3rd, one of the 12th was punished for sleeping on his post. Nine battalions, about 7000 men it was thought, were in the enemy's camp.

Their operations now began to engage our serious attention. They had been permitted to pass and repass unmolested for some time, but the Governor did not think it prudent to allow them to proceed any longer with impunity. A council of war was therefore held on the 11th to confer on the measures to be pursued, with the result that perpetual cannonading on our part became the order of the day.

In the beginning of October, the enemy's army was reported to consist of 15 battalions of infantry and 12 squadrons of horse, which, if complete, would amount to about 14,000 men.

On the 3rd October, a soldier of the garrison (not of the 12th) was reported for having said :—" If the Spaniards came, d—— him that would not join them " ; whereupon the Governor, considering the man must be mad, ordered his head to be shaved, blistered, and bled, and the man to be sent to the Provost, kept on bread and water, to wear a tight waistcoat, and to be prayed for in church.

Towards the end of the month, smallpox was discovered at Gibraltar amongst the Jews. The Governor, apprehensive of its spreading, ordered all those who had not been affected by it, to go southward until it should disappear. The Grenadier Company of the 12th and Light Company of the 72nd, were also ordered to the town.

Provisions of every kind were now becoming very scarce, and exorbitantly dear, in the garrison. Not only bread, but every article, was hard to procure. Veal, mutton, and beef sold from 2*s.* 6*d.* to 4*s.* a pound ; fresh pork, 2*s.* to 3*s.* ; salt beef and pork, 1*s.* 3*d.* a pound ; a pig's head, 19*s.* ; fowls, 18*s.* a couple ; ducks, a guinea, also a goose ; firewood 5*s.* per cwt. ; a pint of milk and water, 1*s.* 3*d.* Vegetables were extremely scarce, a small cabbage costing 1*s.* 6*d.*, whilst a small bunch of the outer leaves sold for 5*d.* Irish butter, 2*s.* 6*d.* per lb. ; eggs, 6*d.* each ; and candles, 2*s.* 6*d.* a lb. Notwithstanding the garrison being almost surrounded by the sea, the best fish was sold at exorbitant prices ; our fishermen being, as foreigners, under no regulation, they exacted most extravagant sums.

It was about this period that the Governor made a trial as to what quantity of rice would suffice a single person for twenty-four hours, and actually himself lived eight days on four ounces of rice per day.

The middle of November found the garrison with a great scarcity of coals ; the fuel issued at this period was wood from ships bought by Government, and broken up for that purpose, but was difficult to light, on account of having been so much soaked in salt water.

On the 20th, 39 bullocks and a few sheep arrived from Barbary, and several of the former being too weak and poor to land, were drowned in their efforts to get on shore.

On the 2nd December, four of the enemy's deserters were sent to the 12th Regiment, and, on the 6th, a man of the regiment was tried by General Court-Martial for sleeping on his post. On the 14th, bread was reported as very scarce. On the 27th, following a violent thunderstorm the previous night, there was found, floating under the walls of the garrison, a vast quantity of wood, cork, &c., which had been wafted by the wind to our side of the bay, and as fuel had long been a scarce article, this supply was considered a merciful act of Providence.

On the 28th, a packet boat from Barbary brought 42 goats, 8 dozen fowls, and eggs, but no mails.

1780

January did not commence with any interesting events.

As the enemy's works progressed, the pavement of the streets of Gibraltar was taken up, all conspicuous towers of buildings pulled down, the stone sentry-boxes removed, guard houses unroofed, traverses raised, a covered way begun, and every measure adopted to prevent the bombardment of the place being attended with serious results.

On the 8th, a Neapolitan three-masted ship was luckily driven under our guns, and obliged to come in, when about 6000 bushels of barley were found on board, a cargo of inestimable value.

The bakers had long been limited in the quantity of bread for daily issue to the inhabitants, who were not the only sufferers in this scene of distress ; many officers and soldiers had families to support out of the pittance received from the victualling office. One woman actually died through want, and many were so enfeebled, that it was not without great attention they recovered. Thistles, dandelions, wild leeks, &c., were for some time the daily nourishment of numbers. Few supplies arriving from Barbary, and there appearing little prospect of relief from England, famine began to present itself with all its attendant horrors.

All soldiers now convicted of theft were hanged with the Governor's approval.

Desertions on both sides had for some time been numerous, many meeting their death in various ways in their endeavours to escape.

On the 11th January, the enemy showed their dissent at our burying our dead outside the garrison, by firing on a clergyman whilst performing funeral rites over a soldier of the 72nd Regiment. On the following day, they fired ten shots from Fort St. Philip into the town, when a woman, passing near one of the houses, was slightly hurt—a strange circumstance that at this remarkable siege a female should be the first person wounded.

That evening, soldiers of the garrison were informed that the Governor was under the necessity of curtailing their rations, ½ lb. of beef and ¼ lb. of pork being taken, per ration, from their allowance in the following week. This intelligence, though disagreeable, was received with the cheerfulness which was so prominent a feature in all the vicissitudes of this trying period, the garrison submitting at all times, without a murmur, to every necessary regulation, however unpleasing.

A deserter, who came in on the 16th January, informed us that the enemy had everything prepared in their lines to bombard the town. In the evening, there arrived a brig laden with flour, which communicated the joyful news of the capture, off the coast of Portugal, by Admiral Sir George Rodney, of a Spanish 64-gun ship, five of 32 and 28 guns, with seventeen merchantmen, and that he was proceeding to our relief with a fleet of 21 sail of the line, and a large convoy of merchant ships and ansports. On the following evening one of the armed prizes came in thout opposition, and, from a frigate which got in about 11, we learnt a complete victory over the Spanish Admiral, who, with three others his squadron, was taken, and he, having been wounded in the engagement, s brought to lodgings in the town on the evening of the 19th, having ery attention and compliment paid him which were due to his rank.

Early on the 22nd, several men-of-war, coming into the Bay, occasioned general alarm in the enemy's camp, and their artillery opening fire, one an was killed and two wounded on board the "Terrible," all of them anish prisoners. Whilst the Fleet remained in the Bay, the Governor and rrison were often honoured with the presence of the Royal Midshipman, ince William Henry, who later became King William IV.

Sir George Rodney arrived on the 27th, when a council of war was mediately held, and the prizes and remaining men-of-war were now all anchor in the Bay. The same day, the Governor ordered all soldiers' ves and children, who were not provided with twelve months' provisions, prepare to leave the garrison with the fleet; 250 lbs. of flour or 360 lbs. biscuit was stated as sufficient for one person. Thus, many useless nds were sent home, which would later have become a great encumbrance the garrison, situated as they were.

On the 29th, the 2nd Battalion 73rd Regiment, or Lord McLeod's ghlanders, commanded by Lieut.-Colonel McKenzie, disembarked from fleet (strength, 36 officers, 50 sergeants, 22 drummers, and 944 rank d file), and occupied quarters in the casemates of the King's Bastion. ny other officers also joined their corps by this fleet, and, among them, lonel William Picton, the newly appointed colonel of the 12th Regiment, o at once took over the command from Lieut.-Colonel Trigge.

On the 5th February, three deserters who came in were sent to join the ners on board the fleet, for passage to England. These men reported at discontent in the enemy's camp, on account of the great scarcity d dearness of provisions.

On the 11th, the invalids and women embarked on board the fleet; the following day the supplies for the garrison were all landed, and rigging of the Spanish prizes having been repaired, the fleet prepared return. Flags of truce were frequently employed, and one of these, the ne day, brought over some English prisoners.

It is reported that the liberality and politeness shown by our Navy and Governor of Gibraltar to the Spanish Admiral and suite had (on the ter being permitted to return on parole to Spain) made a deep and lasting pression on their minds, and was of great advantage later to English soners in Spain, who were ever after treated with great attention and manity.

On the evening of the 13th, the British fleet got under way, with all the nsports that were ready, and was joined by those from Minorca.

The garrison was now considered in a perfect state of defence. The stores and magazines were full, a reinforcement had joined the garrison, and new spirits were infused into the troops, though scurvy had indeed begun to affect many, and threatened to become more general. The number of deserters from the Spanish lines was now very great.

On the 11th March, the Governor ordered all soldiers of the garrison to be victualled monthly (bread excepted) in the following proportion:—each 1st and 3rd week, 1 lb. of pork, 2½ lbs. salt fish, 2 pints peas, 1 lb. flour, 1 lb. raisins, 5 oz. butter, 1½ pints oatmeal. Second and fourth week, 1½ lbs. beef, 2 lbs. fish, 2 pints peas, 1 lb. rice, 5 oz. butter, 1½ lbs. wheat, ¼ lb. raisins. The salt cod, indifferent of its kind, continued to be issued for seven months, and, there being a lack of vegetables, it undoubtedly became the means of increasing scurvy.

In compliance with terms already arranged for an exchange of prisoners, about 390 sailors, prisoners, were handed over on the 12th, and distributed amongst our men-of-war whose crews had been sent to man the Spanish prizes.

On the 29th, we received more English prisoners, and, the same evening, two men of the 12th, one 39th, and one 73rd deserted, making a total of four desertions in the month.

On the 30th, the 12th Regiment was inspected.

On the 6th April, a further exchange of prisoners took place, the sloop "Fortune" taking over to the enemy 300 Spaniards, returning with 9 British, and two days later took over 280 prisoners. On the 11th, all Spanish prisoners were sent to Spain.

During April, we received supplies by two or three boats from the Barbary coast, and, towards the end of the month, the enemy were more active in their camp, being chiefly employed in making repairs, which, however, were so trifling that our artillery did not disturb them.

On the 18th May, two boats arrived from Tetuan with fowls and oil, and, on the 20th, a Moorish sloop from Malaga, with butter, raisins, and leather, the latter being much wanted; indeed, so scarce had it become in the garrison that several officers and men had been obliged to wear shoes made of canvas, with soles of spun yarn.

A valuable cargo of wine, oil, and other articles from Leghorn arrived on the 25th, and, on the 30th, the enemy's army were again under arms. They had already indulged in a few field days, and their movements that day were the attack and defence of a convoy.

Our naval force now consisted of the "Panther," 60 guns, Captain Harvey; "Enterprise," frigate, Captain Leslie; two armed vessels, commanded by lieutenants, with several armed transports, and other merchant ships.

Shortly after midnight on the 7th June, the "Enterprise" was attacked by six fire-ships, whereupon the garrison was as early alarmed as the navy. For some weeks past, we had been very successful in receiving small and y acceptable supplies ; on the 20th, however, it was reported that a large p with coals and butter, bound for the garrison, had been captured by the aniards two days before under the guns of Tangier.

On the 24th, the "Enterprise" had 19 sailors burnt and wounded by the plosion of some powder.

Early on the 27th June, four Spanish gunboats, under cover of night, d on the "Panther," but, on a brisk discharge being returned, they soon ired.

In the beginning of July, the "Panther" receiving upwards of 100 English soners from the enemy, Captain Harvey sailed for England, and, on the , the "Fortune" brought over more British prisoners.

Early on the 17th, five gunboats and four galleys fired on the "Enterse," and on the shipping in the New Mole. One of the frigate's forecastle ns was dismounted, and her fore-stay cut.

During the remainder of the month our firing became brisker on their rties whilst the latter were principally employed in different fatigue duties. veral empty transport ships left the garrison this month for England.

On the 22nd, the 12th and 72nd Regiments marched to the town, and 58th and 73rd to the South Barracks.

Battle of Minden Day was celebrated on 1st August, by Colonel Picton.[1]

In August few incidents occurred on either side. Our provisions began be bad and very offensive. What few supplies we now received were ther luxuries than substantials ; wine, sugar, oil, honey, onions, &c., mposing chiefly the cargoes which arrived.

Sugar had risen to 2*s.* 6*d.* per lb., and everything else in proportion.

On the 16th, the "Fortune" received from the enemy 64 prisoners, an ficer of the 56th Regiment being amongst them, also several who were ssengers in a brig taken by them on the 12th. Between the 15th and th August one soldier and ten sailors had deserted to the enemy, the tter having forced with them the midshipman who commanded the oat. On the 25th, a boat arrived in eight days from Minorca, with wine, a, and sugar.

Colonel Mawhood,[2] 72nd Regiment, died on the 29th August, and a uel took place to-day between two officers, one of whom was wounded.

On the 2nd September, the enemy finished the pontoon bridge they ad been building over the River Palmones, and, on the 23rd, a flag of truce rought over the midshipman, carried off by the sailors who deserted at e latter end of August.

The blockade was now becoming more strict and vigilant. Chains of mall cruisers were stationed across the Straits, at the entrance to the Bay, nd on every side of the Rock, whilst most unwelcome intelligence had een received that the Spaniards would not allow any boats to leave the

[1] Mrs. Green's Diary at Gibraltar.

[2] He had greatly distinguished himself in January, 1777, at the Battle of Princetown, during he war with America, whilst commanding the 17th Regiment.

Bay of Tangier. What small supplies we now obtained came from Minorca, and sold at such enormous prices that few were able to purchase them. The distresses, in short, of the garrison, with the increase of scurvy, threatened to be more acute and fatal than ever.

The enemy's operations on land had, for many months, been so unimportant as to deserve little attention. It was found, however, on the morning of the 1st October, that, during the previous night, they had erected an epaulement about 700 yards in advance of their lines, the noise of men at work having alarmed our outposts. They also made some attempts to burn our palisades with combustibles, but these were soon thrown off without taking effect, and it was later discovered that they had set fire to the fishermen's huts in the gardens. The epaulement was about 30 yards in extent, composed largely of fascines and sand-bags, and was erected about 1100 yards from our Grand Battery. Their guns being elevated, and batteries manned, together with other preparations in their lines, did not induce the Governor to take any notice of their work by day, but, at night, a few shot were discharged to discover if they were making any additions.

The erection of a work so distant, which had no connection with their established lines, caused some amazement to the garrison, as being so contrary to the ordinary method of approaching a besieged garrison. At daybreak on the 4th, it was observed that the enemy had carried away a great quantity of vegetables from the gardens, and had trampled much under foot. An English sloop of 12 or 14 guns was taken by them at about 11 o'clock, and, on the same day, our outposts discovered that they had been again endeavouring to fix fire faggots on our barriers, which was effectually stopped by a smart discharge of musketry from us, causing them to retire.

Forty British prisoners were handed over on the 5th, and, on the 11th, the sailors from our boats boarded a Danish dogger from Malaga, laden with lemons and oranges, which the Governor immediately bought and distributed to the garrison. Few more welcome articles had ever arrived than this cargo of fruit.

Scurvy had been making terrible ravages in our hospitals, and the beneficial effects from issuing lemons to the sick were almost instantaneous; in a few days, men who had been considered beyond recovery were able to leave their beds. Women, children, and officers had all been, to some extent, affected by this dreadful scourge.

A parley, which came from the Spaniards on the 11th October, informed the Governor that all future correspondence or intercourse was to be conducted by flags of truce in the Bay, and this rule was observed until the settlement of peace in 1783.

The assistance in the way of provisions the garrison had received from Barbary having been suspended for several months, Minorca was the only place they could look to for supplies. Several garrison boats were accordingly sent there, which returned in October, laden with the wine of the island, sugar, and cheese, with sometimes a few live stock. By the Governor's order, these articles were all sold by auction, and, though seldom purchased by the lower ranks, they afforded, on the whole, a temporary relief to the garrison.

On the 30th, it was observed that the enemy had posted an officer's guard in the Mill Battery, which was the name we gave to the new work.

On the 2nd November, two soldiers of the 12th and 56th Regiments attempted to desert by swimming round the Old Mole Head, but, a few days after, the body of the former was washed ashore near the King's Bastion, and it was concluded that his comrade had shared the same fate.

On the morning of the 8th, it was observed that the enemy had captured an English vessel, and were then towing in a gunboat. Our fire about this time became more animated. On the night of the 16th, the enemy's gunboats fired several shots about the Rock, and wounded a lieutenant and four men on board the "Enterprise." On the 18th, the gunboats again visited us, and, in returning their cannonade, one of the 32-pounders on the King's Bastion burst, killing an artilleryman and wounding three men of other regiments.

The man who fired the gun escaped, but was a little scorched by the powder.

On the 23rd, our batteries continued to fire at the enemy's new advanced work, to which they had begun to make an approach from their lines, consisting of fascines with sand banked up in front, and, for three nights, the gunboats continued firing into the garrison.

On the 25th, the enemy determined to entirely expel our people from the gardens, which were neutral ground, and this, in a few days, they accomplished.

The 30th November was noted for the arrival of a ship from Algiers, with soap, oil, wine, and candles—a very valuable cargo.

December was ushered in with bad weather.

A letter from the Adjutant-General's Office, dated 2nd December, called on commanding officers to send in a Return to the Commander-in-Chief of such Lieutenants, of over two years' standing, who were ready to raise Independent Companies of 100 privates in each.[1]

On the 21st, the "Speedwell," cutter, Lieutenant Gibson, arrived with despatches, having been attacked by a small vessel, which she beat off, Lieutenant Gibson being dangerously wounded.

On the 23rd, a brig brought some excellent supplies, and, the following day, two vessels from Liverpool arrived with provisions, the cargoes of which sold, at auction, at 300 per cent. profit.

The enemy, having completed their approach to the Mill Battery, were further employed in raising fascine traverses in the rear, for their greater

[1] *W. O.* 3, Book 26, p. 42.

protection, and, on the 26th, and following nights, their carpenters were so noisy at their work as to induce a brisk fire from our batteries.

On the 31st, another vessel got in from Minorca.

1781

In the beginning of January, our carpenters were busily employed erecting stages and temporary cranes in camp and its precincts, to assist the unloading of provisions, under the apprehension of being annoyed by the enemy's advanced battery, which was now finished, and reported to contain eight 13-inch mortars ; precautions which turned out later to be very necessary, allowing for the range and effect of the enemy's fire.

Our Engineers were also indefatigable in putting everything in the best state of defence.

The failure to obtain any further provisions from Barbary was now accounted for by the news that the Emperor of Morocco had been bribed by Spain with an annual grant of £7500 and the redemption of a hundred African prisoners, on condition that he should deliver up, for a certain period, the three ports of Tangier, Tetuan, and Larache, and banish from his dominions the British Consul and subjects of Great Britain; all of which was duly carried out, the British Consul arriving at Gibraltar on the 11th January.

On the 16th, a brig came in from Madeira with 70 butts of wine, followed, two days later, by the "Tartar" privateer, laden with flour and various articles from England.

On the 28th, a ship from Leghorn brought wine, barley, brandy, and deal boards, and, on the 29th and 30th, two vessels arrived with the produce of Minorca. The difficulties and dangers risked by such ships were always very great.

On the 1st February, the bodies of two deserters were found behind the Rock (one a man of the 56th, and the other a sergeant of the 73rd), who, in attempting to escape from the garrison, had fallen down a precipice about 1200 feet, and been dashed to pieces.

A deserter who came in from the enemy on the 8th reported that they now mounted a captain's guard, besides artillery, in the St. Carlos's Battery, as they had named their new work.

On the 18th, the officers of the garrison sent, through their commanding officers, to the Governor a memorial to the King, praying for some extra remuneration, on account of the losses and heavy expenses they were being put to because of the siege, and it is reported that to this memorial and another which accompanied it, of similar import, no official reply was received.

On the 19th the 12th Regiment took part in a field-day on the Rock, and, on the 26th, the regiment was inspected with others of the garrison ; all, after being reviewed, marched to their alarm posts, where several rounds were discharged from the parapets.

Bread, flour, and fuel now began again to be very scarce, and wood sold by the pound. Many families had not tasted bread for several days. Distress was intolerable amongst the poor soldiers, and still more the inhabitants, who could not afford to buy articles from the Minorcan vessels, the cargoes of which were chiefly luxuries. Biscuit crumbs sold for 10*d*. and 1*s*. per lb. The allowance of bread to the troops was also curtailed, and many Portuguese fishermen left the garrison for want of this article.

On the 14th March, a soldier of the 12th deserted.

On the 16th, a vessel from Minorca arrived, laden with brandy, gin, flour, &c.

In the course of the month several small craft arrived from Minorca, and, on the 29th, the invalids of the garrison embarked on the "Enterprise" and "St. Firmin."

On the evening of the 11th April, many signals were made from the west, and, at midnight, arrived the "Kite," cutter, with the joyful news that a convoy was at the entrance to the Straits, under command of Admiral Darby, with the British Grand Fleet.

At daybreak on the 12th, the long-looked-for fleet was sighted, and, whilst exultations were great on the part of the inhabitants, they gave little heed to the possibility of a severe bombardment of the fortress, which was so soon to follow.

Just as the leading ship dropped her anchor, the enemy's batteries began bombarding the town, ceasing only in the middle of the day. The amount of ordnance playing on the garrison, from different Spanish forts and batteries, consisted of about sixty-four 26-pounder guns, and fifty 13-inch mortars. Eighteen gun and mortar boats also attempted to prevent our ships coming in, but were forced, by a line-of-battle ship and two frigates, to make a hasty retreat. The enemy's cannonade was instantly returned by the garrison, and our artillery had orders, at first, to concentrate their fire principally on St. Carlos's Battery, which soon subdued its firing. The joy of the inhabitants soon turned to terror as they flocked to the southward in the greatest confusion, leaving their property unsecured.

The firing of the enemy's batteries recommenced at about 5 P.M., and continued without intermission throughout the night. The landing of our supplies, however, continued without interruption. Five hundred men, with a proportion of officers, were employed for the duty, and were later considerably augmented, working daily with such diligence, that with the assistance of the Navy, the stores were landed in nine or ten days. Our casualties on the 12th were an artillery officer and several rank and file wounded. The bombardment was continued next day, when several soldiers were killed and wounded in their quarters; the sick in the town were now conveyed to the Naval Hospital at the Southward.

On the 14th, the enemy's gun and mortar boats, firing on our shipping, were soon forced to retire. Officers, whose quarters were damaged, received marquees from the public stores, and the distressed inhabitants were accommodated with tents. The total strength of our picquets at this period was 2 captains, 9 subalterns, 9 sergeants, 9 drummers, and 391 rank and file.

On the morning of the 14th, a shell set fire to a wine house in the Green Market, near the Spanish Church, and resulted in four or five houses being burnt to the ground.

The enemy's shells soon forced open the secret recesses of the merchants, enabling the soldiers at once to avail themselves of the opportunity of seizing upon the liquors, which they conveyed to haunts of their own. Such a scene of drunkenness, debauchery, and destruction on the part of the troops now took place as has rarely been seen before, resulting in most flagrant cases of insubordination and defiance of their officers. Some of the offenders died of immediate intoxication, whilst several with difficulty recovered.

On the 15th, the bombardment was continued with greater activity, and, our ammunition running short, our batteries ceased firing except as occasion offered. A redistribution of quarters now took place, the 12th, 39th, and 56th Regiments being ordered to occupy Montagu's Casemate, with the Galley House and Waterport Gateway, the other troops moving simultaneously to their respective destinations. On this date, two soldiers (one of the 12th) drank aqua-fortis for rum; the 12th man died and the other recovered.

On the 16th, the gunboats attacked the shipping, but were soon repulsed by three of our ships, and, on the following day, Colonels Ross (39th), Green (R.E.), and Picton (12th) were appointed to rank as brigadiers, Captain R. T. Picton (12th) being appointed brigade-major to Colonel Picton.

On the 18th, when the gunboats opened fire on our shipping, two of our frigates had several men dangerously wounded, and a shell, which fell where some men of the 12th and 39th Regiments were quartered, killed two and wounded four privates.

On the 19th, the gunboats renewed their attack, but were soon obliged to retreat, and, the following day, the supplies having been landed, the fleet sailed, some of the officers' wives and a great many of the inhabitants taking advantage of this opportunity for leaving the garrison. That night the town was on fire in four different places.

The enemy's cannonade and bombardment raged throughout the 21st and 22nd April, and, early on the 23rd, the gun and mortar boats fired on our parties ranging provisions at the Southward, when a woman of the 58th was killed and several men wounded.

On the 24th, a soldier of the 12th Regiment deserted from the Land Port guard, and the garrison orders of the 26th gave out that any soldier convicted of being drunk on duty, asleep on his post, or found marauding should be immediately executed.

The Governor having previously arranged for a supply of provisions from Minorca, a convoy of twenty victuallers arrived, without opposition, on the 27th, under charge of four of our frigates and the sloop "Fortune."

Whilst the remainder of this month was celebrated for excessive rains and heavy storms of thunder and lightning, the bombardment continued warm throughout the 28th and 29th, and, on the morning of the 30th, the gun and mortar boats again approached the garrison, their fire varying very little from former attacks.

Early on the 2nd May, two vessels arrived from Algiers, laden with bullocks, sheep, wine, and brandy, and the enemy seemed to have abandoned the idea of blockading us to a surrender. A provost-marshal was to-day appointed to our force, and, in the evening, a shell from the garrison, falling on a part of St. Carlos's Battery, above the magazine, communicated with the powder, and blew it up. Our artillery annoyed the enemy greatly during their confusion, though they kept up a brisk discharge from their lines at the rate of 250 rounds an hour.

On the 5th, a soldier of the 58th was hanged on the Grand Parade, at the door of the store where he was detected plundering, and his body remained until sunset as an example to other offenders.

On the morning of the 7th, the gun and mortar boats fired on the town and the New Mole, staying about an hour, and then retired, our return fire having consisted of four hundred rounds.

On the 9th, Lieutenant Lowe, 12th Regiment, whilst superintending a working party, lost a leg by a cannon shot below the knee, and, on the 11th, Lieutenant Thornton, 12th Regiment, was wounded by splinters of stones, thrown up by a shot, which grazed between his legs. On this date, a shell fell in a store of brandy and rum as two of the 12th were robbing it; it killed one man and blinded the other. At about 9 A.M. the gunboats had advanced, but kept so far off that only one shell and two shots came ashore.

The bombardment, with almost daily visits from the gun and mortar boats, continued, and, on the 17th May, the Jews' synagogue and other buildings were burnt down. The enemy's fire seldom exceeded a thousand rounds in the twenty-four hours; their batteries were now much shaken from the firing, and parties were constantly bringing supplies of ammunition and different materials for repairing their works.

The 19th May, 1781, was the anniversary of the 12th year the 12th Regiment had been in Gibraltar.

On the 21st, a poor woman broke her thigh in endeavouring to get out of the way of the shot from the gunboats.

Before daybreak on the 22nd, a shell burst on the Convent Battery, when a splinter flew to the South Bastion and fired the morning gun, a singular circumstance, which caused no little surprise to those who heard the report and did not know the cause. Whilst our daily number of rounds fired was now increased to 150, the enemy had reduced theirs from 1000 to an average of about 650.

On the night of the 23rd, the gun and mortar boats resumed their attack on the camp, which was more dreadful than any our force had yet experienced. Three Jews (one of whom had lost all he had in the town, nearly £10,000), his clerk, and a female relative were all killed by a shell; a soldier of the 73rd Regiment was killed in his bed, and a Jewish butcher was equally unfortunate (in all seven killed), whilst another shell burst in a house containing 15 or 16 persons, all of whom escaped except a child whose mother had met with a similar fate. A shot also went through the roof of the hospital pavilion where Lieutenant Lowe was.

On the 27th, a ship from Minorca brought a supply of wine, brandy, lemons, and salt, and the "Enterprise" frigate, with 17 ordnance ships and transports, sailed for England.

Towards the end of May the enemy's bombardment had considerably abated.

A shell from the gunboats, at about 2 A.M. on the 1st June, fell in the marquee of an officer, 12th Regiment, and, on the 2nd, the 12th Regiment was ordered to occupy the ground behind the hospital when the gunboats next came.

When the latter attacked on the following morning, two sergeants of the 12th and 58th Regiments were killed, and the drum-major of the 12th was wounded. The enemy's cannonade in the beginning of June decreased to about 500 rounds in the twenty-four hours.

On the 4th, the Governor commemorated the King's birthday with a salute, at noon, of 23 guns and 43 mortars, being the number of guns on the St. Carlos's Battery; the Spaniards fired a great deal at the Royal Standard, and sent one shot through it.

On the 6th, Captain ——, 56th, having abused Ensign S—— (apparently Ensign Sandby), 12th Regiment, on the South Port guard, a duel ensued, when the latter was shot through the leg.

At about 1.30 A.M. on the 12th, the gun and mortar boats bombarded the camp, killed a child, and wounded a woman.

On the 16th, the "Fortune," sloop, brought in 141 English and Jews (men, women and children) from two ships that had been captured when leaving Gibraltar. The enemy's bombardment now decreased daily, their fire being chiefly directed to our upper batteries, as the town was almost a heap of ruins.

On the night of the 24th, the gun and mortar boats fired on the camp with little damage, and renewing their tactics at 12.30 A.M. on the 28th, for about two hours, they made their usual signal to depart, but soon after returned and recommenced a brisker fire than before, killing 2 and wounding 11 soldiers, 9 of whom belonged to the 39th Regiment, the greatest loss which had been yet experienced from the gunboats.

On the night of the 30th, Captain C——, 12th Regiment, sailed in a packet for Faro in Portugal.

The inhabitants of Gibraltar were now obliged to pay ground rent without a receipt, or lose the ground, though their houses were destroyed.

On the 3rd July, three women were flogged through the camp, for buying stolen goods. The enemy continued making gabions and bringing much wood into their camp, whilst the garrison were employed in repairs and additions to the works.

On the 11th, the Spaniards adopted firing by night only, and, unless we provoked them, their firing seldom exceeded thirty rounds.

On the 15th, a man of the 12th whose thigh had been broken by a shell, came out of hospital, got drunk the following day, and broke it again.

The troops in garrison changed quarters on the 21st with the exception of the 12th Regiment, who retained theirs.

By a boat which arrived from Portugal on the 22nd, intelligence was received that the army then before Gibraltar consisted principally of militia regiments, the regular Spanish troops having embarked for the West Indies; also that the Spanish fleet had sailed for Cadiz on a cruise.

On the 25th, streets were made in the 12th Camp, and epaulements raised as a protection against the gunboats. At about 2 A.M. on the 1st August, the latter stayed till near daylight, having fired a great number of shell, but at a much greater distance. By the middle of the month, the enemy's bombardment, except when the boats were firing, scarcely exceeded three shells in the twenty-four hours. On the 15th, the fire from the gun and mortar boats caused a few casualties.

A schooner arrived from Faro, on the 17th, with fruit, onions, and salt, and two days later, another boat from Faro, with tea, &c. This was considered the hottest summer the troops had experienced in Gibraltar.

At the end of August, the enemy's attention to the blockade seemed to be revived. Their cruisers were increased, and constantly on the watch. On the evening of the 30th, their cannonade was pretty smart for an hour or two, chiefly occasioned by our firing on their working parties. During the past six weeks, we had lost four men by desertion.

The first half of September was devoid of incidents of note beyond the ordinary operations of a siege.

At midnight on the 17th, an attack by the gunboats lasted two hours and a half, when they expended 130 shells and 87 shot. The garrison retaliated both on sea and land with a smart fire on the enemy's camp. Amongst the casualties, a private of the 12th Regiment had both his legs taken off by a shell; it was later reported that, when taken to the hospital, he lay for 2¼ hours before the surgeons could venture out of their bomb-proof to dress him, so that the loss of blood would alone have killed him.

On the 18th, a soldier of the garrison was hanged for desertion. At about 9 p.m. a shell fell into a house opposite the King's Bastion, where the Town Major (Major Burke), with two majors of the 39th, were sitting. The shell took off Major Burke's thigh, went through the floor and blew him up with it; he lived to have the wounded part cut off, but died soon after, much lamented, and a great loss to the garrison. The other two officers had time to escape before the shell burst, but were slightly wounded by the splinters.

The next evening, Lieutenant Lowe of the 12th, who had lost a leg, left with invalids for England, and, on the 20th, Captain Foulis, 73rd Regiment, was appointed Town Major.

At the end of September, the firing from the garrison exceeded 700 rounds in the twenty-four hours, and the enemy frequently returned 800, and sometimes more.

On the night of the 3rd October, one of the 12th was found drowned; in an endeavour to desert, he had attempted to swim to the enemy from Waterport, and his body was discovered floating by our sailors. Amongst our casualties on the 7th, a shell from the gunboats wounded two men of the 12th Regiment.

The establishment of the corps now in garrison should have been 5776 privates, but they did not muster more than 3600.

On the 12th, a duel is stated to have taken place between Ensign J—— (apparently Ensign Jones), 12th Regiment, and Lieutenant B——, 72nd, but there is no report of the result.

On the 21st, the regiments changed quarters, the 12th relieving the 56th, on relief by the 39th.

On the night of the 30th, four boats arrived from Portugal with lemons, onions, &c., which the Governor purchased for the sick in the hospitals, and detained some of the crew who were suspected of having come as spies. On the 31st, two men of the 58th died of spotted fever, and one of the 12th shot himself. In the course of the month, three men deserted.

On the 6th November, a native of Gibraltar, who came with the boats containing cargoes of fruit, confessed to have been sent as a spy.

On the night of the 11th, the enemy erected an additional battery of six embrasures to those recently raised, and, on the 13th, the "Phœnix," cutter, arrived with ordnance stores, and Brigadier Ross on board, who had left the garrison some months before.

A deserter, who came in on the 20th, gave the strength of the enemy's advanced guards, and related the general inactivity which prevailed in their camp; he also corroborated the former intelligence that the army opposed to us was now principally composed of militia regiments, adding that the men were greatly dissatisfied, and that they had lately suffered very severe losses from our fire.

The Governor, having formed the opinion that the Spaniards were now lulled into security from their superiority of force, and would never suspect the garrison as capable of taking any bold and hazardous measure against them, resolved to make a sortie on their batteries, and, on the 26th, assembled the field officers to be employed on this service to communicate to them his plan of attack.

The following orders were issued :—

"Evening garrison orders. Gibraltar, November 26, 1781: COUNTERSIGN, STEADY.—All the grenadiers and light infantry in the garrison, and all the men of the 12th and Hardenberg's Regiments, with the officers and non-commissioned officers on duty, to be immediately relieved and join their regiments, to form a detachment, consisting of the 12th and Hardenberg's Regiments complete ; the grenadiers and light infantry of all the other regiments ; one captain, three lieutenants, ten non-commissioned officers and a hundred artillery ; three engineers, seven officers, ten non-commissioned officers, overseers, with a hundred and sixty workmen from the line, and forty workmen from the artificer corps ; each man to have thirty-six rounds of ammunition, with a good flint in his piece, and another in his pocket ; the whole to be commanded by Brigadier-General Ross, and to assemble on the Red Sands, at twelve o'clock this night, to make a *sortie* upon the enemy's batteries. The 39th and 58th Regiments to parade at the same hour, on the Grand Parade, under the command of Brigadier-General Picton, to sustain the *sortie* if necessary."

At midnight, the whole were assembled, and being joined by 100 sailors with two officers, the detachment was divided into three columns as follows :—

Left Column
Lieut.-Colonel Trigge, 12th.

	O.	S.	D.	R. & F.
Grendrs, 72nd Regt,	4	5	0	101
Light Coy, 72nd	4	5	0	101
Sailors with an Engineer .	3	3	0	100
Artillery .	1	4	0	35
12th Regiment	26	28	2	430
Light Coy, 58th	3	3	0	57
	41	48	2	824

Centre Column
(The Reserve)
Lieut.-Colonel Dachenhausen, and Major Maxwell.

	O.	S.	R. & F.
Grendrs, 39th Regt. .	3	3	57
Light Coy, 39th	3	3	57
Grendrs, 73rd Regt. .	4	5	101
Light Coy, 73rd	4	5	101
Engrs. with Workmen	6	14	150
Artillery . . .	2	4	40
Grendrs, 56th Regt. .	3	3	57
Grendrs, 58th Regt. .	3	3	57
	28	40	620

Right Column
Lieut.-Colonel Hugo.

	O.	S.	D.	R. & F.
Grendrs, Reden's Regt. . .	3	7	0	71
Grenadiers, De la Motte's Regt. .	3	7	0	71
Engrs. with Workmen . . .	4	6	0	50
Artillery . .	1	2	0	25
Hardenberg's Regt.	16	34	2	296
	30	59	2	570

In the total of the columns, Brigadier Ross, with several officers who accompanied him as aides-de-camp and men are not included, the deficit amounting to 3 officers, 3 sergeants, and 57 men in the Right Column.

	Colonels	Lieut.-Colonels	Majors	Captains	Lieutenants	Ensigns	Chaplains	Adjutants	Quarter-Masters	Surgeons	Mates	Sergeants	Drummers	Rank and File	
Total out with the Sortie	1	3	3	26	60	14	0	3	0	0	2	147	4	1914	Exclusive of sailors from the frigates.
Sick in Hospital . .	0	0	0	1	1	1	0	0	0	0	0	28	6	557	
Remaining in Garrison .	5	5	5	45	71	31	3	7	8	9	14	266	181	2531	
Total strength of the Garrison before the Sortie	6	8	8	72	132	46	3	10	8	9	16	441	191	5002	Total strength 5952.

The detachment being formed in three lines, the right column in the rear, and the left in the front, tools for demolishing the works were delivered to the workmen, and the following directions for their destination communicated to the principal officers :—

> "The right column to lead and march through Forbes' Barrier, for the extremity of the parallel; keeping the eastern fences of the gardens close on their left. The centre immediately to follow, marching through Bayside Barrier, and directing their route through the gardens for the mortar batteries. The left column to bring up the rear, marching along the Strand for the gun batteries. No person to advance before the front, unless ordered by the officer commanding the column; and the most profound silence to be observed, as the success of the enterprise may depend thereon. The 12th and Hardenberg's Regiments to form in front of the works, as sustaining corps; and are to detach to the right and left, as occasion may require. The reserve to take post in the furthest gardens. When the works are carried, the attacking troops are to take up their ground in the following manner. The grenadiers of Reden's and De la Motte's behind the parallel; the 39th and 73rd flank companies along the front of the fourth branch; and the 72nd grenadiers and light company, with their right to the fourth branch, and left to the beach."

The reserve was to be led by Major Maxwell, 73rd Regiment, supported by a body of seamen in two divisions. When the moon had set, the march commenced at about 2.45 A.M. on the 27th. Nothing could exceed the silence and order of this advance, and so well timed and exact was the combination, that at one and the same time the whole of the front of the enemy's work was attacked. The Spaniards, surprised and dismayed, gave way on every side.

Lieut.-Colonel Trigge (12th), with the grenadiers and light companies of the 72nd Regiment, carried the gun batteries with great gallantry. When our troops had taken possession, the attacking corps formed, agreeably to their orders, to repel any attempt which the enemy might make to prevent the destruction of the works, whilst the 12th Regiment took post in front of the St. Carlos's Battery, to sustain the western attack; the reserve, under Major Maxwell, being drawn up in the further gardens.

By the most wonderful exertions, two mortar batteries of ten each, and three batteries of heavy cannon, with all their lines, communications, traverses, beds, carriages, and platforms, were overthrown and consumed to ashes.

The object of the sortie having been fully effected in an hour, trains were laid to the magazines, when Brigadier Ross ordered the advanced corps to withdraw, with the supporting regiments to cover their retreat.

It was not a little singular that the 12th Regiment and Hardenberg's Hanoverians, who, at the memorable battle of Minden, had fought together, (and in the ordinary course of events could never expect to meet again) were now serving a second time on field service, and were the only entire regiments that day engaged.

The magazines blew up one after another. In the course of these conflagrations the whole Spanish camp continued tame spectators of the havoc that was made, without an effort to save or defend their work.

They had 2 officers and 15 privates taken prisoners, and, little opposition being made, very few were killed in the works. Our casualties were 4 privates killed, 1 officer (Lieutenant Tweedie, 12th Regiment), and 24 N.C.O's and privates wounded, and 1 missing, supposed to have been left wounded in the batteries.

General Eliott declared in Orders that "the bravery and conduct of the whole detachment, officers, soldiers, and sailors, on this glorious occasion, surpassed his utmost expectation."

For several days the Spaniards appeared confounded at their misfortune; the smoke of the burning batteries continued to rise, and no attempt was made to extinguish the flames. It was noticed that several executions took place in their camp—probably soldiers who had fled so disgracefully from the batteries. In the beginning of December they again resumed the siege, and a thousand workmen commenced labouring to restore the batteries. Their work was greatly retarded, however, by the fire of the garrison.

On the 1st December, a flag of truce brought letters from the English prisoners lately captured in the cutters bound to the Garrison. Replies were returned on the 2nd, and letters were also sent from the Spanish prisoners, taken in the sortie, to their friends in camp.

One of the officers taken by us was Baron Von Helmstadt, Captain in the Walloon Guards, who had been wounded in the knee, and only under great pressure submitted to amputation.

On the 4th, a flag of truce brought letters of thanks to the Governor from the Spanish General and the Walloon Guards, for the humanity shown to the prisoners taken in the batteries. The boat also brought poultry for the wounded Baron, with clothes and money for the officers.

Our firing continued to vary as the enemy's operations (apparently now only defensive) were more or less noticeable.

The Garrison, at this period, was extremely sickly, nearly 700 men being now hospital patients, thereby greatly curtailing the strength of the working parties, and necessitating officers' servants and others usually exempt from such duties being ordered to assist.

On the 20th, Brigadiers Ross and Green were appointed major-generals in the Army, and, the next evening, General Ross left for England. The same day, a flag of truce brought money and clothes for the prisoners.

About the 27th December, the Baron became dangerously ill, irrespective of his wounds, and flags of truce were daily passing and re-passing to inform his friends of his condition. On the 28th he died, and, on the following day, his body was carried to the New Mole, escorted by the grenadiers of the 12th Regiment, who fired three rounds over it, two barges being in waiting to carry it to the enemy's camp. The Governor and principal officers of the garrison attended the ceremony, and the most scrupulous care was taken to return every article of the late Baron's, including his money, and the fowls and other refreshments sent by his friends.

The funeral procession was officially recorded as follows :—

The 12th Company of Grenadiers, with reversed arms,
Musick playing a Dirge.
The Town Major, a Lieutenant of Royal Navy. The Governor's Secretary.
Boy carrying a Lanthern with two lighted candles.
Boy carrying a vessel of Holy Water.
Spanish Church-Wardens carrying a small Cross.
Vicar of Spanish Church repeating funeral service.
The Corpse dressed in full
uniform in a Coffin covered with black cloth and white furniture.
The Baron's Sword and Scabbard cross'd over the Pall.
Bearers.
Six Captains.
Don Vincente Vasquez Frèire, Sub-Lieutenant of Artillery, Prisoner of War (to whom the Governor presented a sword, his own being left behind when the Advanced Batteries were stormed) as Chief Mourner.
Sir Charles Knowles, Baronet, Captain of the "Porcupine," and the Adjutant-General.
Captain Curtis, commanding the Squadron, with the Governor and his Staff.
Major-General De la Motte.
Suite of Officers.
Three vollies fired on the Coffin being put into the boat.

ON THE WATER.

In 2nd Barge.

The Town Major, Naval Lieutenant, and Secretary with the coffin which was delivered over to a Captain of Lusitania, who, with persons apparently of Civil Departments, came to receive it.

1782

January found the enemy persevering in the reconstruction of their works. On the night of the 12th, two of our ordnance ships left for England, one of them embarking Lieutenant Tweedie, 12th Regiment (wounded), and also the prisoners taken at our sortie, including the Spanish Lieutenant of Artillery, Don Vincente Frèire.

On the 20th, the 12th Regiment relieved Reden's Hanoverians in the Picquet Yard. At about 2 P.M. on the 24th, our batteries began to play on the enemy's advanced works, and a very hot fire was kept up on both sides until about 4 P.M., when ours slackened. The garrison lost one man by desertion this month.

In the beginning of February, the enemy's St. Carlos's Battery seemed almost completed. On the 23rd, a ship arrived from Portugal laden with wine, lemons, &c.

On the 24th, the 12th Regiment was inspected on Windmill Hill, and the officers dined with the Governor.

Throughout March the operations of the besiegers still continued to be tedious. On the night of the 23rd, the "Vernon," store ship, arrived with

70 drafts of recruits for the Garrison, and, some hours after, two frigates, with four transports, having the 97th Regiment on board, anchored under our guns. The arrival of the "Vernon" was considered fortunate, since there were upwards of 30 Spanish men-of-war out to intercept her and her convoy, the frigate "Sussex."

On the 24th, the 97th Regiment, commanded by Colonel Stanton, disembarked, 700 complete, and soon after became very sickly.

On the morning of the 25th, a shell from the enemy took off the legs of two men of the 72nd and 73rd Regiments, one leg of a soldier of the 73rd, and wounded another man in both legs, so that four men had seven legs taken off or wounded by one shot.

The Spaniards now seldom exceeded firing 200 rounds in the twenty-four hours, though they frequently received from us, in that period, double that number.

On the night of the 31st, a boat came in from Portugal with sheep, fowls, oranges and lemons, two others having arrived during the month.

On the 6th April, Colonel Stanton, 97th Regiment, was appointed a Brigadier-General.

On the 10th, as Lieutenant Whetham (12th Regiment) was marching off the Spur Guard, a shell burst and killed him, his servant lost an arm, and the drummer of the guard had his drum broken to pieces on his back; the remainder of the guard escaped.

On the 17th a General Court-martial sentenced a man of the 12th to stand in the pillory[1] for assaulting a drummer.

A change of quarters took place on the 21st, the 12th, 58th, and Hanoverians proceeding to camp, the 39th and 72nd to the Pavilions, and the 73rd and De la Motte's to the town. The men of the 97th were now very sickly, and dying fast from a contagious fever, of which there were other cases in the Garrison.

From 7 p.m. 4th May to the same hour on the 5th was the first period of 24 hours in which both the Garrison and the enemy had been silent for nearly thirteen months.

On the 22nd, 80 dollars was the price given for a sheep which cost 4*s.* at Portsmouth; fowls were selling at 8 dollars a pair, and bread was scarce.

On the 25th, a privateer, 10 guns, arrived from Leghorn with despatches for the Governor, and to inform him of the preparations being made by Spain and France for an attack on Gibraltar. There also arrived, by the same vessel, a Corsican officer and 12 privates, who offered their services as volunteers during the approaching attack, stating that there were more of them to follow. The Governor accepted their services, and directed them to be entertained by the different regiments until the others arrived.

[1] A wooden engine, on which criminals were formerly exposed to public view, held by the neck and wrists.

The same night, a coasting vessel from Algiers brought about 100 butts of wine, &c.

Every preparation was now being made by the Garrison to give the enemy a warm reception, and, among them, an additional number of grates were made for heating shot. Between the 27th and 29th it was observed that the Spaniards were reinforced by about 12 battalions, approximately 9000 men, if the battalions were calculated to muster 750 in each.

On the 2nd June, Brigadier Stanton died of sunstroke.

On the 4th, a salvo of 44 guns was fired by us to commemorate the King's birthday, (he being 44 years old,) and the enemy's batteries immediately returned our fire.

As an instance of the vagaries of shells, Ensign M'Kenzie, 73rd, was, on the evening of the 7th, surprised by a shell carrying away part of the chair he was sitting on ; it then burst in the room below, and lifted him and the chair from the floor without further injury.

On the 11th, an unlucky shell from the enemy caused the explosion of our magazine at Princess Ann's Battery, which contained about 100 barrels of powder ; one drummer and 13 rank and file were thereby killed, and 3 sergeants, 3 drummers, and 9 rank and file wounded, whereupon the enemy continued cannonading the remainder of the day.

On the 14th, a French frigate with 14 other ships arrived in the Bay.

On the 18th, Hardenberg's Regiment was ordered, in case of alarm, to act with the 58th at Europa, the 12th and 56th to occupy the Line Wall in the town, as far as the Zoca Battery, and the remaining regiments to line the intermediate space of the Line Wall.

On the 20th and 21st, the French troops disembarked, and encamped in the Spanish Lines.

All regimental musicians, besides officers' servants, were now detailed for duty.

The Court of Madrid, whose whole attention was concentrated on the recovery of Gibraltar, now appointed the Duc de Crillon (who had formerly commanded at the Spanish lines before Gibraltar) to undertake this arduous enterprise. Meanwhile, his adversary, General Eliott, surrounded by enemies, and without any hope of relief and assistance, placed every confidence in the united exertions of his small army, which he had already found superior to the greatest hardships.

On the 24th, the Garrison began to practise "parapet firing" with ball, at casks placed at intervals in the Bay.

On the night of the 30th, a soldier of the 56th who attempted to desert was dashed to pieces. The next day his body was exposed at the Quarter Guard, as a public spectacle to intimidate others.

Early on the 3rd July, a duel took place between Captain H——, 12th, and Captain C——, 39th, who was wounded in the thigh. About the middle of the month, additional forges for heating shot were established in different parts of the Garrison, with all the proper apparatus.

On the 24th, a sergeant of the 72nd was hanged for desertion.

At 5 A.M. on the 25th, two ships arrived from Algiers, laden with wine, sugar, &c., and 4 officers and 68 men, Corsicans, the remaining contingent who had volunteered their services to the Governor.

On the 31st, upwards of 100 covered waggons came to the enemy's lines, supposed to be laden with ammunition and stores for the batteries.

Judging by the general alertness now being displayed in their camp, their movements, in the month of August, became daily more important; nor was the Garrison less active in taking advantage of this interval, in making general improvements for defence. The gunboats which used to pay such frequent nightly visits were now absent, having been stationed out as night cruisers.

The Corsican Volunteers were formed into an independent corps, under a Captain Commandant, and consisted of 4 officers, besides an adjutant and chaplain, 4 sergeants, 4 corporals, 2 drummers, and 68 privates. They were armed with a firelock and bayonet, with a horse pistol, carried on the left side, and two cartridge boxes. The Governor quartered them on Windmill Hill, and committed that post to their charge.

About the middle of August, a species of influenza became very prevalent in the Garrison.

On the 18th, His Royal Highness the Count d'Artois, brother to the King of Spain, joined the Spanish force to serve as a volunteer at the siege, and having brought with him letters for the Garrison, a flag of truce, on the following day, came from the Duc de Crillon, forwarding the same, together with a most courteous letter from the Duke to General Eliott, accompanied by a present of ice, fruit, and vegetables, which, having been duly acknowledged in equally courteous and dignified terms, the siege proceeded in its ordinary course, and, on the night of the 20th, our artillery fired with great vigour from the upper and lower batteries in all directions.

On the 31st August this year, orders were issued for the 12th Regiment to assume, in addition to its number, the title of "East Suffolk," and to cultivate a connection with that county, which might at all times be useful towards recruiting.

It was now rumoured that the attack on the Garrison was to take place by the 15th September. Affairs seemed drawing to a crisis, and every appearance indicated that it would not be long deferred.

On the 6th September, the Governor made some new arrangements in the Garrison detail, and our picquets were in future ordered to mount fully accoutred with ammunition complete. The alarm posts were changed, the 12th Regiment to the New Mole Parade; the regiment was at this time furnishing a daily picquet of 1 subaltern, 1 sergeant, 1 corporal, 1 drummer, and 33 privates; and Brigadier-General W. Picton was detailed to command the troops in the town.

After the glorious sortie by General Eliott, in the previous November, the honour of inflicting a second disaster was reserved for the Lieut.-Governor, Lieut.-General Boyd, who, on the 6th, recommended the immediate use of red-hot shot against the besiegers' batteries. Accordingly, with the Governor's approval, steps were at once taken for the suggestion to be carried out. The strength of the Garrison, with the Marine Brigade (including officers), in September, was about 7500 men, upwards of 400 being in hospital.

The following abstract shows the daily number for duty :—

Guards	1,091	Men, including officers.
Picquets (including the additions of the 12th Regiment) .	613	Do.
Working parties under the Chief Engineer and the Quarter-Master General.	1,726	Exclusive of the Engineers and overseers
Total . . .	3,430	

exclusive of orderlies, hospital assistants, and men employed in other departments of the Garrison.

By the morning of the 8th September, the preparations under General Boyd's directions were completed, and at 7 A.M. our batteries began a heavy fire on the Spaniards' advanced works, which continued till 4 P.M. with most satisfactory results, setting fire to their Mahon Battery of six guns, and its adjacent works.

At about 5.30 A.M. on the 9th, the enemy commenced a cannonade with a volley of about 60 shells from all their mortar batteries in the parallel, succeeded by a general discharge of cannon, in all about 170 guns. At about 1 P.M., a ship arrived from Algiers, laden with 4 bullocks, 30 sheep, wine, &c., at which time the Spanish Admiral, with five ships of the line, stood within gunshot, and fired at the vessel and the Garrison. While the land batteries were discharging a heavy fire on our North Front, nine line-of-battle ships, including those of the French, poured forth several broadsides at the works. The men-of-war repeated this attack early on the 10th, the enemy's gun batteries, on shore, recommencing at morning gunfire. In the expenditure of ammunition on the 8th, it was calculated they had discharged 5527 shot and 2302 shells, apart from the number fired by the men-of-war and mortar boats. Our loss in this attack was 8 killed and 17 wounded.

On the night of the 11th, the enemy fired from 120 to about 200 shells per hour, killing two men on the Grand Battery and destroying about 30 of the palisades, our return fire consisting of several rounds of grape and round shot; and about an hour after midnight, their gun and mortar boats commenced to bombard the North Front, continuing about two hours.

Early on the morning of the 12th, a report was received that a large fleet was in sight, which proved to be the combined fleets of France and Spain, consisting of seven three-deckers, and thirty-one ships of two decks, with three frigates and a number of miscellaneous craft and hospital ships, the whole under the command of ten Admirals; and in the afternoon they were all at anchor.

The appearance of such a force, to carry out their final efforts, could not fail to surprise and alarm the garrison. There were now assembled in the Bay (Spanish and French) 47 sail of the line, 10 battering ships (armed junks) carrying 212 guns, innumerable frigates and cutters, gun and mortar boats, and smaller craft for disembarking men. On the land side, the enemy had most prodigious batteries and works, mounting 200 pieces of heavy ordnance, and protected by an army of nearly 40,000 men under a general of the highest reputation. It was not surprising, then, that the Spanish nation should anticipate, from this force, the most glorious results.

On the other hand, it must be said, in favour of the besieged, that, although the garrison of Gibraltar scarcely consisted of more than 7000 effective men, they had now become so inured to the hardships and privations of field service, and hardened to the effects of artillery, that, with British pluck to carry them through, and unbounded confidence in the valour and prudence of their officers, they were prepared for any arduous conflict that might ensue. The enormous success, too, which had attended the recent practice of firing red-hot shot, was a great factor in raising the spirits of the garrison, by giving hopes of curtailing the tedious cruelty of a vexatious blockade.

On the 12th September, the Governor reinforced the picquets of the line, nine of which, in future, were to be stationed in the town, and distributed as directed. The following return shows the strength of the picquets at this period :—

	Sub.	Sergt.	Dmr.	R. & F.
The Artillery, and 3 Hanoverian Regiments (each corps)	1	1	1	39
The 12th, 39th, 56th and 58th Regiments	1	1	1	54
The 72nd and 73rd Regiments	1	1	1	76
The 97th Regiment	1	1	1	56
Total, four Captains { one of the Artillery and three of the Line }	11	11	11	580

The enemy's cannonade that night was at the rate of 70 to 130 shells an hour.

On the morning of the 13th—to anticipate the arrival of Lord Howe's fleet—the ten battering ships took up their station before the fortress in the presence of the combined fleets of France and Spain.

A crowd of spectators was now seen on the beach, and the surrounding hills were covered with people, as if all Spain had assembled to behold the imposing spectacle of this attack on the much-coveted rock. On the other

VIEW OF THE GRAND ATTACK UPON GIBRALTAR,
SEPTEMBER 13, 1782.

(*From a Drawing by Lieutenant Sandby*, 12*th Regiment*.)

HALF OF THE COLONNADE ON THE BASTION WITH THE COLOURS OF THE 12TH REGIMENT DISPLAYED ON THE CENTRE ARCH.

(*From the Journal of Captain Spilsbury*, 12*th Regiment*.)

hand, the town batteries were forthwith manned, and the grates and furnaces ordered to be lighted.

Adequately to give an idea of the scene is beyond the description of pen or pencil; suffice it to say that 400 pieces of the heaviest artillery were playing at the same time.

The prodigious shower of red-hot balls, bombs, and carcasses, from the side of the fortress, now filled the air, with little or no intermission, astonishing the enemy, who could not conceive the possibility that General Eliott could have constructed such a multitude of furnaces within the narrow limits of the fortified place. In truth, the ordnance furnaces not being quite sufficient to supply all demands for the heated shot, large bonfires were kindled, and shot thrown upon them. These supplies were jocularly called "roasted potatoes." The wonderful construction of the battering ships seemed to bid defiance to the heaviest ordnance; shells rebounded from their tops, and a thirty-two pound shot scarcely seemed to make any impression on them.

The effect of the red-hot shot was doubted; sometimes smoke came from the ships, but the fire-engines within soon occasioned it to cease, and the result remained uncertain. The fire was, however, persevered in, and incessant showers of red-hot shot, shells, and carcasses flew through the air.

In the afternoon the effects of the red-hot shot became apparent, and volumes of smoke issued from the flagship; the Admiral's second ship was also perceived to be in the same condition, and confusion prevailed. Rockets were sent up as signals of extreme distress; these signals were answered, and boats from the fleet were seen to row round the disabled ships; it was found no easy matter to move these unwieldy leviathans from their moorings, and all attempts to do so were unsuccessful. A little after midnight seven of the battering ships were completely on fire; their flames produced light enough greatly to expose the enemy to observation, enabling the artillery to be pointed on them with the utmost precision. An indistinct clamour, with lamentable cries and groans, arose from all quarters.

The number of prisoners rescued by us from the burning ships was 9 officers, 2 priests, and 334 private soldiers and seamen, all Spaniards; also 1 officer and 11 Frenchmen, who had floated in, who made the total number saved amount to 357. Many of the prisoners were severely, and some dreadfully, wounded. These were sent to the hospital, and the others to encamp on Windmill Hill.

On the morning of the 14th, the enemy kept up a heavy fire whilst our Marine Brigade were endeavouring to save from perishing such unfortunate objects as had been left in the lurch, and one of our gunboats was sunk by one of the explosions. At about 5 P.M. their tenth battering ship blew up, having been set fire to under naval directions. Two of their launches were brought in, whilst several stranded on the shore, and several bodies, including one of a priest, were found among the wrecks of the ships.

Notwithstanding the efforts of the Marine Brigade in relieving terrified victims from the burning ships, several unfortunate men could not be removed. The scene at this time was pitiful in the extreme. Some were crying from amidst the flames for pity and assistance, and others, where the fire had made less progress, were imploring relief in the last state of despair, whilst some trusted themselves to the chance of paddling to the shore, on various parts of the wreck.

The loss sustained by the enemy could never be ascertained, but from the information of the prisoners, and the numbers seen dead on board the ships, it was estimated at not less than 2000, including the prisoners.

The casualties of the Garrison, however, were so trifling, as to make it appear incredible that such a quantity of firing should not have caused a greater loss of men.

September 13th, 1782.

	Killed.			Wounded.	
Regiments.	Officers.	Sergts.	R. & F.	Officers	R. & F.
Royal Artillery	1		5	3	21
12th Regiment					2
39th ,,		2	2		5
56th ,,			2		2
58th ,,			1	1	4
72nd ,,			2		12
73rd ,,				1	8
97th ,,					2
Hardenberg's					1
Reden's					
De la Motte's					1
Engineers with the Artificer Company					
Marine Brigade			1		5
Total	1	2	13	5	63

The damage done to our works held no proportion to the violence of the attack and the excessive cannonade which they had sustained. The enemy had in this action more than 300 pieces of heavy ordnance in play, whilst the Garrison had only 80 guns, 7 mortars, and 9 howitzers. Our artillery had expended upwards of 8,300 rounds (more than half of which were hot shot) and 716 barrels of powder. The amount of ammunition expended by the enemy could never be ascertained, but, in Captain Spilsbury's Journal, it is computed that they fired, from their batteries only, 1782 shells and upwards of 9000 shot.

Thus did the mighty efforts of France and Spain end in defeat and destruction, and the gallant efforts of the brave soldiers who defended Gibraltar elicited the admiration of the nations of Europe. In England the most enthusiastic applause was universal; illuminations and other modes of testifying the joy of the people followed the receipt of the news of the destruction of the boasted invincible battering ships, and every family that could boast of a member who was engaged in the defence of Gibraltar was proud of the honour.

Although the enemy gave up all hopes of reducing Gibraltar by force of arms, yet some expectation was entertained that, if the blockade was continued, the garrison might be forced to surrender from the want of provisions; the combined fleet remained in the bay and the besieging army continued in the lines.

The kilns which our Engineers had erected for heating shot (in place of grates and furnaces, at first used) were large enough to heat a hundred in an hour and a quarter.

A flag of truce went on the 15th with letters from our prisoners to their camp, to which replies were received on the afternoon of the 16th, and, on this date, a man of the 12th deserted. After firing a good deal early this morning, the enemy sent up rockets at 9 P.M. and fired 23 guns from their ships.

The Spanish officers, prisoners, with the Frenchmen taken from the wreck, were sent to their camp on the evening of the 17th, the remaining Spanish privates being given in charge of the Corsicans on Windmill Hill. A Spanish Captain of Marines, who died this evening in our hospital, was buried with full military honours on the 18th, attended by the grenadiers of the 39th Regiment.

The enemy's fire on the 19th was warmer than during the few preceding days, and at 1.30 A.M. on the 20th, the mortar boats, which had remained inactive for some time, bombarded the garrison for about an hour.

Efforts were now being made to raise our two frigates, the "Brilliant" and "Porcupine" which had been sunk by the enemy's fire on the occasion of the grand attack, and on the 24th, the "Brilliant" was successfully raised after much trouble.

Whilst every precaution was still being taken by the Garrison against another attack, the enemy's daily method of firing, at this period, had developed into commencing a cannonade at about 5 or 6 A.M., which continued until noon. Then, after 2 o'clock, as the day wore on, their firing varied, and at 7 P.M. was resumed by the mortars until daybreak of the following day, the amount expended in the 24 hours sometimes reaching 600 shells, with from 600 to 1000 shot.

On the 27th, our Navy raised the "Porcupine" frigate, and, on the following day, there was a duel between two officers of the 72nd, one of whom was wounded in the knee. On the 29th, a flag of truce brought clothes for the prisoners, and, when the firing from the mortar boats took place on the night of the 30th, three shells fell in the hospital, killing one soldier, and wounding two or three men of the 12th, besides other casualties in the Garrison.

On the 6th October, all Spanish prisoners (262 in number) were sent away who were able to go, except 59 who remained as deserters. At about 6 A.M. on the 11th, a Spanish 70-gun ship stranded near the Rock, hoisting the English colours over the Spanish. The officers and men on board were mostly dismounted dragoons, numbering in all 634, and, as prisoners, they were landed and conducted to Windmill Hill. No sooner were they landed than our Navy began to lighten the ship by removing her powder ashore, &c.

From the Spanish officers we learnt that the British Fleet, under Admiral Lord Howe, was on its way to Gibraltar, mustering, however, 34 sail to oppose 42, which remained in the Bay.

Of the ships that Lord Howe had with him, it was reported that 34 sail of the line included 11 three-deckers ; he also had 6 frigates, 31 ordnance transports, and a reinforcement of 1600 men for the Garrison. At about 5 P.M. on the 11th, the English fleet was sighted, and a frigate and 3 transports arrived and anchored under our guns, followed the next night by the "Panther" and 9 or 10 store ships, which anchored in the Bay.

In Garrison orders of the 12th October, there appeared an extract of a letter from the Earl of Shelburne, Secretary of State, dated July 10th, 1782, which was read with the greatest satisfaction, and the "assurance of their Sovereign's favour and high approbation" made the Garrison more resolute than ever in its defence.

Addressed to General Eliott, it further stated :—

> "No encouragement shall be wanting to the brave officers and soldiers under your command ; and I have the King's authority to assure you, that every distinguished act of emulation and gallantry, which shall be performed in the course of the siege, by any, even of the lowest rank, will meet with ample reward from his gracious protection and favour. His Majesty, feeling for the difficulties they are under, admires their glorious resistance, and will be happy to reward their merit."

Another extract from a letter from General Conway, Commander-in-Chief, dated August 31st, 1782, stated the King's "greatest satisfaction with the brave and steady defence made by your Garrison," and added :—

> "His Majesty, desirous of showing them every mark of his Royal approbation, has been graciously pleased to grant bat and forage money, as a proper indulgence to your officers."

The British fleet was again sighted at daybreak on the 13th, and by 1 o'clock, the Spanish Admiral had put to sea with the combined fleets. Before the enemy had entirely quitted the Bay, Captain Curtis landed from the "Latona" frigate, with £20,000 in specie for the Garrison, having narrowly escaped being cut off by the combined fleets.

On the 14th, the enemy's cannonade was continued with great activity.

On the night of the 15th, the "Latona," with 8 or 10 transports, anchored in the Bay. Since the arrival of the first transports the Garrison had been busy landing supplies; whereas former fleets had principally brought provisions, this one brought only men and ammunition.

Our prize, the "St. Michael" (which had grounded on a sand bank and been heavily fired at by the enemy) was, with the aid of 100 soldiers to assist the sailors, successfully floated on the 17th, and, on the following day at about noon, four or five men-of-war arrived from the fleet with the 25th and 59th Regiments on board. The former encamped behind the barracks, and the latter on Windmill Hill. The men-of-war sailed that night. The brig "Minerva," however, with the women and baggage of these regiments, had been captured by the Spaniards and dismasted.

At daybreak on the 19th, both fleets were in sight, and by 2 P.M. Lord Howe, having effected his mission, was out of sight, returning to England, closely followed by the combined fleets.

On the 20th, our store ships unloaded, and the enemy's fire had much abated. The French camp broke up on the 21st.

Sugar was now selling at 12*s.* 8*d.* per lb; ducks, 17*s.* 6*d.* per pair.

A visit from the mortar boats on the night of the 23rd did considerable damage. On the 25th, bread was selling at 13 pence per lb.

A deserter who came in the previous evening stated that there had been an engagement between the British and combined fleets, but could give no particulars; he also stated that the Spanish camp was breaking up. Sixteen battalions had already left, and others were preparing to start. More battalions left the enemy's camp on the 27th, and, on the 28th and 29th, it was still decreasing. Their cannonade was nevertheless continued.

At about 3 A.M. on the 31st, the "Tisiphone" fire-ship, with 5 ordnance ships, sailed for England, having 160 Jews on board.

On the 1st November, our batteries fired a great deal by night. The following day, powdered sugar, at 13 pence per lb., was served out to the officers and women, and port wine sold at 7 dollars per dozen. A deserter who came in on the 4th stated that the enemy had now about 11,000 men in camp.

Bat and forage money was to-day issued, a captain receiving £40.

A flag of truce, on the evening of the 9th, brought over an officer of the 58th with the Quarter-Master of the 25th (who had been taken in the brig "Minerva" with the baggage of the 25th and 59th Regiments), and other prisoners.

On the 10th, all the Spanish deserters were sent to England, and, on the 15th, another regiment quitted their camp. On the 17th, the regiments in the Garrison changed quarters, the 12th, Reden's, and De la Motte's Hanoverians moving to the Pavilions. All the Spanish prisoners taken in the "St. Michael" were, on the 20th, sent to the Spanish camp, excepting a few who chose to remain as deserters. Two boats, laden with sugar, fruit, coffee, &c., arrived from Portugal on the 23rd ; the Spaniards chased them both. Two sheep were to-day sold at 10 guineas each, and raised mutton to *5s. 3d.* a lb. The enemy's fire now scarcely exceeded 150 rounds in the 24 hours. On the 24th, the frigate "Porcupine" was made a hospital ship for the sailors. On the 27th, apples sold at *2s. 7½d.* per lb. Towards the end of this month, the enemy's fire became more faint and ill-directed, whilst ours was more brisk and effectual.

In the early part of December tea sold at 11 dollars per lb., Dutch cheese at *3s. 11d.*, and wine from *7s. 5d.* to *10s. 6d.* a gallon. Up to the 4th, an ordnance ship had arrived from England and two or three boats from Portugal. On the 6th, 300 Bibles were sent to the Garrison as a present. A deserter who came in on the night of the 7th stated that the Spaniards were mining the Rock. On the 14th, cheese was selling at *5s. 3d.* per lb. At about 10 A.M. on the 18th, 29 gun and mortar boats commenced a spirited attack on the "St. Michael," killing 4 and wounding 11 sailors, 3 mortally ; about 80 barrels of powder she had were now thrown overboard.

By a flag of truce on the 22nd, about 150 women arrived belonging to the 25th and 59th Regiments, who had been taken in the brig "Minerva." On landing, they were conducted to the Naval Hospital, where 17 were detained.

At about 4.30 P.M. on Christmas Day, 18 gun and 11 mortar boats began a warm bombardment of our camp, their fire being in a great measure diverted from the shipping by 11 of our gunboats, which opposed them. Seven or eight of their shells exploded in a hospital ward, killing and wounding several of the sick ; another shell burst in the cabin of the "Porcupine," whilst a blind shell fell into the ward room of the "St. Michael." There were 1 killed and 8 wounded in the camp, and several houses and sheds destroyed.

In the course of December, several vessels and boats arrived with supplies, and flags of truce frequently passed between the Governor and the Duc de Crillon, the purport of which was not publicly known.

1783

On New Year's Day, a 26-pounder iron gun (which had been recovered from the wreck of a battering ship) was drawn in procession, with the Spanish Colours affixed, to the Mole Battery, attended by the band of the 12th Regiment, playing "God save the King." Many more of these guns, mostly of brass, on being later recovered, were sold, and the sums accruing therefrom, together with other spoil, was apportioned as prize money. The enemy ceased firing on the 2nd and 3rd, but, on the 4th and 10th, their gun and mortar boats did considerable harm, killing, on the latter date, one, and wounding 17 of the Garrison.

A boat from Faro, on the 11th, brought favourable prospects of peace, and the regiments began to prepare for an impending review. Captain Spilsbury reports in his journal that, on the 18th, the 12th Regiment had a field day. On the 24th, one of the 12th was blown up with the guardhouse at Middle Hill, through an ammunition box taking fire. At 8 A.M. on the 27th, the 12th Regiment was inspected. On the 29th, at about 5 P.M., the enemy's gun and mortar boats came, and their land batteries opening fire at the same time, our artillery replied with great vigour. We had 3 killed and 11 wounded. On the 31st, they fired a good deal, and the 1st February was celebrated by a specially animated fire from the Garrison. On the 2nd, letters from the Duc de Crillon informed the Governor that the preliminaries of a general peace had been signed at Paris on the 20th January, between Great Britain, France, and Spain, and, when the boats met, in delivering the letters, the Spaniards cried out with joy "We are all friends." The blockade by sea being now discontinued, the port of Gibraltar was again open, and, about noon on the 5th, we wantonly fired an elevated gun over their works, which was the last shot fired in this siege.

Thus ended the Siege of Gibraltar, which is celebrated in the military annals of the 18th Century, and the successful defence of that fortress ranks amongst the noblest efforts of British arms. During a period of 3 years, 7 months, and 12 days (from the commencement of the blockade to the declaration of peace) the defenders had experienced a continued term of watchfulness and fatigue, the dangers of famine, and every harassing and vexatious mode of attack which a powerful and revengeful enemy could devise.[1]

The feelings of relief and joy, and the social happiness throughout the garrison, now baffled all description, and operations on both sides being suspended, only the official accounts of peace were awaited from England. Innumerable ships and boats now arrived from England and Portugal, so that provisions became every day more abundant, and cheaper in consequence.

Parleys were now almost daily passing between the Governor and the Duke, and, as the Spanish guards were at this time greatly diminished, a number of deserters daily came over to the Garrison.

On the 16th February, the regiments changed quarters, the 12th moving to the King's Bastion.

On the 22nd, fine loaf sugar sold at 1*s.* 6½*d.* per lb., and writing paper at 2*s.* 2*d.* per quire. On the 26th, the 12th Regiment had another field day for those off duty.

The surgeon's mate of the 12th died on the 2nd March, and on the following day the Duc de Crillon sent the Governor a present of a grey Andalusian horse.

[1] Captain Drinkwater's *History of the Siege.*

On the 5th, a schooner arrived from Larache, with a present of bullocks for the Governor, and a letter from the Emperor of Morocco, the subject of which was conjectured to be a request for a renewal of our friendship with the Moors. Two officers and 24 Corsicans also arrived in this boat to join our force, having been prevented doing so by the Spaniards at an earlier date.

Our prize, the man-of-war "St. Michael" sailed for England on the 22nd, where she arrived safely.

On the 23rd, the Governor and his staff visited the Duke in the Spanish works, and, dining with him at San Roque, returned in the evening. The return visit was made by the Duc de Crillon and his suite on the 31st. On entering the Garrison, he was heartily cheered, and was received with a salute of 17 guns and a captain's guard of honour. All the officers were introduced to his Grace at the Convent by regiments, and he then rode to inspect some of our most important batteries. After dinner at the Convent (at which all the Generals and Brigadiers in the Garrison were present), he returned through the camp at about 8 P.M., the regiments falling in, in review order without arms, as he passed, and giving three cheers, the same salute being fired as at his entrance.

On the 2nd April, the Duke left for Madrid, and was succeeded in command by Lieut.-General the Marquis de Saya, who had served at the previous siege of Gibraltar in 1727.

On the 8th, beef was selling at 5¼*d.* per lb.

A man of the 12th was to-day found who had been absent two or three days. On the 15th, the Governor sent the Emperor of Morocco, in the frigate "Brilliant," a present of four brass 26-pounder guns (which had been recovered from the wreck of the battering ships), complete with carriages, and ammunition in proportion.

Preparations were now made for the investiture of General Eliott, His Majesty having been pleased to confer on him the most honourable Order of the Bath, as a mark of his royal approbation for the defence of Gibraltar. Lieut.-General Boyd was, by the King's desire, to be His Majesty's representative, in investing General Eliott with the insignia of the order, and the ceremony was to be performed in as grand a manner as the state of the Garrison would permit. The engineers began to erect a colonnade on the rampart of the King's Bastion, and, by the 23rd April (St. George's Day), the colonnade was finished; it was composed of masts and yards from the enemy's battering ships, with fascines and sand-bags. At 8 A.M. on that date, detachments from all the regiments and corps in garrison paraded in three lines on the Red Sands, when, the usual compliments having been paid, the Governor placed himself in front of the centre and addressed the troops, conveying to them the high sense His Majesty entertained of their meritorious conduct in the defence of Gibraltar, the King's satisfaction on the event having been imparted to the whole world by his gracious speech to both Houses of Parliament. The following unanimous thanks were then communicated by the Governor.

"13th December, 1782.

"Resolved, by the Lords Spiritual and Temporal, in Parliament assembled, that this House doth highly approve and acknowledge the services of the Officers, Soldiers and Sailors, lately employed in the defence of Gibraltar, and that General Eliott do signify the same to them."

"12th December, 1782.

"Resolved, That the thanks of this House (Commons) be given to Lieut.-General Boyd, Major-General De la Motte, Major-General Green, Chief Engineer, to Sir Roger Curtis, Knight, and to the officers, soldiers and sailors lately employed in the defence of Gibraltar."

The Governor then warmly congratulated the troops "on these united and brilliant testimonies of approbation, and asked their acceptance of his grateful acknowledgments, as having been a constant witness of their cheerful submission to the greatest hardships, their matchless spirit and exertions, and, on all occasions, their heroic contempt of every danger."

A feu-de-joie was then fired by the line, each discharge commencing with a royal salute of 21 guns, and three cheers closed the ceremony. The troops then lined the streets, and the following procession went from the Convent to the colonnade on the King's Bastion, General Boyd being the King's Commissioner.

MARSHAL.

Music, 12th Regiment
Playing "See the conquering hero comes."

ARTILLERY.

Quarter-Master General and Adjutant-General,
Town Major and Deputy,
With other staff of the Garrison.

First Division of Field Officers, youngest first.

Music, 58th Regiment

THE COMMISSIONER'S SECRETARY,

Bearing on a velvet cushion the Commission.

THE COMMISSIONER'S AIDES-DE-CAMP.

Lieut.-General Boyd, the King's Commissioner.
The Governor's Secretary,
Bearing on a crimson velvet cushion, the Insignia of the Order of the Bath.

The Governor's Aides-de-Camp, as Esquires.

GENERAL ELIOTT, THE KNIGHT ELECT;

Supported by Generals De la Motte and Green.

Aides-de-Camp to the Major-Generals.

MAJOR-GENERAL PICTON (12th) and His Aide-de-Camp
The Brigadier Generals, eldest first, and Their Brigade Majors.
Music, De la Motte's.
Second Division of Field Officers, eldest first.
Music, 56th Regiment
The Grenadiers of the Garrison.

No compliment was paid to the Knight Elect, but as the Commissioner passed, each regiment saluted. When the procession arrived at the colonnade, the field officers placed themselves on each side of the throne with the Generals; the Artillery formed under the colonnade, and the Grenadiers, fronting the bastion, along the Line Wall. The proper reverences being made to the vacant throne, the Commissioner desired his secretary to read the Commission, which being done, he addressed the Knight Elect in a short complimentary speech, and then placed the riband over the Governor's shoulder. All then took their seats on each side of the throne, the Governor on the right, and a band played "God save the King." The Grenadiers then fired a volley, which was followed by a salute of 160 guns fired from the sea line. The procession then passed through the colonnade, and returned in the same order. The Field Officers and Staff dined at the Convent, and each soldier received a pound of fresh beef and a quart of wine. In the evening the colonnade was illuminated, and Sir George Eliott, with his company at dinner, assembled at the King's Bastion at about 9 o'clock, when there was a display of fireworks.

On the 3rd May, commanding officers submitted another petition to the Governor for a further allowance of bat and forage money, which he received very graciously and forwarded.

On the 6th, a store ship arrived with shot, shell, and camp equipage, and, on the 11th and 16th, several of the three years' men of the 12th Regiment re-engaged for a further term of service.

Since the declaration of peace, neither Spaniards were allowed to visit the Garrison, nor were our officers and men allowed into Spain. In July, the 72nd, 73rd, and 97th Regiments sailed for England. On the 2nd August, a man of the 12th was found drowned.

On the 15th October, the Governor asked the captains of the 12th if they would assist in sharing the duties of captains in the 39th, owing to so many of the latter being away; the captains of the 12th refusing, the duties were directed to be shared throughout the Garrison.

On the 29th, at about 8 A.M. there arrived, in four ships, the 2nd "Queen's" and the 18th Regiments, and landed on the following day.

On the 31st, 4 companies of the 11th Regiment arrived, with the women and baggage of the 2nd and 18th.

The 32nd Regiment and remainder of the 11th arrived on the 1st November, and, on the 8th, the 12th and 39th Regiments were taken off duty. Before daylight on the 10th, the 12th and 39th Regiments formed up, and waited till 8 for the boats at Water Port, when they embarked for England, on board the ships "Goliath," "Ardent," "Ganges," and "Diadem."

The loss of the regiment during the Siege of Gibraltar was :—

	Officers.	Sergeants.	Drummers.	Rank and File.
Killed	1	3	1	13
Died of Wounds	..	..	..	10
Disabled by Wounds	1	..	..	10
Wounded, that recovered	2	4	7	89
Died of Diseases	..	3	..	32
Total	4	10	8	154

The following is an abstract of the total loss of the Garrison.

Killed, and died of wounds	333
Disabled by wounds (discharged)	138
Died of sickness (exclusive of those who died of scurvy in 1779 and 1780)	536
Discharged from incurable complaints	181
Deserted [1]	43
	1231

The following were the proportions of prize money distributed to the Garrison of Gibraltar from sums granted by Act of Parliament arising from destroying the Battering Ships and the sale of the "St. Michael" man-of-war.

	£.	s.	d.		£.	s.	d.
The Governor $\frac{1}{16}$th	1,875	0	0	Captain	43	10	1
Lieut.-Governor	937	10	0	Lieutenant	25	5	6
Major-General	468	15	0	Second-Lieutenant and Ensign	22	0	6½
Brigadier-General	267	10	0	Sergeant	3	6	9
Colonel	156	1	0	Corporal	2	0	11½
Lieut.-Colonel	80	16	0	Private	1	9	1
Major	57	15	6				

A second Act of Parliament was passed later, authorising another issue of prize money to the extent of £16,000, for brass and iron cannon recovered from the wrecks of the battering-ships, and it was expected that it would be followed by a third.

The old soldier of the 12th, James Campbell (mentioned in the Minden Chapter), followed this success by serving with the regiment in the battles of Warburg, Fellinghausen, and Wilhelmstahl, in 1760–61, under Prince Ferdinand of Brunswick. He then served 14½ years at Gibraltar, taking his full share, under the brave Eliott, of the dangers and hardships of

[1] To this number the 12th Regiment contributed 3.

that memorable siege, and was one of the party that made the sortie from the garrison, and destroyed the Spanish batteries, spiking their guns and mortars, and setting fire to seven magazines. Besides Colonel Trigge and his aide-de-camp (Lieutenant Miles Thornton) he was the last who came into the garrison at Bayside.

It is recorded of James Campbell that "a more loyal, or more honest man, or braver soldier, never partook of the royal bounty."

Medals for the defence of Gibraltar were struck in gold, silver and copper.

The following are descriptions of two regimental medals commemorative of the siege.

No. 2 was made to the order of Major-General William Picton, and presented by him to officers of the 12th Regiment present at the siege, and a specimen is in possession of the officers of the 2nd Battalion.

(1) *Obverse.*—The Castle and Key of Gibraltar; to the left, a cannon; on the right, a man-of-war, rounding a headland; below, on a ribbon, MONTIS INSIGNIA CALPE; above, a trophy of flags and arms, and on an oval shield is the regimental number—12.

Reverse.—A laurel-wreath, in which is inscribed, November 27th, 1781, September 13th, 1782; beneath, the word GIBRALTAR. A large silver medal; all engraved. 2 inches diameter. (See Plate 10.)

(2) A large silver medal, 2¼ in. diameter.

Obverse.—A bird's-eye view of Gibraltar and the adjacent coast. Above, on a scroll, BATTERING SHIPS DESTROYED; below, on a raised field, SEPTEMBER XIII, MDCCLXXXII.

Reverse.—BY A ZEALOUS EXERTION OF PATIENCE, PERSEVERANCE AND INTREPIDITY, AFTER CONTENDING WITH AN UNPARALLELED SUCCESSION OF DANGERS AND DIFFICULTIES IN THE DEFENCE OF GIBRALTAR DURING A BLOCKADE AND SIEGE OF ALMOST FOUR YEARS, THE GARRISON UNDER THE AUSPICES OF GEORGE III TRIUMPHED OVER THE COMBINED POWERS OF FRANCE AND SPAIN. Below, the British Lion supporting a shield charged with the Arms of Gibraltar; in the exergue, BLOCKADE COMMENCED JUNE XXI., MDCCLXXIX: SIEGE TERMINATED, FEBRUARY II. MDCCLXXXIII., the whole surrounded by a wreath of olive-branches. Many examples are met with, framed and glazed, and with a ring for suspension. (See Plate 10.)

The following is an approximate list of officers of the 12th Regiment serving at Gibraltar, during the siege.

Colonel—William Picton.

Lieut.-Colonel—Thomas Trigge.

Major—William Barlow.

REGIMENTAL MEDAL.
1782.

MINDEN MEDAL.
1759.

GENERAL WILLIAM PICTON'S
GIBRALTAR MEDAL TO HIS OFFICERS.
1782.

Captains—Henry Ormsby, John Kay, Samuel Montgomery, Joseph Collins, John Perryn, John Spilsbury, Charles Cottrell, Richard Turberville Picton, Thomas Adams (Brevet-Lieut.-Colonel).

Captain-Lieutenant Christopher Ludwig Forch (Aide-de-Camp to the Governor).

Lieutenants—Thomas Bate, Charles Hastings, John Chalmers, James Lowe (lost his leg by a shot, May 9th, 1781), John Byron, John Perryn, Robert Adair, George Bembrick, Harry Lee, Alexander Graham, Thomas Hitchbone, Thomas Whetham (killed, April 10th, 1782), Prestly Thornton (wounded, May 11th, 1781).

Ensigns—Frederick Draught Odlum, Adam Tweedie (wounded in the sortie, November 27th, 1781), Alexander Munroe, William Sandby, Edward Jones, James Allen, Knyvet Wilson, Thomas Craigie, Alexander Legertwood, Alexander Smith.

Chaplain—Robert English.

Adjutant—Thomas Hitchbone.

Quarter-Master—Robert Smith.

Surgeon—David McNair.

CHAPTER VI

Great Britain, Ireland, West Indies, Flanders, Holland, Germany, England, France, East Indies
1783–1798

1783–84

The regiment, on disembarking at Portsmouth, marched to Hilsea Barracks, and, in the ensuing month, a reduction to the peace establishment followed, as authorised from the 25th June, 1783.

On the 19th and 20th December a move took place to Alton and Farnham, and thence to Windsor, while the invalids were directed to move from Hilsea to Fulham and Putney.

The quarters of the regiment (now 8 companies) were :—4 companies to Old and New Windsor, and 4 distributed between Datchet, Upton, Slough, Salthill, and Cluner.

The War Office books show that contingent allowances to captains of cavalry and infantry were granted from the 25th December, 1783, and also an annual allowance of £30 to each Foot regiment, for postage, stationery, guard rooms and store rooms taken together.[1]

On the 19th March, 1784, the regiment, at Windsor, was directed to furnish parties, in relief of the 7th Light Dragoons, at Slough, to escort 66 convicts to London, who, having embarked on board the ship "Mercury," for conveyance to North America, had taken possession of the vessel by force, and steered her into Torbay, where 67 of them got on shore, and made their escape, but 66 were later apprehended.[2]

King George III was highly gratified at having a corps, which had distinguished itself during the memorable siege of Gibraltar, employed near his person, and on the 1st and 8th of June, 1784, His Majesty reviewed the 12th Regiment in Windsor Park in the presence of the Royal Family and many distinguished personages, and expressed, in very gracious terms, his

[1] *W. O.* 3, Book 26, p. 145. [2] *W. O.* 4, Book 125, pp. 317–18.

high approbation of its appearance and discipline, and of its conduct during the siege of Gibraltar.

The battalion was inspected at Windsor, on the 17th June, by Lieut.-General George Warde ; 349 effective rank and file ; and was reported on as :—" well-formed, quite conformable to order, and, when completely recruited, will be a very fine battalion, and perfectly fit for service."

The regiment received orders on the 2nd November to move from Windsor and out-stations to Leeds, which was later altered to Chatham.

The *London Gazette* of November 16th gave notice of the issue of prize money for Gibraltar (for the sale of a ship stores, &c.) in 30,000 shares, being made to certain of His Majesty's ships and regiments who were present throughout the siege. The amounts were to be paid at " The French Horn," in Crutched-Fryars, London, commencing on the 7th December, and shares not demanded on the date mentioned, were to be recalled at the above place, on the first Friday in every month, for three years to come. One date for application was :—

" 12th December . . . . Royal Artillery and 12th Regiment."

1785–89

A Route, dated 2nd April, 1785, directed a move to Newcastle, and, on the 5th, this was countermanded, and the regiment was ordered to march in two divisions as follows :—2 companies to Sunderland, and 6 to Tynemouth Barracks, so as to arrive on the 9th and 10th May respectively.

While on the march, an order was received to proceed, with all haste, to Newcastle, Gateshead, and Westgate, in aid of the civil power, and, on the 3rd December, the companies at Newcastle proceeded to Tynemouth.

On the 20th January, 1786, the regiment marched from its present quarters to Berwick and Tweedmouth, and thence to Musselburgh, where it was inspected, on the 1st March, by Major-General the Honourable Leslie, and reported on as " a very fine, steady, well-disciplined corps." It then marched to Ayr.

Before leaving Scotland in the autumn of 1787, the regiment moved to Edinburgh, arriving 1st May, where it was inspected, on the 8th June, by Lieut.-General the Honourable Alexander Mackay, and was reported on as " a steady regiment, fit for any service."

On the 22nd September, an augmentation was directed to be made in the establishments of several battalions of Foot in Great Britain, including the 12th, by adding to each of the eight companies :—1 sergeant, 1 drummer, and 14 privates. Also two of the companies were each to consist of 1 captain, 1 lieutenant, 1 ensign, 3 sergeants, 3 corporals, 2 drummers, and 56 private men, and one company (for the special purpose of recruiting) to consist of 3 officers, 8 sergeants, 8 corporals, and 4 drummers. It was

M 2

further stated that it was His Majesty's intention to take the officers for the 11th Company from the Half Pay List.[1]

A General Order, dated, 28th September, laid down that recruits were to be taken from 16 to 35 years, and those for India were not to exceed 30 years of age.

On the 29th September, 1787, the regiment moved from Edinburgh to the south of England, and was directed, on arrival at Carlisle, to march by the shortest and most convenient route to Plymouth Barracks. On reaching Warrington, an order was issued, that on arrival at Lichfield, Stafford, and Newcastle-under-Lyme respectively, a halt for one week was to be made at each of these places, and the regiment was then to proceed to Plymouth.

By an Army Order, dated 31st December, the two fifers, who, for some length of time, had been discontinued for the grenadier company, were now restored to it.[2]

On the 10th January, 1788, the regiment was inspected at Plymouth by His Royal Highness the Prince of Wales, then in his seventeenth year, who afterwards became King George IV, and, on the 12th April, the 12th embarked for the Channel Islands, and were quartered (5 companies at each) at Jersey and Guernsey, throughout the years 1788–89. The headquarter detachment at Jersey was inspected at Fort Henry on the 29th May, by the Lieutenant-Governor, who highly approved of its appearance and the behaviour of the men.

1790

In March, the regiment was relieved from duty in the Channel Islands, and, disembarking at Portsmouth on the 24th, marched to Hilsea Barracks, the usual stoppage of 3*d*. a day, for provisions during the passage to Portsmouth, having been made from officers and men.

On the 1st April, 8 companies were ordered to Portsmouth, and 2 to Forton Barracks, Gosport, and, on the 24th May, the regiment was inspected at Portsmouth by General Scott, and was reported on as "in very good order, very steady, and fit for service."

In 1786, a company of merchants residing in the East Indies had formed a settlement at Nootka Sound—a bay of the North Pacific Ocean, on the west coast of North America—with the view of obtaining furs. This settlement was seized by the Spaniards in 1789, and two ships were detained. To chastise this violation of British enterprise and liberty, a fleet was fitted out, and on the 28th May, 600 men, in all, from the Land Forces, were definitely appointed to do duty as Marines on board His Majesty's ships.

The 12th Regiment had been ordered to be in readiness, since the 26th, and to be completed to its augmented establishment by a temporary transfer of 100 men from the 3rd and 37th Regiments, who were to be

[1] *W. O.* 4, Book 133, p. 406. [2] *W. O.* 3, Book 27.

volunteers, sound and able bodied, if so many offered, a bounty of 1½ guineas being allowed to each man prior to embarkation, and the men to leave their arms and accoutrements with their own corps.

Instructions were issued that the " band of musick " was not to embark, but the sergeants and drummers were.

Embarkation orders were issued on the 7th June in such proportion as the commanding officer of the Squadron should direct, and, on the 15th, the regiment embarked on the following ships of the fleet :

On H.M.S. " Barfleur "—Captain Picton, Lieutenants Allen and Craigie, 3 sergeants, 1 drummer, and 109 rank and file.
,, " Impregnable "—Captain Perryn junr., Lieutenant Lord Falkland, Ensign Hunter, 5 sergeants, 2 drummers, and 106 rank and file.
,, " Carnatic "—Captain Spilsbury, Lieutenants Sandby and Legerwood, 5 sergeants, 1 drummer, and 97 rank and file.
,, " Bellona "—Captain Perryn senr., Ensigns Pogson and Jardaine, 5 sergeants, 1 drummer, and 97 rank and file.
,, " Magnificent "—Captain Tweedie, Lieutenant Tinling, Ensign Wogan, 4 sergeants, 1 drummer, and 98 rank and file.
,, " Edgar "—Captain Martin, Ensigns Reeves and Lyster, 3 sergeants, 1 drummer, and 61 rank and file.
,, " Bellerophon "—(embarked on 23rd October) 1 sergeant and 10 privates.

Field Officers of the regiment, and others who did not embark, proceeded to Hilsea, and, on the 16th November, the remainder of the regiment, on disembarking, joined them there. The drafts from the 3rd and 37th Regiments were now transferred to their former corps from the 10th November inclusive, and the supernumerary arms of the 12th were returned into store.[1]

An official list of quarters of the troops in Great Britain, dated 7th August, 1790, showed the 12th, 17th, 29th, and 31st Regiments on board the fleet.

On the 17th December the 12th were ordered to march to Portsmouth and embark on board the transports provided for their reception, for conveyance to Cork.

1791–92

The regiment landed at Cork on the 3rd January 1791, from on board the ships " Eliza " and " Trio," with a total of 34 officers, 1 surgeon's mate,

[1] *W. O.* 4, Book 141.

22 sergeants, 12 drummers and fifers, and 400 rank and file, and marched the same day to Kinsale.[1]

The annual inspection took place on the 18th June, by Major-General R. Whyte. In the Inspection Report, under the heading "Orderly books," it was noted:—"the regiment has a code, and each officer is furnished with a copy of it." The Inspecting Officer further reported on the corps as "a remarkably fine body of men, well appointed, well regulated, and in every respect fit for any service."

On May 24th, 1792, the regiment marched to Dublin, and arrived on the 7th June.

On the 4th August, it was inspected by Major-General David Dundas, and reported on as "having fired well, marched well, and formations and movements exact; great steadiness and attention both in officers and men; a very fine regiment, and fit for any service."

1793

The disturbed state of affairs in France, occasioned by the Great Revolution, necessitated an augmentation of the forces in England, and this, combined with a bill brought forward by our Houses of Parliament (to suppress revolutionary intrigues by foreign agents), was made, by France, a pretext for declaring war, on the 1st February, against England and Holland.

In March, the 12th Regiment moved to Drogheda, arriving on the 11th.

From the 26th July, 400 men were ordered to be raised by each regiment serving in Ireland, by which two captains, two lieutenants and two ensigns were added, and from the 1st December, 11th and 12th companies were added to each of these battalions, the additional not to be flank companies; and also a second lieut.-colonel and a second major to each battalion, special terms of promotion for officers being stated.[2]

In a Commander-in-Chief's book (Irish) a memorial from Major Bowes, commanding the 12th, shows that, by the 13th September, the regiment had raised 200 men for the 11th company, agreeably to the terms of government, and petitions to recommend certain officers for it.[3]

A Return, dated 16th September, shows present with the headquarters, 16 officers, 726 of other ranks, and 107 "wanting to complete."

The regiment was inspected at Drogheda, on the 1st October, by Lieut.-General the Honourable E. Stopford, and owing to the large number of recruits (519 had enlisted since the previous inspection), there was no drill. (Irish Inspection Returns.)

The spread of violent republican principles to the French West India Islands having caused the European planters in those parts to solicit the

[1] *Irish Government Correspondence Book*, and *W. O.* 8, Book 8.
[2] *W. O.* 8, Book 8, p. 391, and *Anthologia Hibernica*, Domestic, for December, 1793, folio 470.
[3] Book 91.

protection of the British against the fury of the mulattoes and negroes, the deliverance of those islands was now undertaken, and an expedition was sent this year to the West Indies, under General Sir Charles Grey and Admiral Sir John Jervis.

In addition to the despatch of one or two regiments, it was decided to send the flank companies of 14 regiments quartered in Ireland, and a letter from Lord Westmorland to the Honourable Henry Dundas (Secretary of State), dated 23rd September, stated that, on the 20th instant, he had given orders for the flank companies of the following 14 regiments, viz., 8th, 12th, 17th, 22nd, 23rd, 31st, 33rd, 34th, 35th, 38th, 40th, 41st, 44th, and 55th, to be completed to 4 sergeants and 100 rank and file each, the same when completed to march to Cork to embark; also 10 supernumerary rank and file per company, in order to embark complete. Also this force to be properly officered; and he added :—"I shall divide it into four battalions, and appoint field officers, with a proper staff, to each."

The 1st Battalion Grenadiers, commanded by Lieut.-Colonel Sir C. Gordon, embarked at Monkstown on the 13th November, and consisted of the grenadier companies of the 8th, 12th, 17th, 22nd, 23rd, 31st, and 41st Regiments.

(The 12th, under Captain Tweedie, mustered 1 captain, 2 lieutenants, 4 sergeants, 4 corporals, 2 drummers and fifers, 67 privates, 6 women, and 1 child.)

The 2nd Battalion Grenadiers, commanded by Lieut.-Colonel Cradock, embarked with the 1st Battalion, and consisted of the grenadier companies of the 33rd, 34th, 35th, 38th, 40th, 44th, and 55th Regiments.

The 1st Battalion Light Infantry, under Brevet-Major Ross (31st), embarked at the same time and place, and consisted of the light companies of the 8th, 12th, 17th, 22nd, 23rd, 31st, and 35th Regiments.

(The 12th, under Captain Perryman, mustered 1 captain, 2 lieutenants, 4 sergeants, 4 corporals, 2 drummers, 64 privates, and 7 women.)

The 2nd Battalion Light Infantry embarked at the Cove of Cork on the same date, and consisted of the light companies of the 33rd, 34th, 38th, 40th, 41st, 44th, and 55th Regiments.[1]

By a Royal Warrant, dated 15th November, an increase was ordered in the 14 regiments in Ireland, and, from the 1st December, a further augmentation brought the total of the 12th Regiment to 1099 of all ranks.

1794

The above troops, which left Ireland with General Sir Charles Grey, arrived at Barbados on the 6th January.

The operations commenced with an attack on the island of Martinique, which had been captured by the British in 1762, and ceded to the French by treaty the following year.

[1] *Irish Government Correspondence Book*, No. 136.

The entire armament of 19 ships of war, escorting 7000 soldiers, sailed from Barbados, on the 3rd February, in three divisions, for Martinique, and amongst the troops were the four battalions of grenadier and light companies from Ireland.

The 1st Division, under Major-General Dundas and Commodore Thompson (with whom were the 2nd Battalion Grenadiers and 3 Light Infantry Companies) landed successfully on the island, on the 5th and 6th February, at a place called La Trinité, and, under a heavy fire of musketry, advanced against Morne le Brun, which soon fell into our hands. The fort of the mulatto leader was evacuated by him in the night, after his setting fire to the town, and leaving all his artillery behind ; but, with the aid of seamen, the flames were extinguished, and, thereby, large stores of provisions saved. The division reached Gros Morne on the 7th, and, by the 11th, had occupied Fort Matilde, and the strong port of Lemaitre.

The 2nd Division, under Major-General Sir Chas. Gordon, with Colonel Myers and Captain Rogers, R.N., landed at Caisse de Navire, to leeward of La Trinité, and by capturing Fort Catherine, various batteries, and Pigeon Island, had laid open to our shipping the bay and harbour of Port Royal, thus securing the safety of the troops.

The 3rd Division, under Lieut.-General Prescott, with whom was Sir Charles Grey, effected a landing on the 8th February near Trois Rivières, successfully completing the disembarkation of the expedition without loss, the object in landing at three points having been to divide the forces of the enemy.

With a view to capturing the town of St. Pierre, an advance in two columns was made on the 14th, when a body of 500 or 600 of the enemy, who were advantageously posted on a height covering the town, were forced to give way. This was afterwards followed by a complete rout, by General Dundas, of 600 of the enemy, whom he observed moving across his front towards Morne Rouge. At daybreak, he found they had abandoned their position, leaving 2 guns and some stores.

In concert with the two columns advancing on St. Pierre, a force was embarked under Colonel Symes, with orders to land north of the town, whilst another force, consisting of 4 companies of 1st Battalion Grenadiers and 4 of Light Infantry, under Colonel Myers, marched to co-operate with General Dundas.

The main body, advancing at daybreak, had not gone far when they found that Colonel Symes was already in the town, the operations having been quite successful ; towards the end of February, the mulatto leader and his second in command surrendered to Sir Chas. Grey, and were sent as prisoners on board the fleet.

An attack was now made on Port Royal and Fort Bourbon, our land forces consisting of the 1st and 3rd Battalions Grenadiers, and 1st and 3rd Battalions Light Infantry, under Colonel Symes, the naval operations being under command of Commodore Thompson. The surrender of both was achieved without any difficulty, and a large store of guns and ammunition was found in the forts and town.

From Martinique, the battalions of grenadiers, under command of Prince Edward (afterwards Duke of Kent), with some artillery and three other regiments, embarked, on the 30th March, for St. Lucia, and its conquest being promptly effected with no loss on the British side, this expedition returned to Martinique on the 5th April.

The flank companies of the 12th were afterwards engaged in the capture of Guadeloupe and its dependencies, the fleet sailing thither on the 8th, and, on the morning of the 11th, a landing was effected at Gosier Bay with part of the 1st and 2nd Battalions Grenadiers, 1 company of the 43rd Regiment, and 500 seamen and marines. Our advance was made, as before, in three divisions: the first, under Prince Edward, consisted of the 2nd Battalion Grenadiers, and 100 men of the Naval Brigade; the second, under Major-General Dundas, consisted of two battalions Light Infantry and 100 seamen; the third, commanded by Colonel Symes, consisted of the third battalions of Grenadiers and Light Infantry.

The several divisions, which had timed their marches so as to arrive simultaneously at points indicated for attack, rapidly advanced, and the place became ours in an incredibly short space of time.

Our loss was only 13 rank and file killed, and 39 wounded, that of the enemy amounting to 67 killed and 55 wounded, besides 14 whites, 18 mulattoes, and 78 blacks, prisoners.

The capture now followed of the important strongholds of Palmiste and Morne Magdalen, and the French general, seeing that he had no chance of saving the island, capitulated, surrendering Guadeloupe and the islands of Marie Galante, Desirade, Les Saintes, and others.

The Commander-in-Chief, in his despatch home on the capture of Guadeloupe, stated that he could not find words to convey an adequate idea, or to express the high sense he entertained, of the extraordinary merit evinced by the officers and soldiers in this service.

Operations against the island of Dominica by Brigadier-General Whyte had been equally successful in subduing the island with an insignificant loss.

Whilst the flank companies were engaged in the capture of the French West India Islands, the regiment was withdrawn from Ireland and embarked at Drogheda, on the 7th March, in the ships "Leland," "Mary," "Endeavour," "Three Sisters," "Lord Kilwarlin," and "Isabella." An Embarkation Return shows that it mustered 17 officers and 771 of other ranks, 122 women, and 77 children; the Return also shows 2 brevet majors, 2 captains, and 2 subalterns on recruiting duty, and 2 captains and 4 subalterns with the flank companies in the West Indies.

The regiment was directed to disembark at Liverpool, and remained there until the 31st, when it marched in two divisions to Buckingham and Aylesbury, five companies to each, moving on the 23rd April to Gravesend, and embarking, on the 26th, for Ostend, to reinforce the troops under the Duke of York in Flanders.

A letter, two days later, from the Secretary at War, notified that the establishment per company was to be 4 sergeants, 4 corporals, 2 drummers, and 81 privates, and its embarkation strength showed a total of 21 officers, 861 of other ranks, also 60 women and 32 children.

The regiment landed at Ostend on the 6th May, and on joining the division under the command of the Austrian general, Count Clarfait, was posted to Major-General Hammerstein's brigade, with the 38th and 55th Regiments and the 8th Light Dragoons. Our allies were Hanoverians and Hessians.

The 12th Regiment, commanded by Major Frederick Bowes, consisting of 809 rank and file, took part in numerous operations, and was engaged in the general attack on the French positions on the 17th and 18th of May. On the latter day, the 12th were engaged in driving the enemy from Werwick, and in forcing the passage of the river Lys, on which occasion they highly distinguished themselves; but the operations on the above two days were not successful, from the want of a more perfect combination in the movements of the several divisions, and from the superior numbers of the enemy.

In division orders, dated, Camp near Tournay, 20th May, 1794, Major-General Whyte stated:—

> "he had great pleasure in informing the British troops, that General Count Clarfait has highly approved of their spirited conduct in the field, and great exertions in going through such excessive fatigues, as they necessarily have had since their first movement from Ostend. . . . He desires his thanks may be accepted by the commanding officers, and all the officers and men of the 38th and 55th Regiments; and also by Major Bowes and the officers and men of the 12th Regiment, whose conduct has been highly approved of by Major-General Hammerstein, under whose immediate command they served."

In an attack by the Austrians under Count Kaunitz, on the 24th May, the French lost over 5000 men, 3000 of whom were prisoners, and nearly 50 guns.

With a view to regaining their lost possessions in the West Indies, the French had now fitted out an armament, with about 1500 troops, whom they sent to Guadeloupe, for the recovery of that island, under the republican commander Victor Hugues, whose force, escaping the notice of the English cruisers, arrived at Point à Pitre, Guadeloupe, on the 3rd June, and was welcomed by a multitude of mulattoes and blacks.

On account of the extremely sickly state of the British troops, owing chiefly to yellow fever, the French commander took Point à Pitre by storm, and having succeeded in driving out the feeble garrison from Grande Terre, the English commander took up a new position at Basse Terre.

Sir John Jervis and Sir Chas. Grey went at once to Guadeloupe on hearing of the arrival of the French armament, only to find the place so strengthened, and the republican chief's army of mulattoes and blacks so efficiently drilled and disciplined, that he was ultimately able to defeat the combined attack of all the English forces.

In Sir Charles Grey's Return of killed, wounded and missing at Grande Terre, Guadeloupe, from the 10th June to 3rd July, 1794, Captain Tweedie, grenadier company of the 12th, is shown as wounded, and Lieutenant Lyster, of one of the flank companies, killed.

The total casualties of our force in the French West India Islands between the above dates were :—Killed: 1 lieut.-colonel, 4 captains, 7 lieutenants, 7 sergeants, 2 drummers, 94 rank and file; wounded: 1 major, 3 captains, 7 lieutenants, 13 sergeants, 8 drummers, 298 rank and file.

It is much to be regretted that the above Return gives no particulars as to the regiments, or detachments, in which these casualties occurred.

The prisoners taken by the French were, according to the stipulation, to have been allowed to go on board the English ships, but the agreement was broken, and they were kept prisoners for over a year, during which many died owing to the severity of their confinement.

General Sir Charles Grey had written home for the immediate despatch of some troops, which, however, was not complied with, so he wrote, on the 9th July, requesting to be relieved of his command.

Meanwhile, on the 13th June, Major-General Hammerstein, with the 8th Light Dragoons and 38th and 55th Regiments, failed in an attack on a very superior force of the enemy at Ghits. After the action, he retreated to Torout, falling back himself with the Hanoverians to Bruges, and ordering the British troops to Ostend.

The 12th continued to serve under General Clarfait in Flanders, and a despatch from Army Headquarters, dated Tournay, 14th June, gives an account of this general's failure to raise the siege of Yprés (an operation in which the regiment was engaged), the French besieging the place with 30,000 men, and a covering army of 25,000.

A despatch from the Duke of York, on the 22nd June, announced :—

> ". . . his great appreciation of the Home Government for sending a reinforcement to his assistance, under the command of Lord Moira, the Duke's force 'being totally insufficient to act by itself in a country in which the enemy is at least 80,000 strong.'"

The Earl of Moira's force, consisting of Artillery, part of the 8th and 10th Light Dragoons, and 10 line regiments (totalling about 8500), embarked at Southampton on the 18th June, and sailed from Spithead on the 22nd, passing through Lord Howe's fleet, which had just returned from his ever memorable and glorious victory of the 1st June, with 6 captured French line-of-battle ships; and the hearty cheering which arose on both sides had a very fine effect, and greatly exhilarated the spirits of our troops, who arrived at Ostend on the 26th June; every individual, on disembarking, being served with a blanket, canteen (or small wooden cask, holding about 3 English pints), and a haversack, or linen bag, slung over the right shoulder, for provisions, &c. These, with the knapsack and its contents, arms, accoutrements, and 60 rounds of ammunition were the sum total of what the soldier carried, no baggage having been allowed for either officers or men, except what each carried for himself.

Before daybreak, on the 29th June, Lord Moira's force marched towards Bruges (about 18 miles), which was reached about noon, and bivouacked on ground which had that morning been vacated by a detachment of the French. The next day's march to Ghent was to enable this force, if possible, to effect a junction with the Austrians, under General Clarfait, and the Hanoverian and Hessian troops, under Count Walmoden, who were within two miles of Ghent, as the French army in that neighbourhood was known to exceed ours by more than treble the number.

To every British regiment of Foot, there were now attached two light brass 6-pounder guns in charge of 12 artillerymen, and three horses with their drivers; also an ammunition waggon and tumbril to each regiment, which were very heavily laden. Owing to very stormy weather, Lord Moira's force did not reach the ground intended for it until the 1st July, when the necessary junction with the other troops was effected.

On the 4th, the force arrived at Dendermonde, and, on the 6th, at the town of Alost. As the troops now had no protection from bad weather, huts or wigwams, from the boughs of trees (after the Indian fashion) were erected whenever the opportunity offered, and the Hanoverians having just quitted this station, their wigwams were still remaining.

On the evening of the 7th, Lord Moira quitted Flanders, and, entering the province of Brabant, continued the march towards Malines to effect a junction with the army under H.R.H. the Duke of York, which was duly carried out on the 9th, the French army, at this time, being on the move from the neighbourhood of Soignies.

The very superior numbers of the enemy gave them so great an advantage, that the allied army was forced to commence retrograde movements. The 12th Foot remained with Major-General Hammerstein's Brigade until the 9th of July, when the following paragraph appeared in the Division Orders issued at the camp at Contiche,—

> "As the 12th British Regiment is going to leave Major-General Hammerstein's Brigade, he takes this opportunity to assure the regiment of his best acknowledgments for the good and gallant behaviour it has shown during the time the general has had the honour to command it; he likewise thanks it for the readiness and good will with which it has borne so many and great fatigues."

On removal from Major-General Hammerstein's command, the regiment was formed in brigade with the 33rd, 42nd, and 44th Foot, under Major-General Balfour.

On the 13th July, the French attempted to pass the river Nethe by a bridge which was defended by a part of the Hanoverians, under General Walmoden, when Lord Moira was detached with two brigades in support of them, but before he could come up, the French had retired with considerable loss, the Hanoverians having had 14 killed and wounded. Two days later, the enemy were more fortunate when advancing in great force against our left wing, which consisted of Hanoverians and Hessians; they then took possession of Malines before two brigades, detached from our army, could arrive. On the 16th July, the French again attempted to cross the river Nethe, at the bridge of Walheim, but a British brigade stationed there, under General Stuart, consisting of the 12th, 33rd, and 44th Regiments, with six guns and some howitzers (2 companies of Irish artillery), very bravely and successfully opposed their passage. A very heavy cannonade from across the river caused us the loss of some artillerymen, but we made great havoc amongst the French from our well-directed fire, driving them from Malines and capturing 18 guns.[1]

Our hospital, with the sick from Antwerp, nearly 1800 men, was now daily expected at Middelburg, where a house was prepared for their reception.[2]

The combined armies of British, Austrians and Dutch, now formed a complete chain from Namur to Antwerp, any part of which, if attacked, could be supported by the others. General Clarfait was at Namur, the Prince of Orange at Louvain, the Prince of Coburg at Tirlemont, General Walmoden at Lierre, and the Duke of York covering Antwerp.

In consequence of intelligence received on the 18th July, that part of the French, under Pichegru, were pressing hard on the Dutch army, and forcing them to fall back towards Breda, the Duke of York set off to have an interview with the Prince of Orange. The outposts of our troops at Duffell were skirmishing daily with those of the enemy, with losses on both sides, our casualties being chiefly amongst our light dragoons, who were principally engaged.

On the 21st, our outposts were withdrawn from Duffell, and the troops proceeded towards Antwerp, in which direction the main body of the Duke's army had already moved.

On the 22nd, to the great grief of his force, Lord Moira took his departure for England, the assigned reason being, that, on his lordship joining the Duke of York's army, he found himself one of the youngest generals, and, consequently, instead of retaining command of his army, would only have been entitled to the command of a brigade.

On the arrival of our troops at Rosendael, on the 25th July, it was

[1] *Foreign Office Book*, No. 25. [2] *Ibid.*

ascertained that the French had, on the previous day, entered and occupied Antwerp.

Owing to the nature of the country in these parts, no good water could be got near the camps, and, though wells were dug immediately on arrival, the water they produced was very indifferent. Bread also became a very scarce commodity here, as well as most of the necessaries of life, giving it the name of the *starvation camp*.

The sick had gone to Bois-le-Duc.

An advance from Rosendael to the village of Nissien took place on the 3rd August, and, next day, our whole army moved towards Breda, and occupied ground near the village of Etten. The Army again moved on the 5th, and passing Breda, took up a position near the village of Osterhout, where they had the great pleasure of having tents and camp equipage issued, which had been brought in waggons from the Scheldt. Whilst here, the Prince of Orange, accompanied by a numerous suite, came to review the British Army with great splendour.

General Return of the Army, actually present and fit for duty under the command of His Royal Highness, the Duke of York.

Camp at Osterhout, 5th August, 1794.

Nations.	Officers.	Sergeants.	Trumpeters or Drummers.	Rank & File.
British Cavalry	165	231	72	4,247
Hanoverian Cavalry	112	184	44	1,395
Hesse Cassell Cavalry	46	116	23	860
Hesse Darmstadt Cavalry	10	36		281
TOTAL CAVALRY	333	567	139	6,783
British Infantry	583	924	511	19,734
Hanoverian Infantry	143	273	213	3,284
Hesse Cassell Infantry	135	403	160	3,029
Hesse Darmstadt Infantry	44	93		1,327
TOTAL INFANTRY	905	1,693	884	27,374
GRAND TOTAL	1,238	2,260	1,023	34,157

Skirmishes between our outposts and those of the French were almost of daily occurrence, and, from two of their deserters who came in on the 19th August, we learnt that their army now amounted to 100,000 men, and was within a day's march, whereupon all our outposts were promptly visited by the Duke of York. Contrary to expectation, however, the French made no immediate move, whilst everything was being prepared for their reception.

At daybreak, on the 26th, they made their grand attack on the villages of Alphen, Chaen, Roysberg, and Gelder, which they assaulted with great vigour, bringing several guns, and, after an obstinate contest, succeeded in gaining possession of the two latter places, but were repulsed at Alphen and Chaen, which was, in a great measure, due to a timely reinforcement of ours.

On the 27th and 28th August, our army moved off, and passing Bois-le-Duc, took up their ground about nine miles from it, on a heath near the village of Nisterlode.

A despatch from the Duke of York, at headquarters, Uden, dated 31st August states :—

"The army is at present encamped behind the River Aa, from whence we can advance to the protection of any of the Dutch fortresses which may be attacked, whilst the position covers the only passage into Holland which is not defended by the situation of the fortresses, and re-establishes our communication with the Austrian Army."

On the 1st September, the headquarters of the Duke were established at Berlicum, a village close to the camp.

A Monthly Return of British Troops under command of His Royal Highness, the Duke of York, dated 1st September, 1794, shows the 12th Regiment mustering 21 officers, 27 sergeants, 17 drummers, and 653 rank and file fit for duty, 103 sick, 3 on command, 18 recruiting, 7 prisoners of war, 19 attached to Royal Artillery, 51 batmen, 17 sick in England—total rank and file 871.

Our allies, the Hanoverians and Hessians, were now posted near the village of Ordenrog, on the small river Dommel.

On this date, the Duke of York, Prince of Orange, and several generals held a council of war at Bois-le-Duc, which lasted several hours, presumably to concert a plan for the ensuing operations, whether the Allies should advance on the French, or continue on the defensive only, and maintain the chain at present formed until all their reinforcements arrived, or until the opening of another campaign, as both armies would soon be glad to go into winter quarters.

The following was the state of the French army from the best information procurable at this time :—

Army under Pichegru and Martha	50,000
,, Jourdan	70,000
,, Vandamme, Suillard, and Dandelo	35,000
Total	155,000

Of these, the two first were advancing in three columns on our army; the remainder in the neighbourhood of Breda, and ready to co-operate with their commander-in-chief, Pichegru, if required.

On September 14th, a sudden attack was made by the French in great force on all posts of the Allies' right; and that of Boxtel, the most advanced, was assailed, with considerable loss to the Hesse Darmstadt troops who occupied it. As the line of outposts on the Dommel could not be maintained while the enemy were in possession of Boxtel, it appeared necessary to regain it. Lieut.-General Abercromby was therefore directed to march with the reserve during the night, and reconnoitring the post at daylight, to act as he should judge best. He found the enemy in such strength as left little doubt of the proximity of their army, and accordingly retired without loss.

On receipt of information, which was confirmed, as to their strength being scarcely less than 80,000 men, the Commander-in-Chief decided on retreating across the Meuse, which was carried out in two days without any loss, the army crossing the river, on the morning of the 16th, by a bridge of boats constructed for the purpose.

The loss in the attack on the outposts fell chiefly on the Hesse Darmstadt troops, and the principal loss of the British fell on the 12th Foot, who, at one time, were supposed to have been entirely cut off.[1] This was doubtless the post which, Cannon relates, had been surrounded and assailed by superior numbers, and defended with great gallantry. The 12th had 1 man killed, 1 wounded, and Lieutenant Eustace, 3 sergeants, 1 drummer, and 44 rank and file taken prisoners, our total loss having been about 150 killed, wounded, and prisoners, and the capture of two guns.

This action is recorded as interesting to the British as being the first battle of our illustrious Wellington, who was then Lieut.-Colonel Wellesley, commanding the 33rd Regiment.

Prior to crossing the Meuse, the army passed through Grave, and finally occupied ground half-way between it and Nimeguen.

On the 22nd September, army headquarters were established at Cranenberg, and moved on the 25th to a small village called Mook.

On the 3rd October, the Duke of York's headquarters were at Groesbeck, and a despatch of that date intimated that on account of an attack in great force by the enemy on General Clarfait's left flank (causing him to retire behind the Erfft), the Duke's position at Groesbeck was no longer tenable, and he was, in consequence, taking measures to pass the Waal, at Nimeguen, as soon as possible, whence he next wrote on the 6th.

Since the re-capture of Guadeloupe in June, by the republican forces, no reinforcements had, by October, been sent to the assistance of Sir Chas. Grey, whilst our numbers rapidly diminishing, from the great amount of sickness and mortality which prevailed, rendered our troops incapable of regaining any lost honours.

The command of the troops at Basse Terre had devolved on Lieut.-Colonel Graham, 21st Fusiliers, who took up a final position at Bexville Camp, which he defended with the utmost gallantry and spirit until the 6th October, when finding his provisions nearly exhausted, and that he was

[1] *Journal of the Campaign*, p. 72.

cut off from all communication with the shipping, and without hope of relief, he was obliged to surrender, his force being reduced to 125 rank and file fit for duty.[1] By this unfortunate event, the whole of the island of Guadeloupe except Fort Matilda, where Lieut.-General Prescott commanded, fell into the hands of the enemy.

Sir Chas. Grey, who had applied to be relieved of his command, retained it until the 18th November, when his successor, Sir John Vaughan, arrived at Martinique.

In the meantime, a despatch from the Duke of York in Flanders, dated 12th October, mentioned the surrender to the French of Bois-le-Duc, which was, at the time, garrisoned by a corps of Dutch emigrants.

A despatch, dated Nimeguen, 20th October, notified that the enemy had, on the previous morning, attacked the whole of the advanced posts of the Duke's right wing, in very great force; particularly that of Doulen, held by the 37th Regiment.

On the 23rd the Duke's headquarters were established at Arnheim.

Another attack on our outposts, at Nimeguen, took place on October 27th, all of which were driven in, after a very severe conflict, but the French suffered very considerably in this affair, as our troops were for the most part under cover, and could not be dislodged without great loss to the enemy.

On the 28th, a sortie was made from the town on some advanced parties of the enemy, when Captain John Picton, 12th Regiment, was wounded.

On November 4th, another sortie was made in force, when our troops consisted of the 7th and 15th Light Dragoons; the 8th, 27th, 28th, 55th, 63rd and 78th Regiments; the Hanoverian Horse Guards, besides 3 squadrons of their Horse, and the Emigrant Legion of Damas; also 2 battalions of Dutch, covered by the British 59th Regiment.

The French loss on this occasion was computed at upwards of 500, that of the Allies having been 20 killed, 176 wounded, and 20 missing. In this sortie, there was not a single prisoner made by us, and the bayonet decided everything.

The sole object now of the British troops was the preservation of Holland, by preventing the French from crossing the Waal.

Our force at present consisted of 16 regiments British Cavalry, 30 British Infantry, besides Artillery and foreign troops in British pay, consisting of Hanoverians, Hessians, Emigrants, York Rangers, &c., forming, with the Dutch troops, an immense army, which covered the banks of the Waal, and would, it was thought, be a complete check to the French penetrating into Holland across that river.

The prospect now before our army of winter quarters was very dis-

[1] *London Gazette*, December 13th, 1794.

The troops that capitulated were the flank companies from Ireland of the 8th, 12th, 17th, 31st, 33rd, 34th, 38th, 40th, and 55th Regiments, the 39th, 43rd and 65th Regiments, and three companies of the 56th. Their loss in the different actions between the 27th September and the 6th October amounted to 2 officers and 25 rank and file killed, and 5 officers and 51 of other ranks wounded. (Sir C. Grey's despatch, Bulletin, 1794, p. 357, and *W. O.* 1, Book 83, P.R.O., London.)

couraging, there being no other expectation than to pass a dreary winter in a cold country in tents, on the banks of the Waal, and to undergo the severest duties without the common comforts of life.

Whilst the headquarters of our Army were at Arnheim, five of our British line regiments were encamped before Nimeguen until as late as the 7th December, and one or two men had been frozen to death; by the above date, the regiments in camp had built themselves huts of sods and thatch. Our picquets now remained 48 hours on duty.

The distressed situation of the Army, for clothing, was at this time very great, some regiments having lost the whole of their necessaries, owing to the transports they were shipped in having fallen into the hands of the enemy, and many had hardly a coat to wear. It was no uncommon thing to see an officer with the skirts of his regimental coat cut off, to repair the body of it, or to keep it whole at the elbows.

The officers of several regiments set on foot a subscription to furnish their men with a few of the comforts of life. It was curious to see what shifts the men were put to, to keep themselves warm.

One would have on a huge pair of Dutchman's wide breeches to his regimental jacket; another would have a large, fully trimmed Burgomaster's coat, by way of an overcoat, and, another, to his red jacket, would have sewn a pair of wide black or brown sleeves, with long hanging cuffs, and there were other curious contrasts, so that, from the motley appearance of some, it was difficult to distinguish them as English troops. The main object being to keep arms and ammunition in good order, very little attention was paid to dress.

Shoes and stockings were also much wanted, as the weather was very wet, and the roads and fields exceedingly dirty. Upwards of 40 and 50 men in a battalion might now be seen barefooted, but it was chiefly the recruits who were in the most wretched state imaginable, hundreds of them having only linen trousers, without drawers or stockings, and only their slop jackets without waistcoats, whilst the benefit of great-coats was unattainable.

His Royal Highness, the Commander-in-Chief, left Army Headquarters at Arnheim for England on the 2nd December, and the command of the British Army devolved on Sir Wm. Erskine.

The number of sick in our general and regimental hospitals was now very great, amounting, on the 11th December, to 6947, and many deaths occurred daily. Experience also proved too forcibly how very few might be expected to return to duty, owing to want of a system, and a deficiency of hospital supplies, particularly port wine, which could not be procured in that country, together with other necessaries, and even linen.

On the 18th December, on the suggestion of Lieut.-General Harcourt, the proposed establishment of a hospital corps was put into force, at a daily cost of £21 6*s.* 8*d.*, which consisted of :—1 military commandant of hospitals, at £1 a day; 4 officers, to act as paymasters, quarter-masters and adjutants, at 10*s.* each, daily; 4 sergeant-majors and 4 quarter-master sergeants, at 2*s.* 6*d.* each; 10 sergeant ward-masters at 2*s.* each daily; 20 corporals and 20 cooks at 1*s.* 6*d.* each; 200 private attendants at 1*s.*, and 100 washerwomen at 8*d.*

About the middle of December, the allowance of beef to our troops was increased by ½ a lb. a day, owing to their hard duty and the excessive cold. A soldier's daily allowance now was :—1lb. beef, 1¼ lbs. bread, and half a pint of spirits, and with potatoes and greens, &c., for which they foraged, they made a tolerably good living, as coffee being very plentiful, it was sold by the soldiers' wives, so that, for a man's breakfast, he might, for a penny, have an ample amount of coffee, but had to find his own bread and butter. The frost, which had set in rather severely for several days, continued with unabated rigour, and, throughout December, all remained tolerably quiet on the Waal with the enemy, who seemed as anxious for a rest as we were.

In a survey sketch of the country, with a list of cantonments, dated, 19th December, 1794, showing quarters occupied by British troops of the Duke of York's Army, the 12th Regiment is posted to the 3rd Brigade, under Major-General Balfour (with the 33rd, 42nd, and 78th Regiments) in garrison at Echteld, and quartered near the Dyke, for the defence of the Waal.

On the 21st December, the very welcome news reached our army, that an adequate supply of warm winter clothing was actually landed on the continent, and on its way to the army, and Christmas Day this year (one of the coldest ever known in the country) was celebrated by the arrival in camp of a quantity of flannel clothing, which was immediately distributed, and the men, in honour of the day, were allowed a week's pay, to drink to the health of the ladies of England, who, in forwarding the warm clothing, had not been unmindful of the hardships of the troops, and their sufferings abroad.

On December 27th, the French crossed the ice, with 8000 men, at the important post of Bommel (an island, occupied by two battalions of the Dutch), to which our attention was now turned, and, on the 30th, the enemy were attacked by a force, under Major-General Dundas, of British and Hessians (two columns), and driven back across the ice with the loss of a considerable number of men, and four guns. Our loss, amounting to 54 killed and wounded, was trifling in comparison to theirs. In other respects, our troops suffered very severely, not only from their heavy march of thirty miles previous to the action, but also from excessive cold, and the want of shelter at night.

1795

On the 5th January, General Dundas again moved to attack the enemy, who were driven back with considerable loss, ours having been 3 killed, 58 wounded, and 7 missing; he then fell back upon Buren during the night, which was effected unperceived by the enemy, our camp fires having been left burning in order to deceive them.

As our troops, for some time past, had undergone a very great stress of fatigue, and had suffered much from their distressing situation, for want of cantonments, by which they were under the necessity of passing these cold and dreary nights without any shelter whatever, steps were taken to withdraw the Army from its present situation, and a position was taken up behind the River Leck (leaving picquets on the Waal), where the troops were accommodated with the cantonments which were so much needed.

A reconnaissance from Buren on the 8th January was made by Major-General Cathcart, which resulted in a successful action, in which he manœuvred with part of his troops on the ice, and drove a party of the French through Buremalsen, across the Lingen, and thence from the village of Gildermalsen, capturing their only gun, a fine brass 8-pounder, which was unfortunately lost by us the next day when crossing the river, owing to the ice breaking under it. Our loss in this engagement was 15 killed, 119 wounded, and 7 missing.

At daybreak on the 10th, the long projected attack by the French began, when they crossed the ice in six columns, three of which were repulsed, and three, passing at different points (in conjunction with another very strong column, passing near Nimeguen), attacked the whole of our line on that side, which resulted in a retreat of our force, owing to their superiority of numbers, when our troops took possession of the village of Elden, where they bivouacked that night.

A journal of the war says:—

> "It would be doing great injustice to the women of our army not to mention with what alacrity they contributed all the assistance in their power to the soldiers while engaged, some fetching their aprons full of cartridges from the ammunition waggons, and filling the soldiers' pouches at the risk of their own lives, while others, with a canteen of spirit and water, would hold it to the men's mouths, when half choked with gunpowder and thirst; they would also assist wounded men, as much as possible, to the nearest house or waggon, in which friendly offices it was not uncommon for them to get wounded, as well as the men, numerous cases of this happening in the course of the campaign."

On the 11th January, our troops crossed the River Leck, at Arnheim, strewing sand and straw on the ice, to facilitate the passage of the horses.

The whole of our army was now on the north side of the Leck, and the enemy in possession of the country between the rivers Waal and Leck. Our sick and wounded were all sent to Deventer, where, from want of accommodation on the march, and the extreme severity of the weather, great numbers perished, and, what was more distressing, the ground was frozen so hard, and to such a depth, that it was impossible to bury the bodies, which lay on the commons and roads over which the waggons had gone, a most melancholy spectacle to the Army as they passed.

It having been decided to retire from Holland, part of our army marched by different routes, on January 14th, towards the river Ysel; the frost continued severe, and dreadful was the prospect of that retreat, the greatest difficulty being found in obtaining some kind of shelter for the troops at night, which resulted in the Army being divided into brigades, and sometimes single regiments moved independently, in order to obtain some kind of quarters in the wretched villages in which the troops were obliged to shelter.

The severity of the frost was so great, that many who could not keep up were frozen to death on the dreary heaths, whilst others fell into the hands of the enemy, but others again, having found shelter, rejoined their regiments.

Many also were frost-bitten, losing legs, arms, feet, toes, fingers, &c., sufficient indeed, to make one reflect with what a burden on her shoulders England maintains her weight of empire.

January 16th was devoted to rest, after the thirty mile march from Wageningen, and, the enemy not having yet advanced in any great force, it was resolved to indulge in another day's rest. On the 18th, the force proceeded to the small village of Hackforth, where the accommodation was so limited that one entire regiment took up its residence in a large mansion, each company occupying a room. The mansion, the property of Baron de Westerholt, had been deserted by him on the approach of the armies, leaving his larders and wine cellars plentifully stocked, and as wine cellars are not generally the last places visited by British soldiers, (and this day, January 18th, happening very appropriately to be the Queen's birthday,) it may be imagined that the occupants of the house did not neglect the opportunity which the fortune of war had cast in their way, to drink to the Queen's health, and also that of the Baron.

The march was continued for days through very severe weather, with snow lying deep on the ground, and, on the 29th January, the force had the pleasing prospect of passing the frontiers of Holland, when, on entering Germany, they joyfully arrived at Bentheim, receiving from the inhabitants a very friendly ovation, a great contrast to inhospitable Holland.

The hardships and sufferings of our army, in this memorable retreat through Holland, were such as were perhaps never exceeded, for we had not only to make good our retreat before a numerous and victorious enemy, who were continually harassing our troops, but also were under the necessity of being constantly on our guard against the perfidy of the treacherous Dutch, who let no opportunity slip of practising the foulest ingratitude to the scattered remains of those troops who had been so long employed in the defence of their country.

A War Office book, headed "Claims for Losses on the Continent, 1794–95," shows that, on arrival of the British troops in Holland in 1794, a general service depot had been established at Helvoetsluis, for the storing of officers' baggage, regimental necessaries, &c.; also, that owing to raids by the enemy the losses from it had been very great, though some were incurred at different hospitals, by men who had been taken prisoners.

The claims of the 12th Regiment, in 1794–95, for the loss of Regimental Necessaries alone (apart from baggage of regimental officers), amounted to £1508 1*s.* 10*d.*, of which £1368 18*s.* 7*d.* was allowed, one of the items disallowed having been £50 for loss of officers' mess plate. Officers' claims were quite separate.

In a "Return of Cantonments of British Troops," dated 11th February, 1795 (showing 1 Brigade of Cavalry and 6 of Infantry, under Major-General Dundas), the 12th Regiment, in the 3rd Infantry Brigade, appears quartered at Gramsberg, and a Return of the same date, headed "Dislocation of the British Troops," shows it in the same brigade, and the regiment itself occupying quarters at Giersen and Basvinkel, on the east bank of the Ems.

Our marches continued until the 12th February, when we arrived at Schutrope, where a halt was made until the 25th, while the French army had been possessing themselves of Holland.

In a letter, dated, Army Head Quarters, Osnabruck, 7th March, arrangements were made by Lieut.-General Harcourt for the embarkation, at Embden, of our sick for England, and for sending by land to Bremen those who could not be embarked.

March 9th found our army at the village of Laten, whilst the weather continued very cold for the time of year. In this part of the country, provisions were very scarce; beef was not procurable at any price, and the bread, besides being black, was so bad that all who partook of it were immediately taken ill. The troops thus lived on potatoes and other roots, and halted at Laten until the 20th, when the long-wished-for march was made towards the city of Bremen.

On March 21st, the whole of the sick embarked for England on the vessels allotted to them, and proceeded under escort of H.M.S. "Dædalus."

Peace was concluded at Basle on the 5th April, between France on the one side, and Prussia, Spain, Holland, and Tuscany on the other, which necessitated the British troops in Germany returning home. The regiment lost so many men during the campaign and retreat through Holland, that its numbers were reduced from 815 to 425 rank and file.

At a General Court-Martial, assembled at Bremen, on the 10th April, Captain John Picton, 12th Regiment, was tried, for having, without orders, while on duty, quitted his detachment, the advanced post of Beer-Malsen, in front of Buren, where he commanded, and for having fallen back several miles, crossed the Rhine, passed through other posted corps, and joined his regiment, then at Wyck, or in that neighbourhood, before any report of the transaction was made to Major-General Dundas. He was found not guilty of any part of the charge, and was most honourably acquitted. He had been in confinement over three months before obtaining a court-martial.[1]

The regiment embarked at Bremenlee on the 11th April, landed at Gosport on the 12th May, and marched to Porchester, every effort being now made to recruit it as speedily as possible. Here, the regiment was joined by Lieutenant O'Brien (who later became Marquis of Thomond) one sergeant and one private, the only surviving individuals of the grenadier and light companies of the 12th, which proceeded from Cork to the West Indies in 1793, as fine a body of men as ever quitted the shores of Ireland.

These flank companies, it will be remembered, had served at the captures of the Islands of Martinique, St. Lucia, and Guadeloupe, the remaining officers and men having perished in action, or by the fatally insalubrious effects of the climate.

Richard Cannon states, in his history of the regiment (page 56) that the regiment (at Porchester) was reviewed on the 2nd July by His Royal Highness the Duke of York, but in this, it would appear from the following, he is mistaken.

The "Times" correspondent at Portsmouth, writing from that station on July 2nd, states :—

> "About 11 o'clock, H.R.H. the Duke of York arrived here (Portsmouth), and . . . after inspecting the new fortifications around Gosport, reviewed the 28th and 80th Regiments. About 4 P.M., on landing at the Dockyard, he walked into Portsmouth, and, on the parade, reviewed the 128th Regiment, after which, he dined, &c. . . . , and the next day, was to visit the camp on Southsea Common."

On the 7th July, the regiment was ordered to march from Porchester Barracks to Nutshalling Common, near Southampton, in readiness to embark for foreign service.

A letter from the Secretary at War, dated 15th August, directed that the ten companies of the regiment, prior to embarkation, were to be completed to the augmented establishment of 100 rank and file per company, by drafts from the 112th, 124th, and 127th Regiments, and that two companies be added thereto, each consisting of 5 sergeants, 5 corporals, 4 drummers, and 55 private men (with the same number of non-commissioned officers as the other companies), to be constantly employed on recruiting service. The men thus drafted were to take with them the arms of their former corps, and each man transferred was to receive 1½ guineas.[2]

A Horse Guards' letter, dated 11th August, had directed Lieut.-General the Earl of Moira to forthwith cause the following regiments, viz. :—the 12th, 1st Battalion 78th, 80th, and 90th, with a detachment of Artillery, and such proportion of stores and ammunition as he might consider necessary,

[1] *Book of Cuttings, R.U.S.I.*, Vol. ii., p. 231.
[2] *W. O.* 4, Book 160, p. 162.

to be embarked on a sufficient number of transports at Southampton, and placed under command of Major-General Doyle, with secret written instructions to proceed with the troops, under convoy of the Squadron, to the islands of Houat and Hedic, and there to communicate with Sir John Warren, commanding a squadron off the coast of France, for instructions; and to receive from him such intelligence as he had obtained, respecting the state of the Royalist and Republican forces in Brittany, Poictou, and the adjacent provinces, the forces then proceeding to the Island of Noirmontier, with a view to taking possession of it. If satisfied that it could be done without material loss, or risk to the troops, General Doyle was then to capture it, and, if successful, to strengthen it with such works as might be necessary for its defence, reporting duly on the same. If finding, after possession, that it could not afford entire security against probable attack, he was to return with his troops to Cawsand Bay, or such other part of the British coast as circumstances should permit. In the event of taking possession of it, he was to let it be publicly known that it was taken in His Majesty's name, in the endeavour to open a communication with the Royalists in the adjoining provinces, and to afford them such succour and supplies as possible, to enable them effectually to resist the forces of the pretended National Convention of France, and to promote the restoration of order and regular government.

Further instructions left it to the discretion of General Doyle, in concert with Sir J. Warren, to capture, in addition to or instead of Noirmontier, the Isle D'Yeu, its possession being held in the same manner, and with the same objects as referred to Noirmontier.

Lord Moira, in a letter to the Secretary of State, requested that another general officer might be posted to his force in addition to General Doyle, and commented on the quality of the drafts being so bad that he found the number fit for duty with those corps would fall far short of 4000.

The four regiments embarked on the 19th August, and, before sailing, a medical board, convened by Earl Moira, found the greater part of the drafts unfit for service.

The Embarkation Return shows the 12th embarking with 2 lieut.-colonels (Lieut.-Colonel Aston in command), with a strength of 28 officers, 758 of other ranks, 80 women and 45 children, the total rank and file of the four regiments amounting to 2888, instead of 4000 as originally intended.

Lord Moira resigned command of the expedition on the 24th August, when it was offered to Sir Ralph Abercromby.

On the 5th October, the regiment landed in the Isle D'Yeu, in conjunction with a force of 1200 men, under Major-General Needham, who had sailed from England on the 28th, accompanied by a body of French emigrants, and the Count D'Artois, brother to the King of France. L'Isle D'Yeu surrendered to him on the 30th.

The island is six miles long, and three in breadth ; water good, but no forage, and scarcely any pasture, and did not produce sufficient corn for the inhabitants, who numbered 1800, and manifested the most favourable and friendly disposition towards the British troops.

A consensus of opinion being now taken of Major-General Needham, the Adjutant-General, Quarter-Master-General, and the officers commanding regiments, they were unanimous in deciding that it would be advisable to abandon the projected attack on Noirmontier, and to retain possession instead of L'Isle D'Yeu, in the belief that the latter would ultimately prove of more value from the greater facility of defending it with a small force.

A despatch from Major-General Doyle, dated L'Isle D'Yeu, 7th October, showed his troops to be destitute of fresh provisions, the little they could get being kept for the sick.

Sickness had also set in for want of straw, or fresh bedding, whilst application to the home authorities was at the same time made for material for building barracks or huts ; also fuel (partly wood) for the ovens, and forage for the horses of nearly 2000 cavalry, who had joined the force.

Money also, for payment of the troops, was becoming scarce, and the want of halfpence, in particular (there not being any on the island), necessitated the cutting of dollars for immediate circulation.

The salt provisions also in possession were estimated to last only thirty days, and provisions which had been expected from Quiberon had not yet arrived, except for three days' consumption. Beef and pork were in particular request, and also candles, of which the force was entirely deficient.

On the 26th October, the following indent for the most necessary articles was rendered by the Quarter-Master-General, and sent home, viz. :—vinegar (1 tun), 40 (10 lb.) jars mustard ; 100 lbs. pepper, 40 lbs. ginger, and 1600 lbs. each of French barley, rice, candles, soap, cheese, and butter ; 800 lbs. tea, 4000 lbs. sugar, 40 hogsheads port wine, 400 lbs. portable soup, and 60 head of live cattle.

The troops were from the first employed in fortifying the island.

No circumstances having occurred to favour any further attempts connected with the expedition on this enterprise, it was found expedient to embark the British troops for home, early in December, and an Embarkation Return shows the following as the state of the 12th Regiment between the 5th and 16th December, viz. :—36 officers and 850 of other ranks, 90 women and 49 children—total 1025, embarking in the following ships of the fleet, viz. :—H.M.S. "Anson," "Queen Charlotte," "Robust," and hospital and commissariat transports.

On the passage home, the regiment was exposed to several violent storms at sea, but arrived safely at Southampton on the 15th, and marched to Iron-hill Barracks.

1796

A letter from the Secretary at War, dated 7th January, directed that the 12th Regiment was to be sent, with all practicable expedition, to the West Indies, and to be held in readiness to embark accordingly, which was, however, changed later in the year to the East Indies.

On the 11th January, the regiment marched from Iron-hill Barracks to Lyndhurst, and, on the 8th February, in three divisions, to Newbury, Speen, and Speenhamland. On the 24th March, orders were received to march to Portsmouth, for conveyance to the Isle of Wight, where it was quartered at Newport until the 8th June, when it embarked at West Cowes, on board the ships "Rockingham" (Colonel Aston, with the two flank companies, band and colours), "Hawkesbury," "Airlie Castle," and "Melville Castle," Indiamen (800 tons each), for service in the East Indies, convoyed by the "Fox" frigate, as the Dutch fleet was at sea looking out for them.

Colonel Bayly, in his diary, describing the march from Newport to embark, says :—

> "the regiment was accompanied by at least 500 women, wives of the soldiers, only 60 of whom were allowed, by regulation, to embark with their husbands, the scenes which followed (as is usually the case) being distressing beyond description."

The regiment remained on board off St. Helens, Isle of Wight, until the 27th June, and, in that interval (duelling being now much in vogue), two of its young officers (Lieutenants Price and Willock) proceeded one day, as the result of an altercation, with pistols and their seconds, to the nearest land at St. Helen's, and, having landed, fired a couple of shots at each other without effect, when the affair was amicably adjusted, and they returned to the ship.

Sailing from St. Helens on the 27th June, the ships anchored on the 19th September in Table Bay, Cape of Good Hope, which had been captured from the Dutch a short time previously.

The regiment did not disembark, but the officers were allowed to go on shore, and a few of the men by turns.

During a stay of nearly two months at the Cape, a good deal of duelling took place, and in one of the encounters Ensign Jordan was shot through the breast by Lieutenant Willock, both of the 12th.

On the 9th and 10th October, the Indiamen, whilst in Table Bay, encountered a terrific gale, which injured every vessel in the fleet, several losing their anchors; and the greatest anxiety prevailed on shore for the safety of the regiments, which, however, providentially escaped.

Our fleet, which consisted of the "Fox" frigate and eleven large Indiamen (many of them having thirty-six 18 pounders) with three British regiments on board (the 12th, 86th, and 94th), sailed for Madras on the 10th November, when all was hurry and bustle in exercising the great guns, as it was expected our ships would now encounter the French frigates cruising in the Bay of Bengal.

The regiment left at the Cape Lieut.-Colonel T. Grey (its second lieutenant-colonel) who died shortly after of an abscess.

1797

After clearing for action several times, our ships at length anchored in the Madras Roads, without further impediment, on the 19th January, and next morning, the regiment (870 rank and file) marched into Fort St. George, vacated by the 74th Regiment.

A Madras General Order, dated 13th June, directed that the pay of soldiers of the 12th and 74th Regiments, whilst practising the Gun Exercise, was to be made up to that of the Royal Artillery.

Referring to troops on home service, a Horse Guards' Circular, dated 10th July, authorised that such part of them as could be spared from military duty might be employed by farmers, to assist in getting in their harvest, not exceeding one-third at a time, to go to harvest work in the neighbourhood of their respective quarters, and not beyond ten or fifteen miles from them.[1]

The 12th continued to do garrison duty at their new station until August, up to which time the regiment was drilled, without intermission, throughout the hottest part of the season, becoming sufficiently expert in military evolutions to justify its being in that respect (as Colonel Bayly relates) " perhaps superior to any King's regiment then serving in India."

A general Order, issued on the 11th August, directed the regiment to be held in readiness to embark, with other troops, for foreign service at the shortest notice, and by the 27th the whole regiment was on board ship in two detachments, 6 companies, on the 17th, in His Majesty's ships "Trident" and "Victorious," for passage to Penang (or Prince Edward's Island), and the remaining 4 companies on the 27th.

The officers' mess, at this period, appears to have been conducted on an economical principle, satisfactory to all, a contract with two natives having been made for the supply of an excellent dinner, with dessert and a pint of Madeira to each officer, for 10 pagodas (about £4) a head monthly ; also, twice a week, Thursday and Sunday, a superior dinner, including European articles, such as hams, tongues, cheese, &c. The dinner hour was at 3 P.M. and Thursday and Sunday were the days set apart for guests, who are stated to have been often three times the number of their hosts.[2]

On the usurpation of the Mysore throne by Hyder Ali, the British territories on either side of that state (the Carnatic on the east, and Malabar on the west) were continually harassed and threatened, their security varying from time to time, in proportion to the strength maintained by our military establishments in particular districts, and their preparedness for war.

[1] *W. O.* 3, Book 31, p. 46.

[2] Captain Eler's *Memoirs*, pp. 60–1.

On the death of Hyder Ali, his son Tippoo pursued a similar policy of reprisals and intrigue towards the British, against whom he had even a greater hatred than his father.

The treaty of 1792, which followed the investment of Seringapatam in the previous year (when Tippoo had come to terms), only weakened his power and resources for a time, whilst his revengeful feelings (at the loss of a part of his territory, and the infliction of a fine, equivalent to 3½ millions sterling) made him eager to seize the first opportunity to recommence his policy of aggression and intrigue. His design was to drive the British out

Col Thomas Grey 12th Inf.
Ob 1797

of India, and to induce others to join in the enterprise, he sent missions to the Isle of France (Mauritius), to Constantinople, Delhi, and to the Mussulmans beyond the Indus, and also corresponded with Napoleon in Egypt.

In 1796, his intrigues and movements caused a concentration of troops to be made in India, but no expedition to Mysore was undertaken, and, in 1797, his doings gave rise to such alarm, that it was not considered advisable to despatch a force out of India, which had been detailed for an expedition to Manila.

The greater part of this force had, however, embarked by the 27th August this year, and the following troops from Madras had arrived at Penang before they could be recalled, viz. :—

Detachment Royal Artillery; 6 companies (461) 12th Regiment; 74th Regiment (680), besides a total of 1355, made up from the 2nd Battalion 5th and 33rd Regiments. Penang had been an English settlement not more than ten years, and was inhabited chiefly by Malays and Chinese.

On the 27th August, the four remaining companies of the 12th and the remaining R. A. embarked on board His Majesty's Ships "Suffolk," and "Arrogant," but did not sail.[1]

A large fleet was now assembled at Penang, with a considerable force of troops from the Bengal, Bombay, and Madras Presidencies, ostensibly for an expedition against the Philippine Islands, until the order arrived, towards the end of September, for their immediate return. The six companies of the 12th sailed a few days after its receipt, and encountered such severe weather that the transports, finding it impossible to contend against the north-west monsoon, were compelled to return to the island in October.

In the meantime, the 4 companies at Madras had been directed to furnish a detail of 50 men, together with those who had been trained to the Gun Exercise, to embark, at gun fire on the 20th September, to act as marines on board the "Britannia."[2]

On the 15th November, the 6 companies at Penang again sailed for Madras, and landed at Fort St. George on the 12th December; during their absence, the remaining 4 companies of the regiment had exchanged a few shots with a French Squadron, which had appeared in the Madras Roads, and succeeded in driving an Indiaman on shore, under the works of the fort.

1798

Early in January, there was a most dangerous and secret plot to mutiny existing amongst the East India Company's troops, and, on the 18th, a Court of Enquiry assembled at St. Thomas's Mount, Madras, to ascertain the causes which had led to it in the 1st Battalion of Artillery, and whether there were grounds for believing that any combination existed between the Artillery and the King's regiments in the neighbourhood. The 6th item in the finding of the Court was :—

> "That no imputation whatever rested upon H.M's 12th, 73rd, and 74th Regiments, as being connected in any way with the disaffected men at The Mount."

On relief by the 52nd Regiment, at Fort St. George, the 12th marched, on the 28th January, to Tanjore, the capital of a well-cultivated province in the Carnatic, where it arrived on the 1st March, the flank companies being detached under Major John Picton to Fort Vellum, 10 miles off, a wild and cheerless place, situated on a large sandy plain, the few rocks in the neighbourhood being infested with large snakes of the most dangerous

[1] General Order, Fort St. George, dated 26th August, 1797.
[2] *Ibid.*, dated 19th September, 1797.

description.[1] Three times a week, the eight companies at Tanjore met this detachment, on a fine open plain, equidistant from the two stations, for drill and manœuvres ; an arduous duty which could not last long, consisting as it did, in five miles march to the drill ground, two hours' incessant evolution, and five miles back, under the fierce rays of a tropical sun. Many men died from sunstroke, and, in hospital, there were 400 afflicted with dysentery and other severe complaints. After the deaths of two officers and upwards of 100 men, either from climate or excessive fatigue, the Government issued a peremptory order preventing any future meeting of the separated portions of the corps.[2]

On the 4th May, at Tanjore, the regiment was inspected by Major-General Floyd, who expressed, in orders, to Colonel Aston, his officers and men, the satisfaction he felt on inspecting the eight companies of the 12th Regiment at that station, and added :—

> "In the masterly hands of their commanding officer, there is every reason to expect that His Majesty's 12th Regiment of Infantry, will, whenever they are called upon, be ready and disposed to renew in the East the glories of Minden and Gibraltar."

Tippoo's army, this year, was reorganised and officered by selected men, all Mussulmans, and the position became more acute.

To strike an effectual blow at the naval, commercial, and colonial greatness of the British had long been an object of primary importance with France, and it was well known that Tippoo was tendering to the French offers to enlist, in his own service, any men they might send him ; also, that French troops, having landed at Mangalore, had proceeded to Seringapatam.

Lord Mornington, Governor-General of India, accordingly issued orders, on the 20th June, to the Governor of Bombay to assemble a force on the coast of Malabar, and to Lieut.-General Harris, commanding in Coromandel, to concentrate his scattered army, whilst efforts were also made to induce the Nizam of Hyderabad and the Peshwa of the Mahrattas to give auxiliary aid.

A War Office Circular, dated 14th July, directed that the regiment was now to consist of ten companies, each of 120 rank and file, for service abroad, and one company to be formed at home for recruiting service, the latter mustering 1 captain, 2 lieutenants, 1 ensign, 8 sergeants, 8 corporals, and 4 drummers, without any private men.[3]

On the 22nd July, the 12th marched from the fortress of Tanjore, *en route* to join the army assembling under the command of Lieut.-General Harris, and arrived during the rainy season at the Fort of Arnee, where it was destined to lose its colonel.

He had proceeded on leave to Madras, on private affairs, and during his absence had received a note from Lieutenant Hartley of the regiment complaining of the Officiating Paymaster, Major Allen, relative to some

[1] Captain Eler's *Memoirs*, pp. 75–6. [2] Colonel Bayly, p. 58.
[3] *W. O.* 4, Book 299, p. 431.

pecuniary transaction, in reply to which the Colonel wrote that Allen was an " illiberal fellow," or words to that effect. There is no doubt that when Colonel Aston made use of this term he never imagined that Hartley would abuse his confidence in reading his letter to different officers of the regiment, which, however, he did, and this coming to the ears of Major Allen, a man of a high sense of honour, he immediately consulted his friend Major Picton, who at once issued an order for all the officers to assemble at his quarters, to investigate the difference between Major Allen and Lieutenant Hartley. The result of the Enquiry proved that Major Allen's accounts with Lieutenant Hartley were perfectly clear, and that Major Allen did not merit the observation made by Colonel Aston, on Lieutenant Hartley's representation. The minutes of this court of Enquiry were forwarded to Colonel Aston, who resolved on rejoining his regiment to issue a severe order against Major Picton for presuming to take advantage of his temporary absence, in convening a meeting of the officers without his sanction. It appears, however, that prior to the colonel's return in November, Major Picton had written to him, and the letter was returned from Madras by the Colonel in a blank cover. Irritated by this contemptuous mode of treatment, and deeming it a palpable insult, Major Picton resolved on seeking satisfaction for the affront, and, on the colonel's return, sent him a challenge. They met, when Colonel Aston was attended by Captain Craigie as his second, and Major Picton by Lieutenant Crawford. The distance of eight paces was regularly measured, and the signal given, when Major Picton's pistol missed fire, and he threw it on the ground in a rage. The Colonel told him to " try again," but this the seconds, very properly, would not allow, whereupon Colonel Aston fired in the air. No apology was made, and the respective parties returned to their quarters.

According to established etiquette in the then reigning system of duelling, Picton would not have been justified in firing a second time, and retired very ill-satisfied with the result of the *rencontre*.

The next day was Sunday, when it was customary for captains of companies to attend on the commanding officer, with their respective company " states." Major Allen, who was a major by brevet, presented his at the breakfast-table, and remained until every officer had gone. When alone with Colonel Aston, he said :—" I wish to consult you, Colonel, about exchanging out of the regiment. From certain unpleasant circumstances that have lately occurred, I find my situation, after the reflections you have applied to me, not what it was." Colonel Aston replied :—" If you ask my opinion, I think, as senior captain, you would be wrong to quit the regiment ; and as to your feelings, I am ready to atone in any way you wish." " Will you then give me a meeting ?" demanded Allen. " Certainly, instantly, but allow me to say, you have been very tardy in demanding it, as I have been some days with the regiment, without assuming command of it, in order to give you and others who feel aggrieved, an opportunity of satisfying themselves."

The meeting was very shortly arranged, and the Colonel and Captain Craigie happened to arrive on the ground a few minutes before Major Allen and his second, an assistant-surgeon, by the name of Erskine. Major Allen apologised for keeping him waiting, adding :—" I am sorry, upon my soul, Colonel Aston, it should ever come to this." Colonel Aston merely said :—" Take your ground, sir." The distance was measured ; Allen fired, and from the circumstance of the Colonel standing perfectly upright, with his pistol levelled, the seconds concluded that the ball had passed him. The Colonel dropped his pistol arm and said, " I am wounded, but it shall never be said that the last act of my life was that of revenge." Then turning to his second (Captain Craigie) he faintly observed :—" Support me ; I am going to fall." The Colonel had fainted, and, in this state, was conveyed to his house. The despair of poor Allen was inexpressible ; he threw himself on the ground, and was quite overcome by sorrow and remorse, saying he had killed his best friend, nor was it ascertained what had become of him for many days after. Even Erskine, his second, was ignorant of the place of his seclusion.

Colonel Aston's wound was so deeply situated that probing was adjudged dangerous, but, from his cheerful demeanour, the most favourable anticipations were entertained of his ultimate recovery, until the seventh day, when, desiring to be assisted out of bed, he had scarcely touched the floor with his foot, when he fell dead on the bed. Three days after his decease, his remains were conveyed, with all respect and military pomp, from his house to the Fort, and interred in the little cemetery in the eastern quarter ; his own regiment, a native regiment, and a company of artillery attended. Minute guns were fired ; his beautiful Arab charger was hung with crape, and his boots pendant from the holsters. A handsome monument was erected to his memory, which, some years later, was most generously repaired at his own expense, by Major Munro of the East India Company's service. The unfortunate Major Allen had shut himself up in his quarters and fastened all the doors and windows, everyone supposing, from the deserted appearance of the rooms, that he had escaped to Madras. He, however, joined the funeral procession of his lamented friend, and never was such a change in a man produced in so short a time ; pale and emaciated, and so weak that he was scarcely able to follow to the grave. His head servant reported that the Major had never touched an atom of food since the fatal event. He only asked for water, continually repeating the expression :— " Oh, God ! I have destroyed my dearest and best friend." Majors Picton and Allen were both placed in arrest, sent to Madras, and tried by court-martial. They were acquitted, and Allen was tried by the Civil Court and acquitted. They both returned to duty with the regiment, but poor Allen was never seen to smile again, and died in less than three months of a raging fever, being the first officer who fell a victim to climate, at the commencement of the siege of Seringapatam.

With reference to this duel, the following autograph letter from His Royal Highness, the Duke of York, Commander-in-Chief at this period, is inserted by the courtesy of a descendant of Colonel Aston's.

"From H. R. H. the Duke of York to Mr. Windham :—

"Horse Guards, June 7th, 1799.

"DEAR SIR,

I return many thanks for the narrative you were so good as to send me. It clearly proves Colonel Aston (and, of that indeed I had no doubt before) to have been perfectly right, and his whole conduct throughout the melancholy affair, does infinite credit, both to his head and heart.

This must surely be great consolation to his family.

Believe me &c.

(Signed) Frederick."

CAPTAIN GEORGE ELERS.
(Lieutenant 1796, to Captain 1803.)

CHAPTER VII

Siege of Seringapatam, East Indies, Isle of Bourbon, Mauritius, Ireland, England, France.

1799–1817

1799

The following appears in " The Gentleman's Magazine " for January :—

> " *January* 1*st. Deaths.* At Dublin, James Bellairs Esqre. . . . late Second Major of the 12th Regiment of Foot. During a period of 18 years, few officers have gone through more severe service, and none with more bravery or fortitude."

On the 1st January, the regiment joined the camp of the army advancing towards Mysore.

This army, consisting of nearly 21,000 of all arms, had assembled near Vellore, under the command of General Harris, the Commander-in-Chief, and was composed as follows :—

Two Cavalry Brigades, with artillery mustering 60 guns and 40 battering cannon ; the Infantry, in two wings, were formed into 6 brigades of 3 regiments in each, and there were also 1000 of the Madras Pioneers.

The 12th Regiment (693) was in the 1st Brigade, commanded by Major-General Baird, with the 74th and 94th Regiments.

Colonel Bayly gives a description of a subaltern's requirements at this period for six months' field service, which may convey a pretty correct idea of the number of the followers of an army of 30,000 men at that period. He had two bullocks laden with biscuits, two with wine and brandy, two with his trunks, and four for the marquee, and, in addition, a dubash (head servant), maty boy, and six coolies, to transport his couch, chairs, and various other belongings. He was thus accompanied by ten bullocks and eight servants, the majority of whom were followed by every individual of their family—grandfathers, grandmothers, uncles, aunts, nephews, nieces, with whole generations of children. This may appear an exaggerated statement, but (as he says) " no less extraordinary than true," every officer in the army having been then similarly encumbered, and sometimes even more so.

The army arrived at Ryacottah on the 1st March, and, on the 5th, a detachment, consisting of 1 troop native cavalry, the 1st Battn. 6th N.I., and the light companies of the 12th and 74th Regiments (the whole under command of Major Cuppage, 6th Regiment) took possession of the small hill forts of Neeldroog and Auchittydroog, which lay on the route, two other forts surrendering a few days later. The first of these was captured, without opposition, by the light company of the 12th, under Captain Woodhall; it was occupied by a company of Sepoys, and the light company returned to camp. On the 7th, the 12th Regiment, with numerous flank companies of other corps, marched all night, in order to surprise a large cavalry cantonment, fifteen miles off; the enemy, however, on obtaining intelligence of the design, made a hasty retreat.

Lieut.-Colonel Robt. Shaw was now temporarily transferred from the 74th Regiment to command the 12th, in succession to the late lamented Colonel Aston, and he immediately issued the following order:—

"Camp near Killamungahim.
8th March, 1799.

"As the 12th Regiment, from having the honour to be the oldest King's regiment with the army, is more liable to be called on for immediate service than other corps, the Commanding Officer expects the Officers, Non-Commissioned Officers, and Private Men, will be ready, night or day, to turn out at the shortest notice, and to parade under arms without noise or confusion. On all sudden alarms, the light company is instantly to accoutre without waiting for orders, and to be in readiness to march whenever their services may be required."

Nothing of importance occurred for a while except the surprise of one of our picquets of Sepoys, eighty of whom, in a night attack, were literally cut to pieces by the enemy's cavalry. The Sepoy officer was left among the heap of dead, a figure too horrible to describe. He had been wantonly and barbarously mutilated, with about thirty sabre wounds in his body, yet he recovered and returned to his native country, to linger on in a wretched existence.

At a place called Amboor, previous to our mounting the Ghauts, the army had been joined by a considerable number of the Nizam's forces—a disorderly set of savage, undisciplined barbarians, clothed in stuffed cotton and steel chain armour, but, at the same time, excellent horsemen.

They were certainly an addition to our numerical strength, but otherwise frequently in the way, in constantly dashing through the infantry columns at full gallop; and being often mistaken for the enemy's irregular horse were fired at accordingly.

As the march was continued, towns and villages were in flames in every direction, and several of our unfortunate camp followers, being entrapped at various times, by the enemy, were sent into camp with noses and ears cut off. We retaliated by hanging these barbarians whenever they were taken prisoners. Not an atom of forage was procurable, and every tank or reservoir of water was impregnated with a species of poison called milk-hedge, many men becoming dangerously ill, and several horses and bullocks falling victims to its deleterious effects.

On the 21st, a night march was made to secure two tanks which the enemy had begun to destroy. On the 24th, they were in position, with horse, foot, and guns, to dispute the passage of the Madur River, but withdrew on the advance of the British. On the 26th March, the army encamped at Malleville, and received information that Tippoo intended to attack them as they left the jungle country, through which, until then, the route had led.

Action at Malleville.

At about midday on the 27th March, as the Quarter-Master-General's Department were fixing the place of encampment, and the lascars commencing to pitch the tents, on an extensive plain of sand (a few hundred yards behind the mud fort of Malleville), they were suddenly interrupted by a heavy cannonade from a height, 1½ miles in front; at the same time, the advanced picquets, with two 12-pounder guns, under Captain M'Pherson, 12th Regiment, were briskly attacked, but bravely repulsed a vastly superior force. The right wing of the British army, arriving shortly after on the plain, immediately advanced in close column of regiments in the direction whence the cannonade proceeded, where large bodies of cavalry and infantry were plainly discernible. The enemy's rockets considerably annoyed the advancing columns of the British, but, on a nearer approach, they retired with their heavy guns, which had only done execution at the commencement of the action.

The British columns, when within a few hundred yards of the summit of the height, deployed, advancing steadily until they had passed it. When the enemy's immense army appeared a short distance in front, just opposite to the 12th Regiment, a large body of cavalry, formed in the shape of a wedge (and headed by two elephants with howdahs on their backs) were observed in the act of charging, but found it impossible to make any impression on H.M's 12th and 94th, who received them with the greatest steadiness, and by a close, continued, and well-directed fire, repulsed them with considerable loss.

The British line halted to receive the attack. Two other masses of cavalry were also discovered in topes or woods, just in their rear, consisting of ten or twelve thousand, preparing to support the first charge. The Commander-in-Chief, General Harris, sensible of the danger that menaced the regiment, placed himself in its rear, and frequently called out "Steady Twelfth," "Steady old Twelfth." The wedge of cavalry had now approached within a hundred yards of the line, discharging carbines and pistols, but without execution. The 12th Regiment had made ready, and was, of course, at the "recover," when General Harris gave the word "Fire." To the credit of their coolness and unexampled discipline be it recorded, that not a shot was fired, nor even a movement made that indicated indecision, so accustomed were they to the voice of their commanding officer, who allowed the enemy to approach within about thirty yards, when a volley was poured into the wedge of cavalry, followed by rapid and well directed file firing. A rampart of men and horses strewed the front of the 12th Regiment. The rear of the wedge, embarrassed by the heaps of slain, were actually incapable of continuing the charge, and the elephants, maddened by pain, were making off, swinging their chains in the midst of the cavalry. The howdahs were smashed to atoms, the leading chiefs, who had directed the charge, falling headlong from the backs of the enraged animals. The elephants now becoming more excited from numerous wounds, and deprived of their conductors, turned all their fury on the Mysoreans, overturning all who opposed them in their retrograde movement. A few horsemen only cut through the regiment, but were instantly shot in its rear.

The defeat was complete, and Tippoo withdrew with all possible haste. After a short pursuit, the want of water not permitting our troops to encamp on the field of battle, the force returned to its original encampment in the vicinity of Malleville.

The 19th Dragoons, the 12th, 33rd, 74th, and 94th Regiments, which alone of His Majesty's Corps were engaged, were equally distinguished by their steadiness and gallantry.

The loss of the enemy was estimated at about 2000 killed and wounded, our casualties amounting to 66 men and 48 horses, killed, wounded and missing, of which there were none in the 12th Regiment.

In General Orders issued that evening, it was stated :—

> "The Commander-in-Chief congratulates the army on the happy result of this day's action, during which he had various opportunities of witnessing its gallantry, coolness, and attention to orders ;"

In Brigade Orders :—

> "Major-General Baird, with the most heartfelt satisfaction congratulates the Brigade on the victory obtained this day over the enemy ; it is sufficient for him to say that the valour of the corps fully answered his expectation."

Immediately after the action at Malleville, General Harris crossed the Cavery on the 29th and 30th March, and resuming his march on the 1st April, the army encamped on the 3rd within four miles of Seringapatam.[1] Here, escorted by the 12th, he reconnoitred the enemy's position from

[1] The proper name of that city is Siri Runga Patan.

the apex of some adjacent hills. This evening, the 12th Regiment, with the flank companies of the 74th and 94th, proceeded at 6 P.M. to surprise the enemy's cavalry encampment, but were out all night without any surprise being effected. At about 3 A.M., however, on the 4th, they suddenly came upon a body of Mysorean Cavalry, when they rushed forward and bayoneted nearly every man before the Mysoreans could mount their horses, which were led into the British camp at 6 o'clock, just as the army was about to commence its march.

General Harris, practically unmolested by the enemy since the battle of the 27th, encamped before Seringapatam on the 5th, about two miles from the south-west face of the fort, when the following General Order was issued :—

> "The Commander-in-Chief takes this opportunity of noticing the high sense he has of the general exertion of the troops throughout the long and tedious march, with the largest encampment ever known to move with any army in India ; and, in congratulating them on a sight of Seringapatam, he has every confidence that a continuance of the same exertions will very shortly put an end to their labours, and place the British colours on its walls ! "

TAKING OF SHAW'S POST.

On the evening of the 5th April, two separate bodies of troops were ordered to parade at 6 o'clock, one for the purpose of taking possession of the dry bed of a nullah, or watercourse ; the other, to occupy a wood at some distance on the right of the river, but it was not known beforehand that both these positions were completely occupied by strong and select columns of Tippoo's troops.

The 12th Regiment was detailed for the first of these enterprises, and, with two battalions of Madras Sepoys, proceeded, under Lieut.-Colonel Shaw (12th), at sunset, to attack a post occupied by the enemy in a ruined village, about 2000 yards from the fort, and in front of our left.[1]

Scarcely had this force cleared the outposts, when thousands of rockets and fire-balls appeared in the air, followed by long peals of musketry, as if from an extensive line of infantry. The night was exceedingly dark, but the numerous fire-balls, thrown by the enemy, afforded sufficient light to discover the approaching British force, on whom a well-directed fire of rockets and musketry was showered without intermission, and with considerable effect. Throughout the night, the 12th did not fire a shot, an order having been given that all must be done with the bayonet. Scarcely had this order been conveyed through the ranks, when an increased and tremendous peal of musketry for several minutes was distinctly heard from the wood on our right, a certain indication that our other advancing body of troops was also seriously opposed. This soon ceased, but immediately afterwards the rear of our right flank was turned, from whence

[1] Colonel Wilson's *History of the Madras Army*, Vol. ii. pp. 319–20.

the enemy poured in deadly volleys of musketry. Thus situated, it became a paramount object to shelter, and an order was given to lie down. The enemy supposing from our recumbent posture, which was plainly exposed by the light of the fire-balls, that the majority were annihilated, a heavy column of their troops ventured a desperate attack at the point of the bayonet, and actually drove our Sepoys in confusion on to the 12th, killing their commandant, and wounding many officers. As soon as the regiment was separated from the flying Sepoys, who scampered pell-mell over our prostrate line, the command "Up 12th, and charge" was obeyed with alacrity, and the regiment plunged headlong into the ranks of the swarming foe, springing on them like lions. The effect was magical; they were seized with a general panic and scoured over the plain much more rapidly than they had advanced, and were scattered in all directions. The murderous rockets and musketry still showering from the other quarters, we were soon compelled to resume our prostrate manœuvre. At about 2 A.M. the enemy's fire ceased, and Colonel Shaw resolved to await the dawn of day. His object was to take possession of the nullah, about 1½ miles in front of the encampment.

The attack was renewed at daylight on the 6th, by the same troops, reinforced by the 94th and two Madras battalions.[1] At 4 A.M., the 12th moved forward, and at daybreak found themselves in rear of a long mud wall, and fragments of a ruined village, three hundred yards from the nullah, which was occupied by thousands of Mysoreans, with a body of French troops, supported by large masses of infantry on both sides. Thus outflanked, mounds of earth were instantly thrown up to secure the men from enfilade. In executing this work, the force had only been partially discovered, but broad daylight brought with it incessant showers of musketry. Guns from the fort also played on the same spot, but being upwards of 1½ miles from it, very few shots took effect.

At length, the 12th Regiment, supported by the Sepoys, dashed into the nullah, when there was not a single idle bayonet; oaths, shouts, and carnage presented a terrible scene of human ferocity, and never did men more heroically perform their duty. The conflict was excessively murderous and obstinate, as Tippoo's troops were brave, numerous, and well disciplined. On the enemy being driven from the nullah, they formed on the other side of the high bank, and made a show of retaking the post, but, from the smart file firing of the 12th, and a battalion of Sepoys, they were compelled to retire. The post thus captured was called "Shaw's Post" in honour of the commanding officer of the 12th, and became a landmark in the campaign.

We had scarcely taken possession of it, and sheltered under the embankment from the guns of the fortress, when the enemy in the wood

[1] Colonel Wilson's *History of the Madras Army*, Vol. ii. p. 331.

delivered a heavy fire on the position, and poured in volleys of musketry, which would have expelled our force from this exposed situation, had not a brigade arrived from camp in time to occupy the wood, which was accomplished with little or no loss, the enemy quietly abandoning it on the approach of the British, and the 12th had the honour of making the first direct attack on the strong fortress of Seringapatam.

From Shaw's Post, the enemy retreating from the wood were saluted with a sharp fire on their flank, when they lost one of the bravest chiefs of their army. This evening (6th April) the 12th were relieved in the captured position by the 74th Regt., and were heartily glad to rejoin the encampment after 24 hours' hard fighting, fatigue, and fasting. In this brilliant affair, 11 officers and 180 men were killed and wounded, the casualties in the 12th being Lieutenants G. Nixon, T. Falla, and 10 men killed; Captain Whittle, Lieutenants R. Nixon, Percival, King, Ensign Neville, and several men wounded. Lieutenant Falla, who died of his wounds, had been struck by a 26-lb. wrought iron shot, in front of the right hip, which broke the thigh, and lodged between the bones.

Storming of Seringapatam.

The army halted in the position taken up on the 5th April, and entrenched, preparatory to commencing the actual siege of the fortress. The position was uncommonly strong against counter-attack. The left rested on the Cavery,[1] the right on a high commanding knoll sloping towards the river, and in front was a winding stream covered by a chain of advanced posts. Within the position was an ample supply of good water, and large quantities of excellent materials for siege works.

On the 5th and 6th April, operations began by determined attacks on the enemy's outposts on the west side of the river. These were successful, and the army took up a final position for the siege on the 7th. A detachment was also sent towards Periapatam, to effect a junction with the Bombay army. This army eventually arrived before Seringapatam on the 16th, and crossing the Cavery, took up a position opposite the north face of the fortress.

The point of assault having now to be decided on, it was determined to attack a prominent bastion, forming an acute salient, and pointing up stream, where the Cavery divides. This point in the north-west bastion appearing to be the weakest in the fortress, it was decided to attack the western face at this spot, and to breach the curtain near the angle.

From the 11th April, the siege was advanced by regular and ordinary stages. No particular difficulties presented themselves, and a number of ravines and nullahs proved most favourably situated.

The batteries were constructed at night, and, by the 24th, the enemy's

[1] Sometimes written "Cauvery."

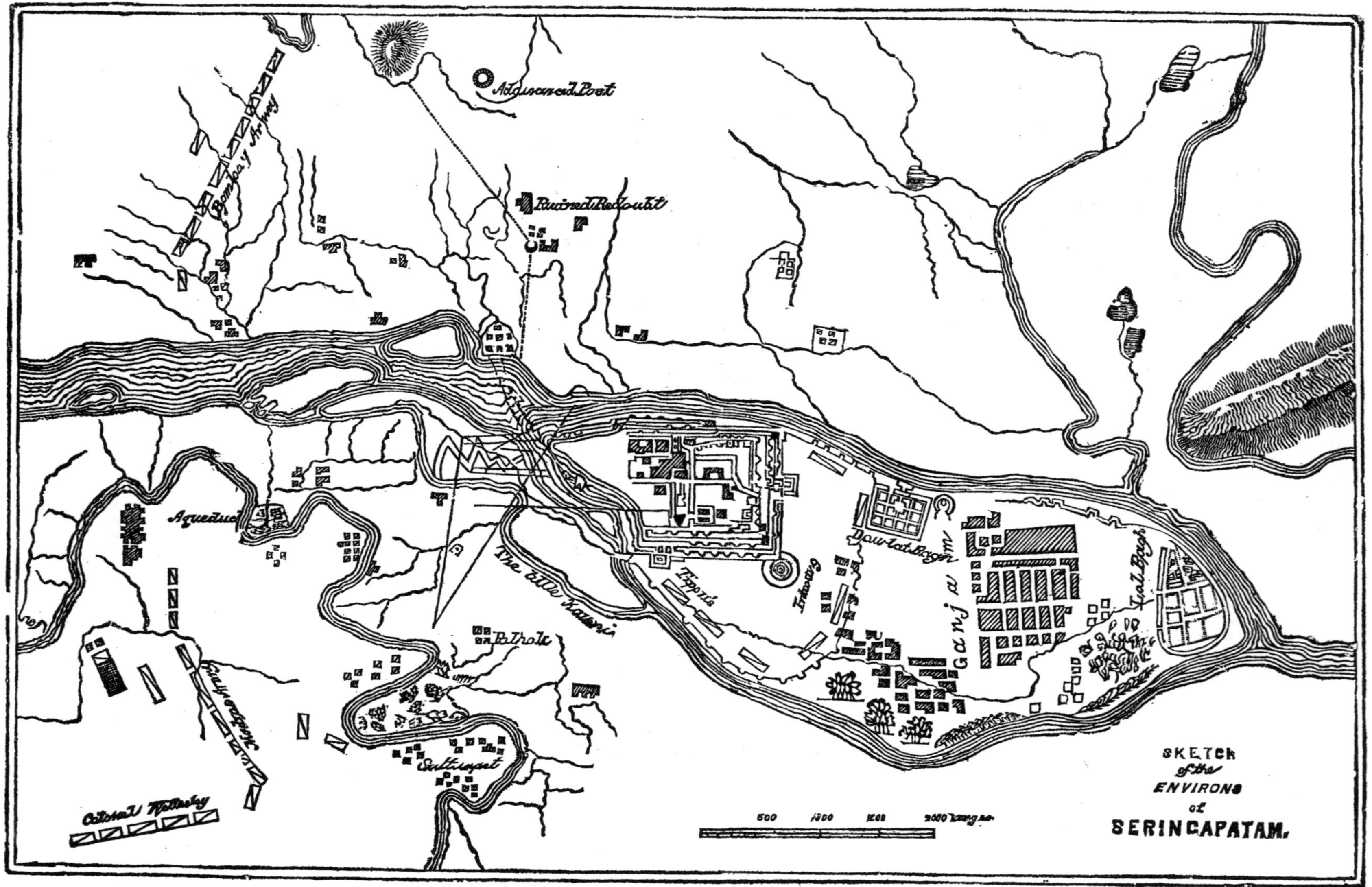
SKETCH
of the
ENVIRONS
of
SERINCAPATAM.
Advanced Post
Ruined Redoubt
Bombay Army
Aqueduct
The little Kaveri
Tippu's
Daulat Bagh
Ganjam
Lal Bagh
Sultanpet
Colonel Wellesley

guns on the western face were practically silenced. On the 28th, their last remaining foothold on the western bank was captured, and, with the exception of some cavalry still in the open country, the defenders were now all driven within the fortress. The breaching batteries were in position on the 1st May, and fire was kept up against the point to be breached until the 3rd, when it was reported practicable. The General decided to storm the place the next day, for his army was on the verge of starvation. There was only two days' rice left in camp.

By daylight on the 4th, troops were massed for the assault in the advanced trenches, the batteries keeping up a well-directed and incessant fire on the breach and remaining defences of the fort, which was warmly returned by the enemy till noon, when their fire slackened.

N.-W. Corner of Fort, near Breach, Seringapatam.

The following troops were warned for the assault :—

Six flank companies of European regiments from the Bombay Army (the 75th, 77th, and 103rd) ;

4 flank companies from the 94th, and the Swiss Regiment De Meuron ;

10 do. of Bengal, 8 of Madras, and 6 of Bombay Sepoys ;

H.M.'s 12th, 33rd, 73rd, and 74th Regiments ;

100 Madras Artillery, with a proportion of gun lascars ;

200 of the Nizam's Infantry ;

The corps of European and Native Pioneers.

The whole, under the immediate orders of Major-General Baird, consisted of nearly 2500 European, and 2000 Native troops.

Major-General Baird, who had solicited the command, divided this force into two separate columns, which, on mounting the breach, were to file off to the right and left. The right column was destined for the attack of the southern rampart. The forlorn hope of each attack consisted of a sergeant and 12 Europeans, who were followed by two subalterns' parties; that of the right was commanded by Lieutenant Hill, 74th, and, the other, of the left column, by Lieutenant Lawrence, 77th Regiment. Colonel Wellesley remained in the advanced trenches in command of the reserve, to support the troops in the assault, in case it should be necessary.

At 1 o'clock, the troops began to move from the trenches. The width and rocky channel of the Cavery, though it contained little water, its exposure to the fire of the fortress, the imperfection of the breach, the strength of the place, the numbers, courage, and skill of the defenders, constituted such an accumulation of difficulties that nothing less than unbounded

Outer Ramparts, North Face, Seringapatam.

confidence in the spirit and courage of his men could have inspired a prudent general with hopes of success. The troops descended into the bed of the river, and advanced, regardless of a tremendous fire, towards the opposite bank. In less than ten minutes from the time of issuing from the trenches, the British Colours were planted on the summit of the breach. Tippoo Sultan, though advised by the most judicious of his officers, had neglected to cut a trench, so as to insulate the angle of the post in which the breach had been effected; and the assailing parties, under Major-General Baird and Colonel Sherbrooke, taking the directions ordered, soon cleared the southern rampart. General Baird then halted his men at the eastern cavalier without having met with any serious opposition except near the Mysore Gate, where many more were killed and wounded, and was here joined by the other column.

The left attack met with more resistance, the traverses on the north rampart having been resolutely held until the defenders became exposed to a flanking fire from a detachment of the 12th Regiment, which had got on the inner rampart, and had advanced parallel with the main body of the column. With this assistance, Captain Lambton, who had assumed command *vice* Lieut.-Colonel Dunlop, disabled in the breach, forced the traverses one after another, and drove the enemy to the north-east angle of the fort, where, having perceived the near approach of the right column, they fell into confusion and great numbers were killed. Immediately after this, Captain Lambton joined General Baird near the Eastern Gate.

In the left attack, Tippoo was himself defending the traverses with the best and bravest of his troops.[1]

It was here that the gallantry of Captain Woodhall, commanding the light company of the 12th, became conspicuous.

He had rushed down a rugged confined pathway into the ditch, and, ascending the second or inner wall by an equally difficult road, he mounted to the summit, followed by his company. Before attaining a footing there, he had clasped a tuft of grass with his left hand, and was on the point of surmounting the difficulty, when a Mussulman, with a curved scimitar, made a stroke at his head, which completely cut the bearskin from his helmet without further injury. Woodhall retaliated, separating the calf of the fellow's leg from the bone. He fell, and the gallant officer was on the rampart in a moment, surrounded by a host of the enemy, whom, with the assistance of his company, he soon drove before him, thus relieving General Baird and his column on the outer wall from the destructive fire from the interior rampart, and thereby saving hundreds of lives. Tippoo, finding his troops fired on from the inner ramparts, hastened to the sallyport. Here Woodhall and his men were already in the interior of the town, and a sharp melée ensued. The gateway was filled to the top of the arch with dead and dying. The column under General Baird had pursued the flying enemy to the sallyport, and whilst Woodhall was bayoneting and firing in the front, the enemy were also attacked in the rear. The body of Tippoo was afterwards found amongst a promiscuous heap of slain. The fortress now became a wild scene of plunder and confusion. Captain Woodhall, with his company, had been ordered to extinguish the flames of some burning houses in the vicinity of the Grand Magazine, which, had it ignited, would have blown the whole garrison, friends and foes, into the air. He performed this arduous duty effectually, and, although first in the town, his company was the only part of the regiment who did not reap any pecuniary reward for such daring heroism. The rest of the troops had filled their muskets, caps, and pockets with zechins, pagodas, rupees, and ingots of gold.

Although all accounts concur in describing the resistance to the right column as having been much less vigorous than that opposed to the left

[1] Colonel Wilson's *History of the Madras Army*, Vol. ii.

attack, yet the casualties in the former somewhat exceeded those in the latter.

As soon as the ramparts were occupied, a detachment was sent to secure the palace, and protect the Sultan's family from insult. The 8th Madras N.I. were already formed in its front, and after some parley, conveyed through Major Allan, the Deputy-Quarter-Master-General of the army, General Baird (escorted by part of the 12th and 33rd Regiments, and the 2nd Battn. 9th Madras N. I.), was admitted into the palace by two sons of the Sultan, on his promise of their personal safety. Tippoo's body, having been found amidst a heap of slain, was removed to the palace, and identified by his family and servants. He, from all accounts, had fallen under the fire of a soldier of the light company of the 12th, after having been previously twice wounded. (See Appendix VII.)

Within two hours, the whole of the works were occupied by our troops, with the British Colours flying, and, on the enemy surrendering after the death of Tippoo, the capture of this important city and fortress was achieved. His cruelty had known no bounds, and it is recorded that twelve soldiers of the 33rd Regiment, who had been taken prisoners in the attack on the wood, were put to death in his presence, in a most barbarous manner, by having nails driven into their skulls.

During the night of the 4th, almost every house in the town was plundered, and it was not until the 6th that Colonel Wellesley (who later became Duke of Wellington), having been appointed to command in the fort, reported that plundering had been stopped, the fires extinguished, and that the inhabitants were returning to their homes. In the interim, several soldiers had been executed, and a number flogged, for looting.

In Vol. i. of the "Wellington Despatches," it is recorded that the 12th, 33rd, and 73rd Regiments having, on the 5th, been without food, and exposed to heavy rain, Colonel Wellesley, on the 6th, ordered extra biscuits and an extra dram to be supplied to them.

The force of the enemy amounted to about 21,800 men, of whom 13,750 regular infantry were in the fort, and the remainder in the entrenchments on the island. The loss of the latter was not accurately known, but has been computed at about 40 men killed and wounded, per diem, during the siege.

The General Return of casualties (European and Native) before Seringapatam, from the 4th April to 4th May 1799, shows a total of 300 killed, 1042 wounded, and 122 missing. (Of the last named, 100 were natives.) The losses of the 12th Regiment during the above period were 17 killed, 49 wounded, and 1 missing. Lieutenant M. Shawe (posted to the regiment in the early spring of 1799) was shot through the leg during the operations, and the following officers died of wounds, illness, and fatigue:—Major Allen, Captain Buckeridge, Lieutenant Percival, Ensign Walter Gahan, and Assistant-Surgeon Bacot.

Referring to the bravery of Captain Woodhall, of the light company of the 12th, one writer says :—

> "Had it not been for his noble attack, and consequent success, by which the rampart of the second wall was cleared of the enemy, the capture of Seringapatam would have been at least retarded, if not abandoned for that season, as, on the following day, torrents of rain fell among the distant mountains, inundating the river, and rendering it totally impassable ; nor was there any other possible mode of obtaining possession of the inner wall, except by this fortunate discovery of the pathway, which was carried under such very appalling circumstances. Fifty thousand dead bullocks polluted the British camp, rendering it completely pestiferous, and not a day's rice remained for the consumption of the troops, and had the rains filled the river one day sooner, the British must have burnt their camp equipage, destroyed their battering train, and been compelled to make a hasty and disastrous retreat, followed and harassed by at least 60,000 of the enemy's cavalry, during the whole march to the Carnatic, about 200 miles. Had this second wall remained in the enemy's possession twenty-four hours longer, a still more disastrous fate might have attended the army ; the elite of the force, who gained the outer ramparts, would probably have been made prisoners, and the remainder in camp exposed to almost inevitable destruction. It must therefore be justly inferred that the success of Captain Woodhall's heroism was the salvation of the British army.
>
> The poor fellow was not long fated to enjoy his triumph. He was promoted, by purchase, to a majority in the 12th Regiment in the year 1800 ; he then married, had one child, and died from dysentery in 1801. His body, and that of his infant, who died immediately after, were brought to England from Madras, and interred in Redcliffe Church, Bristol, his native place ; and there rest the remains of as brave a spirit as ever tenanted the frail structure of mortality—an honour to his profession, a credit to his city, and a glory to his country. It may be truly added, that he was a hero, *sans peur et sans reproche.*"

The value of the treasure and jewels, captured in the fort ofSeringapatam, was estimated at £1,143,216, and an order from the Governor-General in Council directed an immediate distribution of the above to the allied troops, in such proportion as conformable to the usage of the British service in cases of a similar nature, the ordnance and military stores being held in reserve, pending instructions from England.

Nine hundred and twenty-nine pieces of ordnance were found in the fort, of which 287 were mounted on the works. There was also a very large quantity of gunpowder, round shot, small arms, and military stores of different kinds.

Of the artillery, however, there were no fewer than 436 guns throwing shot under 5 lbs. Out of 373 brass guns, 202 were from Tippoo's own foundry, 77 were English, and the rest French, Dutch, and Spanish ; of the 466 iron guns, only 6 were from Tippoo's foundry, 260 having been of foreign, and 200 of English, make. Of 60 mortars and cohorns, 22

THE STORMING OF SERINGAPATAM.

1. Sergeant Graham, of the Forlorn Hope—killed.
2. Colonel Sherbrook leading the Right Attack.
3. Captain Cormick, Madras Pioneers—killed.
4. The Sally Port. (Where Sultan Tippoo fell.)
5. Major Wallace, 74th Regt.
6. Major Skelly, Scotch Brigade.
7. Major Craigie, 12th Regt.
8. Lieut.-General Baird, who commanded the attack.
9. Captain Grant, his aide-de-camp.
10. Captain Faulkner, aide-de-camp.
11. Captain Whittle, 12th Regt.
12. Captain Woodhall, 12th Regt.
13. Lieutenant Gawlor, 73rd Regt.
14. Lieutenant Prendergast, 74th Regt.—killed.
15. Captain Farquhar, 74th Regt.—killed.
16. Captain Owen, 77th Regt.—killed.
17. One of the Brass Tiger Guns at the Fort.
18. Tippoo's Tiger Grenadiers.
19. Lieutenant Harris, 74th Regt.
20. Lieutenant Brutton, 75th Regt.—wounded.
21. The Sirdar who wounded Col. Dunlop—killed.
22. Leader of the Tiger Men.
23. Highland Sergeant.
24. Captain Molle, Scotch Brigade.
25. Major Moneypenny, Scotch Brigade.

were Tippoo's, the rest English and foreign. The howitzers, 11 in number, had, with one exception, been cast in Seringapatam.

A few days after the storming, the sons of Tippoo, and most of the Sultan's principal officers, came in and surrendered.[1]

On the 5th May, 9800 of the enemy's dead were buried by the Provost Marshal.

On this date, it was stated in orders :—

> "The Commander-in-Chief congratulates the gallant army he has the honour to command on the conquest of yesterday ; the effects arising from the attainment of such an acquisition as far exceed the present limits of detail, as the unremitting zeal, labour, and unparalleled valour of the troops surpass the power of praise for services so incalculable in their consequences : he must consider the troops well entitled to the gratitude of their country. . . .
>
> On referring to the progress of the Siege, so many occasions have occurred for applause to the troops employed, that it is difficult to particularise individual merit, but the gallant manner in which Colonel Shaw, 12th, &c., &c., . . . conducted attacks entrusted to their guidance, on the several outworks of the enemy, deserves to be recorded."

On the 13th, the Army encamped on the island of Seringapatam.

On the 15th May, in General Orders by Government it was stated :—

> "The Right Hon'ble the Governor General in Council having this day received from the Commander-in-Chief of the allied army in the field, the official details of the glorious and decisive victory obtained at Seringapatam on the 4th May, offers his cordial thanks and sincere congratulations to the Commander-in-Chief, and all the officers and men composing the gallant army which achieved the capture of the capital of Mysore on that memorable day.
>
> His Lordship views with admiration the consummate judgment with which the assault was planned, the unequalled rapidity, animation, and skill with which it was executed, and the humanity which distinguished its success.
>
> Under the favour of Providence, the justice of our cause, and the established character of the army, had inspired an early confidence that the war in which we were engaged would be brought to a speedy, prosperous, and honourable issue ; but the events of the 4th May, while they even surpassed the sanguine expectations of the Governor General in Council, have raised the reputation of the British arms in India to a degree of splendour and glory unrivalled in the military history of this quarter of the globe, and seldom approached in any part of the world.
>
> The lustre of the victory can be equalled only by the substantial advantages which it promises to establish, by restoring the peace and safety of the British possessions in India, and a durable foundation of genuine security.

[1] Colonel Wilson's *History of the Madras Army*, Vol. ii. p. 331.

The Governor General in Council reflects with pride, satisfaction, and gratitude, that in this arduous crisis the spirit and exertions of our Indian army have kept pace with those of our countrymen at home ; and that in India, as in Europe, Great Britain has found in the malevolent designs of her enemies an increasing source of her own prosperity, fame, and power."

On the 17th May, at Seringapatam, the Commander-in-Chief published he following to the Army, from a letter from the Governor-General :—

"With the warmest sensation of admiration, affection, and attachment, I offer my cordial thanks, and zealous congratulations to you and all the officers and privates composing the gallant army, which has achieved this glorious and decisive victory, with a degree of energy, rapidity, and of skill, unparalleled in this quarter of the globe, and seldom equalled in any part of the world."

On the 26th May, the army crossed the River Cavery, and encamped near the French Rocks, about 3½ miles from Seringapatam, and, on the 28th, marched to Yarriagoranelly, 12 miles south of it.

On the 3rd June, the following General Order was published :—

"The Colours or Standards taken by the following corps from the enemy, during the late service, to be sent on without delay to the Adjutant-General, in order to their being lodged at Seringapatam until they can be forwarded to the Presidency.

By His Majesty's 12th Regiment—8 Colours.
,, ,, ,, 74th ,, —3 Colours."

On the 12th, Colonel Shaw, commanding the regiment, accompanied by a battalion of Sepoys, and a heavy train of artillery, marched from Seringapatam, and proceeded towards the strong fortress of Ghooty, which still held out for the Mysore chieftains, many of whom had assembled there, on the dispersion or disbanding of their army. After about 100 miles' march in that direction, and whilst encamped at a place called Sera, information was received that General Harris, then on a tour of survey, was surrounded by a large body of Mysorean cavalry, in a small mud fort, about forty miles off. The light company of the 12th, with every available officer mounted, numbering thirty, instantly proceeded to the rescue. The banditti had, however, obtained information of their approach, and silently retired, thus relieving, from his dangerous position, the besieged general, who soon met the relief force with cordial expressions of approbation for their prompt succour; the party reached their camp on the following day, complete in all respects, after a march of 80 miles in two successive days, a remarkable occurrence in a tropical climate.

On the 18th June, the captive princes, with their families, left Seringapatam for the fort at Vellore, which had been prepared for their reception. Their escort consisted of two troops of native cavalry, a detachment of the Madras European Regiment, and the 2nd Battn. 5th N. I. The remainder of the families of Hyder and Tippoo were removed to the same place shortly afterwards.

The headquarters of the army left the neighbourhood of Seringapatam on the 10th July, reached Chittledroog on the 24th, and, early in August, advanced to Hurryhur, on the east bank of the Toombuddra, a fort which had surrendered on the 10th July.

Meanwhile, the 12th Regiment was proceeding on its march, over sandy plains, through dense jungles, and various nullahs and rapid rivers, until finally encamping in a wood about fourteen miles from Ghooty; here a courier brought despatches announcing that Colonel Gowdie had succeeded, with a small force, in compelling the Kellidar of the fort to capitulate, and consequently the aid of Colonel Shaw's force was not required. The Kellidar had refused to give it up until the 9th August, when a practicable breach had been made. The regiment, on marching from Seringapatam, had taken with it all the Colours captured by the army during the war, which Colonel Wellesley ordered to be delivered to the commanding officer at Sera.

In September, the regiment arrived at Bangalore, and encamped until the 14th October, when the march was resumed towards the fortress of Seringapatam, which it occupied on the 23rd, and remained so quartered until the 17th November, when an order was issued to march as soon as the necessary carriage and baggage animals could be supplied; three brigades of 6-pounders, with their proportion of tumbrils, were also to be in readiness to accompany the corps to Vellore in the Carnatic, which was reached on the 8th December. Advancing on the 10th to Arcot, the regiment proceeded to Wallajabad to be quartered, a station surrounded by an immense extent of paddy or rice fields, and where the finest snipe shooting was to be had.

By order of the Court of Directors, Madras, a medal for the capture of Seringapatam (in commemoration of the brilliant success of the British arms in Mysore, in 1799) was struck for distribution amongst the officers and soldiers (European and Native) employed on that occasion, which is described as follows :—

Obverse.—The storming of the breach of Seringapatam, from an actual drawing on the spot, with the meridian sun denoting the time of the storm, and the following inscription, in Persian, beneath: "The Fort of Seringapatam, the Gift of God, 4th May, 1799."

Reverse.—The British Lion subduing the Tiger, the emblem of the late Tippoo Sultan's Government, with the period when it was effected, and the following words in Arabic, on the banner: "Assad-oollah-ul-Ghalib," signifying "Lion of God, the Conqueror." Size 1·9 inches. (See Plate 3.)

OFFICER'S BREASTPLATE (*exact size*).
1799–1815.

OFFICER'S SHAKO PLATE (*exact size*).
1815–28.

1825–42. OFFICERS' BREASTPLATES 1843–55.
(*reduced sizes*).

Of these medals, gold ones were struck for His Majesty the King, the Right Honourable Lord Melville, the Governors in India at the time, the Marquis Cornwallis, the Nizam and his two ministers, the Peshwa and his minister, the Nabobs of Arcot and Oudh, and the Rajahs of Tanjore, Travancore, Mysore, Coorg and Berar, Dowlut Row Scindiah, the Commander-in-Chief, General Officers on the staff, employed on the service, and the Oriental Museum.

Silver-gilt medals for the members of Council at the three Presidencies, the Residents of Hyderabad and Poona, the Field Officers, and the General Staff on the service.

Silver for the Captains and Subalterns on the service.

Copper bronze for the Non-commissioned Officers, and pure gravin tin for the Privates.

The medals were not received at Madras until January, 1811, when orders were issued for their distribution to survivors, whether effective or otherwise, and also to the heirs of deceased persons, who had been entitled to them.

The following is a list of officers in the 12th Regiment for the year 1799, showing casualties during the operations in Mysore.

Colonel Wm. Picton.
Lieut.-Colonel James Taylor.
Majors: John Picton.
Thomas Craigie.
Captains: Jas. Allen (Brevet Major. Died from illness during the operations.)
Joseph Moore.
Thos. Woodhall.
Thos. James.
Jno. Wilson.
Wm. Whittle. (Wounded during the operations, 6th April.)
Keith Young.
Mark D. Buckeridge. (Died from illness during the operations.)
Captain-Lieutenant Wm. O'Brien.
Lieutenants: Jonas Crawford.
Robert Nixon. (Wounded during the operations, 6th April.)
Nicholas Eustace.
Patrick Moyna.
John Parker.
George Nixon. (Killed in action, 6th April, 1799.)
Wm. Erlam.
Hon. Jno. Meade.
Saml. Percival. (Died from wounds received in the action of 6th April.)
Wm. Morris.
Richd. Ashton.
Jas. Cavendish.
Matthew Price.
George Elers.
Wm. Gahan.
Jas. Seaton.
Wm. Langford.
Richard Bayly.
Thos. Hartley.
Chas. King. (Wounded during the operations, 6th April.)

Lieutenants: Wm. Firth.
Robert Hen. Sale.
Thos. Falla. (Died from wound received in action, 6th April.)
Thos. Wm. Edwards.
Ensigns: Walter Gahan. (Died from illness during the operations.)
Henry M'Keady.
Chas. Rist.
Wm. Lewis Hurford.
Michael Grace.
John Rist.
Edmund Nevill. (Wounded during the operations, 6th April.)
Thos. Tarrens Morgan.
Henry Kater.
Adjutant Wm. Langford.
Quarter-Master Alexr. Stewart.
Assistant-Surgeons: Robt. Erskine.
Henry F. Bacot. (Died from illness during the operations.)

In the churchyard of Seringapatam will be found the following inscriptions to Officers who died during the time it was garrisoned by British troops:—

SACRED
TO
THE MEMORY
OF
CHARLES RIST
LIEUTENANT H.M. 12TH
REGT. OF FOOT
WHO DEPARTED THIS LIFE
8TH JANUARY, 1807,
AGED 28 YEARS

CHAS KING JENKINS, ESQRE.
PAYMASTER, H.M. 12TH REGIMENT
WHO DEPARTED THIS LIFE
3RD DECEMBER, 1805
AGED 69 YEARS.

ALEXANDER DALZELL CAMPBELL ESQ.
SURGEON IN
H.M. 12TH REGIMENT
WHO DEPARTED THIS
LIFE ON THE 25TH DAY
OF AUGUST, 1805.
AGED 20 YEARS.

Outside the Fort at Bangalore is a monument with the following inscription:

THIS MONUMENT
WAS
ERECTED BY THE OFFICERS OF
H.M. 12TH & 74TH REGTS.
TO THE MEMORY OF THE UNDERMENTIONED
OFFICERS WHO WERE KILLED OR DIED
DURING THE SIEGE OF SERINGAPATAM.

H.M. 12TH FOOT
LIEUT. GEORGE NIXON,
KILLED
5TH APRIL, 1799.

LIEUT. THOMAS FALLA,
KILLED
6th APRIL, 1799.

Drawing from a photograph by Captain G. H. Walford, Suffolk Regt.

MAJOR JAMES ALLEN,
DIED
22ND APRIL, 1799.

ENSIGN WALTER GAHAN,
DIED
7TH MAY, 1799.

Then follow the names of 5 officers of the 74th Regt., killed between the 25th April and the 4th May, 1799.

1800

At a levée held by the King, at St. James's Palace, on the 23rd January, Lieutenant Harris, of the 74th, was introduced by the Adjutant-General, and had the honour to present to His Majesty the 11 Colours taken from the Sultan Tippoo at Seringapatam.

Six of them were deposited in the Royal Hospital at Chelsea, where their remnants can still be seen, but it is not known what became of the other five.

The 12th having captured 8 out of the 11, it is clear that of the six here shown, at least three must have been taken by the regiment, but no evidence is obtainable as to which they were, whilst the possibility exists that all six may have been amongst the eight captured by the corps.

The spear-heads of the large elephant standards (Plate 15) of silver-gilt, measuring 1 foot, are handsome and curious, having on them inscriptions from the Koran : "The Battle Array. There is no God but God. Mahomed is the Prophet of God, Victory from God, and a speedy victory" &c. These large standards (one measuring 13 by 11½ feet, and the other, 12 by 9 feet, next the pole) are made of cotton with gold ornaments. The innermost part of No. 31 (torn in the centre) consists of gold ornaments on a pink ground, and similar ornaments outside it on a light green ground, the whole edged with a yellow border, the staff having green and yellow stripes alternately. The centre triangle of No. 32 consists of a gold star and ornaments on a dark blue ground, surrounded by similar ornaments on a light blue ground, and both inner and outer parts edged with a yellow border.

The two pendants of sunflower device (Nos. 29 and 33) consist of yellow stars and ornaments on a crimson ground, and gold fringe.

The flags numbered 20 and 28 were evidently companions, as both have silver spear-heads in the shape of a crescent, inside of which are a sword and sprigs of laurel, and a portion of one is missing. Being French tricolours, the combination of the Mahomedan Crescent and the tricolour is exceptional. (See Plate 15.)

In a Regimental Order, dated Wallajahbad, 27th March, Lieut.-Colonel Shaw notified a free issue of two drams of arrack to each of the non-commissioned officers and men of the regiment to celebrate the anniversary of "the first opportunity he had of observing that attention to orders, steadiness, and discipline, for which he trusts the 12th Regiment will, as heretofore, ever be distinguished when in the face of an enemy."

Duelling, which was too often resorted to on the most frivolous occasions, had not yet died out of the army, and, whilst at this station, Lieutenant Bayly and Captain Crawford, both of the 12th, met in one of these encounters. It is related that the former fired without aim, and, of course, wide of the mark, and the pistol of the latter flashed in the pan, resulting in apologies on both sides for supposed intemperate conduct.

On the 3rd May, the regiment marched from Wallajahbad, and embarked on the 23rd at Marmalong for Poonamallee, where it encamped. Whilst here, Colonel Shaw was transferred to his former corps, the 74th, and was succeeded in the command by Lieut.-Colonel G. W. R. Harcourt, who had been gazetted to the 12th on March 1st, 1799, on transfer from the 40th Regiment, which he had commanded in the West Indies.

Two companies of the regiment, under Captains Eustace and Ashton, embarked on frigates at Madras on the 28th June, and proceeded with other troops to Batavia, and three companies, under Major John Picton, were sent southward against some turbulent Polygar Rajahs.

On the 29th August, the remaining 5 companies and headquarters, under Colonel Harcourt, marched to an encampment near Seringapatam, and, arriving on the 15th November at Nandygherry, were joined by the 77th Regiment, some battalions of Native Infantry, and a proportion of Native Cavalry, forming a small force, under Colonel Pater, for the purpose of subduing the warlike tribes of the Wynaad country,—a mountainous district abounding with woods, and comprising about 1200 square miles, situated in the province of Malabar.

Colonel Pater's force marched from the encampment near Seringapatam, on the 26th December, to the Mysore country. During the march, and after entering this territory, an unfortunate dispute took place between Lieutenant Price and Captain R. Nixon of the 12th, which resulted in the former being tried by court-martial and cashiered. The sequel to it was that a Mr. Washington Price, an Assistant-Surgeon of the regiment (no relation to Lieutenant Price, just named) became so violent against Captain Nixon that a fierce quarrel ensued, some months after, at Seringapatam, which ended in a duel at Warriore, near Trichinopoly.

Captain Elers relates that, after marching about the Mysore country, the regiment returned to Seringapatam, prior to a force of about 5000 men being formed, in three brigades, under Colonel Stephenson, Madras Cavalry, to proceed to the Cotiote country against the Rajah of that name. Colonel Harcourt (12th) commanded the 1st Brigade, which consisted of the 77th Regiment and 5 companies of the 12th, with Captain Elers as Brigade Major. The Cotiote country was 60 to 80 miles distant from Seringapatam, and this force was despatched to avenge the barbarity of its Rajah, who, in 1797, had surprised and massacred a battalion of native infantry; no favourable opportunity for this having until now presented itself.

From the year 1800 to 1867, the following were the approximate rates of pay issued in the ranks, after deducting 6*d*. a day for rations :—Sergeant-major and quarter-master-sergeant, 1*s*. 6¾*d*. ; sergeant, 1*s*. 0¾*d*. ; corporal, 8¾*d*. ; drummer, 7¾*d*. ; private, 6*d*.

1801

Starting from Mysore Chattamully on the 1st January, the force, in a few marches, crossed the frontier of the enemy's country, and the roads being very bad, the pioneers were constantly at work, the progress of the march being very slow, not more than two miles an hour. A broad and rapid river ran on the left of the force, and, owing to the thickness of the jungle, it was not possible to see more than a few yards from the flanks.

On the 9th January, hostilities commenced, when, from the opposite banks of the river, the enemy opened fire from the tops of trees on the advanced guard, which was supported by the grenadiers of the 12th Regiment. They were so well concealed that only their firing was discernible, which was silenced by the discharge of some grape shot. The casualties in the attacking force were not above thirty. On arrival at the village of Manantoddy, the force, halting for three weeks, established a strong post, and built a formidable stockade, sending out parties to scour the country, and otherwise subjugate it. By the 8th March, the country had been traversed in nearly every direction, and, after four months on this service, the European soldiers became very sickly from this unhealthy district, where the miasma arising from rotting vegetation proved very prejudicial to their health. The province became completely subdued, and the Rajah, who had opposed the English for so many years, was taken prisoner, whereupon the 5 companies of the 12th employed on this service returned to Seringapatam on the 31st May.

In the meantime the 3 companies of the regiment, under Major John Picton, had, early in March, joined a force under command of Colonel Innes, operating in the hill tracts of Dindigul, against the Polygar chiefs, the most refractory of these being the Polygar of Verapatchy, two of whose strongholds were taken on the 21st, without loss on the British side, followed, on the 27th, by the capture of most of his horses, baggage, and elephants, he himself taking refuge with another chief, until the 4th May, when he was captured. Precautions had been adopted to cut off any attempted retreat, by an advance, on the 12th April, to Dhullee, which was taken next day with trifling loss, dispersing the Polygars and their adherents. Those who escaped fled to Jellipatty, where they made a last stand on the 22nd, after which no further opposition was attempted.

Colonel Innes and the troops were thanked by the Government, in an order, dated Fort St. George, 22nd May, 1801, in which the Governor in Council "viewed with the highest satisfaction the undaunted and persevering ardour, with which the difficulties opposed by the united force of the rebels, the severity of a fatiguing service, and of an unfavourable climate, had been surmounted."

The 3 companies, 12th Foot, with detachments of artillery, 5th Cavalry, 2nd Battn. 13th N. I. and pioneers, were further detailed under Colonel Innes, early in June, to co-operate with the troops in Shevagunga, in suppressing rebel chiefs, and, after continuous marching and fighting in a close and difficult country, the 12th Regiment, in an action on the 26th June, had 4 rank and file killed, Lieutenant Firth and 9 rank and file wounded, and Lieutenant Parker died of jungle fever. On the 29th the united detachments encamped at Serruvial, after having been fired on nearly the whole day, and on the 30th, the place was abandoned by the Polygars, after firing a few shots. The 3 companies of the 12th then marched to Vellore.

The dates in Lieutenant Bayly's diary are not very clear, but, it would appear that he, having arrived home in 1800, was this year appointed a Recruiting Staff Officer at Bath and Wells.

At the latter place, he had collected obout 200 substitutes for the Reserve Army, each of whom was to receive from £40 to £60 for his services. He relates that, as a very small proportion was allowed to the men at the time of enlistment, he consequently became possessed of several thousands of pounds, in bank bills, which he had carefully sewn in the lining of his regimental cap, considering it risky to confide it to country bankers. An appeal to the War Office, pointing out the dangerous responsibility to which he was subjected, failed to elicit any reply, so that the cap had to be worn or watched morning, noon, and night, and it was this vigilance that attracted the attention of an old rogue of a sergeant. The moment came when the cap was left unguarded, and the sergeant seized the bills and absconded. Bayly's horror, on discovering this was great.

His landlady said that she had seen a fat sergeant enter his room, and further enquiries told him that the same man had left by coach for Bath; so mounting a beautiful fleet young mare, a present from his father, he galloped thirteen miles in thirty-five minutes. The mare, on entering Bath, gave a sudden plunge forward, and fell dead at his feet. The object of his search was not there; he had quitted the coach a mile from Bath, and walked in the direction of the Bristol Road.

It was whilst arranging to continue the pursuit, that the Bristol coach rattled through the town for London, and there in the back seat was the sergeant. "Stop coach! Stop coach! A thief! A thief!" soon brought the miscreant down, when all the missing notes were found secreted in his neckcloth.

The sergeant was brought to a court-martial, broken, and flogged. The dead mare—valued at least at 100 guineas—was hard to replace.[1]

On the 16th October, the headquarters and 5 companies of the regiment marched out of Seringapatam, and encamped on the island. The march was continued on the 18th towards Trichinopoly, which was reached on the 14th November, the detachment proceeding on the 17th to Warriore, a large cantonment near the fort of Trichinopoly.

1802

In January, the remains of the 2 companies that had proceeded to Batavia rejoined the regiment, 3 officers and 63 men only having returned, out of 5 officers and 125 men who had embarked at Madras. The others fell victims to the baneful climate of the Island of Java. Lieutenants Gordon and Neville died of fever. The three-company detachment, under Major Picton, also rejoined here from Vellore, after their successful employment against the Polygar chiefs.

In the spring, the duel already referred to between Assistant-Surgeon Washington Price and Captain R. Nixon took place at Warriore.

When they met, Captain Nixon's shot took effect in the fleshy part of Price's hand, hit the stock of his pistol, and the ball, lodging near the top of

[1] Colonel Gardiner's *Centurions of a Century*, pp. 177–8.

the arm, was cut out directly. There the business ended. Captain Nixon soon after obtained leave and went home.

In May the regiment left Warriore, and moved to the barracks formerly occupied by the 19th Light Dragoons, on Trichinopoly Plain, and, there being no barracks for the officers, they had to live in detached bungalows wherever they could get them. The regiment had not been long in its new quarters when scarcely a night passed without robberies, for which Trichinopoly has ever been famous, committed by thieves in the hills, called Bhils. They stripped entirely and oiled themselves all over, so that if caught, they slipped through one's fingers like an eel, and so made their escape.

The regiment was inspected in May and October by Colonel Pater, who on each occasion expressed his entire approbation of its appearance and steadiness.

By War Office letter, dated 30th June, the following establishment for the 12th had been authorised, viz.:—10 companies, to consist of a total of 34 sergeants, 22 drummers, and 750 rank and file.

On the 1st July, an Armourer-Sergeant was added to the establishment of every regiment of cavalry and infantry.

1803

A War Office letter to General William Picton (Colonel of the Regiment), dated 11th February, 1803, appointed, for the first time, a paymaster, to be borne on the establishment from the 25th December, 1802.

Among the officers appointed this year to the Perthshire Militia was Major Thomas Craigie, who commanded the grenadier company, 12th Regiment, at the storming of Seringapatam in 1799.[1]

On the 19th May, 1803, the installation as a Knight of the Bath took place in King Henry VII's Chapel, Westminster Abbey, of General Sir Thomas Trigge, K.C.B., who had served in the 12th Regiment throughout the Siege of Gibraltar, and had distinguished himself in the grand sortie on the 26th November, 1781.

The Duke of Wellington was the last knight installed there, before the disuse of the chapel for this purpose,[2] and, up to that time, every knight had three squires as attendants.

Those of Sir Thomas Trigge were: Henry Garratt, James Waller, and Benjamin Bathurst. The chapel contains Sir T. Trigge's banner, and also his coat of arms on a brass plate, together with those of his three squires on brass plates. The General's motto was: "Dulcis pro patria labor." He was buried in Westminster Abbey. (See Plate 45.)

A General Order, dated 6th June, authorised that in future "each

[1] *History of Perthshire*, Marchioness of Tullibardine.
[2] This installation in the chapel was revived in 1913.

1800–06 (*reduced size*). OFFICERS' SHAKO PLATES. 1806–15 (*exact size*).

troop and company throughout the army, shall have an effective captain, and consequently the field officers of the several regiments of cavalry, foot guards, and infantry of the line shall no longer have troops or companies."

By the same General Order, the rank of Captain-Lieutenant was abolished.[1]

During August and September, an expedition was formed at Ganjam, for the purpose of taking possession of the province of Cuttack, then belonging to the Rajah of Berar.

The force was placed under command of Lieut.-Colonel Harcourt, 12th Regiment, and consisted of 200 of the 22nd Regiment, 50 Madras Cavalry, 30 Madras Artillery, and detachments of native infantry, mustering 2491. The troops left Ganjam on the 8th September, and occupied Juggernauth on the 18th without opposition. On the 24th, the advanced guard had several skirmishes with the enemy, and, on the 10th October, the town of Cuttack was entered without resistance, the force taking by storm, on the 14th, the adjoining fort of Barabutty, around which the country was very luxuriant, abounding in rich pasture land. After that, the greater part of the province submitted. The casualties during the operations, from the time of leaving Juggernauth, were small, viz. 6 killed and 47 wounded.

Lieut.-Colonel Harcourt, 12th, and his troops were thanked by the Governor-General in Council, in an order, dated at Fort William, on the 22nd October, 1803.

On the 18th December, 5 companies of the regiment encamped on Trichinopoly Plain, prepared for field service, under Lieut.-Colonel John Picton, and, marching on the following day, arrived near Golpettah on the 31st, *en route* to the Travancore country.

1804–06

Continuing the march on the 1st January, the regiment joined a considerable number of sepoys on the 3rd, and the force advanced under Major-General McDowell. Before the end of March, however, orders were issued to return and encamp on Trichinopoly Plain, and, on the 31st July, the regiment marched to Seringapatam, which was reached on the 25th August.

The battalion was inspected on the 25th March, 1806, by Major-General McDowell, who, in expressing to Lieut.-Colonel Picton his entire satisfaction at its appearance, directed that, as a mark of approbation, 2 drams of arrack were to be issued to each man present at the inspection.

Great mortality occurred in the regiment this year, owing to a severe epidemic of fever which broke out, involving a move to an encampment some distance from the fortress of Seringapatam.

[1] This rank had hitherto been conferred on the senior subaltern as lieutenant of the Colonel's company. He ranked as the junior captain, and had a prescriptive claim to the first vacancy which occurred amongst the captains.

1807

The epidemic still prevailing in the spring, with more than half the men in hospital, an order arrived from Madras, on the 17th April, to move to Cannanore on the Malabar Coast, where the regiment arrived on the 4th May, in such a deplorable state of sickness, that, on the march from Seringapatam, upwards of 400 men had to be conveyed in doolies.

A War Office Order, dated 18th September, directed that volunteers for the 12th Foot were to proceed to the Army Depot, Isle of Wight.

A Regimental Medal as here shown was this year instituted :—

Obverse.—The Castle and Key of Gibraltar; above it "12" and "Gibraltar"; below, the motto "Montis Insignia Calpe."
Reverse.—"A Reward for Military Merit, 1807."

By the courtesy of Major A. Murray, Polmaise Castle, Stirling.

A thin silver engraved medal (1·55 inches diameter) with ring for suspension. The colour of the ribbon worn with it cannot be ascertained.

1808

At Cannanore, the health of the men was restored, and, when inspected in 1808, by Colonel Cuppage, a district order was published, in which the colonel expressed

> "his thanks to Captain Eustace and the officers and men of His Majesty's 12th Regiment, for their handsome appearance at the review. The dress, steadiness, and general appearance of the men, marked the great attention paid to their discipline, and their uniform good conduct and friendly disposition towards the natives reflect every credit on the corps."

War in Travancore.

A dispute had arisen this year between Lieut.-Colonel Macaulay, the Resident, and the Rajah of Travancore, owing to the subsidy payable by the latter having fallen considerably in arrear, whereupon the Resident urged upon him, as a measure of economy, the expediency of dismissing a body of regular infantry in his service, known as the "Carnatic Brigade," but this proposal was not acceptable.

The Resident, believing that the Dewan, or prime minister, was the principal cause of the opposition, insisted upon his removal from office, which was agreed to, but, during the interval of selecting a successor, the minister employed himself in organising an insurrection of the Nairs, with a view to accomplishing the murder of the Resident, and he induced the minister of the Rajah of Cochin to join in the plot.

This design having become known at Madras, additional troops were immediately ordered into Travancore. H. M.'s 12th Regiment sailed from Cannanore for Quilon on the 26th December, followed, a few days later, by the 2nd Battn. 18th N.I., and a detachment of artillery (Europeans and Natives) with four 6-pounders and two 5½-inch howitzers, whilst the 1st Battn. 17th N. I. were to proceed to Cochin.

The battalion companies of the 12th embarked on twelve patamars (open boats of 40 to 100 tons), and the flank companies on an old country-built brig. With the exception of four boats, these were all leaky, and consequently unseaworthy; but in them the regiment had to proceed upwards of 300 miles, to Quilon. Owing also to a proportion of native cooks, lascars, and officers' servants having to accompany the troops, the boats were excessively crowded.

Rice, salt fish, and arrack (a native liquor) were the only provisions taken, and an unequal distribution of arrack (two or three boats only having been supplied with it) caused much distress throughout the voyage. Captain Bayly describes the situation of the troops as "distressing in the extreme," from the confinement to one position, being absolutely wedged together, without the means of any refreshing slumber, whilst *mal de mer* was not entirely absent; being also without any covering, exposed to the scorching rays of the sun by day, and the baneful and heavy dews of night, with the deadly land winds blowing off the shore, were enough to injure the most robust constitution in a few hours. It is recorded, however, that whilst labouring under the effects of such complicated miseries, these gallant sufferers refrained from the expression of the least complaint, except a little regret at the loss of the accustomed dram of arrack, which might have cheered their spirits amidst the evils they endured.

Four of these boats, with 4 companies of the regiment on board, reached Quilon on the 29th December, and were disembarked the same day, but with the greatest difficulty, as the small canoes employed for the purpose were only calculated to hold three or four soldiers at a time, and were managed awkwardly by the lascars who accompanied the regiment. The troops fortunately were unmolested and landed without opposition, which was somewhat surprising, as this proceeding was in direct violation of the treaty existing between the Rajah and the East India Company, which specified that the landing or marching of Europeans in the Travancore dominions would be considered an open declaration of war.

In the meantime, the house of the Resident (Colonel Macaulay) who was then at Cochin, had, on the previous night (28th), been surrounded by a number of armed men, and he had barely time to conceal himself, when they broke into the Residency. Not finding him, they left the place, and he got on board one of the vessels coming down the coast with troops (probably travelling with part of the 12th) which, most fortunately, happened to pass Cochin, about daybreak on the 29th December.

On the Europeans disembarking at Quilon, there was a general howl amongst the inhabitants, who abandoned their dwellings in every direction and quickly moved away. The detachment of the 12th at once joined the Sepoy force, under command of Colonel Chalmers, hitherto employed in the service of the Rajah of Travancore, but now threatened with destruction by his troops. Colonel Chalmers, that night, detailed 4 companies of his Sepoys, with a 6-pounder gun, to occupy a height commanding the Dewan's palace, in order to ascertain any hostile movements of his forces.

The party was attacked by the Travancoreans, and a sharp conflict ensued. Ensign Keappock and 40 men of the 12th were then ordered to support the Sepoys, and the enemy were forced to retire with the loss of 100 men, 60 of whom lay dead on the spot. The Travancoreans were excited to such rage and fury against the British, who had thus gained a footing in the heart of their country, that they assembled in immense multitudes before the camp, kept up an incessant fire on the picquet, and with heavy columns menaced the encampment; the soldiers were thus kept constantly ready for action, and they lay on their arms night and day.

On the 29th December, a tremendous gale of wind dispersed the patamars containing the remainder of the regiment in various directions, and some were wrecked off the coast of Cochin, but, by incredible exertions, the men were saved, whilst some of the boats, more fortunate, succeeded in entering the harbour.

One patamar, containing the second sergeant-major and 33 men anchored in the roads of Aleppi, a seaport belonging to the Rajah, unfortunately mistaking it for Quilon, which was thirty miles off. Canoes pushing off from the shore, they landed without hesitation or suspicion, being glad of relief from the miserable and dangerous confinement they had undergone, but only to find themselves soon overpowered and secured by the Travancoreans, who first cruelly broke their wrists, and then, tightly tying their arms behind them, and neck and knees together, plunged them headlong into a deep unwholesome dungeon. In this shocking condition they remained four days and nights, and, on the fifth morning, surrounded by an exultant populace, they were taken separately, in a deplorable state of exhaustion, to the Backwater, about three miles distant, where it was many fathoms deep; heavy stones being then attached to their necks, they were hurled into the water amidst the barbarous shouts and music of the remorseless natives.

The second sergeant-major (Tilsey) was the last victim of this awful tragedy ; he repeatedly called for a sword, that he might die like a soldier, but all in vain ; he was precipitated, in spite of cries and struggles, into the watery grave already shared by his unfortunate comrades. These particulars were communicated by a cook boy, who had accompanied the detachment, and had been an eye-witness of the whole inhuman transaction, having been himself similarly threatened. Another patamar, with nearly a whole company of the 12th on board, commanded by Lieutenant Gray, with whom were the adjutant and his clerk, approached the coast a few miles distant from Quilon, and were deciding on the advisability of landing (in consequence of the shattered state of the vessel) when a volley of musketry from the shore announced the hostile intentions of the natives. The vessel was bound together with some large tents to prevent its splitting, and it arrived safely at Cochin, but went to pieces in the harbour. All the patamars being damaged, other vessels were procured to continue the voyage.

At the end of December, the troops at Quilon consisted of 270 men of the 12th Regiment with 1500 Sepoys. On the morning of the 31st, the Dewan's palace was taken possession of, and eight guns, all pointing towards the principal entrance, each doubly charged with round and grape shot, were captured, which, however, proving useless (as the calibre was not suitable to our shot), were later spiked by our troops. The same evening, an insulting message was received from the Dewan, stating that, unless the Europeans were re-embarked immediately, he would, that night, drive the English force into the sea, and, if any were taken prisoners, they should be trampled to death by elephants. This was followed by our picquets being driven in, and numerous bodies of the enemy appeared on all sides, indicating a combined attack, whilst others entirely surrounded the palace.

Under these circumstances, a retirement from the encampment was directed with a view to occupying a strong position four miles off, in the remains of an old Dutch fort, which had the advantage of a commanding situation, and, in order to deceive the enemy, the camp was left standing.

The force took up its position at about 9 P.M., without impediment, on the ruins of the rampart, and had the misfortune to bivouac here, unsheltered, in an incessant torrent of rain, which lasted throughout the night, so that they were soaked to the skin, without even the excitement of being attacked.

1809

Captain Bayly relates that the 1st January was consequently ushered in by the most lamentable scenes that can be imagined. The troops being without a single dry cartridge, it was resolved to gain the ground of the encampment at the point of the bayonet, and, parading at 5 A.M., they moved off in torrents of rain, led by four companies of the 12th, which pushed forward towards the Dewan's palace, and it was re-occupied immediately.

When within a few hundred yards of their encampment, the little force was greeted by about a dozen fishermen, who rushed towards them, with their noses and ears cut off, and their faces streaming with blood; the Dewan's spies having detected them selling fish to our troops, this mutilation was the brutal punishment that had been inflicted.

From these wretched objects, it was ascertained that the British manœuvre of the previous evening had deceived the Rajah, who, suspecting the sudden departure from camp as a ruse to attack his army in some other quarter, had concentrated his whole force on the opposite bank of the Backwater, about five miles distant.

On arrival in camp, a scene of desolation presented itself, every tent and marquee being level with the ground, whilst boxes and liquor cases had been broken open, and wearing apparel and empty bottles were scattered in all directions, the work of our own camp followers, many of whom were lying about in a beastly state of intoxication and insensibility; the principal part of the lost clothing was recovered from them, but the wine and spirits were irrecoverable. The magazine and all public stores remained uninjured, and precisely as abandoned on the previous evening.

On the 8th January, the remainder of the regiment, accompanied by the 18th Madras N.I., arrived at Quilon, with the exception of 50 men of the 12th, under command of Lieutenant Thomas Thompson, who had been left at Cochin, for the protection of the Resident, Colonel Macaulay, as an attempt had been made to assassinate him. Six hundred Sepoys, with several guns, had also been left for the defence of Cochin.

The above reinforcement at Quilon, together with four 9-pounder guns and a howitzer, had infused fresh courage into our exhausted troops, who had been ten days and nights under arms, surrounded by a large force in the heart of the enemy's country, exposed to all the fury of a numerous and enraged population, and threatened with an attack on the encampment by heavy columns. The British force now amounted to upwards of 3000 effective men, 700 of whom were Europeans. As the remaining companies of the 12th, with the Sepoys and guns, were landing, the enemy made some movements in front to prevent the operations, but the whole force was safely on shore by 6 in the evening. Unfortunately, however, they were without the means of carrying on active operations.

Never were troops more scantily attended or equipped than the army in Quilon, without a single bullock, or conveyance for guns or baggage of any description; they were thus exposed to the united efforts of the armed population of Cochin and Travancore, without the possibility of retaliating in the event of obtaining any partial success.

The force was obliged to remain encamped on a sandy plain near the sea, enclosed by an almost impenetrable forest of cocoanut trees.

The Resident prohibited the felling of trees to conciliate the natives; but this produced no advantage, and the wood afforded shelter to the Travancorean marksmen, who annoyed the camp with their fire, keeping the troops in a constant state of alarm; the outposts were also frequently attacked by parties of the enemy.

The troops were fully occupied until the 12th January in landing stores and artillery, which latter, from the small size of the canoes, would have been impracticable; but, with the assistance of a frigate which had fortunately arrived, the object was accomplished. The enemy had now assembled in the neighbourhood to the number of about 30,000 with 18 guns, but only affairs of outposts occurred until the morning of the 15th January.

Before daylight on that date, they were observed approaching in force from several directions, and the troops were immediately got under arms.

Lieut.-Colonel John Picton, with a wing of H.M.'s 12th, 8 companies 18th N. I., and 2 guns, advanced against a large body which had occupied some heights in front of the cantonment, while Major Hamilton of the 13th N. I., with the other wing H.M.'s 12th, the 1st Battn. 2nd N. I., and his own battalion, moved against another considerable body with guns, which had taken up a position on the left, about half a mile from the cantonment, upon which they had opened fire. Colonel Chalmers accompanied this column, and Captain Newall, with the 1st Battn. 4th N. I., was left to defend the camp.[1]

As the advance commenced, our picquets came rushing in, closely followed by the enemy, who had with them 40 guns against 5 small field guns of the British, whilst our artillerymen present did not exceed thirty. The Dewan had advanced his guns during the night, in profound silence, to within a quarter of a mile of the British encampment, and their fire was so rapidly directed on the Europeans, that before the 12th could form into line several men fell dead. The two forces soon came to close quarters, when Colonel Picton, commanding the right attack, directed a bayonet charge, but, on entering a wood, two of the enemy's guns killed 20 of our grenadiers; the guns were, however, taken, the gunners bravely defending them until bayoneted on the spot.[2] The battle now raged on all sides, and amongst the weapons used by the enemy were barbed arrows, which inflicted most cruel wounds.

Whilst the right attack had carried everything before it, entirely defeating the Travancoreans (or Nairs), and killing a great number, the left attack was equally successful, having dispersed two separate bodies of them, one

[1] Colonel Wilson's *History of the Madras Army*.
[2] Colonel Bayly's *Diary*, p. 156.

of which was gallantly pursued by 5 companies of the 12th through the woods.

During the pursuit, part of the Carnatic Brigade, with guns, appeared on the heights to the west of the cantonment, and opened fire. It was immediately charged, forced to retire, and the guns taken.

In the meantime, three regular corps of Nairs, which had advanced on the right of the camp, had been attacked and dispersed by Captain Newall, 4th N. I. A number of them were killed, and four of their guns captured. Captain Newall was wounded, but his battalion did not lose a man.[1]

Captain Bayly relates that during the action, the Nairs had taken possession of the encampment. The 12th, however, had not the trouble of retaking the camp, which the Nairs had precipitately abandoned in their retirement after four hours' hard fighting, leaving about 1500 killed and 2000 wounded on the field of battle.

So closely were the contending forces engaged that several officers distinguished themselves by personal prowess, Lieutenant Keappock of the 12th cutting down two men who were opposed to him, and both wings of the regiment obtained an equal share of glory. "Remember Aleppi!" were the rallying words of the old 12th.

The Sepoys emulated the Europeans in this day's glorious action by various remarkable instances of bravery, especially the battalion defending the right flank of our line, which captured five guns, and repulsed the enemy with considerable slaughter.[2]

The British were unable to follow up the advantage from the want of stores, which prevented their quitting the coast.

The enemy appear to have been very confident of success on this occasion, and to have been intent on the annihilation of the Europeans. Several Travancoreans of their Carnatic Brigade were taken prisoners, and ropes being found in their possession, they were questioned on the subject, when they confessed that the cords were brought for the purpose of hanging British soldiers, and that the British officers were to have been trampled to death by elephants.

The total casualties of the British force amounted to 141 killed and wounded, of which the 12th Regiment had 8 killed and 47 wounded, the latter including 3 officers, Captain Bayly, slightly, in the head, and Lieutenant Molloy and Surgeon Erskine, severely. Fifteen guns were taken.[3]

A heavy fall of rain the same night destroyed many of the wounded, who were left on the field. So improvident had the Government been that the force was not supplied with a single doolie, so that the wounded were all borne to the hospital tent on the shoulders of their comrades.

[1] Colonel Wilson's *History of the Madras Army*.

[2] Colonel Bayly's *Diary*, pp. 158–9.

[3] Colonel Wilson's *History of the Madras Army*. (R. Cannon states 18 guns captured, and Capt. Bayly, p. 162, states 26.)

The following order was issued on this brilliant occasion, dated Quilon, January 16th, 1809 :—

"It is with the greatest satisfaction that Lieut.-Colonel Chalmers congratulates the troops he has the honour to command, on the glorious success obtained yesterday against the attack of an enemy whose force did not amount to less than thirty thousand men. He begs to offer his most particular thanks to Lieut.-Colonel Picton, who commanded the right wing of this little force with a wing of the 12th Regiment, and to the officers, non-commissioned officers, and privates, whose gallantry and high discipline have on all occasions appeared conspicuous. Lieut.-Colonel Chalmers has to offer his thanks to Major Hamilton, who commanded on the left with a wing of His Majesty's 12th Regiment, and to the officers, non-commissioned officers, and privates, whose gallant conduct needs no further comment than that they belonged to His Majesty's 12th."

The Political Resident, Colonel C. Macaulay, stated in a letter to Lieut.-Colonel Chalmers :—

"I have received the details of the victory over the united forces of the Dewan, an achievement that reflects signal honour on the discipline and animated valour of the troops under your command, and sheds fresh lustre on the British arms."

The following order was also issued on this occasion, dated Fort St. George, 4th February, 1809 :—

"The Honourable the Governor in Council has recently received a particular account of the action which took place at Quilon on the 15th ultimo, between the British troops and the troops of Travancore, in which, after a severe and long contest, the latter were defeated with heavy loss. From the extent of the combined force which was opposed to the British troops, this signal victory reflects the highest honour on their discipline and valour, and the Governor in Council has great satisfaction in expressing his strongest approbation of their meritorious conduct.

The Governor in Council accordingly conveys to Lieut.-Colonel Chalmers, his public thanks, and he is requested to convey the thanks of the Governor in Council to Lieut.-Colonel Picton, H.M's 12th Regiment, &c., &c. . . . with the other officers and troops who bravely signalised themselves on the occasion."

The British now divested themselves of all false delicacy for the religious prejudices of the natives, and every cocoa-nut tree was cut down that impeded our operations. Three large batteries were erected, and the captured guns placed in them ; these trees thus became one of our principal sources of defence, in addition to which, they afforded nutriment to the troops.

On the 18th January, a cannonade was heard in the direction of Aleppi, and, a few hours after, the "Piedmontaise" frigate was again anchored off Quilon with the Resident on board.

As all communication had been rejected at Aleppi, Captain Foote had bombarded the town, as an intimation that the barbarous murder of 34 men of the 12th was not forgotten. He announced to Colonel Picton, that, with the aid of Lieutenant Gilmore, commanding a small cruiser, he had completely destroyed all the enemy's vessels in the port of Aleppi, upset the guns in their batteries, and set fire to part of the town, the troops drawn up on the beach receiving several broadsides of grape from the frigate, dispersing them in the greatest confusion, with the loss of some hundreds killed and wounded.

At about 6 A.M. on the 19th January, the detachment of 50 men of H.M.'s 12th Regiment, and 6 companies, 1st Battn. 17th Native Infantry, under command of Major Hewitt of the latter corps, at Cochin, were attacked by three columns of the enemy at three different points. The defence was made with much skill and gallantry, and, after a severe engagement of three hours, the assailants, after several determined attempts, were repulsed on all sides, and compelled to retreat with the loss of about 300 men and two guns.

From Major Hewitt's report, the enemy appeared to number about 3000 well-disciplined troops, but from rumour, they amounted to much more. A battery of two guns, erected by them on the other side of the river, had also played on the defenders with some effect, on their charging the enemy on the south side.

At one period of the attack, the Sepoys showed every appearance of wavering, but the undaunted firmness of Major Hewitt, and the conspicuous prowess of the remaining Europeans infused confidence through their ranks. The gallant 12th, leading every charge, cleared the streets of Cochin, and nothing could resist the charge of the English bayonet. At length, Lieutenant Thompson, of the 12th, fell, pierced with wounds; one ball entered his face just below the left eye, four fingers of his left hand were shot off, and a ball through each thigh. Fifteen men of his detachment also fell.

The plan of the Cochin Rajah was to have annihilated every soul in the town. The inhabitants gratefully acknowledged the services of the 12th, openly attributing their preservation to the distinguished heroism of the company of the regiment, which had saved them from the contemplated massacre. Barricades were now thrown up at the entrance to each street, to prevent the effects of enfilade, as the enemy continued a furious cannonade; and in this situation they remained several days until the arrival of the "Piedmontaise" frigate and a Bombay cruiser, which relieved them from their difficulties.

The Rajah of Cochin now making pacific overtures, a negotiation was commenced and speedily terminated, whereby the confederacy between the two Rajahs at once ceased, and the Cochin troops being immediately disbanded and separated from those of Travancore, had the effect of relieving the British at Quilon of a force of 10,000 men opposed to it.

A subsequent report from Major Hewitt, dated 21st January, concludes as follows :—

> "The small detachment H.M.'s 12th Regiment behaved with great gallantry, and showed a noble example to the 1st Battn. 17th Native Infantry, who followed it entirely to my satisfaction."

The detachment received the thanks of Government in a General Order, dated 4th February, in which the Governor in Council expressed :—

> "his warm approbation of the conduct of a detachment of troops stationed at Cochin, who . . . . bravely repulsed a numerous and united force of the troops of Travancore and Cochin, in an attack which they made on the British detachment, on the 19th ult., and has particular satisfaction in expressing to Major Hewitt and the officers and troops under his command, his public thanks for their highly deserving conduct."

The gallant conduct of Lieutenant Thompson of the 12th is not alluded to in this Order, though he charged nearly 4000 of the enemy seven distinct times with only 50 Europeans, and at length fell covered with wounds and glory. This omission of justice to the deserts of so brave a man had an immediate effect, for after joining the force at Quilon, and listening silently to the reading of the Order, his agitation became so great that a fever seized him, and he was a corpse the next day. Thus perished a gallant officer, a martyr to the neglect of form in an official document.[1]

In the above engagement at Cochin, the 12th had 1 private killed, Lieutenant Thompson and 14 rank and file wounded; the 17th N. I. having 10 privates killed, one officer and 45 rank and file wounded.[2]

The Nairs at Quilon continued to attack our picquets daily, driving in the working parties; and the excessively severe duties involved on our troops (from being unceasingly harassed by day and night, combined with disease from constant exposure in an unhealthy climate), now began to tell heavily, and largely increased the number of hospital patients.

Moreover, the scanty provision of stores landed with the troops had been soon consumed, and rice in the husk being the only staple food now served out, there was every appearance, at this period, of the force being threatened with famine, but for the timely arrival, on the 28th January, of a supply of various descriptions, which had been expedited, on board some patamars, by the indefatigable exertions of Mr. Baber, Collector of Tellicherry, who (as Captain Bayly says) "may have been justly esteemed the

[1] Colonel Bayly's *Diary*, pp. 171–2.
[2] Colonel Wilson's *History of the Madras Army*, Vol. iii. p. 209.

saviour of this little deserted army, thus preserving them from the most terrible of all disasters."

At daylight, on the 31st January, the Dewan again advanced to an attack, when our troops were assailed by the usual salutation of rockets, grape, and musketry. On the approach of the enemy across the open sandy plain in front, our batteries began to play on them with great execution. Continuing their advance, however, they commenced a smart cannonade on the encampment from the summit of some sandhills, but their guns were quickly dismounted by the superior force of our batteries. Their next effort was to mount the entrenchments, rushing forward with the intention to charge, but our grape shot had so depleted their ranks that they were thrown into the utmost confusion. One long line, further to the left, beyond reach of our batteries, was seen advancing rapidly to charge our battalion of Sepoys stationed there. Their attack on the right was more successful, in driving back a battalion of Sepoys on to the 12th Regiment, the left wing of which, under Captain W. H. Forssteen, immediately charged, and in five minutes took four of the Dewan's guns, and drove the enemy into the woods, where they kept up a galling fire on our troops until about 2 P.M., when they slowly retired in all directions, carrying off their dismounted guns and innumerable wounded men.

The Nairs suffered so severely that they began to desert in considerable numbers.[1] It was estimated that their loss that day was as great as on the 15th instant, 1500 of their dead having been buried on the plain where they commenced the action, the casualties on the British side amounting only to 5 men killed and 20 wounded. The British were unfortunately still unable to pursue their advantage owing to the force being so miserably equipped.

Many French families, on hearing of the second defeat of the Dewan's army, now settled at Quilon, and were pleased to denominate the British troops "The Band of Heroes."[2]

On the following day, the troops were thanked in orders for the steady and cool manner in which they met and repulsed the attack of the enemy.

The gallant little army at Quilon still maintained their ground in spite of the Dewan having assembled an armed population of 100,000 men in the neighbourhood of the encampment, and our troops, now reduced to little more than 2000, were fully occupied as working parties in felling trees, working the batteries, and digging trenches, apart from repelling partial attacks.

When in command of one of these working parties, Captain Bayly, of the 12th, was wounded, and remained an invalid for many months.

On the 13th February, there was a welcome reinforcement to the garrison, when the 19th Regiment, accompanied by a detachment of artillery, with several guns, arrived in patamars from Colombo, bringing with them a

[1] Colonel Wilson's *History of the Madras Army*, Vol. iii. p. 216.
[2] Colonel Bayly's *Diary*, p. 170.

seasonable supply of provisions, which were badly needed, and army stores. Another supply following later, with a scanty number of bullocks for the conveyance of baggage, Colonel Chalmers was enabled to undertake an attack on the enemy's entrenchments at Killianoor, about 2½ miles from cantonments, after having fortified the old triangular fort of Trangacherry, about 4 miles off, for the reception of the sick and wounded, leaving 4 companies of Sepoys for its defence. Killianoor was protected by batteries, having a deep nullah full of water in their front, and defended by about 5000 men.

The 1st Brigade, composed of the 12th and native troops, under Lieut.-Colonel Picton, was ordered to turn the left of the position, while the 2nd Brigade, composed of the 19th and native troops, attacked in front. At daylight on the 21st, the force moved out, by separate roads, and both attacks were completely successful, the enemy taking to flight and leaving their guns behind.

Our casualties, amounting in all to about 22, were confined exclusively to the 2nd Brigade.

On the 3rd March, a General Order was published, conveying the thanks of the Honourable the Governor in Council to Lieut.-Colonel Picton (12th) and Lieut.-Colonel the Hon. Stewart (19th), and to the officers and men who served under them, for their meritorious exertions on this occasion.

In the meantime, a despatch having been received that another British force, under Colonel St. Leger, had penetrated the Travancore country, and was encamped a few miles north of Trivandrum, (the capital of the Rajah's dominions), the Quilon troops advanced on the 22nd February, with four days' provisions, to effect a junction with Colonel St. Leger's troops. The combined force encamped at the village of Attingwery until the 31st March, pending peace negotiations between the Resident and the Rajah of Travancore, the result of which was that the Rajah should pay the expenses of the war, and that the Carnatic Brigade and Nair battalions in his service should be dismissed.[1]

The 12th Regiment was now ordered to Quilon. The Dewan committed suicide, but his brother was taken and hanged for his participation in the murder of Assistant-Surgeon Hume of the Company's service, and 34 men of the 12th, which had been treacherously perpetrated at Aleppi.[2] A few weeks later, two of the Dewan's chiefs who had been implicated in it were hanged directly over the spot where the cruel massacre took place.[3] The body of the Dewan was hung in irons and exposed on a gibbet by order of the Resident, which called forth severe censure from the supreme Government.[4]

[1] Colonel Wilson's *History of the Madras Army.* [2] *Ibid.*
[3] Colonel Bayly's *Diary*, p. 190.
[4] Colonel Wilson's *History of the Madras Army.*

The regiment, having received orders to proceed to Seringapatam, marched on the 23rd May, and, owing to the Malabar monsoon having set in with great severity, four short marches were at first effected, with extreme difficulty and danger. Even in the short distance achieved, the provisions, tents and new clothing had all been destroyed, and many of the officers lost all their baggage, owing to the inundated state of the country. In the endeavour to pass over a river, on the last of the above marches, part of the regimental baggage and camp equipage was swept away, and several men were drowned. A three weeks' halt now took place in a few old huts, the only shelter available in this deplorable situation, surrounded by floods of water, and exposed to incessant torrents of rain. Tents and provisions were at length supplied, but not before 30 or 40 deaths in the regiment had taken place from disease.

After five weeks' unexampled exposure and suffering, the regiment, by a forced march, succeeded in reaching the Coromandel Coast, to which the influence of the Malabar monsoon did not extend.

Another halt then ensued, when, on a representation being made to Madras, of the exhausted state of the corps, and the sickness pervading it, an order was received, changing its destination to Trichinopoly, where the regiment arrived on the 24th July.

About the 12th August, the flank companies of H.M.'s 12th and 30th Regiments, with a company of Artillery, the 6th Madras Cavalry, and the 2nd Battns. 13th and 24th N. I., under command of Colonel Wilkinson, left Trichinopoly, by order of Government, for the purpose of enforcing obedience throughout the Division, and arrived at Dindigul on the 17th. This move was in connection with what was known as the mutiny of European officers, of the Madras Army, who, for the most part (during the term of office of Sir George Barlow, as Governor of Madras) refused to sign a declaration of obedience in support of his authority as the Governor in Council of Fort St. George, which was in course of circulation.

1810–11

The regiment having moved to Wallajabad, orders were received, in March, 1810, for the flank companies to be completed to 100 rank and file each, for the purpose of being employed in an expedition against the French colonies, the Islands of Bourbon and Mauritius. The companies accordingly marched to Madras, and, embarking on the 4th May, arrived at the Isle of Rodrigues on the 20th June, and sailed thence on the 3rd July.

The whole of the King's troops having, on the 6th, been transshipped on board the men-of-war (at the place of rendezvous, twenty leagues to windward of Bourbon), the fleet made sail, and anchored on the 7th, within gunshot of that island, between St. Marie and the Cutoor River, but it was found that the surf was so high as to render a landing dangerous. An effort was, however, made. The troops had been formed into four brigades, under Colonel Keating, and the 4th Brigade consisted of the flank companies, 12th and 33rd Regiments, a detachment of the 56th, and a party of pioneers, under Lieut.-Colonel Campbell, 33rd.

The disembarkation began at 2 P.M. on the 7th, and about 300 men of the 3rd and 4th Brigades, under Colonel McLeod, together with a party of seamen, had landed, with the loss of a few men drowned, when the surf became so heavy that nothing more could be done. Upon this, Colonel McLeod moved as far as St. Marie, where he stormed the batteries, and, taking possession of St. Marie, remained there the night.

The 1st Brigade, under Colonel Fraser, succeeded in landing at Grand Chaloupe without loss, and, early on the 8th, attacked the enemy, who were drawn up in two columns on a plain with 2 guns, and supported on the flank by a redoubt. They were quickly driven back with the bayonet, and retreating into the town, left the redoubt in our hands. The guns were then turned on the place, and the rest of the troops coming up about 4 P.M. the commandant surrendered, and a capitulation was signed, by which the whole island, together with all public property, was ceded to the British. The British loss was 18 killed and 79 wounded, exclusive of 5 seamen; and, to this, the 4th Brigade contributed 1 killed, of the flank companies (corps not stated) and 1 of the 56th Regiment; wounded, 2 officers and 5 men, 12th Regiment, 3 of the 56th, and 4 pioneers. The officers were Lieutenants Spinks and Whannel.

Lieut.-Colonel Campbell, with the 4th Brigade (including the 12 flank companies), was detached, on the 10th, to St. Paul's, and took possession of that place, where 1500 men laid down their arms. A large amount of ordnance, gunpowder, and a quantity of small arms and military stores fell into our hands.

An Order was later issued by the Governor-General in Council, in which the officers and men engaged in the expedition were thanked for the zeal, courage, and perseverance, by which a conquest of so much importance to the national interests had been achieved.

A letter from Army Headquarters, dated Madras, 19th August, 1810, directed the battalion companies of the 12th to be in readiness to march to Fort St. George, on or about the 28th instant, and the corps being under orders for foreign service, any men unfit to accompany it were to proceed to Cornamallee during the march of the regiment from Wallajabad.

On the 27th, the regiment marched, and arrived on the 30th at St. Thomas's Mount, halting until the 21st September, when the march was continued to Madras.

The embarkation of troops for the expedition against Mauritius began on the 17th September, and was concluded by the 24th.

The regiment embarked as follows :—

	Ship	Officer	Sergeants.	Men.
Frigates.	"Cornwallis"	Lieut.-Colonel J. Picton and headquarters	5	130
	"Cornelia"	Captain King	5	75
	"Hesper"	Major O'Keefe	4	34
	"Bucephalus"	Captain McKeady	5	75
	"Clorinde"	Captain Ashton	6	90
74	"Russell"	Captain Bayly	5	90
		Total	30	494

The Embarkation Return shows a total of 22 officers and 524 of other ranks.

A Horse Guards General Order directed that, from the 25th September, 1810, a Trumpet-Major in regiments of cavalry, and a Drum or Bugle-Major, in regiments of infantry, were to be borne on the establishment of each regiment with the rank of sergeant.

On the 6th November, 1810, the Squadron and transports reached the island of Rodrigues, which is about twelve miles long and five broad, several hundred feet above the sea, extensively wooded, and, at that time, inhabited by three or four French families. It had been in our possession ever since the expedition to Bourbon, and was now garrisoned by a company of Sepoys. The Bombay Contingent had already landed there, but that from Bengal did not arrive until the 21st.

On the 20th November, the battalion companies of the regiment transshipped from the men-of-war to the "Lord Castlereagh," Indiaman, the grenadier and light companies having embarked at St. Paul's, to join the force. The fleet anchored in Grand Bay, on the northern extremity of Mauritius, on the 29th, and the troops were landed without opposition the same afternoon.

The force was divided into five brigades ; the 1st Brigade consisted of the 12th and 22nd Regiments, and right wing of the Madras Volunteer Battalion, under command of Lieut.-Colonel John Picton, 12th Regiment, with Captain Bayly (12th) as his Brigade Major. In addition to the five, there was a Reserve Brigade, under Lieut.-Colonel Keating, composed of the flank companies of the 12th and 33rd Regiments, the 84th, and two companies of the 30th.

On disembarking, the column, led by General Warde (with the Reserve Brigade in advance) moved off towards Port Louis, the capital of the island, and, after passing Canonier Point, marched by a pathway through a wood, skirted on both sides by almost impenetrable jungle. Hitherto no opposition had been offered, but when half-way through the jungle, the light company of the 12th under Captain Forssteen suddenly came on one of the enemy's posts, and a short skirmish ensued with a French picquet, resulting in their being driven from their position at the point of the bayonet. Two men of the 12th were killed; Colonel Keating and Lieutenant Ashe (12th), and three men wounded. Lieutenant Ashe was shot through the thigh and arm, but had vigorously cut down two of his opponents before he fell; he had also received a bayonet wound across the forehead. On debouching at dusk, from the wood, the troops bivouacked in an extensive maize field, and having exhausted the contents of their canteens, there was extreme distress throughout the night for want of water, which, during an excessively hot march, had not been procurable; a small well, now within reach, affording only a scanty supply.

The march was resumed on the morning of the 30th, but the heat was so great that, after proceeding five miles, the force was obliged to halt and encamp in the immediate vicinity of some powder mills, called "Moulin au Poudre," through which flowed a beautiful transparent stream of water.

About midday, a reconnoitring party, under General De Caen, attacked our picquets, and compelled them to retire, but, being reinforced, they rallied, and drove the enemy back with some loss. The picquet of the 12th stood the brunt of the attack, and, being joined by the light company of the 59th, soon dislodged the enemy's marksmen from the old houses and barns in front. General De Caen had a shot through his boot, when riding off towards Port Louis.[1] Our troops were now amply supplied with provisions from the fleet.

Before daylight on the 1st December, 1810, Colonel McLeod (69th), was detached with the 4th Brigade (composed of the 69th Regiment, 300 Royal Marines, and flank companies, 6th, and 12th Madras N.I., to take possession of the batteries at Tortue and Tombeau, about a couple of miles to the right of the line of march, which service was successfully performed.

At about 5 A.M. the main body marched towards Port Louis, the grenadier company of the 12th, under Captain Frith, leading, with a section, under Lieutenant Keappock, detached a short distance ahead. The light company of the regiment, under Captain Forssteen, was close on their flanks, in extended order.

After proceeding about two miles, it was found that the passage of the River Tombeau was disputed by about 300 of the enemy, with two guns, and that they had partially destroyed the bridge. They were soon driven back, and, on their retiring without further opposition at this point, the troops passed rapidly over the bridge, after which, by the active exertions of a body of sailors who had been landed from the fleet, our guns were dragged through the river at a ford lower down.

The troops were again opposed, about two miles further on, at the River Sêche, when a battery of the enemy's, on the other side, showered volleys of grape up the road by which our column was advancing, with great execution. Lieut.-Colonel Campbell (33rd) commanding the flank battalion, was killed, and Captain Bayly relates that "just at this period, the 12th had 60 men mowed down."

[1] Colonel Bayly's *Diary*, p. 231.

It was now that Major O'Keefe of the 12th was killed, whilst with the leading division of the regiment, a cannon ball having carried off the upper part of his skull.

Lieutenant Keappock was wounded in the side, but continued at his post until a shot in the head forced him to retire. His honourable, though dangerous post was taken by Lieutenant Jenkins, who received a severe contusion of the breast by a ball, but continued at the head of the leading section.

The enemy were driven back with the loss of about 100 men, and pursued to the River Latamers, near the outskirts of the town, having abandoned several guns and a howitzer, and fled in all directions.

The 12th Regiment were now ordered to ascend the Long Mountain, and storm the fortification at its summit, a manœuvre which took about an hour, a few straggling shots having been all the opposition offered; and the French fled, leaving a gun behind them. The 12th remained on the mountain, and the rest of the troops, having been withdrawn beyond the range of the batteries, encamped for the night.

A flag of truce was sent to the camp by the enemy the next morning (2nd December), and terms of capitulation having been agreed on, they were ratified on the 3rd, when the whole island was surrendered, together with a large quantity of military and naval stores and ammunition; 209 pieces of heavy ordnance were found in the works.

A number of British ships were recaptured, and about 2000 English seamen and soldiers, who had been taken prisoners, were rescued from confinement.

The French troops and seamen were not made prisoners of war, but were embarked on some of our transports for France, with their arms and Colours, and all personal effects, at the expense of the British Government, far more favourable conditions than they had any right to expect, considering the great disparity of force.[1]

Our loss was only 28 killed, 94 wounded, and 45 missing, 167 in all, of which the casualties in the 12th Regiment were:—Major O'Keefe, 1 drummer and 16 rank and file killed; Lieutenants Keappock and Ashe, 3 sergeants and 28 rank and file wounded, and 5 men missing.

The gallantry displayed by the flank companies of the 12th Regiment (serving with the flank battalion) was admirably conspicuous, and shows how nobly they exerted themselves in this short but brilliant and decisive attack, from their losses alone.

In a General Order, dated Headquarters, Port Louis, 1st December, 1810, on the success of that day's operations, Major-General Abercrombie was "happy to acknowledge the steadiness shown by H.M.'s 12th and 22nd Regiments," and the services of the officers and men of the

[1] Colonel Wilson's *History of the Madras Army*, Vol. iii. p. 311.

expedition having been duly acknowledged by him in his report, the force was thanked on the 9th February following, by the Governor-General in Council.

The principal part of our army that had been engaged returned to the Indian Presidencies, to be employed in the contemplated expedition to Batavia, leaving the 12th, 72nd, and 87th Regiments to garrison Mauritius.

On the 4th December, the 8 battalion companies of the 12th Regiment descended the Long Mountain, and embarked from Tortue Bay on board the "Psyche" frigate for Grand Port, which was reached on the 7th, when the regiment disembarked, and was joined on the 12th by the flank companies, after eleven months' absence.

The depot of the regiment, which had been moved from the Isle of Wight to Ipswich, was this year ordered to Stowmarket.

On the 25th April, 1811, the depot was directed to march to Hadleigh, and to remain until the volunteering from the Militia was over. In September, the depot moved to Maldon.

General William Picton died on the 14th of October, in his 88th year, and was succeeded in the colonelcy of the 12th Foot by Lieutenant-General Sir Charles Hastings, Bart., from the 77th Regiment.

The 1st Battn. 12th was stationed at Mauritius during the years 1811 and 1812.

1812–14

A distinguished officer, General Sir Richard England, K.H., G.C.B., who entered the service in February, 1808, exchanged, as Captain, on the 1st January, 1812, from the 60th Rifles to the 12th Regiment, and was borne on its rolls for over eleven years. He joined the 2nd Battalion in Paris, after the Battle of Waterloo, and, on appointment to the staff remained in France until the withdrawal of the Army of Occupation in 1818, when, after a further term of staff service, he was promoted Major into the 75th Regiment, with which corps he served in South Africa in 1836–37. He commanded the 41st Regiment in Candahar in 1839–41, and finally commanded a division in the Crimea, having been present at Alma, Inkerman, and the Siege of Sevastopol. (See Plate 18.)

The war with France was now approaching to a crisis; Napoleon Bonaparte had attained the summit of power, and the efforts of Great Britain were commensurate with the importance of the contest; the army was augmented.

Formation of a Second Battalion.

A Horse Guards' letter, dated 18th January, 1812, directed that a 2nd Battalion was to be added to the 12th Regiment from the 25th December, 1811. It was to consist, in the first instance, of 1 field officer and 4 companies of 100 men in each, with the usual staff; and, on its effective strength exceeding 400 rank and file, the battalion was to be augmented to 6 companies, of 100 men in each, and another field officer; on completion of that number it was to be further augmented to an establishment of 10 companies, with the usual complement of field officers. The recruiting company of the 1st Battn., and such non-commissioned officers and men as were then in England, were to be incorporated with it.[1]

An Army Circular, dated 25th March, 1812, limited the award of corporal punishment by regimental courts-martial to 300 lashes.

The 2nd Battn. was inspected at Maldon on the 26th May, and mustered only 4 companies with a total establishment of 400 rank and file; no Colours had been provided for it. There were no parade movements, and, as a newly formed battalion, it was well reported on.

A War Office letter, dated 2nd June, to the Officer Commanding 2nd Battn., directed that the recruiting company of the 1st Battn. having been discontinued on the establishment, steps were to be taken to procure the transfer to the 2nd Battn. of the non-commissioned officers and drummers of the said company, and appoint them to the first vacancies which might arise in it.[2]

On the 23rd September, 1812, the 2nd Battn. was ordered to march from Witham to Gravesend, Northfleet, and Chalk, and await embarkation for Ireland, which took place on the 29th, in the ship "Atlas," mustering 436 all told, including 28 women and 27 children.

A War Office book shows the 2nd Battn. 12th taken on the Irish establishment from October 25th[3] inclusive, mustering 6 companies, with 25 officers and 647 of other ranks.

The battalion arrived at the Cove of Cork on November 13th, and, disembarking on the 15th, marched to Fermoy the following day.

The 1st Battn. was inspected at Port Louis, Mauritius, on the 26th October, 1812, by Lieut.-Colonel John Picton. Captain William Frith was in command, and the establishment showed 760 privates. From the Inspection Report, it appears that the junior major (Ashton) was at the time rendered totally unfit for service, owing to a paralytic stroke. A Return of Courts-martial, since the previous inspection, shows that punishments of upwards of 600 lashes were still being awarded, though not in every case inflicted.

In October this year the depot was at Horsham.

By the 6th April, 1813, 6 companies 2nd Battn. had marched from Fermoy to Clonony, King's County.

On the 28th April, the 1st Battn. embarked at Port Louis, Mauritius, for the Isle of Bourbon, where the regiment arrived on the 1st May.

In consideration of the meritorious services of the non-commissioned officers of the army, and with a view to establishing in the infantry a corresponding rank to that of Troop Sergeant-Major in the cavalry, a

[1] *W. O.* 3, Book 55, p. 138. [2] *W. O.* 4, Book 214. [3] *W. O.* 8, Book 11. p. 276.

General Order, dated, 6th July, directed that certain selected sergeants, with an increase of pay, should be called "Colour-Sergeants," the duty of attending the Colours in the field being at all times performed by them, whilst it should in no way interfere with their other duties, the said sergeants to be distinguished by an honourable badge of a Union Jack supported by two crossed swords above a double gold chevron.

On the 21st August, the depot was ordered to march from Horsham to Bristol, and embark for Ireland.

On the 20th and 21st September, the 1st Battn., under command of Major Eustace, was inspected at St. Denis, Isle of Bourbon, by Lieut.-Colonel Keating, and favourably reported on, but the accoutrements, in general, were in a bad state, and required renewing.

By Royal Warrant, dated 16th November, 1813, the 2nd Battn., formed as in 1812, was discontinued on the Irish Establishment, from the 25th September this year, and, on the same date, was reinstated on the same establishment, with a considerable increase, so that its strength was now :—1 lieut.-colonel, 2 majors, 10 captains, 22 lieutenants, 8 ensigns, 1 adjutant, 1 paymaster, 1 quartermaster, 1 surgeon, 2 assistant surgeons, 56 sergeants, 21 drummers, and 1000 rank and file. Total of all ranks 1126.[1]

On the 6th December, the battalion marched to Athlone.

Large reductions of the infantry regiments took place in 1814. The majority of the 2nd Battalions, raised about the year 1782, were disbanded by War Office Order, dated 12th October.

An Embarkation Return, dated 11th November, 1814, shows that the 2nd Battn. 12th embarked on that date, at Monkstown, Ireland, in the ships "Crown No. 16," "Durham H.L.," "Albion A.T.," "Thomas 308"; strength, 26 officers, 39 sergeants, 17 drummers, 391 rank and file, 2 officers' servants (not soldiers), 108 women, and 108 children—total 691; and proceeded to Plymouth.

1815

The depot was ordered in February to Portsmouth, *en route* to the Isle of Wight.

In the previous year the tyrannical power of Bonaparte had been overthrown, and the Bourbon dynasty was restored to the throne of France. On the re-establishment of peace, the Island of Bourbon was restored to the French monarchy, and, in consequence, the 1st Battn. 12th returned to its former station, Mauritius, embarking at St. Denis on the 3rd April.

French troops, who arrived to take possession of the island, landed as the British went on board the ships prepared to receive them.

The 1st Battn. sailed on the 4th, anchored in Port Louis Harbour, Mauritius, on the 7th, and disembarked the same morning.

[1] *W. O.* 8, Book 11, p. 344.

On the 11th April, the 2nd Battn. marched to Dartmoor, to take charge of the American prisoners confined there.

Soon afterwards, Bonaparte quitted the Island of Elba, in violation of his engagements, and regained the throne of France, when the powers of Europe took arms against the usurper, and his veteran legions were overpowered on the field of Waterloo by the allied army, under Field-Marshal the Duke of Wellington, on the 18th June. The allies then advanced to Paris.

To replace the losses of the British army at Waterloo, additional troops were sent to the continent, and the 2nd Battn. 12th (which had returned a few months previously from Ireland) embarked on board H.M.S. "Tigre" at Plymouth for Flanders, on the 27th June, under command of Colonel Julius Stirke. The battalion landed on the 7th July at Ostend, and, arriving at Paris on the 18th, joined the army under the Duke of Wellington at St. Denis on the 25th, forming, with the 36th and 64th Regiments, the 12th British Brigade, under command of Colonel the Honourable Sir C. Greville.

On the 5th July, Montmartre had been occupied by British troops, and, after the capitulation of Paris had been signed, the greater part of the British army was encamped in the Bois de Boulogne, where the 2nd Battn. was inspected, on the 23rd October, by Colonel Bilson.

From the 24th August the battalion had been posted with the 1st Battn. "Buffs," 2nd Battn. 30th, and the 33rd Regiment, to the 5th British Brigade.

Referring to the 2nd Battn. 12th, the following is reported from the "Naval and Military Gazette," of September 13th, 1838 :—

> "One day an Irishman by the name of Ryan was digging for pipe-clay on Montmartre, the citadel of Paris, when he happened to poke his way into one of Napoleon's wine-cellars. Pat Ryan, who was a Tipperary boy, lost no time in telling his wife, Molly, the secret. She began to vend the wine, some of it twenty years in wood, at the rate of two sous per soldier's mess kettle. Most of the soldiers in garrison were kept in a state of continued intoxication for about six weeks before the secret was found out, and Molly Ryan set sail for the 'Land of Potatoes,' and purchased a small farm near the Gaulty Mountains, at the expense of the Emperor Napoleon. The Duke of Wellington suspected that the 88th, who were quartered at the foot of the hill, were the thieves, and had them confined to barracks in their cantonments. There were a few Birmingham lads serving in the Light Company, 2nd Battalion 62nd Regiment, on the hill, who by means of the cook's ladle, applied as a crucible, and many impressions of chalk, converted the Britannia metal buttons of the 'Springers' (62nd) into counterfeit dernier francs, and for every piece thus fabricated Molly Ryan, in her hurry of business, gave the Light Bobs five mess tins of wine. The secret was found out at the attesting of a recruit belonging to the 81st Regiment. Every officer

DISTINGUISHED OFFICERS WHO SERVED IN THE 12TH.

LIEUT.-GENERAL SIR RICHARD ENGLAND, K.H., G.C.B.

MAJOR-GENERAL SIR ROBERT SALE, K.C.B.

and soldier of the 12th doing duty in Paris at the time was ordered to pay the ex-Emperor's cellarer one day's pay each. The company to which Ryan belonged was, up to that period (1838), called the 'Mining Company' by the old 12th in commemoration of finding out the cellar and the ex-Emperor's wine at Montmartre."

A Horse Guards Order notified the Duke of Wellington, as early as the 27th October, that the 2nd Battn. 12th (610 rank and file) was one of the many regiments to return from France, and a General Order, dated Paris, 30th November, directed the battalion to be brigaded with the 2nd Battn. 30th and 33rd Regiments, for the march to the port of embarkation, under Colonel Stirke, 12th Regiment.[1]

The battalion remained in the neighbourhood of Paris until the 4th December, when it commenced its march to Calais, to embark for England, and, arriving on the 24th, disembarked at Deal.

In the Monthly Return of Mauritius for June, 1815, an Embarkation Return shows that the light company of the 1st Battn., under Captain Turbervill, with Lieutenants Burrowes and Lawson, embarked in the ship "Jessy," on the 16th of that month, for service in Bengal, numbering 3 officers, 5 sergeants, 2 drummers, and 74 rank and file. Troops embarking with them consisted of the 87th Regiment, and the flank companies of the 22nd. The whole, numbering a total of 1107 officers and men, including staff, were under the command of Colonel Keating, and joined a force under the Earl of Moira, but not for long.

The light company of the 1st Battn. rejoined the regiment from Bengal, at Mahéburg, Mauritius, on the 14th November, and the battalion marched from Mahéburg to Flacq on the 21st.

In the Duke of Wellington's despatches, the name appears of Lieutenant Robert Nixon (Lieutenant 12th Foot, 21st August, 1794; Major, 1st Royals, 8th February, 1810, and Lieut.-Colonel in the army, 4th June, 1814), awarded the 4th Class of King Wilhelm's Order of the Low Countries.

1816–17

On the 2nd January, 1816, the 2nd Battn. embarked at Deal for Ireland, and marching on arrival to Fermoy, was taken on the Irish establishment from the 25th January, with a strength of 10 companies, and a total of 906 of all ranks.

At two inspections in May and October, the 2nd Battn. was highly complimented in every way by the Inspecting Generals.

On the same date, the depot embarked at Cowes for Ireland in the brig "Ladies' Adventure," including, all told, 143.

By the 26th March, the 2nd Battn. had marched to Galway.

On the 17th April, orders had been issued from Army Headquarters at

[1] Wellington's *Despatches*.

Madras, for the abolition, at Fort St. George, of the Military Institution which had been established there since 1804, for the improvement of military education, by the instruction of a certain number of officers in geometry, military drawing, &c. In promulgating this order, the Governor-General in Council "took the opportunity of expressing his entire approbation of the mode in which the duties of Mathematical and Drawing Instructor to the Institution had been discharged by Captain Troyer, 12th Regiment."

A Horse Guards' letter, dated 21st December, notified that the ship "Tonnage" had been sent out to bring the 1st Battn. 12th home from Mauritius.

The battalion continued to form part of the garrison of Mauritius throughout the year 1816, and the first six months of the following year.

On the 1st of July, 1817, a serious fire broke out at Port Louis, when the exertions of the garrison to extinguish the flames called forth the admiration and thanks of the inhabitants, which were communicated to the troops by the Governor. The regiment was, at this period, at Flacq, and Major Bayly, arriving at Port Louis, a day or two after the fire, describes it as a "scene of melancholy desolation, such as he had never witnessed."

Transports having arrived to convey the regiment to Europe, a General Order was published, in which it was stated :—

> "Major-General Sir Edward Butler, in taking leave of the 12th Regiment, feels himself highly gratified in stating its conduct, during its services in this island, has, in every particular, been such as to meet with his highest approbation."

The regiment embarked, on the 21st July, in the transports "Ocean" "Alexander," and "London," and, sailing from Port Louis on the 25th, arrived at Portsmouth on the 10th November. Two days later, orders were received to proceed to Ireland, without landing in England. The regiment was then transshipped into one large transport, the "Brailsford," in which they lay over six weeks, detained by contrary winds, and some tremendous gales, and did not arrive at Cork until the 27th December, disembarking the same evening, after an absence from Europe of nearly 22 years. The disembarking strength was :—14 officers, 29 sergeants, 22 drummers, 266 rank and file, 17 women, and 25 children—total 373.

By the 23rd December, the 2nd Battn. had proceeded in two divisions to Athlone, and the 1st Battn. marched there from Cork on the 29th.

CHAPTER VIII

Ireland, England, Gibraltar, England, Ireland, Mauritius.
1818–1847

1818–20

The 1st Battn. arrived at Athlone on the 9th January, 1818, and joined the 2nd Battn., which was stationed there.

In compliance with instructions from Army Headquarters, the 2nd Battn. was disbanded on the 15th January, transferring 660 men to the 1st Battn. The officers were placed on half-pay from the 25th January, with the exception of 2 captains, 5 lieutenants, and 5 ensigns, who, having been permitted to withdraw from the battalion, prior to its reduction, were to be placed on half-pay within two months from the period on which they ceased to do duty. Officers, at the following rates :—1 lieut.-colonel at 11/- daily, 2 majors at 9/6, 8 captains at 7/-, 1 lieutenant at 4/6, and 10 at 4/-, 7 ensigns at 3/-, adjutant and two assistant-surgeons at 4/- each, paymaster 7/6, quarter-master 3/-, and surgeon 7/-. The effective non-commissioned officers and privates were to be transferred to complete the 1st Battn. to the establishment of 800 rank and file, any supernumerary men being invited to enlist in corps whose establishments were incomplete.

Major Robert Sale, the senior major, now placed on half-pay, was restored to full pay as major in the 13th Regiment on the 28th June, 1821, and, later, obtaining the command, made a great name for himself in Afghanistan, and became a highly distinguished officer. He was duly knighted, in addition to receiving many other honours, and, in 1844, was appointed Quarter-Master-General in the East Indies, retaining the appointment during the Sikh War in 1845, in which he was mortally wounded at the battle of Moodkee, and died on the 21st December of that year.

General Sir Robert Sale, K.C.B., was known as a brave soldier, and was nicknamed "Fighting Bob." Wherever there was fighting, he was always in the thick of it; his men followed him anywhere.[1] (See Plate 18.)

[1] *Dictionary of National Biography.*

In June 1818, the regiment, now a single battalion, marched to Limerick, and, throughout the year 1819, continued to do duty in the counties of Limerick, Cork, and Clare, until early in the summer of the following year.

King George III died on the 29th January, 1820, and was succeeded by King George IV.

The regiment marched in June to Dublin, where it arrived on the 17th, and was ordered, on the 12th October, to embark in three divisions for England. On arrival of the headquarters at Liverpool, routes were issued for the march of 6 companies to Macclesfield, and 4 to Manchester, where the headquarters joined on the 13th November.

Since its return home, the regiment had been most highly reported on by all Inspecting Generals, and, prior to leaving Ireland, Major-General Egerton, at an inspection in Dublin on the 2nd October, declined to see it perform any evolutions, "in consequence of the knowledge he had acquired of its precision of movement, during three months' previous acquaintance with it in that garrison."

1821–24

On the 11th February, 1821, routes were received for the march of the regiment in three divisions to Portsmouth, where they had arrived by the 5th March. Prior to leaving Lancashire, Major-General Sir James Lyon, K.C.B., addressed the following letter to Lieut.-Colonel Forssteen :—

> "Although the 12th Foot has been stationed but a short time in this place, I cannot refrain from expressing to you that no military change could have given me more concern than their departure.
>
> I have had every opportunity of observing their uniform good conduct and strict attention to every branch of discipline, and nothing but satisfaction has ever been manifested to me by the civil authorities, and the inhabitants in general, on the very exemplary behaviour of the men. I beg of you to make known to the corps the value I attach to the honour of having had a regiment of such high character placed under my orders, and that I must ever take an interest in its welfare and success."

At noon, on the 7th March, the regiment embarked at Portsmouth for the Channel Islands, as follows:—headquarters and 5 companies, under Major Bayly, in the ship "Fanny Fox" for Jersey; 5 companies, under Brevet Lieut.-Colonel Hare, in the ships "Zephyr," "Integrity," and "Crown" for Guernsey; and arrived at their respective destinations on the 16th March.

While stationed at these islands, the appearance of the regiment, the conduct of the men, and the excellent system of interior economy which existed in the corps, elicited the commendations of Major-General Sir Colin Halkett; at the inspections in October 1821, May and October 1822, and when the 12th were about to return to England in May 1823, the Major-General repeated his expressions of approbation, and of the warm interest he took in the welfare of the corps. The conduct of the four companies at Guernsey, under Major Bayly, was also specially commended by the Lieut.-Governor, Colonel Sir John Colborne.

On the 28th March, 1823, an order was received for the removal of the regiment (now 8 companies) to England, and, on the 8th May, the head-quarters and 4 companies embarked at Jersey, on the transports "Cato" and "Zephyr." On the 15th, two companies landed at Sheerness, and occupied barracks; the other two proceeded in the "Cato" to Chatham. The remaining 4 companies, under Major Bayly, arrived on the 25th, in the transport "Loyal Britain," at Sheerness, to be quartered.

On the decease of General Sir Charles Hastings, Bart., the colonelcy was conferred on Lieut.-General the Honourable Robert Meade, from the 90th Regiment, by commission dated 9th of October, 1823.

On the 25th, routes were received for the march of the regiment to Fort Cumberland, near Portsmouth, where it remained until the 5th November, and, on the 8th, embarked by wings on His Majesty's Ships "Ganges" (84 guns) and "Superb" (74) for service at Gibraltar, the scene of its former triumphs, and arrived at that celebrated fortress (strength, 700 rank and file) on the 25th of the same month.

The Governor of Gibraltar, at this period, was General Sir George Don, G.C.B.,[1] and the garrison consisted of a battalion of artillery and six regiments of infantry, with a proportion of engineer officers, sappers, and miners, amounting in all to 4500 effective men.

1825–27

In 1825, the establishment of the regiment was augmented from 8 to 10 companies; 6 to be considered service companies, with 86 rank and file in each, and were to remain at Gibraltar, mustering a total of 26 officers and 550 of other ranks.

Four depot companies, of 56 rank and file in each, were to be stationed in the United Kingdom, mustering 14 officers and 240 of other ranks. In consequence of this arrangement, the officers and non-commissioned officers of two companies serving abroad, were sent to England; Brevet Lieut.-Colonel R. Bayly, as senior Major, returning home at the same time to command the depot.

The four depot companies were being formed at Plymouth, and, scarcely had 200 youths been enlisted, when they were ordered in December to Ireland, and, reaching Cork Harbour in three days, proceeded in a steam vessel to Middleton, where they disembarked and marched to Youghal, a dreary town on the sea coast. Lieut.-Colonel Bayly relates that, as the men were marching through the town, the poor girls, with whom the cottages

[1] Sir George Don was appointed Lieutenant-Governor of Gibraltar on the 25th August, 1814. As the nominal Governor of Gibraltar was an absentee, General Don was practically the Governor of that fortress until his death, on the 1st January, 1832.

were swarming, looked out of their half-formed doors, exclaiming:—"Plase, your honour, give us some of your men for husbands!" And, in the course of the three months the depot remained at Youghal, these cottage Venuses (many of whom, though wretchedly clad, were really beautiful) contrived to marry several of the undisciplined youths then serving at the depot.

The highest flogging sentence on record in the British Army was passed this year on a private soldier, who was awarded one thousand nine hundred lashes by a court-martial, and received one thousand two hundred.[1]

In February 1826, the depot proceeded to Fermoy, and thence to Bantry for three months, which was followed by moves to Kinsale, Cork, and Tralee.

On the 1st May and 4th December, the regiment was inspected at Gibraltar by General Sir G. Don, G.C.B., and reported on in each instance as "in high order in every respect." In both of these Inspection Reports, the Colours were reported on as "unserviceable."

Particulars of the presentation of new colours to the regiment, in May 1827, will be found in Chapter XIV, "Notes on Colours."

1828–29

Brevet Lieut.-Colonel R. Bayly was, in September 1828, promoted lieutenant-colonel, in succession to Colonel Forssteen, retired, and had scarcely arrived at Gibraltar, to assume command of the regiment, when a most dreadful epidemic broke out, early in September, of fever of a most virulent type, its symptoms somewhat resembling the yellow fever peculiar to the West Indies.

The garrison (consisting of the 12th, 23rd, 42nd, 43rd, 74th, and 94th Regiments, with a proportion of artillery and engineers) had upwards of 500 men, women, and children affected by it, whilst the civil population suffered in a far greater measure. The troops were moved to the Neutral Ground, but many deaths continued to occur until the 10th of December, after which no fresh cases were admitted into hospital, and those already suffering soon resumed their wonted health.

As instances of the severity of the disease, it may be mentioned that the men selected to attend the patients in the naval hospital died so rapidly that, at length, no volunteers offered themselves for this service, as the employment was attended by certain death. Twenty-three soldiers of the garrison, who had acted as orderlies in the hospital, ceased to exist in the course of three weeks. Scarcely a family in Gibraltar escaped the direful effects of this desolating scourge; in about six weeks from its commencement upwards of 4000 bodies were buried in the long trenches on the Neutral Ground.

On the 27th December, His Excellency the Lieut.-Governor inspected the regiment, and was pleased to express his entire approbation of its

[1] *The Army Book for the British Empire* (Lieut.-General Goodenough).

BATTALION COMPANY OFFICER'S SHAKO PLATE
(*exact size*).
1829-45.

LIGHT COMPANY OFFICER'S SHAKO PLATE
(*exact size*).
1829–45.

appearance, perfect state of interior economy, and correctness of the books, which reflected infinite credit on those entrusted with the discipline of the corps during this trying crisis.

The regiment sustained a loss of two officers (Lieutenants H. G. Forssteen and O. K. Werge) and 53 soldiers, from the distressing malady; and 4 officers and 218 men had been affected by it.

The following is a summary of the effects of the disease :—

84 officers were attacked, of whom			15 died.	
1494 rank and file	,,	,, ,,	436	,,
426 soldiers' wives and children	,,	,, ,,	61	,,
4701 civilians	,,	,, ,,	1281	,,

Enlistment from 1829 to 1847 was for life.

The regimental metal seal, here shown, was found amongst the effects of Lieut.-Colonel W. H. Forssteen, 12th Regiment, who died in 1828. By the courtesy of his grand-daughter (Miss Craig) it is now in possession of the officers, 1st Battalion.

Regimental Seal, 1828.

1830–34

In 1830, the first Charter of Justice was given to the City of Gibraltar. A magistracy was established, and the advantage of civil liberty accorded to the inhabitants. Nor had its military interests been neglected; the fortifications, always strong, had been vastly extended and improved, the heaviest ordnance replaced lighter guns of former days, and the number was greatly increased; 550 guns were then in position, while, during the Great Siege, the number was 96 only; also, immense stores of supplies and ammunition were constantly maintained in the greatest efficiency.

On the 20th June, 1830, at 3 A.M., His Majesty King George IV expired at Windsor Castle, and His Royal Highness the Duke of Clarence succeeded to the Throne, with the name and title of King William the Fourth. As the King had no heir, this made the Princess Victoria the immediate heiress to the throne.

Corporal punishment awarded by regimental courts-martial was, in 1830, limited to 300 lashes, and in 1831, by district court-martial, to 500.

On retirement from the service, in 1831, of Lieut.-Colonel R. Bayly, Major Gervas Turbervill was promoted Lieut.-Colonel.

The death occurred on the 1st January, 1832, of General Sir George Don, G.C.B., Governor of Gibraltar, whose service in the army exceeded 61 years; he was buried with full military honours on the 4th, in the Garrison Church, where a monument is erected to his memory.

The regiment continued at Gibraltar throughout the year 1833, and embarked for England, by detachments, on the 4th and 5th April, and 6th May, 1834, in the transports "Hope," "Sylvia" and "Stentor," and, on disembarking at Portsmouth, on the 30th April and 10th June, marched to Winchester.

Prior to leaving Gibraltar, where the corps had been stationed ten years and six months, His Excellency the Lieut.-Governor published in a Garrison Order that he

> "could not suffer the 12th Regiment to embark without expressing his satisfaction at their regular and orderly conduct during the period of serving under his command in the garrison, and, more particularly, he acknowledged his sense of the zealous and constant exertions of Captain French, to maintain the discipline of the corps whilst under his command."

In the middle of September 1834, the regiment marched from Winchester, in three divisions, to Weedon, and, arriving on the 24th, remained until the 6th October, when 5 companies marched to Manchester, and were detached as follows:—2 to Bolton, and 1 to Burnley, Rochdale, and Blackburn respectively. On the 13th October, further changes of companies took place, one proceeding to Nottingham, and, on the 23rd, the headquarters and 2 companies marched to Blackburn, where, on arrival, one was detached to Burnley.

1835–36

The 12th remained at the above stations until the 25th May, 1835, when the headquarters and several detachments received Routes to march for Salford Barracks, Manchester, where the corps was concentrated on the 2nd June.

On the 19th November, the regiment proceeded by rail to Liverpool, embarked, on the following day, on the steamers "Express" and "Ennisfall," and disembarked on the 21st at Dublin, occupying quarters at Beggar's Bush, George Street, Portobello, and Royal Barracks, until the 28th December, when the headquarters and four companies moved to Richmond Barracks, where the whole regiment was together shortly after.

Between the 30th September and the 3rd October, 1836, the regiment marched from Dublin in three divisions, and had arrived at Athlone by the 7th, detaching one company to Shannon Bridge, and the grenadier company to Roscommon.

1837–41

The next move was directed on the 30th May, 1837, when routes were received to march to Cork, where the corps was concentrated on the 16th June, preparatory to embarking for Mauritius, in relief of the 29th Regiment.

On the 20th June, King William IV died, and Her Majesty Queen Victoria succeeded to the throne.

The regiment was inspected at Cork on the 22nd July, by Major-General Sir A. Norcott, K.C.H., when the establishment was directed to again consist of 6 service and 4 depot companies, the former mustering 25 officers and 520 of other ranks; the depot, of 4 companies, to consist of 14 officers and 196 of other ranks, under a field officer.

On the 9th August, the headquarter division of the regiment, under Lieut.-Colonel J. Jones, marched from Cork Barracks, to embark at the Cove of Cork, on the troopship "Athol," for conveyance to Port Louis, Mauritius, and, arriving on the 30th November, disembarked on the following day.

The depot companies proceeded in August to Kinsale.

The 2nd division of the regiment, under command of Captain Sir R. Douglas, Bart., embarked at the Cove of Cork on the 27th September, in the transport "Permailia," which sailed on the 12th October, and, with reference to this detachment, the following is recorded (dated 23rd September, 1837), prior to its departure from Cork.

> "There was on Saturday and Sunday last, in Cork, a most dreadful conflict between some soldiers of the 12th, and another regiment stationed in the city. The cause of the trouble arose with regard to the bravery and valour of the respective parties. So serious has been the *rencontre* between them, that 18 are in hospital, desperately wounded. Both regiments were ordered out of the town on Wednesday."

On the 23rd January, 1838, the 2nd division of the regiment arrived at Port Louis, Mauritius, and disembarked the following day.

The depot companies, this year, were removed to Cork.

It was this year ruled, that no sentence of corporal punishment in the Army was to exceed 200 lashes.

On the 7th May 1839, the depot companies embarked in the steamer "Victory" at Cork for Aberystwyth, and were stationed at Newtown, Builth, and Brecon. Strength all told, 498, including 66 women, 129 children, and 3 officers' wives.

On the augmentation of the army in August, the establishment of the 12th was increased to 47 sergeants, 14 drummers, and 800 rank and file.

The depot companies proceeded, in May 1840, to Scotland, and occupied the barracks at Paisley, and in May 1841, they were stationed at Sunderland.

1842

Formation of a Reserve Battalion.

An augmentation of the regiment having been authorised in April, the establishment was now to consist as follows:—1 colonel, 1 lieut-colonel, 2 majors, 12 captains, 14 lieutenants, 10 ensigns, 6 staff, 67 sergeants, 25 drummers, and 1200 rank and file, a total strength of 1338.

It was further directed, on the 7th May, that the regiment, numbering 540 rank and file, was to have an addition of a Reserve Battalion, mustering the same numbers, the depot being made up to 120 rank and file.

The present depot to be completed by volunteers, and formed into six equal companies, to which the men were to be posted without reference to size or appearance, and to be nominated the "Reserve Battalion," the present service companies being considered the 1st Battn. (In compliance, 255 volunteers were received from the 6th, 16th, 19th, 32nd, and 59th Regiments.)

The Reserve Battalion was not to have flank companies, and the band to remain with the 1st Battn., the Reserve Battn. not being permitted to form one.

The post of the lieut.-colonel was to be with the 1st Battn., and, assisted by the junior major, he was to be considered as having the general charge and superintendence of the whole regiment, and, when both battalions were together, to be responsible for its discipline and efficiency.

The senior major was to command the Reserve Battn., with which the whole of the Depot Staff were to continue to serve, with an additional officer acting as quarter-master.

In the senior major's absence, the junior major was to command the Reserve Battn.

When both battalions were stationed together, there was to be one officers' mess, the Reserve Battn. carrying with it the whole of the Depot Mess establishment.

When the two battalions were separated, the mess plate and utensils were to be divided as under the depot system then in vogue.

The Depot to consist of 1 captain, 3 subalterns, 6 sergeants, 2 drummers, and 120 rank and file.

At least three of the depot non-commissioned officers were to be expert drills, capable of instructing recruits, and the men to be selected, in the first instance, from the oldest soldiers.

On the 2nd July, the Reserve Battn. was completed in appointments, and 600 stand of new arms and 20 fusils were issued for its use from the Tower; immediately after, the battalion was distributed in several detachments in the Northern Counties, in aid of the civil power, furnishing companies to Stafford, Wolverhampton, Newcastle-under-Lyme, Burslem, and Coventry. These detachments were finally concentrated at Weedon on the 4th September.

A Horse Guards' General Order, dated 23rd September, announced the Queen's command, to express Her Majesty's high approval of the good conduct, exemplary forbearance, and steadiness of the military force in support of the civil authorities, during the disturbances which had unhappily prevailed in many of the northern and midland counties—the same to be communicated through General Officers to all concerned.

OFFICERS' SHAKO PLATES (*reduced sizes*).

1846–55.

GRENADIER COMPANY OFFICER. BATTALION COMPANY OFFICER. LIGHT COMPANY OFFICER.

By the 23rd September, the Reserve Battn. had arrived in two divisions at Chatham, where it was quartered until the 19th October, when it moved to Winchester.

On the 29th October, Major Sir Robert Douglas, Bart., succeeded to the command of the Reserve Battn. *vice* Major O'Neill retired.

The battalion was inspected on the 1st November, by Major-General the Honourable Sir H. Pakenham, K.C.B., and orders were received this month to tell off a depot company of 1 captain, 2 subalterns, 2 sergeants, 2 drummers, and 115 rank and file, completing them with all the necessary appointments, fusils, muskets, etc.

On the 9th November, the Reserve Battn. proceeded to Portsmouth to embark on the transport "Java," and sailed on the 16th for Mauritius.

1843–45

On arrival at the Cape of Good Hope, on the 13th January, 1843, the battalion was detained by the Governor, Major-General Sir G. Napier, K.C.B., until the disturbance caused by the Boers on the northern frontier, and in the vicinity of Port Natal, should be quelled. The battalion, on disembarking, furnished small detachments at Simon's Bay and Robben Island, and, on the 18th May, re-embarked in H.M.S. "Thunderer," arriving at Mauritius on the 11th June. During the detention of the battalion at Cape Town, final arrangements were made (in accordance with the instructions received) for the separation of the battalions, the Officer Commanding 1st Battn. transmitting a report of the same, for the information of the Commander-in-Chief, to which a reply came, dated 25th July, approving of all the arrangements that had been made (in connection with the posting of officers and men to the 1st and Reserve Battns. 12th Regiment) "all of which appeared to His Grace to be perfectly satisfactory."

On the 1st July, the Reserve Battn. was augmented by a transfer of 104 men from the 1st Battn., with 60 stand of arms and sets of appointments, and in August, a further increase of 50 men from England.

Lieut.-Colonel Jones retired on the 18th August, and was succeeded in command of the 1st Battn. by Lieut.-Colonel Patton.

Captain William Bell was, on the 1st November, ordered to assume command of the Reserve Battn. in succession to Major Sir Robert Douglas, Bart., who had died after a lingering illness.

On the 3rd June, 1844, the headquarters of the Reserve Battn. marched from Port Louis to Mahébourg, to relieve the headquarters of the 1st Battn., and furnished detachments to Flacq, Grand River, S.E., Black River, Souillac, and Canonier Point, in relief of detached companies.

The Reserve Battn. was inspected, on the 28th November, by His Excellency the Governor, Lieut.-General Sir William Gomm, K.C.B., and was then under command of Major Bell, who had been promoted.

On the 23rd January, 1845, Major Glover was appointed to the command of the Reserve Battn. *vice* Major Bell, transferred to the 1st Battn., and in May, the headquarters Reserve Battn. returned to Port Louis, where the detachments rejoined.

1846–47

By a circular memorandum, dated 20th February, 1846, the 1st Battn. was increased to 560 rank and file, to provide for musicians, and, on the 5th May, a lieut.-colonel and an adjutant were posted to the Reserve Battn. By this arrangement, Major Glover was gazetted to the lieut.-colonelcy. On the 7th August, a surgeon was also appointed to it, and, on the 11th December, a quarter-master.

On the 4th and 6th May, 1847, the Reserve Battn. marched, in two divisions, to Mahébourg, to exchange quarters with the 1st Battn., which was proceeding to Port Louis, to embark for England.

A Horse Guards' letter, dated 12th July, having directed that the former should be completed by volunteers from the 1st Battn., prior to the embarkation of the latter, 89 non-commissioned officers and men were transferred to the Reserve Battn. on the 9th December, and all invalids and inefficient men were, under the above order, transferred from it to the 1st Battn.[1]

On the 2nd November, the 1st Battn. 5th Fusiliers arrived at Mauritius, in the troopship "Resistance," to relieve the 1st Battn. 12th, which embarked for England on the 15th December, under command of Lieut.-Colonel Patton.

Enlistment for ten years was this year introduced.

[1] *W. O.* 3, Book 105, p. 289.

CHAPTER IX

England, South Africa, Ireland, Australia, Great Britain, New Zealand, Ireland, East Indies. 1848–1867

1848

The 1st Battn. arrived at Spithead on the 1st March, and disembarked on the 3rd at Portsmouth, where it was joined by the depot company from the Isle of Wight.

A Horse Guards' letter, dated 4th March, directed that, from the date of the battalion disembarking, its establishment was, until further orders, to be 660 rank and file, together with the prescribed number of thirty supernumeraries. Moreover, the battalion was henceforth to serve as the depot for the Reserve Battn., and to be prepared to supply it, from time to time, with such drafts of non-commissioned officers and men as might be required to keep it complete, during its term of service with the Colours.[1]

Apprehensions were entertained that the public peace would be disturbed by the several meetings of Chartists in the vicinity of the Metropolis on Monday, the 10th of April, 1848; and as they appeared determined to unite on Kennington Common, in order to proceed, in procession, to the House of Commons with their petition, the Government took the usual precautionary measures to prevent tumultuous assemblages of the people.

Accordingly the 1st Battn. was ordered, by rail, to London on the above date, but, as the meetings dispersed more quietly than was anticipated, the regiment continued its route to Weedon on the same day.

The headquarters and detachments of the Reserve Battn. were again concentrated at Port Louis, Mauritius, on the 16th May.

On the 12th June, under the temporary command of Major Bell, it was inspected by Lieut.-General Sir William Gomm, K.C.B., when the Inspecting Officer (having already expressed his satisfaction at the soldierlike manner

[1] *W. O.* 3, Book 108, p. 19.

in which the march from out stations to Port Louis was performed) was "now equally satisfied with the general appearance of the corps, both in quarters and in the field," remarking particularly on the "orderly arrangement of the packs (knapsacks), and the cleanliness of the barrack rooms."

1849

The 1st Battn. continued at Weedon, furnishing detachments during the winter, in the northern and midland counties of England, until the 19th March, when the regiment moved to Northampton, the barracks at Weedon being required for the augmentation of the 87th Regiment, on its proceeding to India. The 12th returned to Weedon on the 19th April, when the several detachments rejoined headquarters, except those at Wellington (Salop) and Northampton. Lieut.-Colonel Glover, commanding Reserve Battn., having retired on the 30th March, was succeeded in the lieut.-colonelcy by Major Bell.

On the 14th July, new Colours were presented to the 1st Battn. at Weedon. At 12 o'clock the regiment, under Lieut.-Colonel Patton, paraded in review order, and was inspected in line by Colonel Arbuthnot, commanding the District, after which the new Colours were consecrated by the Rev. G. R. Gleig, and presented by the Honourable Mrs. Arbuthnot.

Colonel Arbuthnot narrated, in complimentary detail, the distinguished achievements of the regiment in the field, and commended their good conduct and discipline since return from foreign service.

The Rev. Mr. Gleig then eloquently vindicated the rite of consecration, and entreated those who heard him to honour the Colours entrusted to their keeping, as much by their moral conduct in quarters as by their bravery in the field.

After the ceremony, luncheon was served at the Officers' Mess, of which more than 150 partook. The sergeants also gave a ball; and a good dinner, outside the barracks, was provided for the men, General the Honourable R. Meade (the Colonel) having sent £100 to the regiment for the occasion.

The correspondence which followed, as to the disposal of the old set of Colours, will be found in Appendix XII.

1850

On the 2nd April, Lieut.-Colonel Patton, on appointment to a Recruiting District, was succeeded in command of the 1st Battn., by Lieut.-Colonel Rumley, transferred from the 6th Regiment, and Major Perceval was promoted to command the Reserve Battn., on the retirement of Lieut.-Colonel Bell.

Between the 3rd and 5th April, the 1st Battn. marched, in three divisions, from Weedon to Chatham, detaching one company to Harwich.

At the end of May, the headquarters Reserve Battn. moved to

Mahébourg, in relief of the 5th Fusiliers, furnishing detachments to Flacq and Souillac, the latter rejoining headquarters on the 16th September.

As the result of an inspection of the Reserve Battn. on the 3rd October, by Major-General the Honourable Wm. Sutherland, commanding troops at Mauritius, the Inspecting Officer was pleased to express his approbation of the clean and soldierlike appearance of the battalion, the cleanliness of the barracks and hospital, and the highly creditable state of the men's kits.

1851

On the 15th and 17th July, the Reserve Battn. marched in two divisions to Port Louis, where the detachments rejoined, and on the 12th August, under Lieut.-Colonel Perceval, embarked in the steam sloop "Hermia," for conveyance to Port Elizabeth, Cape Colony, where the regiment landed on the 24th, to reinforce the troops employed under Lieut.-General Sir Harry Smith, G.C.B. (in active operations against the Kaffir tribes), the regiment's services being required in Lower Albany, where the Kaffirs were threatening Grahamstown, the chief town of the district.

On the 27th August, the battalion marched to Grahamstown, which was reached, without any incident of interest, on the 3rd September, *en route* to join the camp of Major-General Somerset, commanding the 1st Division.

The Diary of Lieutenant and Adjutant E. Foster relates that, on marching from Port Elizabeth:—

> "every officer of the regiment was on horseback, carrying a gun, pair of pistols, sword, &c., and that the water, *en route*, was very bad, consisting chiefly of stagnant, odoriferous pools, the days being extremely hot, and the nights very cold. The Kaffir country was intersected with ravines, or kloofs as they were called, and by far the greater part of it consisted of bush or low jungle."

At Grahamstown, the men's white cross-belts and large pouches were taken into store, and replaced by white waistbelts with pouches affixed. A jacket of red cloth was also worn, and trousers of moleskin.

Details of the actual strength of the battalion are not forthcoming, but, besides Lieut.-Colonel Perceval and Major Horne, it appears to have consisted of six companies, commanded by Captains Bagnell, Espinasse, Crofton, Palmer, Hearne, and Gillman ,and seven subalterns ; also Lieutenant and Adjutant Foster, Acting Paymaster Herrick, and Surgeon-Major Dick.

On the 6th September, the headquarters marched towards Bathurst, leaving one company at Grahamstown, and furnishing detachments to Riebeck and Fort Brown. From Bathurst, they marched, on the 10th, to join the camp of Major-General Somerset, commanding 1st Division, at Rietfontein.

The following is given by Lieutenant Foster as the list of troops then serving here.

Detachment Royal Artillery (strength not mentioned); the headquarters, 2nd "Queen's," 6th Regiment, 12th (Reserve) Regiment, 4 companies 60th Rifles, 1 company 73rd, 74th, and 91st Regiments, 3 or 4 troops Cape Mounted Rifles, and a large number of Fingoes, the whole mustering upwards of 4000.

The difficulties to be encountered in the fighting in this country were, that owing to the enemy's natural adroitness in concealing themselves in the bush, they were seldom seen, whereas the attacking force had to advance across open country or up narrow defiles, and a novelty for the troops was their being confronted by assegais.

On the 2nd October, the detachment at Fort Brown, under Major Horne (Reserve Battn. 12th) had a smart engagement with the enemy, which was commented on in a highly favourable report from Major-General Somerset, as "a very gallant affair, in which Major Horne, by the judicious and soldierlike disposition of his force, defeated a large body of combined Hottentots and Kaffirs, killing ten of their number, and capturing several head of cattle."

To the above (published in a General Order, dated 7th October), Sir Harry Smith added his "gratification in noticing the spirited conduct of the officers and men engaged on this occasion."

On the 3rd October, the whole of the troops at Rietfontein, under Major-General Somerset, moved to Fort Beaufort, and, on the 13th, proceeded on patrol, under command of that officer, to attack the enemy's strongholds in the Water Kloof, Blinkwater, and Fullers-Hoek; the detachments from the Bathurst district having rejoined the headquarters from that date, they were actively engaged with the enemy until the 6th November.

In an engagement on the 4th November, in which the 12th, 74th, and 91st Regiments were engaged, the 12th had no casualties, but the horse Lieutenant Foster rode was shot in the neck. The 74th had about a dozen casualties, and Lieutenant Foster relates that the bodies of our troops who fell at this period, had to be buried in solemn silence without full military honours, "ammunition being too precious to expend on funerals."

The Water Kloof proved a very great difficulty, and was attacked several times before the bush was finally cleared of the enemy. In spite of the fighting taking place in kloofs, the want of water was seriously felt.

On the 19th November, the battalion marched from Blinkwater to Lower Albany, and arrived on the 22nd, the headquarters being encamped at Governor's Kop, and detached camps posted at Collingham, Botha's Hill, and Driver's Farm.

Kaffirs and Hottentots were constantly crossing the frontier to raid farms in Albany.

These several detachments seem to have formed a barrier between the seat of war on the east, and Grahamstown and the rest of the Colony on the west; their object was to prevent the rebel Kaffirs and Hottentots from crossing into the colony, and, when they had been successful in getting across, in preventing them from getting away with their loot.

Lieut.-Colonel Perceval was placed in command of the Lower Albany district, and, for the energetic measures which he adopted for the protection of the colony, was thanked by Sir Harry Smith, who expressed his marked approbation of the colonel's zealous exertions.

The Reserve Battn., 12th, continued to guard the frontier from the time of their arrival at the above posts, until the following March, when they again took the field against the Kaffirs in the Great Fish River Bush. Large quantities of looted cattle were thus recaptured, and returned to their owners, and a number of the enemy killed.

1852

The Wreck of H.M.S. "Birkenhead."

H.M.S. "Birkenhead," Commander Salmond, an iron paddle-wheel transport, sailed from Cork on the 7th January, 1852. She had on board 731 souls, of whom 132, including officers, belonged to the Royal Navy, and 360 officers and men to different regiments, while Lieut.-Colonel Seton, 74th Highlanders, commanded the troops.

The second in command was Captain Wright, of the 91st Regiment, who was one of the two, out of eleven, officers on board, who escaped, and to whose narrative is due most of the information received.

The following was the Embarkation Return :—

Officer Commanding—Lieut.-Colonel Seton, 74th Highlanders.

	Officers	Men
12th Lancers	1 officer,	4 men.
2nd Queen's	1 officer,	35 men.
6th Regiment	1 officer,	47 men.
12th Regiment		55 men.
43rd Light Infantry	1 officer,	29 men.
45th Regiment		3 men.
60th Rifles		30 men.
73rd Regiment	2 officers,	54 men.
74th Highlanders	1 officer,	48 men.
91st Regiment	1 officer,	44 men.
Army Medical Dept.	2 officers,	
Total	11	349
Grand total of officers and men,		360

Most of the officers were very junior, and of the men, the largest proportion were recruits. It should also be noted that the non-commissioned officers were few in number, and that three of the drafts had no officer of their own regiment with them. These facts enhance greatly the splendid behaviour of the men.

The "Birkenhead" had a fairly prosperous voyage, and arrived at Simon's Bay on the 23rd February, when orders were received that she was to proceed at once to Algoa Bay, and the Buffalo River, in order to land the drafts. Accordingly, on the evening of the day of arrival at Simon's Bay, she resumed her voyage. The night was calm, the sea smooth, the stars were shining, and the shore was about three miles distant. Look-out men were posted, the lead was kept going, and everything presaged a safe passage. Shortly before 2 A.M. on the 24th, the man who was casting the lead called out, "twelve fathoms," but the words had hardly passed his lips when a violent shock was felt. The ship had struck on some sunken, jagged rocks, not marked on the chart. Water at once rushed in through a hole in the fore part of the vessel, and drowned many soldiers, who were lying in their berths on the lower troop-deck. As soon as the shock was felt, the Captain and the officers not on watch rushed on deck. Colonel Seton at once addressed the officers, who gathered round him, on the necessity of preserving discipline and silence among the men, and then ordered the troops to draw up on both sides of the quarter-deck. The men silently and promptly obeyed, as calmly as if they were falling in for inspection. Parties were then told off to different duties; one party being directed to work the pumps, another to assist the sailors to lower the boats, and a third to throw the horses overboard. Captain Wright says, "Everyone did as he was directed; all received their orders and had them carried out, as if they were embarking, instead of going to the bottom. There was only this difference—that I never saw any embarkation conducted with so little noise and confusion." In the meantime, the Captain had directed the engines to be reversed, in the hope of backing the ship off, and getting nearer the shore. The result was disastrous, for the vessel came on the rocks again with another bump, which knocked a great hole in her bottom, and admitted tons of water, which extinguished the engine-room fires. In the meantime, three boats were successfully lowered; the others had been stove in, or could not be launched. Into the three boats the women and children were passed down, the soldiers remaining steady in the ranks, though they saw their last chance of safety disappearing. Colonel Seton himself stood in the gangway with a drawn sword, to see the embarkation conducted, and to prevent any rush on the part of the men. None, however, was attempted, and all the women and children being thus placed in safety, orders were given that the boats should lie off at a distance of 150 yards from the doomed ship. About ten minutes after striking, the Birkenhead broke in two at the foremast, and the funnel fell, crushing some, and throwing into the sea others who were trying to clear the paddle-box boat. Nevertheless, the men, notwithstanding that many were mere lads, and that the horror of their situation was aggravated by the fact that it was night, kept their places. Owing to the breaking off of the fore part, the ship began to settle down forwards, and the survivors took refuge on the poop without any disorder. At length, Commander Salmond, seeing that all hope was lost, advised all those who could swim to jump overboard, and try to reach the boats. Colonel Seton, however, exhorted the men to do no such

WRECK OF H.M.S. "BIRKENHEAD."
(26TH FEBRUARY, 1852.)

(*By the kind permission of Messrs. Henry Graves & Co., Pall Mall, London, S.W.*)

thing, as if they succeeded in reaching the boats, they would be sure to swamp them, and send the women and children to the bottom. Nobly they responded to the appeal, and not a man stirred. The officers, now seeing that the end was near, and the vessel rapidly sinking, shook hands, and bade each other farewell. Immediately after, the ship broke again abaft the mainmast, twenty minutes after striking, and threw the noble survivors into the water, as well as the main-top-mast, and a few spars. Many were carried down by the sinking ship, entangled in the rigging; others were drowned at once. Some managed to obtain a temporary refuge on the main-top-mast and other spars which were above water, and attached to the wreck. Others again, who could swim, struck out for the shore. An eye-witness says, that up to the disappearance of the ship, "there was not a cry or a murmur." When, however, the ill-fated "Birkenhead" made its final plunge, we can well imagine that there were many cries sent up to heaven by sinking men, though, no doubt, the survivors were all too much occupied by efforts for their own safety to hear much, or to think of anything but themselves. The disciplined body of British soldiers and sailors had been dissolved, broken into units.

It is impossible to say how many of the four or five hundred, still left alive when the ship sank, could swim. Probably not many could accomplish more than a few yards, for there were then fewer swimmers, both in the Army and Navy, than now. Even the strong swimmers were almost numbed with despair. It is said that the shore was about three miles from the wreck, but judging by the comparatively large number who reached it (though all did not succeed in landing) it seems reasonable to suppose that the interval was not more than a few hundred yards. This handful of men must have undergone a seeming eternity of torture, as they struggled on their weary way through the darkness, for sharks abounded, and imagination must, at every moment, on any contact with seaweed or a piece of wreckage, have suggested that one of these monsters of the deep was on them. It was, indeed, no imaginary danger, for many of the swimmers were seized and carried down by these ferocious creatures. Even of those who did succeed in reaching the shore, some failed to find escape from death, for the coast was fringed with a thick and impenetrable growth of seaweed; many also were dashed against the rocks. No doubt the majority sank from exhaustion. Captain Wright (91st), and seven men of his regiment, were more fortunate than their comrades. They succeeded, though in a state of extreme exhaustion, in effecting a landing at Point Danger. Weak as they were from the physical and mental trial which they had gone through, they struggled over fifteen miles of rocky and barren coast, till they reached a house belonging to Captain Small, where they were treated with the utmost kindness, until conveyed to Cape Town by H.M.S. "Rhadamanthus." Captain Wright experienced a most remarkable escape. When the ship sank he had a life-belt on, but was carried down by the suction. His jaw was broken, and he was entangled in some of the rigging. Providentially, he had on his person a clasp knife, with which he cut himself free, when he rose to the surface, and struck out for the shore, which he reached in safety.

When the "Birkenhead" went down, the boats picked up as many as they could of the survivors—some eight or ten. When they reached the shore, they found it impossible to land, so they pulled towards Simon's Town.

They had not gone far when they came across the coasting schooner "Lioness," which took them on board, the captain, his wife, and the crew treating the hapless people with the greatest humanity and tenderness. The "Lioness" then made for the scene of the wreck, being joined by H.M.S. "Rhadamanthus," and found, at 2.30 P.M., some 45 soldiers and sailors, who had clung to the mast for about twelve hours, and who were all rescued. Of the total number of 731 souls on board, all the women and children, and 164 officers and men of the Army and Navy were saved; 376 officers and men of both Services, including seven naval and nine military officers, were drowned, or devoured by sharks.

In a General Order, issued by the General Officer Commanding the Forces in South Africa, on the occasion, it is stated that Captain Wright merited every encomium, that true valour was never better exemplified than on similar awful occasions, and that the bravery, gallantry, and soldier-like conduct with which the men had met their fate would be reported to the Duke of Wellington, their Commander-in-Chief.

Captain Wright was promoted to a brevet-majority, and the Queen caused a memorial to be erected in Chelsea Hospital, to the memory of those who perished, every officer's and every man's name and regiment being inscribed. The King of Prussia was so struck by the magnificent behaviour of the troops, that he brought the glorious episode to the notice of his whole Army, by ordering a statement to be read at the head of every regiment. The story speaks for itself, and needs no embroidery of praise; but we may be permitted to remark that the British Army may with justice regard the wreck of the "Birkenhead" as an undying subject of pride. Nothing grander has ever been recorded in the history of any nation, in the annals of any army.

Napier, the Peninsular historian, writing on the event, said:—

> "The records of the world furnish no parallel to this act of self-devotion; men like these do not perish; their bodies may be given to the fishes of the sea, but they live on immortal in the minds and hearts of their fellow men."

Thackeray, the author, in an impressive speech, paid eloquent tribute to this act of heroism, and said:—

> "No saint ever died more simply, no martyr more voluntarily, no victim ever met his fate in a more generous spirit of self-renunciation than those who were content to be engulfed so long as the women and children could be saved."

Referring to the Kaffir War, the following extract from a letter, received by Lieut.-Colonel Perceval from Sir Harry Smith, was published in Regimental Orders, on the 8th January, 1852, for the information of the Reserve Battn.:—

> "The conduct of the Reserve Battn. 12th, under your command, has most deservedly acquired my approbation upon all occasions, when employed on the harassing and most fatiguing duties of a desultory warfare like this."

The Reserve Battn. was now constantly employed in patrols, and, on the 26th January, one of them captured 78 head of cattle. But, in spite of all the precautions taken, looting continued until the end of the war. Cattle were being constantly stolen from settlers, and considerable numbers were actually conveyed across the frontier, evading the constant watch that was kept. Lieutenant Foster states that patrols were ordered to search for men only; as, however, it had been customary to dispose of the proceeds of the sale of captured cattle among the troops, it is not improbable that capturing cattle became more interesting than searching for Kaffirs, who were very difficult to find, and, when found, seldom gave a fair fight.

Throughout the months of January and February, troops, under Colonels Eyre and Michel respectively, were employed in scouring the country, and destroying the enemy's crops. Lieutenant Foster relates, that on one of these expeditions, 30,000 head of cattle were captured, and herds of goats, besides horses and stores.

Mention is made more than once in Lieutenant Foster's letters, of ammunition being sold by Dutch colonists to the rebels, and, somewhat later in the war, letters were found incriminating persons of no mean position. It was believed that the action of these traitors had resulted in prolonging the war, for without ammunition it could not continue.

About the middle of January, the Governor created a service known as the Rural Police, consisting of 2 officers, 10 sergeants, and 400 privates, half of which were to be mounted. This was followed, early in February, by a Proclamation issued by the Governor, calling out all Burghers between 20 and 50 years of age; the 8th of March being fixed for assembling.

It having been decided by His Excellency the Commander-in-Chief, Sir Harry Smith, that a general attack should be made upon the various camps and strongholds occupied by the Kaffirs and rebel Hottentots, along the whole line of the Eastern Frontier, and in the Amatola range of mountains, on the 8th March, the Reserve Battn., 12th Regiment, reinforced by a detachment of Royal Artillery, with a 12-pounder howitzer, and a supply of rockets; a detachment of Cape Mounted Rifles; Armstrong's Horse, under Major Armstrong, C.M.R., some mounted Burghers from Lower Albany, and some of the Grahamstown Fingoe Levy, in all, a force of about a thousand fighting men, under the command of Lieut.-Colonel Perceval, 12th Regiment, proceeded to attack the camps occupied by the followers of the Kaffir Chiefs Stock Tola and Seyolo, and that of the rebel Hottentots in the Fish River Bush.

In the course of seven days' operations, the enemy, who were found to be numerous, certainly some two thousand fighting men (strongly posted in three distinct camps, with connecting kraals, extending for some miles along the left bank of the Great Fish River), were completely driven out of their several fastnesses with the loss of many of their numbers, and the capture of five hundred head of cattle, a portion of which was the property of the Chief Stock Tola. Several horses and many hundred goats were also captured.

It was reported that among the slain were Stock Tola and Pakati, his councillor.

These camps, from which, since the commencement of the war (fifteen months previously), so many marauding parties had issued to devastate the frontier farms, having been completely destroyed, the Division followed the enemy to the Keiskama River, which they crossed by the drift near the now abandoned post of Fort Wiltshire.

Having scoured the country for two days along the banks of the Iquibeka, the Division then proceeded, via Fort White, to Fort Cox, where orders were found for the force, under Lieut.-Colonel Perceval, to co-operate with that of Lieut.-Colonel J. Michel, 6th Regiment, in scouring the Amatala Basin, and devastating the enemy's crops.

On the 24th March, the various patrols which had proceeded at the commencement of the month to attack the enemy's strongholds, rendezvoused at Fort Cox; they had all been eminently successful.

The following were the troops now assembled at Fort Cox:—six gun detachments "Royal Artillery," 12th Lancers, 6th, 12th (Reserve Battn.), 43rd, 60th, and 73rd Regiments, and detachments of the 2nd "Queen's," 45th and 91st Regiments, Cape Mounted Rifles, sundry European and Native Levies, and mounted Burghers.

On the 26th, the various divisions again marched off; Colonel Perceval, on this occasion, co-operated with Colonel Eyre, 73rd Regiment, in scouring the dense forests of the Buffalo Poorts and Peri Bush, which was most successfully performed, during the course of three days' operations.

Some cattle fell into our hands, and a large quantity of crops were destroyed.

Two companies of the battalion, and a party of Cape Mounted Riflemen, which had been detached for the last two days under the command of Captain Gillman, 12th Regiment, whilst on their march to rejoin the Division, were attacked by an overwhelming number of Kaffirs near the Tab-Indoda Mountain; the firing having fortunately been heard, the remainder of the division pressed rapidly on with the Artillery and succeeded in rescuing them.

On the 31st, the Division reached the Gubula River, and, from thence, scoured the country along the banks of that river to its junction with the Keiskama, as also the dense forest land in which the Ischaka and Gubula take their rise. The Division then moved into the Keiskama Hoek. From this camp a patrol was taken out by Lieut.-Colonel Perceval for three days, in the course of which, the whole of the mountainous and forest land between the Nyami and Gologha Mountains was most thoroughly searched, and about six hundred head of cattle and several horses were captured.

On the 7th April, the Division returned to Gubula, the 73rd and 43rd Regiments (Colonel Eyre's Brigade) replacing it at the Keiskama Hoek.

Before commencing further operations, it was considered advisable to send the cattle already captured into King William's Town, and, with this intention, the Reserve Battn., 12th Regiment, proceeded to Bailey's Grove, where fresh supplies met them, a company and a detachment being sent on to King William's Town with the cattle, returning the following day.

On the 11th April, an attack made on the Kraal of Sandili's half-brother Clu-Clu (or Xo-Xo) was successful; the huts of that Chief, which were most strongly posted, were burnt, and his cattle taken, as well as that of many of his followers. Xo-Xo himself narrowly escaped; his horse, from which he threw himself, was taken. During the following days, the regiment was constantly occupied in scouring the country and devastating the crops in the neighbourhood of the Tab-Indoda Range, and the sources of the Gubula and Isckaka Rivers. In the middle of April orders were received from General the Honourable George Cathcart (who had recently arrived to relieve Sir Harry Smith), for the 12th Regiment to return to Lower Albany, searching, on their way, the countries of Stock Tola and Seyolo, and examining the positions of the old camps in the Fish River Bush. This service was duly performed, and on the 21st April the battalion encamped at Botha's Hill, after a long and most arduous patrol of forty-six days' duration, which had been most successful, while the loss of life, on the part of the Division, was but trifling. The patrol was supposed to have marched 1014 miles, killed 124 of the enemy, captured 1476 head of cattle, including 60 sheep and goats and 16 horses.[1]

The casualties during this Patrol were two men killed, one man wounded, one man accidentally killed by a Burgher, and one man lost in the Bush. There were no casualties among the officers. Governor's Kop was finally reached on the 23rd.

Lieutenant Foster mentions having seen a copy of the last despatch written by Sir Harry Smith before leaving; in this he refers to Colonel Perceval "several times, praising his conduct and unwearied exertions in the highest degree."

The usual expeditions after Kaffirs began very soon after returning to Governor's Kop. Before the 1st May, Captain Hamley, who was with the

[1] Diary of Private Thomas Scott.

headquarters, had a brush with the enemy, in which six Kaffirs were killed.

In the month of May, a company under the command of Captain Espinasse proceeded to Bloemfontein, in charge of waggons containing arms, ammunition, &c., and rejoined the headquarters in Lower Albany, having marched a distance of nearly nine hundred miles.

Captain Espinasse did not get back until early in July. While waiting at Bloemfontein, he had some success with big game shooting, three lions having fallen to his gun.

The weather in Lower Albany, especially at Governor's Kop, was at this time very variable; the sudden changes from heat to cold were trying. A flight of locusts crossed the Kop in May. "It was a most extraordinary sight; they keep in a line, and move on without turning right or left, as if bound for some place; they are so close together that you can hardly see through them, and the line extends for miles." (Lieutenant Foster's Diary.)

On the 4th June, a column, composed of the headquarters of the Reserve Battn., 12th Regiment, detachment of Cape Mounted Rifles, and Fingoe and Hottentot Levies, under the command of Lieut.-Colonel Perceval, marched from Governor's Kop, to open the communication between Grahamstown and King William's Town, acting in co-operation with a division of the army from the last-named station, under the command of Major-General Yorke, commanding second Division, South African Army.

At daylight on the 12th June, Captain Moody, Royal Engineers, commanding an escort of 34 Sappers, left Grahamstown with a convoy of 9 waggons for Fort Beaufort. The waggons conveyed, besides stores, thirty-six Minie rifles and 3000 rounds of ammunition; there were also some women and children. Shortly after crossing the Koonap River, they were attacked by a body of Hottentots and Kaffirs, some of the former being dressed as Cape Mounted Riflemen. Nine of the Sappers were killed and six wounded. Captain Bagnell, at Fort Brown, hearing the distant firing, sent out 50 men of the 12th, and a few Cape Mounted Rifles. At the same time news was sent to Governor's Kop; Colonel Perceval was absent at the time, so Captain Prior of the Lancers, the senior officer, started off with a troop of his corps, and Lieutenant Durant. Lieutenant Studdert, with 30 men of the 12th, followed immediately. Colonel Perceval, returning the same evening, started off at once with the other troop of Lancers. Captain Moody was relieved, and the escort and convoy were got back to Fort Brown. Only one of the women was killed, and this was done by a Kaffir, in spite of the Hottentots, some of whom spoke English, promising that the women and children would not be injured. The leader of the attacking party was supposed to be a Hottentot, and a very brutal man.

This was looked upon as a very serious affair. The Minie rifles were then a new invention, and a few had been sent to South Africa as an experiment. All the 36 rifles and the whole of the ammunition were lost.

This incident proved beyond doubt that the rebels had spies in Grahamstown who kept them informed of the departure of convoys, and, more important still, of the strength of the escorts. Some of the Minie rifles were subsequently recovered.

Referring to the wreck of the "Birkenhead," in February this year, Colonel C. H. Gardiner (late 12th Regiment) relates the following, at page 318 of his book "Centurions of a Century." "An old bay horse swam from the wreck of the 'Birkenhead,' and enlisted as a charger with one Captain Moody, R.E., and again ran the gauntlet well, avoiding all the bullets as she had shied off all the sharks." Presumably this horse was with Captain Moody in the Kaffir War, as there was no other captain of the same name in the Royal Engineers in 1852.

On the 20th June, the headquarter camp, with the 12th Lancers, moved from Governor's Kop to Driver's Farm, which was only 17 miles from Grahamstown; here they encamped until late in September, when Governor's Kop again became the site of the headquarters camp.

In the observations on the half-yearly Inspection Reports by the Commander-in-Chief, for 1852, His Grace the Duke of Wellington was pleased to express his greatest satisfaction at "the admirable order in all respects of the 1st Battn., and its good state of discipline and efficiency."

On the 22nd July, Lieut.-Colonel Perceval, Reserve Battn. 12th, was invalided to England by a Medical Board.

General Somerset issued the following order on his departure :—

"Grahamstown, 30th July, 1852.

"Major-General Somerset desires to express to Lieut.-Colonel Perceval, Commanding the Reserve Battalion 12th Foot, the regret he feels at losing the services of an Officer, whose zeal and activity in command of the District of Albany, and in command of his Brigade in the field, have been so conspicuous, and which have merited the high encomiums of the late Commander-in-Chief, Sir Harry Smith."

The residents of Grahamstown also showed their appreciation of the Colonel by presenting him with an address, expressing their gratitude for the services rendered by him to the Town.

On the 29th July, Lieut.-General Sir Richard G. Hare Clarges, K.C.B., was transferred from the colonelcy of the 73rd to that of the 12th Regiment, in succession to Lieut.-General the Honourable Robert Meade, deceased.

On the 3rd and 5th August, the 1st Battn. (now 6 companies) proceeded by rail and steam conveyance from Chatham to Liverpool, and, embarking for Ireland, the regiment disembarked at Dundalk. The headquarters proceeded on the 7th to Newry, detaching 3 companies to Crossmaylen, Rostrevor, and Newtown Hamilton.

Major Horne had succeeded Colonel Perceval in command of the Reserve Battn. and he seems to have been kept very busily employed. For a short time he had also command of the Albany District. Throughout August and September, the Gaika Kaffirs gave trouble, raids into the colony and robberies were incessantly occurring, and numerous punitive expeditions were sent to the Fish River Bush, and to the Keiskama River, besides the usual patrols, with varying success. Captain Gillman, 12th, on one of these expeditions, was the hero of a brilliant little affair. With only 15 men, he encountered 80 or 100 rebels, killed a number, and captured 300 head of cattle.

Towards the end of August, Major-General Somerset, commanding the 1st Division, was ordered home, on promotion to the rank of Lieut.-General, and was succeeded by Colonel Buller, Rifle Brigade.

During the very early hours of the 23rd September, the Orderly Room Tent of the Reserve Battn. was burnt down. Fortunately the fire was discovered in time to save the Regimental Records, but the Orderly Room Clerk, who occupied the tent, slept so soundly, that he was rather seriously burnt before being rescued. The fire was supposed to have been caused by his leaving a candle burning on the ground, which set fire to the straw he used for his bed.

Camps were formed and held by the battalion at the Line Drift, Fort Peddie, Trumpeter's, Fraser's Camp, Driver's Farm, Botha's Hill, Neimanos Kraal, Riebeck, and Koonap, from which posts almost daily patrols were made, frequent recaptures of cattle taking place.

On the 11th October, Captain Hearne, his servant, and a civilian were killed within a mile of Driver's Farm Camp.

It was one of the duties of the above Camp detachment to send out a patrol every Monday to clear the Trumpeter's Road for the usual weekly convoy. On this occasion, the patrol started as usual, but Captain Hearne, his servant and a civilian, who were mounted, followed after a short interval. Soon after they started, Captain Hearne's horse was seen by a soldier of the regiment galloping about riderless. This soldier ran out from the camp to catch it, but was captured by the rebels, stripped, and shot.

News was sent to Governor's Kop. Lieutenant Foster, who was alone at the time, started post haste with the Cape Mounted Rifles for Driver's Farm. The bodies of Captain Hearne, the servant, and the civilian, were all found in the bush within a mile of the camp. They had been stripped and murdered. The enemy of course had all disappeared.

It would appear that this affair was the direct result of Seyolo, the Kaffir Chief, having given himself up. He was being conveyed to Grahamstown in a waggon under a small escort of the Rifles. Two of the escort sat in the waggon with muskets at full cock, with orders to shoot the prisoner if any attempts at rescue were made. The rebels seem to have become aware that the escort transporting Seyolo was to pass Driver's Farm, and they are supposed to have lain in wait for it. They allowed the Driver's Farm Patrol with a convoy of 15 waggons and 200 head of cattle to pass by unmolested.

The temptation offered by three men, unsuspicious of any one being near, no doubt instigated them to commit the deed. It thus came about that when Seyolo and escort arrived, the rebel ambush had disappeared, and Lieutenant Foster and party were in possession. Whereas, had Captain Hearne accompanied the Patrol, he would not have been murdered ; the rebels would have remained, and the waggon in which Seyolo was being transported would have been attacked, and the prisoner himself immediately shot. Seyolo therefore owed his life to that half-hour which Captain Hearne spent in Driver's Camp after his Patrol had started. Captain Hearne was buried in the Grahamstown Cemetery on the 13th October.

Seyolo was imprisoned in the Drosdty Barracks, Grahamstown, where he seems to have been kept some time in comparative comfort, with a view to encouraging other rebel chiefs to surrender. Events however showed that this treatment of a rebel chief was not sufficient inducement to encourage others to do so.

Two or three small engagements with the enemy had occurred before the Driver's Farm incident. Major Horne, patrolling on the 8th, fell in with the enemy, killing 12 or more. Another patrol captured 12 head of cattle and killed one or two Kaffirs.

About the middle of October Colonel Napier was given command of Lower Albany, which Major Horne had held for a short period.

The Governor made every effort to induce the rebel Chiefs to surrender. Knowing, as they must have known, that the promise of their lives being spared would be fulfilled, it is strange that they still persisted in defying authority. Large numbers of Hottentots were known to be starving in the Bush. They had lost their cattle, and all their crops had been destroyed —yet they held out. Finally, the Governor, late in October, issued a proclamation pardoning all rebels except Withaalder, the Hottentot Chief, on whose head there was a reward of £500, and a few others of the ring-leaders, for each of whom £50 was offered.

Macoma and Sandilli still continued to hold out, but there was ample evidence to show that the war was practically over. Small disturbances, however, continued.

On the 4th November, the headquarters of the Reserve Battn., 12th, at Collingham, moved into Grahamstown and were quartered in Fort England.

1853

On the 8th January, a detachment of the Reserve Battn., under the command of Captain Espinasse, fell in with, in the Fish River Bush, near Jantjees Kraal, a party of rebel Hottentots, under their Chief Hans

Brander, whom they engaged for nearly two hours, in which time the enemy lost eighteen of their number, there being but one man of the regiment killed, and one wounded.

On the 8th March, peace was proclaimed, and some of the detachments rejoined the headquarters.

The 1st Battn. proceeded by rail, on the 23rd March, from Newry to Belfast, detaching a company to Carrickfergus, and, on the 24th, detachments from Newtown Hamilton, Carrickmacross, and Crossmaylen, rejoined headquarters.

As the result of the half-yearly inspection of the 1st Battn. at Belfast, on the 18th May, by Major-General Thomas, C.B., an Adjutant-General's letter, dated, Horse Guards, 29th August, 1853, intimated that the battalion was "admirably commanded, and altogether in a very satisfactory state of discipline and efficiency."

The battalion was also inspected this year by Lieut.-General the Right Honourable Sir Edward Blakeney, G.C.B., commanding the forces in Ireland.

1854

One company of the 1st Battn. which had proceeded to Cork, embarked there on the 18th January in the freight ship "Gloucester," and sailed on the 20th for Van Dieman's Land, and by the 31st May, the headquarters and remaining five companies of the battalion had arrived at Cork.

By the exchange of Lieut.-Colonel Rumley, 1st Battn., to the 27th Foot, and the retirement of Lieut.-Colonel St. Maur, the command of the 1st Battn. now devolved on Lieut.-Colonel T. Brooke, who was at once transferred, to assume command of the Reserve Battalion at Cape Colony.

On the 1st July, the headquarters and 3 companies 1st Battn., under command of Major Kempt, embarked on board the transport "Camperdown" for Australia, and disembarked at Melbourne on the 18th October.

The second division of the regiment, consisting of 2 companies, under Captain Atkinson, embarked at Cork on board the transport "Empress Eugenie" on the 28th July, and disembarked at Melbourne on the 6th November.

A nominal depot was left at Cork, to proceed to Chatham, under command of Captain Segrave.

On the arrival of the headquarters at Melbourne, the company which embarked for Van Dieman's Land in February was stationed at Castlemaine and Sandhurst in the colony of Victoria, and, on the 21st October, one company proceeded to the Gold Fields at Ballarat.

On the 27th November, one company was despatched, express in cars, to Ballarat, to reinforce detachments of the 12th and 40th Regiments, in support of the civil power, to suppress a disturbance amongst the gold diggers. When entering the gold fields, the rear car was attacked, and Drummer Eagan of the battalion was wounded.

On the 1st December, Major-General Sir R. Nickle, K.H., with the headquarters of the 1st Battn. 12th, and 40th Regiments, and a strong mounted police force, marched, accompanied by 4 guns (manned by sailors of Her Majesty's ships "Electra" and "Fantome") with a proportion of officers, for Ballarat.

On the following day, 2 companies of the battalion, and 1 company 40th Regiment (the whole under command of Captain Thomas, 40th) attacked and destroyed a stockade at the Eureka, where the disaffected diggers were in force. In this affair, Lieutenant Paul, 12th, and an officer of the 40th were severely wounded, and several men were also wounded, amongst whom were seven privates of the 12th, two mortally.

1855

Lieut.-Colonel Brooke assumed command of the Reserve Battn. 12th Regiment on the 16th January, in succession to Colonel Perceval, C.B., who had proceeded to take command of the 1st Battn. in Australia.

On the 5th March, the headquarters and 2 companies, 1st Battn. returned to Melbourne, and were joined on the 31st by 1 company from Castlemaine and Sandhurst; the remaining 3 companies were detached to Ballarat, Castlemaine, and Sandhurst respectively, and, on the 15th August, Colonel J. M. Perceval, C.B., assumed command of the battalion.

On the 30th the headquarters of the Reserve Battalion, under Lieut-Colonel Brooke, moved from Grahamstown to Kaffirland, and with the detachment that had rejoined, the battalion was stationed as follows:—headquarters at the Keiskama Hoek, and detachments at the extreme frontier posts of Dolme, Kabourie, and Baillie's Grave.

The companies of the 1st Battn. on detachment at Ballarat and Sandhurst (Victoria) rejoined headquarters on the 30th August and 20th September respectively.

On the 1st November, one company embarked on a coasting steamer for Adelaide, South Australia, to relieve a company of the 40th, and, on the following day, another company embarked for Launceston, Tasmania, in relief of a company of the 99th Regiment.

On the 14th, one company rejoined headquarters from Castlemaine, and, on the 27th, the 1st Battn. was inspected by Colonel E. Macarthur, Deputy-Adjutant-General, commanding the forces.

The headquarters and 4 companies 1st Battn. remained at Melbourne until the 20th December, when they embarked on board the transport "Windsor" for conveyance to Hobart Town, Tasmania, where they arrived on the 29th. Three companies (with headquarters) disembarked on the 31st, and moved into barracks, in relief of the 99th Regiment, ordered home, whilst the fourth proceeded in the "Windsor" to Swan River, West Australia, in relief of a company of the 99th.

The system of the supply of clothing by colonels of regiments (in accordance with the Clothing Regulations, issued in the time of Queen Anne, 14th January, 1707) had held good up to this date, and the colonels had already arranged for the annual supply for 1856–57. Consequently the clothing supply for 1857–58 was the first undertaken under the new system, which consisted of supply by contract. The supply of clothing for the foot guards, infantry, and cavalry had, up to this period, been a close monopoly in the hands of a few firms, and great opposition was made by these firms to the abolition of the monopoly. For the clothing required during the year 1857–58 the first contracts under the new system were made in 1856.

1856

Two detachments of the 1st Battn., under Ensigns Fitzgerald and Williams, proceeded on the 1st January to Port Arthur and Eagle Hawk Neck, Tasmania Peninsula, and the company which had left for Swan River reached its destination on the 9th February.

The 1st Battn. now received 191 volunteers, transferred from the 99th Regiment, prior to the latter embarking for England.

In March, this year, the only regiments that possessed Reserve Battalions were the 12th and 91st.

A Horse Guards Circular Memorandum, dated 13th October, announced the formation of Depot Battalions, each regiment of infantry (excepting those serving in the East Indies and local corps) being divided as follows, viz:—8 service and 4 depot companies, regiments at home being similarly divided to those abroad, and the depot companies were to be formed into battalions under officers specially appointed to command them. The 4 depot companies to each infantry battalion were to muster a total of 12 officers, 10 sergeants, 4 drummers, and 200 rank and file. All officers were to join these depots on first appointment, and the formation of bands of music at quarters of depots was strictly forbidden.

Eleven infantry regiments, however, on foreign service at this period (including the 12th) were to be exempted from this new Depot Battalion system, and to remain temporarily unaltered.

A Horse Guards' letter, dated 31st October, to the General Officer Commanding at Melbourne, notified that it was intended shortly to send two companies of the 12th Reserve Battn. to Tasmania, and to bring to England the remaining four, forming them, on arrival, into a depot, thus assimilating the 12th to other regiments.[1]

In December, the Reserve Battn. was armed with the Enfield Rifle.

[1] *W. O.* 3, Book 119, p. 355.

1857

In February, the depot of the 12th was at Colchester. On the 30th March the Reserve Battn. was relieved by the 73rd Regiment, and proceeded to the Tamacha Post, British Kaffraria, furnishing detachments to Galube, Baillie's Guard, Line Drift, and Fort Peddie.

The 1st Battn. was this year armed with the Enfield Rifle.

On the 14th April, Major-General C. A. F. Bentinck was appointed Colonel of the 12th Regiment, in succession to Lieut.-General Sir R. G. H. Clarges, K.C.B., deceased.

A War Office letter, dated 14th September, to Major-General Bentinck, notified that the Queen had been pleased to order the establishment of the 1st Battn. of the regiment to be raised from the 1st September, to consist of 12 companies, and to include 46 officers and 1081 of other ranks—total 1127.

FORMATION OF THE SECOND BATTALION.

Twenty-five new battalions being now added to the first 25 regiments of Infantry of the Line, an order, dated 19th September, was issued for a second battalion of the regiment to be raised, from the 2nd September, the establishment to consist of 8 companies, including 32 officers, and 695 of other ranks—total 727.

1858

On the 17th February, a letter was received from the Horse Guards directing that the formation of the Second Battalion was to be proceeded with independently of the Reserve Battalion, but, on the 23rd March, it was decided that the Reserve Battalion, 12th Regiment, at Cape Colony, would form the headquarters of the Second Battalion, retaining its staff, and would be brought home for the purpose ; also that the portion of the corps being raised by Lieut.-Colonel Hamilton, was to consist of six companies, and be termed the "left wing."

In March, the quarters of the Reserve Battn. were extended, by furnishing detachments to Keiskama Hoek and King William's Town.

In the "Personal Reminiscences" of Lieut.-Colonel Walter J. Boyes, late 12th Regiment, at this period, he states :—

> "The Regimental Depot at that time, at Parkhurst, Isle of Wight, formed the nucleus of the 2nd Battalion which was there raised by Colonel Meade Hamilton, an Officer who had served with distinction in the 47th, and on the Staff in the Crimea.
>
> There was, I may here mention, considerable jealousy, amounting almost to friction, between the *quasi* 2nd Battalion (of mushroom growth) and the old Reserve Battalion. Colonel Hamilton prided himself on what he was pleased to call the 2nd Battalion, but which

the old officers of the Reserve Battalion spoke of as the Left Wing. Difficulties arose at times over promotion to be made in the N. C. grades, and in the filling up of various minor posts in the Regiment. The Drum-Major of the Reserve Battalion happened to be a very small man; whereas the Acting Drum-Major of the Left Wing was a very tall fine-looking man, and presented a very imposing appearance, marching at the head of the regiment—Drum-Major Haggan was however a splendid musician, and the Reserve Battalion prided themselves on their Drums and Fifes, which were particularly good. A compromise was however arrived at by arranging that the Drum-Major proper should be in charge of, and always train and play with the Drums and Fifes, whilst the Acting Drum-Major of the Left Wing (dressed also as a Drum-Major) should march in front of the regiment with the Drum-Major's Staff. He was accordingly dubbed 'the walking Drum-Major.'

The Band, under Herr Rix, had become so far proficient as to be able to play a few marches, and to head the Regiment when out route-marching. The big-drummer, Silvey by name, was a tall West African Native, with curly hair; so, with our fine 'walking Drum-Major' at the head of the Regiment, and a well-bearded body of Pioneers, all wearing the Kaffir Medal, we caused no little sensation when marching through Campbelltown, and along the country roads in the Highlands."

In the "reminiscences" of Colonel W. C. S. Mair, late 12th Regiment, who was gazetted to it in March this year (and posted to the 1st Battn. at Sydney, New South Wales), he states :—

"that, on joining, he was ordered to report himself to the Officer commanding the Depot, which then formed part of the Depot Battalion at Deal. The battalion was about one thousand strong, and was commanded by Colonel Whitmore C.B.

Our Depot consisted of three Captains, eight Subalterns, and some two hundred and fifty men. The first night I joined,—we had to dine in full dress in those days—we sat down to mess 63 officers, which was about the usual number of dining members.

At the time of my joining it was not the custom to breakfast at mess, and few lunched there. We were charged, by General Order, two shillings a day for our dinner, but if any officer was dining out, and warned the mess man before one o'clock, his account was re-credited with sixpence.

No wine was allowed at mess except port and sherry, there being a Horse Guards' order to that effect."

On the 8th April, the headquarters, 1st Battn., and the grenadier company, embarked for Sydney, New South Wales, in the steamer "Tasmania," in relief of the 77th Regiment, arriving on the 12th April, and disembarking on the same day. Three companies, under Major Hutchins, were left to do duty in Tasmania.

Prior to the departure of the headquarters from Hobart Town, Tasmania, the inhabitants presented the following address to the battalion on parade.

"To Colonel Perceval, C.B., Commanding Her Majesty's 12th Regiment.

We, the undersigned inhabitants of Tasmania, regret to learn that the headquarters of the regiment, which you have the honour to command, have been ordered to Sydney, to replace the 77th Regiment, about to proceed to the seat of war in China.

During the period you have commanded the regiment in this colony, it has been eminently distinguished for its good conduct and discipline, and we are sensible of the pains you have taken, and the sacrifices you have made, to promote temperance and good order in its ranks; you have, at the same time, shown a careful solicitude for its officers, have rightly discountenanced extravagance, and encouraged them to preserve an honourable and independent bearing.

The troops under your command have always been at the call of the community in every emergency, and the inhabitants of Hobart Town are deeply indebted for the valuable assistance rendered by you upon several occasions of alarming fires.

You have been also ready at all times to contribute to social and manly recreations, and have done your utmost to promote public and other entertainments.

The high qualities which you have displayed in the field, and which have been marked by the approbation of your Sovereign, have been, if possible, exceeded by the virtues which have distinguished you at the head of your regiment in time of peace.

You have combined firmness with modesty of demeanour, and have involuntarily won the respect which never fails to wait upon unpretending worth.

Upon the eve of your departure from amongst us, we desire to place on record the sentiments of the inhabitants of this colony towards you, and the officers of the gallant regiment which you command, and we conclude in wishing you every health and happiness in the new sphere of duty to which you are called."

To this, Colonel Perceval made the following reply.

"GENTLEMEN.—The address with which you have honoured me, signed by persons of the highest position and respectability, is a flattering testimony to the discipline and good conduct of the soldiers of the 12th Regiment, and in their name, I thank you.

The eulogium passed on myself has reference only to what is the duty of every officer in command; that duty has been performed to the best of my ability, and I thank you for its recognition.

That the officers of the regiment should have given any impulse to the social and public amusements at Hobart Town will always be the subject of pleasurable remembrance, and those among us now leaving the colony, experience the regret, too familiar to soldiers, of friendships hastily severed, and of, perhaps, a final departure from scenes most agreeably associated with many acts of kindness and hospitality.

In conclusion, the officers and men join with me in the expression of our best wishes for the prosperity of Tasmania."

On the 18th April, one company of the 1st Battn. rejoined headquarters from Adelaide, South Australia, after an absence from the regiment of two years.

A War Office letter, dated 19th April, directed the establishment of the 1st Battn. to be increased by the addition of a School Master (to be appointed by the Secretary of State for War), and a reduction of 2 sergeants, 2 corporals, and 48 privates, making a total of 1074, the above sergeants and corporals to remain as supernumeraries until they shall be absorbed into the new establishment.

On the 8th July the headquarters 2nd Battn. marched from Tamacha, and very shortly afterwards arrived in Grahamstown, *en route* for Port Elizabeth.

The Commander of the Forces at the Cape, General Sir James Jackson, whose headquarters were in Grahamstown, thought this a fine opportunity for a military display; so all the troops in garrison at Grahamstown were paraded to escort the 12th for some miles on its march, and eventually they were all drawn up on a range of hills overlooking Howison's Poort, and fired a salute, as the Regiment, followed by its long train of ox-waggons (12 or 14 oxen to each waggon), wended its way down the kloof. It was a very picturesque sight.[1]

On being joined by its detachments, the 2nd Battn. embarked on the 18th July for England, and arrived at Portsmouth on the 2nd September. Whilst in harbour at Portsmouth, the battalion was ordered to Fort George, North Britain, where it arrived and disembarked on the 10th September.

On the 26th September, the left wing from Chester joined the headquarters at Fort George, and, on the 1st October, the battalion was formed. The depot also joined at Fort George on the same date.

On the 8th October, the 2nd Battn. was inspected by Major-General Viscount Melville, K.C.B., who was very well pleased at its fine appearance, half the battalion consisting of men who had seen service and were decorated with medals.

Enlistment at this time was for ten years—or twelve in case of war—the recruit received a bounty of £2, which was increased to £3 later on; the pay of the private soldier was 1*s.* per day, with an allowance of one penny per day termed "liquor money;" out of which he paid 8*d.* per day messing, washing 1*s.* 3*d.* per month, sheet washing 2*d.*, hair cutting 1*d.*, and barrack damages an average of 4*d.* per month, the soldier getting the balance by daily payments, which usually came to about 4*d.* per day if no other necessaries were required.

1859

The detachment of 3 companies 1st Battn., from Tasmania, rejoined headquarters at Sydney, New South Wales, on the 22nd July.

[1] Lieut.-Colonel Boyes' *Reminiscences.*

Colonel Mair (then an Ensign at the depot) relates that, at the end of March this year, he proceeded with a draft of 2 lieutenants and 100 men to join the headquarters, 1st Battn., at Sydney, New South Wales, where they arrived after a voyage of nearly fifteen weeks.

Colonel Mair states :—

> "Nine companies were at headquarters, and one on detachment at Swan River, a convict station about four hundred miles distant. The Battalion could turn out about seven hundred strong—and a fine-looking lot of old soldiers they were—many of the unlimited and twenty-one years' service men being still in the ranks."

In continuation of this, Colonel H. D. Cutbill, who was then serving as a subaltern in the 1st Battn., writes :—

> "In the ranks of the 12th there were not a few men who had been serving in the old Bengal European Regiments, and who were transferred or re-enlisted after the Mutiny; they were fine men, but the enervating Indian climate had left its mark on a good many of them."

Colonel Mair continues :—

"Sydney was a most delightful quarter, and as it was said to be expensive, Colonial pay was given by the Colonial Government at the rate of *5s.* per day for Company Officers, and *7s. 6d.* for Field Officers.

When the 12th first arrived in the Colonies they had a good deal to do in the way of guards over the convicts, but by the time our draft reached New South Wales, we had to furnish only one convict guard, and that was over a few hundred 'lifers' and long sentence prisoners. The guard was rather an interesting one. It consisted of a subaltern and fifty men, and mounted for 14 days on an island in the middle of the magnificent harbour of Sydney. We had little to do with the convicts, as there was a Governor and regular staff on the island, but we became acquainted with a few of them. Some of the prisoners had been men of good position. Three had been in the Church, one was the son of a Marquis, two were Baronets, and there were many others for whom one felt specially sorry. This guard was useful to some of us, as five shillings a day and allowances went with it.

The Regimental Mess was conducted on what we should now consider old-fashioned lines, but at the same time it was most thoroughly comfortable. The mess man was well pleased if he had fifteen dining members on his books. Colonel Perceval did not consider it necessary to comply with some of the minor rules, such as dining in tunics, and allowing port and sherry to be the only wines drunk at and after Mess. On guest nights, champagne was handed round once, by the mess butler, and officers who did not intend to drink it, merely turned their champagne glasses upside down, then their wine glasses were filled with sherry. After that the wine was handed round no more, but, as each officer desired to have wine, he told the mess butler or corporal to take the wine with his compliments to ——; he would 'be glad to drink wine with him.' One generally drank wine with each guest.

Dinner on guest nights, or, more correctly speaking, the time we sat at table, was apt to be a little long. The band did not commence until the cloth had been removed and the health of Her Majesty proposed. The programme commenced as it ended with 'God Save the Queen.' There we sat, and as we had not a string band, and had much brass in the Regimental one, conversation was rather a difficulty. This was no doubt the cause of so much wine being drunk after dinner—about one bottle and a half per man. Smoking at the mess table had not then been introduced.

It seems a little strange that even so late as the fifties we should carry our mess table and billiard table about with us. Colonel Hamilton decided to sell them, on the headquarters being ordered from Sydney. The dining table was a handsome piece of mahogany, and there were many legends connected with it. On that table were said to have been laid out the bodies of officers of the 12th who fell at Seringapatam, and the morning afterwards, all officers and men who were not on duty marched through the mess tent, saluting the poor fellows, who had been dressed in full uniform. There were marks on the mahogany made by the spurs of a certain very diminutive Adjutant, who, when specially cheerful after mess, would send for his fiddle, and, to his own accompaniment, would perform a 'pas de seul' among the bottles and glasses. One or two old officers of the battalion objected to the disposal of the mess table, but Colonel Hamilton said that he did not intend the Regiment to become an antiquarian museum."

On the 23rd March, the 2nd Battn. moved from Fort George to Glasgow.

A letter was received from the War Office, on the 2nd April, fixing the establishment of the corps at 57 sergeants, 50 corporals, 24 drummers and fifers, and 900 privates.

On the 1st July, the 2nd Battn. proceeded to Aldershot, and joined the Reserve Brigade, and was afterwards transferred to the 1st Brigade, under the command of Major-General A. J. Laurence, C.B.

On the 7th July, the 1st Battn. was formed into ten service companies, and, on the 22nd, 2 companies embarked for Launceston, Tasmania, where, on arrival, one proceeded to Hobart Town.

1860

The natives of New Zealand had for some time been in an unsettled state, and, in April, the Governor of that colony applied to Sir William Denison, Governor of New South Wales, who was also Governor-General, for assistance, as he feared a native rising, and had only one regiment and a few gunners in his Colony. Sir William Denison could only venture to spare two companies of the 12th and half a battery of Artillery, as the gold diggers were giving much trouble. They embarked in the "City of Sydney" steamer, under command of Captain Miller, with Lieutenants

Richardson and Lowry and Ensign Latouche as Subalterns, also Assistant-Surgeon Lynch in medical charge. Their destination was New Plymouth, or Taranaki, as it was called, being the native name. They arrived on the 16th April, and found the little town in a state of siege. It was garrisoned by 600 of the 65th Regiment, under command of Colonel Gold, who was the senior officer in the Colony. The village was crowded, as the farmers in the neighbourhood, in fact from all over the province of Taranaki, had been obliged to leave their delightful homesteads and seek the protection of the military. The natives had been fair and honest with the settlers, warning them in good time that the outbreak would take place, and advising them to leave the out-districts.

Soon after the arrival of the detachment, Captain Miller distinguished himself, by an heroic endeavour to rescue from drowning a young man of the militia, who, on horseback, was trying to ford the Ruatokie Stream which runs into the sea just below the town. Captain Miller chanced to be passing, and seeing the man washed off his horse, immediately plunged in to try to save him. He swam strongly towards the lad, who had been washed down stream, and was carried in among the breakers at the mouth of the swollen river. The surf was heavy, and, more than once, when almost within reach of the drowning man, the current carried them far from each other, and at last Captain Miller was washed up, apparently in a dying state, on to the beach, whence he was carried into the Hawan Pah, and, after a considerable time, and the application of vigorous remedies, he was restored to consciousness.

The character of the Maoris may be briefly described as very complex. Though cannibals, and blood-thirsty to a degree, their sense of honour was high, and their word, once pledged, was considered inviolable. They were by no means devoid of chivalry; their language was full of poetry, their manners were dignified, their laws well defined, and the tenure of land and the ownership of movable property were regulated by customs, enforced by the whole power of the clan. The Maoris' *pahs* were stockaded and entrenched villages, usually perched on cliffs and jutting points overhanging river or sea, and were defended by a double palisade, the outer fence of stout stakes, the inner of high solid trunks. Between them was a shallow ditch. Platforms as high as forty feet supplied coigns of vantage for the look-out. Thence, too, darts and stones could be hurled at the besiegers. With the help of a throwing stick, or rather whip, wooden spears could be thrown in the sieges more than a hundred yards. Ignorant of the bow and arrow, and the boomerang, the Maoris knew and used the sling; with it red-hot stones would be hurled over the palisades, among the rush-thatched huts of an assaulted village, a stratagem all the more difficult to cope with as Maori pahs seldom contained wells or springs of water.[1]

[1] Pember Reeves. *The Long White Cloud*, pp. 48, 49.

Major-General Sir J. Alexander, an officer of great experience, describes a track passing through a New Zealand forest, as follows :—

> "The bush of New Zealand is wonderfully dense and entangled. A European going into it, about twenty yards, and turning round three times, is quite at a loss to find his way out again, unless he is somewhat of an Indian path-finder, and can judge of the points of the compass by the bark of the trees, the sun, &c. Trying to run through the bush, one is tripped up by the supplejack, and other creepers." [1]

The natives had no artillery except three old carronades, which they had got from wrecked ships, and which they only fired three or four times, and they had no better shot than steelyard weights and similar substitutes for cannon-balls.

These they abandoned at the evacuation of Meri Meri, from which time they never had a big gun. Their small arms consisted of old Tower muskets, many flint and steel (*temporis* George III), single and double fowling-pieces, such as are made for Colonial trade, and a very few rifles, perhaps not more than one in a thousand. At the close quarters at which the engagements generally took place, these weapons were actually better than the Enfield rifles of our troops, as being more easily re-loaded, and their double barrels giving two shots for one man. The Maoris had no cavalry.[2]

A considerable number of women go into the field with men in Maori warfare, to assist in various ways; building wharrés or huts, collecting and preparing food, nursing the wounded, besides encouraging the sterner sex to rival the deeds of their ancestors. Female voices would be heard at night, sometimes, with songs and cries.

In Taranaki, in the middle of June, 1860, the hostile natives were collected in some strength, and, near the Bell Block, carried off the stock of the settlers, and broke up and destroyed the insides of their houses. These devastations took place at night.

Portions of the 12th and 40th were sent out to endeavour to surprise the marauders, when several skirmishes occurred, and the energy and enterprise of Lieutenant Richardson (in command of the 12th detachment) were favourably noticed by Colonel Gold, 65th. The rebels, however, seemed to have early intimation of the movements of troops, thereby causing great difficulty in their capture.

A reinforcement being required, another detachment of the 1st Battn. (mustering 6 officers, 4 sergeants, 1 drummer, and 100 rank and file) under Major Hutchins, embarked at Sydney in H.M.S. "Fawn," on the 17th July, and arrived at Taranaki on the 23rd. On the 27th, they marched to Waireka to construct a redoubt, with a view to arresting the advance of hostile natives from the south, in their intended attack on the town of Taranaki. The

[1] *Bush Fighting: the Maori War*, p. 59. By Major-General Sir J. Alexander.
[2] *New Zealand War*, p. 3. W. Fox.

officers with this detachment were Major Hutchins, Captains Queade and Leeson, Lieutenants Dudgeon and Mair, and Ensign Hurst.

The redoubt was partially invested from the 11th to the 23rd of the following month, during which period no duties about the camp could be performed without interruption from the enemy's fire, every wood and water fatigue involving a skirmish. During its occupation, the detachment was favourably mentioned in General Orders, in praise of the manner in which the duties had been conducted. Finding their efforts unavailing, the enemy abandoned their project.

On the 1st August, the 2nd Battn. proceeded by rail from Aldershot to Portsmouth, where it embarked on board the "Himalaya" for Plymouth, and, disembarking on the 3rd, furnished a detachment to Bull Point, and another to Horfield Barracks, Bristol.

On the 10th September, a large expedition was organised at Taranaki, under Major-General Pratt, to advance as far as possible towards Pukerangeora on the Waitara. The force, numbering 1400, was told off into three divisions, No. 2 being commanded by Major Hutchins, but his detachment of the regiment was not included. Pahs were destroyed,and good service done, and, next day, Nos. 2 and 3 Divisions returned to Taranaki.

Major Hutchins was directed to proceed south on the 18th September, with 157, 12th Regiment, under Captain Miller, 270 of the 65th, 17 R.A. with two 24-pdr. howitzers, 13 R.E., 68 Militia and Volunteers, 10 men of the mounted corps, and friendly natives under Mr. Good. The expedition first encamped on the north bank of the Oakura River, and after destroying eight pahs, returned to Taranaki on the 24th.

The 12th detachment then formed part of a field force under Major-General Pratt, which started on the morning of the 9th October, for the reduction of three strong pahs, two on the right, and one on the left bank of the Kaihihi River, and 18 miles from Taranaki. The approaches to the pahs were carefully reconnoitred; two of them had been evacuated, and each, after capture, was found to be very strong, with rifle pits most skilfully contrived with covered passages, and, at the last pah, was an underground hospital, for the wounded from the first pah. The pahs had the usual two rows of palisades. The conduct of the troops of all arms was excellent, the different detachments vying with each other in the field and trenches. Among the casualties were a captain and a sergeant, R.E., severely wounded. It was now determined to attack the strong position of Mahoetahi, where the enemy were in force. Colonel Mair relates :—

"Before daylight, on the 6th November, a force of 1500 men, composed of the 12th, 40th, and 65th Regiments, and some light guns, left Taranaki and crossed the Mangoraka River. We found that the position was not only a strong one, but that it had been well fortified. The 65th, and some of the Militia, formed the storming party, and it was carried with a rush, the natives not having had time to complete the stockading on one of the flanks.

This tribe was armed with well finished English rifles and double-barrelled fowling-pieces, and were able to keep up a continuous fire, whilst their power of concealment was wonderful."

The British loss at Mahoetahi was 4 killed, 2 officers and 13 men wounded. General Pratt pursued the flying enemy with a portion of the 12th, 40th, and 65th Regiments, and 2 guns, and rejoined at Mahoetahi, when, leaving a force of 300 men to occupy the position, the remainder of the troops returned to Taranaki and Waitara Camp, after a long and arduous day's work. The troops behaved with great energy throughout, and amongst the officers specially mentioned was Major Hutchins, 12th Regiment.

Late in November, Major Hutchins was put in command of 500 men and ordered to take Wakoruo Pah. The time and all particulars were left entirely to his own judgment. The men were suddenly turned out at ten minutes' notice, at 11 o'clock at night, and marched nine miles to the position. It was a complete surprise, and the natives gave in without firing a shot.

An expedition started on the 28th December, to reduce Matarikoriko, under General Pratt. The Naval Brigade and 12th detachment, under Captain Miller, a brave and meritorious officer, had charge of the right flank, to keep that clear, whilst the 40th and 65th were thrown out on the left, towards the strong position of Matarikoriko.[1]

Colonel Mair continues :—

"The 12th formed the advance guard, under Captain Miller. The enemy had entrenched themselves in two strong positions, about six miles inland, on what was known as the Puketakawre Block, which was surrounded with scrub and fern six feet high. As the General considered that it would take some time to reduce these strong pahs, situated in such commanding positions, he determined to throw up a redoubt. When the enemy saw what was intended, they advanced in large numbers, and made a most spirited attack on the left flank, which was guarded by the Naval Brigade and our detachment. So determined was the attack that the working parties had to throw down their entrenching tools and join the melée.

By the evening, the redoubt being raised high enough to give a certain amount of protection against a sudden rush, the 12th and 65th were left in charge, the remainder marching back to Waitora. Until 4 o'clock on the morning of the next day (Sunday), a brisk fire was kept up upon our position, when it suddenly ceased, and soon after daylight, when the General visited us, a white flag was flying on the flag-staff at the pah. The Acting Chaplain, who knew the natives well, advanced and met the senior Maori Chief, who intimated to him that it was the desire of the tribes engaged 'not to desecrate the Sabbath by spilling blood,' so our skirmishers were not thrown out, and, though we went

[1] *History of the New Zealand War*. Sir J. Alexander, p. 230.

on with the work of the redoubt, we had a quiet Sunday. The enemy, in full confidence of our good faith, came out of their pahs in numbers, and showed themselves some distance in advance; whilst our men, unmolested, gathered potatoes from a field of some six acres on our flank. On the Monday morning it was found that Matorikoriko was deserted, and that the natives had given up the strong position on our left flank.

No. 1 Redoubt was finished, and the headquarters of the 65th Regiment, the detachment of the 12th, a few Artillery in charge of two guns, and some Royal Engineers, pitched their tents in it, Colonel Wyatt, 65th, being in command. Soon after, another redoubt was erected some distance in advance, and occupied by the 40th Regiment.

The enemy having been defeated, and compelled to retire to a fresh position at Hiurangi, No. 1 Redoubt had been erected on the ground they vacated. The subsequent operations in which Major Hutchins' detachment was actively employed, were those connected with a regular approach on this strong position, by a series of redoubts and a sap, which forced the belt of bush, and the rifle pits of Hiurangi."

1861

On the 9th January, a detachment of the 1st Battn., under Lieutenant Seymour, embarked at Sydney, N.S.W., for Brisbane, Queensland.

Colonel Mair continues:—

"The General's chief object now was to take a pah some few miles in the bush called Te Arie. It could be seen from the redoubts, and native report had it that it was the strongest and best defended pah in the country. It certainly was a well selected situation; in the first place, between our position and the very thick bush, there was a mile of perfectly level ground from which the fern had been removed, and, just on the border of the dense bush, there were numbers of well-constructed rifle pits, covered over and quite invisible, which extended for about a mile. Behind the pits there was a dense bush, so thick with under-growth, that but for some paths, eighteen inches wide, made by the natives, there was no means of penetrating it, except by cutting down the underwood. Round the pah, which stood on a considerable rise, there was a cleared space, and more rifle pits. The river Waitora, with steep banks, almost cliffs, protected the position on the right and on the left, and at the rear there was more thick bush.

Strange to say, up to this time, until the year 1862, I think, each regiment had its own bugle calls (apparently for parade purposes, in addition to regimental calls).

Our 'Advance' was the 65th 'Extend,' our 'Commence Firing' was their 'Close,' and their barrack calls were the same, but conveyed different meanings.

The Maoris also had some bugles, and could imitate the calls of the different regiments, and would send out their buglers at night, and make terrible confusion, until we became accustomed to them."

Soon after the occupation of No. 2 Redoubt, a great stir was observed for two days in and about the Te Arie position, and it was reported that large reinforcements had arrived from Waikato. Shortly before 4 A.M. on the 23rd January, the enemy made a determined attempt to seize the above redoubt, when they were repulsed with great loss. On the detachment of the 12th, under Captain Miller, advancing (in conjunction with the companies of the 65th), and driving the enemy out of the ditch at the point of the bayonet, Captain Miller was wounded, and Lieutenant Lowry, who continued the advance, was favourably noticed by the Major-General.

Subsequently, the whole of the detachment was engaged in operations for the reduction of Pakeranguir, which, on the 10th February, led to a brisk engagement. No. 7 Redoubt was thrown up in the face of a heavy fire, and occupied that night, and part of the next day, by the detachment; a desultory fire having been kept up by the enemy for thirty-six hours.

Colonel Mair states :—

> "Soon after this the natives abandoned the whole line of works, which was about two thousand yards long, and the position gained by General Pratt was an important success. The pits having been abandoned, a force, composed of the 12th, 14th, 40th, 57th, and 65th, with artillery, attacked the bush more than the natives, and in two days managed to force their way through it. Though the Maoris had abandoned their pits, they were full of determination to defend Te Arie; they called to our men to come on, and the women constantly cried out 'Kintoa, kintoa!' 'Be brave, be brave!' The troops had one more hand-to-hand encounter before we reached the open space in front of Te Arie, but as the 12th were in charge of the very limited supply of baggage allowed to be carried, they took no part in it."

On the 22nd February, Brevet Lieut.-Colonel H. M. Hamilton was promoted to the command of the 2nd Battn., *vice* Colonel Thomas Brooke, to half-pay.

On the 25th, a detachment, 1st Battn., under command of Captain Atkinson, marched from Sydney, N. S. W., to Lambing Flats, in aid of the civil power, and rejoined headquarters on the 4th June.

On the 5th March, the Maoris at Te Arie advanced to such close quarters, to intercept progress of the sap, and fired so briskly, that our troops fixed bayonets, expecting an immediate rush at the trenches.

Their defence was most obstinate, and the difficult country abundantly favoured them. The 12th, 14th, 40th, 57th, 65th, R.A., and sailors were all actively engaged, and anxious to be let loose, to charge the pits and Te Arie Pah, and it was difficult to hold them back.

Colonel Mair continues :—

> "On the 18th March, after a hard day's work, and the loss of an Artillery officer and a Captain of the 65th Regiment, the detachment 12th, under Major Hutchins, was on outpost duty, when, about

midnight, he was told that two native chiefs from Te Arie wished to see him. They were sent under escort to the rear, when they informed the General that it was the desire of the natives 'to talk of peace.' On the morning of the 19th, a truce was decided upon, a white flag was run up, and the last shot of the native rising of 1860–1 had been fired.

Within an hour of the terms of peace being signed by His Excellency the Governor, Lieut.-General Sir Duncan Cameron arrived at the Waiteru, and succeeded Major-General Sir Thomas Pratt.

General Cameron decided that his headquarters should be in Auckland, and the 12th detachment, which had received a strong draft direct from England, and the 14th, 40th, and 70th Regiments, were formed into a brigade, a few miles from Auckland, the 65th taking up their old quarters at the latter place.

H.M. Ship 'Niger' conveyed us to Manaukau Harbour, and from there we marched to an old pensioner settlement called Otahuhu, about eight miles from Auckland, where we pitched our camp."

Among the natives, throughout this New Zealand rising, it was understood that 300 had been slain, besides a great number wounded.

Those who have fought with the Maoris are the last to despise them as foes; on the contrary, the British troops who contended against these lusty, active, intelligent tattooed warriors, respect them.

The Maoris, too, have a chivalry of their own, in not taking undue advantage, or striking before they have given warning to their enemies, but, when once the contest is begun, as is usual with other contending parties, take every means in their power to discomfit their opponents. Yet, anxious as they are to be thought civilised, and superior to their ancestors in manners and customs, they had not then understood that prisoners and wounded men should be spared.

As untaught engineers, who had not passed through any military college, their ability was wonderful in choosing and fortifying a position with pahs or stockades, as was their arrangement of rifle pits to fire from, under cover of the picketing and outside the pah, to take in flank an advancing enemy, and, if needed to provide a rapid retreat for themselves down a wooded ravine in the rear. Young Maori women used their fire-arms as well as the men in the rifle pits of Taranaki.

Of endurance and determination in a Maori, there was a remarkable instance at Hiurangi, in the summer of this year.

Natawa, a wild character, tired of firing away all day in his rifle pit, got up into a tree, ten feet above the ground, to fire with better effect, on the 12th, 14th, 40th, and other skirmishers, but he was dropped by a ball in the forehead.

Having perhaps a thick skull, the Minie ball stuck fast over one eye, without passing into the brain, and Natawa, recovering himself, went on fighting for two days afterwards.

The second evening, some of his friends tried to get the ball out, by moving it with their fingers, but, perhaps a portion of bone was dislodged, and touched the brain, and Natawa, after five days raging madness, died.[1]

The 2nd Battn. was inspected at Plymouth, on the 23rd March, by Major-General Hutchinson, who expressed his approbation of their steadiness under arms, and clean and soldierlike appearance.

On the 3rd April, the battalion was inspected at the Citadel, Plymouth, by His Royal Highness, the Duke of Cambridge, Commanding-in-Chief, who was pleased to express his approval of their soldierlike and steady bearing under arms, and his remarks, on perusing the Confidential Report on the prior inspection, were :—" The report on the 2nd Battn. 12th Foot shows a great improvement in every respect, and that it is in a most creditable state."

In consequence of disturbances this year, and part of next, at the Lambing Flats Gold Diggings, the 1st Battn. was again called on to aid the civil power, and, on the 17th July, a detachment of 6 officers and 101 men, with two field guns, under Lieut.-Colonel Kempt, marched from Sydney, New South Wales, to Lambing Flats.

Colonel H. D. Cutbill, late 12th Regiment, states :—

> " The European diggers had fallen out with the Chinese, and had been maltreating them, cutting off their pigtails and otherwise persecuting them ; for this there was no justification.
>
> The gold claims were mostly alluvial, and, if the yield was not up to a certain value, the European diggers would abandon the work, and commence others. The claims thus abandoned would be seized by the Chinese, and the excavated earth, large mounds of which often remained unwashed, would be taken advantage of, and yielded John Chinaman a good profit—hence the trouble! The Europeans did not consider these abandoned claims as worth anything, but objected to the Chinese taking advantage of them, although I believe, from a legal point of view, a claim having been abandoned for a certain time, anyone was at liberty to take possession of it.
>
> We had no trouble, however, with either European or Chinese diggers, for, as soon as the troops arrived, order was re-established."

On the 11th August a letter was received from the Horse Guards directing the 2nd Battn. to be at once divided into service and depot companies, preparatory to the depot being moved to Chatham.

On the 23rd, the headquarters and right wing, under the command of Lieut.-Colonel Hamilton, embarked at Plymouth in the Steamship " Megera," and arrived at Cork on the 26th.

On the following day, the depot companies which had been left behind on the departure of the 1st division, embarked on board the " Megera " for conveyance to Portsmouth.

On the 3rd September, the left wing of the battalion, under the command of Major Hibbert, embarked at Plymouth, and arrived in Queenstown on

[1] *History of the Maori War.* Sir J. Alexander.

the evening of the 4th. On the following day the division was disembarked, furnishing detachments to Spike Island, Haulbowline, and Fort Elizabeth.

A portion of the detachment 1st Battn., which had been on duty in New Zealand, rejoined headquarters at Sydney, N.S.W., on the 16th October, under command of Captain Leeson, mustering 2 officers, 3 sergeants, 1 drummer, and 100 rank and file.

During the next six months, the detachment remaining in New Zealand had a peaceful time, the brigade to which it belonged being under canvas at Otahuhu, near Auckland.

On the 5th November, Lieut.-Colonel E. G. Hibbert succeeded Lieut.-Colonel H. M. Hamilton in command of the 2nd Battn., on transfer of the latter to the 1st Battn.

Colonel Mair writes relative to New Zealand :—

> "In December, it was decided to make a military road through the thirty miles of very thick bush, and over the hills and deep gullies between the capital and the great Waikato River, a distance in all of thirty-seven miles. The 14th Regiment under their Colonel, Sir James Alexander, was sent, with the 12th Detachment under Major Miller—Colonel Hutchins having been appointed Military Secretary to the Lieut.-General Commanding—as far as the river, and from there we worked backwards towards Auckland. The work for the road parties was hard, but the pay was good and the duties not heavy, as, besides our regimental guards, we had only to take the precaution of having a captain, two subalterns, and a hundred men on picquet duty. At the end of six months, the road being completed, we all returned to our old quarters, where huts had been erected."

1862–3

On the 9th April, 1862, a Horse Guards' Circular fixed the establishment of the 2nd Battn. at 10 service companies, consisting of 38 officers, 47 sergeants (exclusive of schoolmaster), 21 drummers, and 770 rank and file ; also 2 depot companies, consisting of 6 officers and 144 of other ranks.

On the 28th May, the 2nd Battn. marched to the Curragh Camp in seven detachments (the last one arriving 9th June), and joined the 1st Brigade, under command of Brigadier-General the Honourable A. H. Gordon, C.B.

The detachment 1st Battn. stationed at Lambing Flats, N.S.W., in aid of the civil power, joined headquarters at Sydney on the 13th August, and that at Perth, West Australia, joined on the 29th March, 1863.

A Horse Guards' Circular, dated 1st April, 1863, directed the establishment of the 1st Battn. to be 12 companies, and to consist of 39 officers, 58 sergeants (exclusive of schoolmaster), 25 drummers, and 900 rank and file. The same circular regulated the establishment of the 2nd Battn. to be the same as in the previous year, with the exception, amongst the service companies, of one extra sergeant, and the total number of privates to be reduced from 730 to 640, whilst, of the depot companies, the number of privates was to be reduced from 120 to 110.

On the 21st April, Captain and Lieut.-Colonel A. E. Ponsonby, Grenadier Guards, succeeded to the command of the 2nd Battalion, on exchange with Lieut.-Colonel E. G. Hibbert.

Referring to the detachment 1st Battn. in New Zealand, Colonel Mair writes :—

> " In the spring of this year, the natives again began to give trouble. The Chiefs of the Waikato tribes informed the Governor, Sir George Grey, that they objected to the military road being carried beyond a certain point, and that if an attempt were made to bridge a certain stream, they would look upon it as a declaration of war."

On the 4th May, they assumed the offensive, marking their hostility by firing on a party of officers and soldiers, whom they shot and tomahawked, except one man who escaped; and committed other atrocities.

By the end of May, the Maoris having collected about 600 fighting men in a strong pah, on the left bank of the Katikaka River, General Cameron determined to attack their position, and strike a decisive blow, which was successful in causing their defeat, with a loss to our troops of 3 men killed and 8 wounded.

On the 9th July, the General assembled a considerable force at Drury, and Colonel Mair relates :—

> " planned a night march to attack some few hundred of the enemy, who had commenced to erect a fortification on a large scale, about ten miles from the Queen's Redoubt (the general's headquarters), and as the detachment 12th Regiment stood high in his estimation, and had had considerable experience in night marching, they were selected to furnish the advance guard."

The General established strong posts along the line of communications, crossed the Mangatawhiri, and occupied the high ground beyond it, an important position on the Koheroa Range.[1]

Colonel Mair continues :—

> " The force, which started from camp about 8 P.M. on a winter's night (July 12th) consisted of only 400 men, 100 of which, with 6 officers, formed the advance guard, under Major Miller. On arriving just before dawn, about a quarter of a mile from the native fortification (which, on this occasion, was not a pah), a ten minutes' halt was ordered to enable the main body to come up, and, on the order to advance, the position was taken at the point of the bayonet."

The engagement at Koheroa commenced at 11 and ended at 1 o'clock. The enemy's loss was estimated at 30 or 40 killed, besides wounded, the casualties of the troops being 2 men killed, 1 officer and 10 men wounded.

[1] *Bush Fighting*. Colonel Sir J. Alexander.

The General spoke highly of the conduct of the officers and men engaged, and of the able way they were led by their officers; Major Miller, 12th, being honourably mentioned.

General Cameron's headquarters being at the Queen's Redoubt, the headquarters 14th Regiment, with 180 of the 12th, were pushed forward to Whangamarino, overlooking the Waikato River, and in sight of the strong Maori position of Meri Meri, which Colonel Mair describes as "a hill well protected with two deep rivers, one in front and one on a flank, and with a swamp in rear, and on the fourth side. Great preparations were made later for attacking this most formidable position."

A strong stockade was erected by the troops at Whangamarino. Single natives used to pay the troops daily visits there, in the most daring manner, to have a shot at the sentries. One night, the camp was alarmed by a sentry of the 12th Regiment, who had been attacked on his post by a Maori, who attempted to seize the sentry's rifle with one hand, and to tomahawk him with the other; he cut off the sentry's thumb, but did not get his rifle, and escaped uninjured into the forest.[1]

General Cameron, having been informed that a body of the enemy had collected at the villages of Paparoa and Paparata (to the east of the Koheroa position) marched on the night of the 1st August, from the Queen's Redoubt, with a force of 700 soldiers, seamen, and marines, with the intention of surprising them, but, on reaching the villages, they were found deserted, the natives having retired into the dense bush behind them, from whence they wounded a soldier of the 12th Regiment. The troops returned to the Queen's Redoubt about 3 P.M., having been under arms since 7 P.M. on the previous evening, and having marched cheerfully about thirty miles.

The General and Quarter-Master-General proceeded, on the 12th August, to reconnoitre the enemy's position in the steamer "Avon," when shells and rockets were thrown into their works, inflicting some loss. On the steamer's return, a running fire was opened on it, and was replied to by the rifles of the "Avon." One seaman was grazed by a buck shot.[2]

The headquarters 1st Battn., consisting of 3 captains, 7 subalterns, 4 staff, 20 sergeants, 9 drummers, and 222 rank and file, under command of Colonel H. M. Hamilton, embarked at Sydney on the 22nd September, on the Steamship "Curaçoa" for service in New Zealand, arriving at Auckland on the 3rd October. Captain Vereker broke his collar-bone on the passage.

Lieut.-Colonel Kempt, with 5 officers and 120 men, were left at Sydney, and small detachments of the regiment at Hobart Town, Tasmania, and Brisbane, Queensland.

On the 25th September, the 2nd Battn. marched from the Curragh Camp to Dublin, halting for the night at Naas, and, arriving on the 26th, was quartered at Richmond Barracks.

[1] *Bush Fighting*, p. 64. Sir J. Alexander. [2] *Ibid.*, pp. 69, 70.

On the 1st October, the battalion was inspected at Dublin, by Brigadier-General Ellice, C.B., who expressed his entire satisfaction at its state.

A diary of the New Zealand War, by Lieutenant J. Boulton, 12th Regiment, states that:—"on arrival of the 1st Battn. at Auckland, 'embarkation practice' in boats, by the regiment, at once took place."

On the 9th October, the headquarters staff of the 1st Battn. under Colonel Hamilton (including Lieutenants Crawhall, Lacy, and Morris, Surgeon Bartley, and Quarter-Master Laver) marched from Auckland, *en route* to the advanced post at Koheroa, where the detachment under Major Miller was stationed, arriving on the 13th October, and, on the same day, the remainder of the battalion, under Captain Downing, marched from Auckland to Otahuhu, where they encamped until November 16th.

On the 19th October, a party, consisting of 1 sergeant, 1 drummer, and 50 rank and file, under Lieutenant Mair, marched from Koheroa to Queen's Redoubt, *en route* to Wairoa, to join a flying column under Colonel Nixon, of the Colonial Forces.

On the 25th, the battalion received orders to move, without tents and in light marching order, at a moment's notice.[1]

Lieutenant Boulton relates that:—

> "on the afternoon of the 27th, the gunboat "Pioneer" arrived at the Bluff, having steamed up the River Waikato, with perfect impunity, under some heavy firing, and that she brought some large bullet-proof boats for the conveyance of troops, and also two 40-pounder Armstrong guns. She looked very grand, being 140 feet long, with accommodation for nearly 500 men, and is quite bullet-proof."

In October, the idea of a Soldiers' Industrial Exhibition was first thought of by Colonel Ponsonby, and a meeting of the men of the 2nd Battn. was convened, and the project explained to them.

Firstly.—The Exhibition was intended to give those who were willing to work for it something to amuse and instruct them during their leisure hours.

Secondly.—Such articles as they contributed were to be sold at a little above their actual cost, as the object of the Exhibition was employment, not profit.

Thirdly.—To show the public generally that the Soldier was not a useless member of Society.

The scheme having been taken up, a Committee was formed.

Materials were purchased for those who were without the means of obtaining them, and everybody set to work.

On the 1st November, a mixed force, from six regiments, of 26 officers and 500 men (including 9 officers and 166 of other ranks, 1st Battn., 12th), the whole under command of Lieut.-General Cameron, C.B., were conveyed from Koheroa up the River Waikato to Meri Meri, where a pah

[1] *Diary of Lieut. Boulton, 12th Regiment.*

and numerous lines of rifle-pits had been constructed by the rebel Maoris, who did not, however, await the attack, but fled southwards across country, which recent rains had made impassable for Europeans, whereupon the General occupied their position, and fortified it.

Lieutenant Boulton writes :—

> " We landed immediately, most of us up to our necks in water, and ran up the hill, as far as the pah and flagstaff. Not a single man was to be found, and so this famous stronghold, with its innumerable rifle pits, and other defences, fell into our hands without a blow ; we found two of the enemy's guns, and another is supposed to be in the river. Dusk now approaching, we proceeded to make large fires, and be as comfortable as possible, with nothing to eat, and no blankets, on a very cold night, with occasional showers. The next morning (November 2nd) whilst anxiously awaiting the arrival of rations and bedding, the 'Avon' came up about daybreak, with a few blankets and some rum, and was followed at 8 o'clock by the 'Pioneer,' bringing a portion of the 12th baggage, lots of provisions, and 400 men of the 18th and 70th Regiments. We now set about pitching our tents on the slope of a hill leading down to the river, a very pretty spot, and there being only two tents for nine officers, three of us commenced to build a hut, which was sufficiently complete to sleep in, by the evening.
>
> On the 3rd, we commenced to build, at Meri Meri, (under the superintendence of an officer of the Royal Engineers,) a redoubt, on the flagstaff hill, for 200 men, which was occupied on the 11th by a detachment of 3 officers and 50 men of the regiment."

A Horse Guards' letter, dated 13th November, directed, that in consequence of a strong detachment of the 1st Battn. 12th Foot being now in New Zealand, with other detachments required at stations in Australia, its establishment is increased to 1000 rank and file.

An expeditionary force, under Captain Downing, 12th, with 6 officers and 178 men, left Otahuhu on the 16th November, and sailed next day in H.M.S. "Miranda" from Auckland, for the Thames River, on the east coast, where they disembarked on the 22nd, and were employed under Colonel Carey (18th) in erecting a line of redoubts between the Thames and Waikato Rivers.

Lieutenant Boulton states :—

> "Two more men-of-war accompanied this expedition, conveying 200 men each from the 18th and 70th Regts., with 50 cavalry, and 300 militia, the whole under command of Colonel Carey, 18th 'Royal Irish.' On the following day, the General proceeded up the river in the gun-boat 'Pioneer,' with the 'Avon' in attendance, as far as Rangariri, which was shelled by our force, the natives replying with musketry, having built a pah on the west side of the river, opposite to Rangariri."

On November 19th, the troops at Meri Meri were reinforced by 200 each of the 40th and 65th Regiments ; also, provisions and baggage for 1200 men were conveyed there by steamers in the course of the day.

On the 20th, a mixed force, comprising officers and men of the Naval Brigade, with 46 officers and 1135 men of Artillery, Engineers, and four infantry regiments (to which the 1st Battn. 12th contributed 5 officers and 107 men) marched from Meri Meri, under command of Lieut.-General Cameron, to Rangariri, where the Maoris had erected a formidable line of earthworks, extending from the River Waikato to the Waikare Lake, thereby impeding the advance of the British troops into the heart of the country.

This line of works showed great engineering skill, comprising a pah in the centre, on the highest ground, well protected by a parapet 20 feet high from the bottom of the ditch, and numerous lines of rifle pits, of most intricate nature, in front.

The works were assaulted again and again by the regiments comprising the attacking force, and, after some hours' fighting, the pah was surrounded by our troops. Under the cover of night, several hundreds of the enemy escaped through the swamp ; at daybreak, the remainder surrendered, and 183 prisoners were taken. The casualties on our side were 132.

Colonel Mair, describing the attack, says :—

> "The crossing of the river and the landing in front of Meri Meri—there being only one very small boat for the transit—was rather a difficult business, but Commodore Seymour managed it, and the 12th, 14th, 40th and 65th Regiments were conveyed to the north side of the Waikato River. The same afternoon the place was attacked. I can't quite remember how the storming parties were told off, but I know that one hundred men each of the 12th and 65th Regiments were provided with scaling ladders, each party being given about twenty. All went well, only one or two men being knocked over in crossing the cleared space in front of the pah. This pah had a most unusual addition to its defences, a deep ditch and earthwork which ran round the stockading. Our party advanced at a good pace, and got into the ditch, where they discovered that the ladders were five feet too short, the distance from the lowest part of the ditch to the top of the ramparts being eighteen feet. From a good flanking angle the Maoris fired on our men, only one of whom succeeded in getting to the top of the earthwork. Lieutenant Murphy, 12th, one of the officers of the storming party, who, with two sergeants, made a desperate attempt to climb the earthwork, was killed, and the others had to retire.
>
> As it was almost dark by the time the last storming party retired, the order was given for the different parties to bide for the night pretty well in the positions which had been allotted to them. Biscuits and the usual ration of rum were sent round, and the men, having their great coats, made themselves fairly comfortable.
>
> At daybreak, next morning, we found the pah empty excepting the bodies of two men, one of the 12th and one of the 65th, being found inside. I think the 12th man must have managed to get on to the top of the defences, and was shot through the head and so fell forward."

Lieutenant Boulton, in his journal, gives the following account of the action, on the 20th and 21st November :—

> "At 7 A.M. on the 20th, 400 men of the 65th, 100 of the 40th, with 100 artillery and 20 engineers left Meri Meri for Rangariri. At about 9 A.M. the General and staff arrived at Meri Meri in the 'Pioneer' with 200 of the 40th Regiment, and 150 sailors. The General then landed and sent the 'Pioneer' on, to land the 40th beyond Rangariri, to cut off the enemy's retreat, whilst he himself proceeded by land, accompanied by 100 of the 12th, and 160 of the 14th. On reaching Rangariri about 4 P.M., they found themselves opposed by a redoubt and earthworks, whose parapets were twenty feet high, with ditches on both sides. On the order to charge and escalade, the men charged and rushed on, but the ladders proved too short; nevertheless, they scrambled on to the parapet only to be shot down, and it was here Lieutenant Murphy of the 12th was killed. Four times were the troops led to the assault, and as many times repulsed. When dusk approached, our troops were in possession of a few of the outworks, but the principal redoubt remained untaken. The men bivouacked on the wet ground, disgusted and disheartened, and were kept awake by a chorus of bullets throughout the night. Unfortunately the 'Pioneer' did not land her troops in time to cut off the retreat of some of the enemy, but enabled the 40th to do some execution among them. The British loss was heavy, but that of the enemy very great, as large numbers were slain crossing the swamp.
>
> The 12th Regiment, in the engagement, had suffered more than any other in proportion to its numbers. Out of 5 officers and 107 men engaged, the regiment had 1 officer (Lieutenant Murphy) and 5 men killed, and 19 men wounded; thus, about one man in every four was hit.
>
> The 182 Maoris who surrendered included all the most influential chiefs of the Waikato Country, and 220 stand of arms were taken."

In a General Order, published the next day, the Lieut.-General Commanding congratulated the force under his command on the success of the attack on the enemy's formidable position at Rangariri, and on the capture of a large number of prisoners, thanking the officers and men by whose valour and conduct the important advantage had been gained, with the promise of bringing their services to the favourable notice of His Royal Highness the Commander-in-Chief.

General Cameron, after consulting the Governor, next decided to invade the King's Country, as some eighty miles of beautiful land beyond the Waikato is called, but as there were no roads, and many rivers and swamps to cross, the preparations for advance took two months.

The following move of the headquarters of the 1st Battn. from Koheroa to Ratinipokeka is described by Lieutenant Boulton as follows :—

> "*December 7th.*—Colonel Hamilton, the adjutant, Major Miller, Dr. Bartley, and the writer of the journal, with 110 men, marched to

Meri Meri and Rangariri, arriving at the latter, at about 5.30 P.M., after a twenty mile march, over a most desolate and hilly country, almost devoid of wood, and the march rendered doubly tedious by the slow progress of the bullock drays.

December 9th.—We were joined at Rangariri to-day by Captain Williams, Lieutenant Featherstonhaugh, Ensign Cooper, and 70 men of the regiment.

December 10th.—Moved forward sixteen miles to Ratinipokeka, where we were joined by Captain Cole and 80 men of the regiment. The country we passed through was flat and open, with numerous deserted native villages, and settlers' houses, with well-cultivated cornfields. We now encamped on rough scrubby ground close to the Waikato, and, the next day, received orders to remain, in order to finish the redoubt and other works.

December 19th.—Captain Vereker rejoined to-day from sick leave, after having broken his collar-bone on the voyage from Sydney, New South Wales."

The headquarters of the army in New Zealand, under Lieut.-General Cameron, had, on the 9th December, occupied Ngaruawahia, the capital of the rebellious country, and, on the 20th, a General Order announced his pleasure in publishing to the troops the following resolutions, adopted by the House of Representatives, Auckland, December 1st, 1863, "feeling sure that this honourable tribute to their gallantry and valour will be fully appreciated by all ranks under his command."

"Resolved:—That the thanks of this House be presented to Lieut.-General Cameron, C.B., Commanding Her Majesty's Forces in New Zealand, for the energy and ability with which he has conducted the military operations in New Zealand, and especially for the decisive defeat of the rebels at Rangariri.

That the thanks of this House be given to the Officers of Her Majesty's Army for their zeal and gallantry, and to the non-commissioned officers and soldiers for the discipline and valour they have displayed in the military operations in which they have been engaged, and especially at the assault and capture of Rangariri.

That the Speaker do communicate these resolutions to Lieut.-General Cameron, and that he be requested to signify the same to the officers and soldiers under his command."

On the 22nd December, the detachment, 1st Battn., belonging to the Thames Expeditionary Force, marched, under command of Captain Downing, to the Queen's Redoubt, and followed the headquarters of the battalion up country as far as Ngaruawahia, where they arrived on the 31st.

Lieutenant Boulton's diary gives the following:—

"Pursuant to an order from the General, Colonel Hamilton took thirty men to the other side of the river, to endeavour to capture a chief and twelve rebels, said to be concealed amongst the friendlies. The party went with their rifles hidden at the bottom of the boat, and immediately surrounding them, captured the whole, and at once despatched them in the 'Pioneer' to the General."

On the 26th December, the headquarters of the 1st Battn., under Colonel Hamilton, were conveyed by steamer from Ratinipokeka to the advanced post at Ngaruawahia, and joined the headquarters of the army, under Lieut.-General Cameron, C.B.

Referring to this move, Lieutenant Boulton writes :—

"At this station were the General and his staff, a battery of Armstrong guns, some R. E., about 750 of the 40th and 65th Regiments, and a large staff of Commissariat. The King's palace here, consisting of one large room with a portico, is now the guard-room, where eleven native prisoners are confined. There are large cultivations of potatoes here, and the camp is pitched in a potato field. The surface of the soil is covered with light sand, and the dust is perfectly blinding."

On December 31st, the headquarters detachment, 1st Battn., consisting of 4 officers, 5 sergeants, and 100 rank and file, under Colonel Hamilton, marched to Wata Wata, where they arrived at 3 P.M., *en route* to join Lieut.-General Cameron, who had again moved.

1864

On the 1st January, the march was continued five miles farther to Tinkaramayo. The General now had with him 1100 men, consisting of detachments of the 12th, 40th, and 65th Regiments.

On the 9th, the remainder of the 1st Battn. arrived at Wata Wata from Sydney, N.S.W.

On the 12th January the Soldier's Industrial Exhibition instituted by the 2nd Battn., which was held in the Rotunda, Dublin, was opened by His Excellency the Lord Lieutenant (Lord Carlisle), attended by Sir George Brown, Commander of the Forces, the Lord Mayor, and a numerous assemblage of the nobility and gentry of the country.

His Excellency the Lord Lieutenant was entertained by the Officers of the regiment on the occasion, and at the conclusion made a suitable speech.

On the 27th January, the field force at Tinkaramayo (New Zealand), under the Lieut.-General Commanding (having been joined by the 1st Battn. 12th and Colonial Forces from Wata Wata), advanced towards the enemy's strongly fortified positions at Pipopoko and Patermigi, and arrived the same evening at Te Rore, having, on the march, detached 4 officers and 136 men, under Captain Vereker, 12th, who proceeded to Nyalimapouri, to erect redoubts on both sides of the River Waipa, and, on relief by a party of the 40th, this detachment rejoined headquarters.

On the 31st January, the 1st Battn. 12th were appointed the General's guard and escort.[1]

[1] Extract from *Diary of Lieutenant Boulton*.

The headquarters of the battalion remained at Te Rore a considerable time without any change, garrisoning three redoubts, and supplying a detachment to the enemy's evacuated stronghold at Paterangi.

On the 31st March, a detachment, under Captain Vereker, left Te Rore to reinforce the troops engaged at the storming of a Maori pah at Orakau. On the pah being surrounded, the natives, although without water, and suffering severe loss, held out in the most determined manner for three days, when, in an escape that was attempted, many were killed or taken prisoners. A sap of some length was successfully conducted by Lieutenant Hurst, 12th Regiment, attached to the Royal Engineers, up to within a few feet of the pah. The detachment of the regiment had 4 rank and file wounded on this occasion.

On the 29th April, a detachment of the 1st Battn. under Captain O'Shaughnessy, attached to the Flying Column, took part in the storming of the Gate Pah, near Tauranga, on the east coast of New Zealand, when the 43rd Regiment suffered almost unparalleled losses among their officers. One man of the 12th was killed, and one wounded, in this memorable engagement.

On the 23rd May, a detachment of 100 rank and file, under Major Miller, left headquarters at Te Rore for Nyalimapouri, and, on the same date, a similar detachment under Captain Downing marched to Raylim, a town on the west coast.

Early this year it was notified that the 2nd Battalion would embark for India during the ensuing summer, and, on the 20th April, orders were received to prepare for embarkation.

The battalion embarked for India in four detachments with the following strength :—

34 officers, 41 sergeants, 37 corporals, 19 drummers, and 782 privates—total 913.

The first detachment left Dublin for Cork on the 15th July, and sailed on the 17th, under command of Major Espinasse, on board the "Alnwick Castle," which encountered a very heavy cyclone in the Bay of Bengal. From 1 to 4.30 A.M. on the 24th October was an anxious time for those on board, but beyond severe damage to the ship from the heavy gale, there was no grounding of the vessel, nor were any lives lost.

The second detachment, under Lieutenant Keough, left Dublin on the 16th July, and sailed from Cork on the 18th, on board the "Trafalgar."

The third detachment, under Major Queade, sailed from Cork, on board the "Calcutta," on the 21st July, and the fourth, under Major Atkinson, on board the "Aliquis," on the 23rd.

The 1st Battn., on the 13th October, under Colonel Hamilton, left Te Rore for Ngaruawahia, furnishing, *en route*, three detachments to stations on the line of communication between Auckland and the headquarters of the army.

DISTINGUISHED OFFICERS WHO SERVED IN THE 12TH.

On the 15th and 27th, the headquarters of the battalion were augmented by the return of the detachments which had been attached to the flying column, and from Wata Wata respectively, but, on the 28th, a party, consisting of 4 officers and 145 of other ranks, left for the Queen's Redoubt, thereby reducing the strength of the headquarters to 7 officers and 282 of other ranks.

On the 29th October, Lieut.-General Henry Colvile was appointed Colonel of the 12th Regiment in succession to Lieut.-General C. A. F. Bentinck, deceased.

The four detachments of the 2nd Battn. had arrived at Calcutta by the 6th November, and were stationed at Fort William until the 10th, on which date, the 1st division proceeded by rail towards Seetapore and Bareilly, followed by the headquarters on the 11th, and the last division on the 17th, Lieut.-Colonel Ponsonby joining the battalion at Allahabad, from England, to assume command.

Captain Vereker's detachment, 1st Battn., rejoined headquarters at Ngaruawahia, on the 17th November.

On the 26th, the right wing of the 2nd Battn., under Major Atkinson, marched from Allahabad to Bareilly, arriving on the 2nd December, and, on the 28th and 30th, the headquarters and left wing, under Lieut.-Colonel Ponsonby, left Allahabad for Seetapore, where they arrived on the 15th December.

A War Office letter had been received in the previous month, fixing the establishment of the 2nd Battn., at 40 officers and 919 of other ranks, to 10 service companies, the depot to consist of 2 companies, mustering 6 officers and 114 of other ranks, to take effect from the 12th March.

1865

On the 2nd January, the 1st Battn. furnished a detachment to Wata Wata, and another, on the 29th April, to Mangawarra Creek, Timpiri, where they erected a redoubt.

Orders were issued from the Horse Guards on the 10th June, that the companies of the battalion were to be distinguished by letters from "A" to "M" instead of by numbers.

On the 21st July, the detachment of the 1st Battn. at Otahuhu was strengthened by a party of 53 of all ranks from Queen's Redoubt, the detachment at Timpiri joining headquarters on the 8th November.

On the 20th October, Surgeon W. G. N. Manley, V.C., a distinguished medical officer, was posted, on promotion, as a regimental surgeon to the 1st Battalion.

As Assistant-Surgeon, on first appointment to the Royal Artillery, in June, 1855, he served in the Crimea, at the Siege of Sevastopol, and, in 1864, whilst serving in New Zealand, gained the Victoria Cross for the following act of bravery.

At the Maori pah, Tauranga, on the 29th April, 1864, he volunteered to accompany the storming party into the pah. Here, Commander Hay, R.N., was mortally wounded, and, when removed, Doctor Manley followed, amid a hail of lead, to attend upon him. This done, he again volunteered to enter the pah in search of wounded, and being successful in finding many, he was the last man to quit it. He also possessed the Royal Humane Society's Medal for saving the life of a man of the Royal Artillery, in New Zealand, on July 21st, 1865, and, for service with the British Ambulance in 1870–71, Doctor Manley received the Prussian (steel) War Medal, the Iron Cross, and the Bavarian Order of Merit.

He exchanged into the Royal Artillery on the 8th November, 1867, finally attained the rank of Surgeon-General, and was appointed a Companion of the Order of the Bath. (See Plate 28.)

The headquarters, 1st Battn., under Colonel Hamilton, left Nagawahia for Otahuhu on the 4th December, *en route* to Napier on the east coast. The detachments at Wata Wata, Rangariri, and Queen's Redoubt rejoined, and arrived at Otahuhu on the 6th.

On the 9th, the headquarters having marched from Otahuhu, embarked at Auckland, in H.M.S. "Esk," and, arriving at Napier on the 11th, occupied barracks there; the remainder of the battalion, under Brevet-Major Miller, embarking at Auckland on the 15th, in H.M.S. "Eclipse," joined headquarters at Napier two days later. A subaltern's detachment was left at Otahuhu.

1866

On the 31st January and 17th February, the 1st Battn. furnished detachments to Wairoa and Tauranga respectively; the headquarters, under Colonel Hamilton, following to Tauranga a week later, and, between the 9th March and 23rd May, officers and men from three outposts (under Ensigns Taylor, Turner, and H. S. Bolton), and the detachment under Captain O'Shaughnessy from Raylim, rejoined headquarters.

Instructions having been issued in May this year that a reduction of 50 privates was to take place in the establishment of the 1st Battn., from the 1st April, the strength of the battalion of 10 companies, as now approved, was 39 officers and 750 of other ranks.

On the 28th August, an officer's party, from the detachment of the battalion at Napier, proceeded, under Lieutenant Hurst, to Waipawamate.

On the 9th and 10th November, the 1st Battn. furnished two strong detachments of 200 rank and file, with a proportion of officers to each, in aid of the civil power. The first of these proceeded on the 9th, over a very hilly and rugged country, and pitched their camp in a commanding position, subsequently named "Minden Peak," about 15 miles from Tauranga, and opposite the Maori village of Waiwhatawata, and the second marched on the 10th to the Wairoa River. The services of the battalion not being required on either occasion, they returned to camp.

In consequence of reports received of reinforcements of hostile natives having joined the rebels, another reserve, consisting of 5 officers and 156 of other ranks, marched to Minden Peak, where it remained.

On the 12th November, a party of 12 officers and 200 of other ranks, under Colonel Hamilton, left for the scene of operations, but returned to camp at Tauranga the same night.

All danger of attack from hostile natives having now ceased, the camp at Minden Peak was broken up the next day, when the detachment returned.

1867

On the 5th January, His Excellency the Commander-in-Chief, Sir William Mansfield, inspected the headquarters of the 2nd Battn. at Seetapore, and expressed his entire satisfaction at the smart and soldier-like appearance of the men under arms, the efficient manœuvring, and the neat and cleanly state of the barracks, and was pleased to ask for the release of the regimental prisoners.

In consequence of a telegram received, on the 23rd January, from the Governor of New Zealand, a party of 230 men, with a proportion of officers, under Colonel Hamilton, again marched to Minden Peak *en route* to Waiwatawata. This village, however, having been burnt down by the militia, the party returned to camp at Tauranga, which was reached at midnight.

At the request of the Defence Minister, that Colonel Hamilton, commanding the 1st Battn. 12th Regiment, should aid the Militia and the Arawas (a friendly tribe), when attacking the villages of Meene, Ake Ake, and Taumata, Colonel Hamilton, with a party of 7 officers and 225 men, marched to Taumata, and, after the destruction of these villages (from which the natives fled at the approach of the troops) returned to Tauranga, leaving 175 men, under Captain Sillery, until next morning, in a redoubt on the Wairoa River.

On the 13th February, the 1st Battn. received orders to be held in readiness to embark for England at short notice.

Accordingly, the headquarters and the detachments embarked at Tauranga at intervals, as expeditiously as possible, for Auckland, where the whole battalion was concentrated by the 9th April.

On the 2nd May, 5 companies, under Captain Sillery, embarked at Auckland, for England, in the transport "England," mustering 11 officers and 285 of other ranks.

A General Order, dated Headquarters Auckland,[1] 16th May, by Major-General Trevor Chute, commanding the force in New Zealand, announced:—

> "The Major-General Commanding cannot allow the 1st Battn. 12th Regiment to leave the command without placing on record the

[1] It is recorded in the *New Zealand Index Directory of Towns* that the town of Hamilton, on the Waikato River, 86 miles from Auckland (which in 1912 contained a population of 3600) was founded by military settlers of the 12th Regiment, and named after Colonel H. M. Hamilton, C.B., commanding it.

very high opinion he entertains of the services, discipline, and good conduct of the corps.

On the occasion of his recent inspection, the appearance of the regiment on parade, their steadiness under arms, and proficiency in manœuvring, as well as the excellence of their interior economy, reflects the highest credit on Colonel Hamilton, his officers, and the regiment. Of their valuable services in the field, prior to the Major-General's arrival in the country, he is well aware, and he cordially thanks them for those rendered since he assumed the command."

Prior to the issue of the foregoing General Order, a letter had been received from the Governor of New South Wales, dated Government House, Sydney, 29th January, 1867, to the Officer Commanding, as follows :—

"As the detachment of the regiment under your command has been relieved by Her Majesty's 50th Regiment, I feel bound to take this opportunity of expressing to you my appreciation of the services rendered by the 12th Regiment while quartered in Sydney. Personally I had every reason to be well pleased with the bearing of all ranks, and I am persuaded that the inhabitants of Sydney generally concur in this favourable opinion."

The headquarters, 1st Battn., and remaining 5 companies, under Colonel Hamilton, mustering 11 officers and 286 of other ranks, embarked at Auckland, in the transport "Mary Shepherd," on the 17th May, for England.

Up to this date, the loss sustained by the 2nd Battn. during its three years' Indian service had been 46 of all ranks by death, including 3 officers ; also 101 by expiration of service and invalided home.

CHAPTER X

East Indies, England, Ireland, East Indies.
1867–1878

1867

The 2nd Battalion having received orders to move from Seetapore and Bareilly to Jubbulpore and Nagode, the march towards the latter places commenced on the 1st November, when the headquarters wing proceeded towards Jubbulpore, and, on arrival at Bareilly, was joined by the left wing, when the whole battalion marched on the 14th to Allahabad, where they arrived on the 22nd and encamped near the railway station. Here they were inspected by Major-General Beatson, C.S.I., who desired that Colonel Ponsonby should convey to his officers, non-commissioned officers, and men, the satisfaction he felt in finding the battalion so smart and soldier-like on parade, and so perfect in arrangements, as regarded neatness and regularity in camp. The battalion proceeded from Allahabad to Jubbulpore and Nagode by train, and arrived on the 2nd December, relieving the 23rd Fusiliers. The wing of the 1st Battalion, under Captain Sillery, landed at Plymouth from New Zealand on the 14th August, after a voyage of 104 days. Lieutenant Woodward and one private died on the voyage. The headquarters, under Colonel Hamilton, landed at Plymouth on the 28th August, after a voyage of 103 days, and proceeded to Tregantle Fort, Cornwall, to be stationed (6 companies), and 4 companies at Devonport.

A War Office letter, dated October 7th, authorised the establishment of the 1st Battalion to consist of 10 companies, mustering 39 officers, 50 sergeants, 20 drummers, and 600 rank and file.

On the 18th November, the headquarters, under Colonel Hamilton, proceeded from Tregantle Fort, Cornwall, to Devonport, to be quartered.

Enlistment for twelve years was this year introduced, and the following new rates of pay:—sergeant-major, 2*s.* 11½*d.*; quarter-master-sergeant, 2*s.* 3½*d.*; colour-sergeant, 2*s.* 1½*d.*; sergeant, 1*s.* 7½*d.*; corporal, 11½*d.*; drummer, 8½*d.*; private, 7½*d.*, after deductions for rations.

1868

On the 5th February, the detachment of 3 companies 2nd Battalion at Nagode, marched to Nowgong, where they arrived on the 14th, and on the 31st March, the headquarters were inspected by Brigadier-General Sir W. Turner, K.C.S.I., who expressed himself well pleased with the steadiness of the men, the celerity of their movements, and the manner in which they performed outpost duty.

On the 16th June, Lieut.-Colonel Ponsonby died at Jubbulpore, of cholera, and was succeeded in the command by Major and Local Lieut.-Colonel R. Atkinson, promoted, who, on the 17th November, exchanged to the 35th Regiment with Colonel J. McN. Walter, C.B. The depot of the 2nd Battalion was now at Gosport.

1869

On the 13th March, the headquarter companies, 2nd Battalion, at Jubbulpore, were inspected by the Brigadier-General Commanding, who remarked, in the most complimentary terms, on their drill and skirmishing (the inspection lasting three days), and commented on the fit state of the battalion to take the field. He had expressed himself as favourably on what he had seen of the detached companies at Nowgong in the previous month.

On the 23rd March, the 1st Battalion, consisting of 29 officers and 596 of other ranks, under Colonel Hamilton, embarked on the troopship "Orontes" for Portsmouth, *en route* to Aldershot, and, arriving on the 25th, was posted to the 1st Brigade, in the South Camp.

On the 3rd April, Colonel Walter, C.B., joined the 2nd Battalion, and assumed command.

On the 7th, the 1st Battalion, under Colonel Hamilton, was inspected at Aldershot, by Lieut.-General the Hon. Sir J. Yorke Scarlett, K.C.B., commanding the Division, who expressed his entire satisfaction with its appearance, and was pleased to say he was "glad to have his command augmented by such a smart battalion."

Under War Office instructions, dated 27th April, a reduction of 40 privates, in the establishment of the 1st Battalion, was authorised, with effect from the 1st April, the other ranks remaining as in the previous year.

On the 9th December, the 1st Battalion left Aldershot by special trains for the Northern District, headquarters and 3 companies proceeding to Carlisle, 4 companies to Sunderland, and 3 to Burnley; all arriving at their destinations on the following day.

1870

On the 21st February, the detachment at Sunderland was broken up, 2 companies proceeding to Preston, 1 to the Isle of Man, and 1 rejoined headquarters at Carlisle. On the same date, 1 company left headquarters for Preston, and, on the 28th, the 3 companies at Burnley were transferred there.

On the 1st March, the depot, 2nd Battalion, arrived at Preston from Parkhurst, Isle of Wight, and was attached to the detachment of the 1st Battalion, the headquarters and 1 company of the latter moving from Carlisle to Preston on the 29th.

Medals for the New Zealand War having been now received, they were presented, at a parade held on the 1st April, by Colonel Hamilton, to officers and men of the 1st Battalion who were entitled to them.

On the 13th April, whilst the headquarter companies, 2nd Battalion, at Jubbulpore, were being inspected by Brigadier-General Forrest, in review order, His Excellency, the Commander-in-Chief in the East Indies, Lord Napier of Magdala (having just arrived from Bombay, on his way to Simla) rode on to the ground and inspected the battalion, which was put through the manual, platoon, and bayonet exercises, followed by movements under several officers, who were called out. An inspection then followed of the hospital and barrack rooms, &c., whereupon His Excellency expressed great satisfaction at everything he had seen, commenting favourably on the drill, the smart appearance of the battalion, and the neat and orderly arrangement of the barrack rooms, and requested that all prisoners and defaulters might be released.

On the 4th May, Colonel J. Walter, C.B., proceeded to take temporary command of the Rohilcund District, and Major G. Walker assumed command of the 2nd Battalion.

Under War Office instructions, dated 14th May, the establishment of the 1st Battalion was, from the 1st April, to consist of 10 companies, mustering 31 officers, 50 sergeants, 20 drummers, and 500 rank and file—total 601, and 2 medical officers attached.

On the 1st July, the establishment of the 2nd Battalion (as fixed for British infantry regiments in India, for the year 1870) was to be as follows:—8 service companies, of 30 officers, 49 sergeants, 16 drummers, and 820 rank and file—total 915; whilst the depot was to consist of 2 companies, mustering 5 officers, 12 sergeants, and 100 rank and file—total 117. In consequence of the reduction in the new establishment of the 2nd Battalion, two companies were now broken up, and the men distributed amongst the other companies.

From the 15th August, the following increase in the establishment of the 1st Battalion was authorised:—rank and file (corporals and privates)—raised from 500 to 800.

On the 26th October, Colonel Walter, C.B., rejoined headquarters, from the Rohilcund District, and assumed command of the 2nd Battalion, which was inspected, on the 29th, by Brigadier-General Forrest, who expressed himself as highly pleased with everything.

Short service was this year introduced into the Army.

1871

The 2nd Battalion having been ordered to Subathu, the regiment, starting on the 9th December, proceeded in three detachments from Jubbulpore and Nowgong, and, by the 27th January, the whole of the companies had arrived there.

Prior to the headquarters leaving Jubbulpore, Brigadier-General Forrest published a District Order, expressing his entire satisfaction at the conduct of the battalion, and adding that "the absence of crime, and the soldierly bearing of the men, had been a great source of pride and pleasure to him, and was, in his opinion, a sure indication of the interest taken by the officers in their men, and the consequent reciprocation of this feeling."

By an Army Circular, dated 1st February, the following decrease in the establishment of the 1st Battalion was authorised this year :—rank and file (corporals and privates) lowered from 800 to 600.

On the 10th March, Colonel Walter, on taking over temporary command of the Sialkot Brigade, was succeeded by Lieut.-Colonel Foster.

On the 27th June, Major J. McKay was promoted Lieut.-Colonel in the 2nd Battalion in succession to Colonel Walter, C.B., appointed to the Indian Brigade Staff. The detachments of the 1st Battalion at Carlisle and the Isle of Man joined headquarters at Preston on the 25th and 29th September respectively.

By Royal Warrant, dated 30th October, purchase in the army, and the ranks of Cornet and Ensign, were abolished, that of Sub-Lieutenant being substituted for them.

On the 11th November, Lieut.-Colonel McKay assumed command of the 2nd Battalion.

1872

The approved establishment of the 1st Battalion, authorised for this year, on the 1st May, was :—33 officers, 48 sergeants, 18 drummers, and 820 rank and file—total 919.

On the 30th May, the 2nd Battalion was inspected at Subathu by His Excellency the Commander-in-Chief, Lord Napier of Magdala, and, on the following morning, 5 companies were sent to Kuckerbuttie, under Colonel Foster, to attack Subathu, which was defended by 3 companies under Lieut.-Colonel Walker. His Excellency addressed the battalion on parade, and was much pleased with the clean and soldier-like appearance of the men, and the cleanliness of the barracks, hospital, &c., and spoke in high terms of the good tone and high state of discipline in the battalion, and the absence of crime, particularly that of a serious or insubordinate nature.

On the 16th July, the 1st Battalion, under Colonel Hamilton, proceeded by rail from Preston to Liverpool, and embarking on the "Orontes," sailed for Ireland. Disembarking at Kingstown on the 18th, the regiment was

OFFICER'S SHAKO PLATE
(*reduced size*).
1855-61.

OFFICER'S SHAKO PLATE
(*reduced size*).
1861–70.

CAP BADGE,
1847–55.

OFFICER'S SHAKO PLATE (*reduced size*).
1871–78.

OFFICER'S HELMET PLATE (*reduced size*).
1878–81.

conveyed by rail to Dublin, and furnished detachments to Tuam, Dunmore, Castlebar, Galway, Ballaghadereen, Ballinrobe, Ballina, Newport, and Westport; the headquarters, with the attached depot of the 2nd Battalion being stationed at Athlone.

During part of the years 1869 to 1872, recruits were raised for the regiment from all districts in Great Britain and Ireland; during this period 635 men were enlisted, chiefly from the Midland counties of England, and a large number from Lancashire.

On the 12th and 13th October, His Excellency the Commander-in-Chief, accompanied by his staff, inspected the 2nd Battalion at Subathu, and again alluded to its good behaviour and soldierly bearing. When inspecting the Regimental School, he addressed the men on the advantage to be derived from regular attendance, and, as an incentive to their acquiring knowledge, His Excellency gave a donation of Rs.50, towards this praiseworthy cause, to be distributed between ten adults, and directed that eight prizes be given to the children who had shown most diligence and attention at school, also that the adult classes be provided with dictionaries at his expense.

The Tuam detachment rejoined headquarters, 1st Battalion, on the 22nd October. On the 24th, the 2nd Battalion proceeded to Dhurmpore, to join a camp of exercise with the 85th Light Infantry, and a mountain battery, under the supervision of Major-General J. M. Tytler, C.B., commanding Sirhind Division, and, on the 29th, the troops marched to Solan, where field movements took place daily, until the camp was broken up on the 2nd November, when the battalion returned to Subathu.

On the same date, the Dunmore detachment, 1st Battalion, rejoined headquarters at Athlone, and the Ballinrobe detachment on the 13th.

On the 15th November, the 2nd Battalion marched from Subathu to Ferozepore, arriving on the 2nd December.

Tuam was re-occupied by a detachment of the 1st Battalion on the 12th December.

The following new rates of pay were introduced this year, with *no deductions*:—sergeant-major, 3*s*. 6*d*.; quarter-master-sergeant, 2*s*. 10*d*.; colour-sergeant, 2*s*. 8*d*.; sergeant, 2*s*. 2*d*.; corporal, 1*s*. 6*d*.; drummer, 1*s*. 3*d*.; private, 1*s*.

1873

Brigade Depots were formed on the 1st April, that of the 12th Regiment being No. 32, at Bury St. Edmunds, but no depot companies joined there for duty until June 1878.

On the same date, Major E. H. Foster was promoted to the command of the 1st Battalion *vice* Colonel Hamilton, appointed to the Brigade Depot, Bury St. Edmunds.

The Ballina detachment, 1st Battalion, moved to Castlebar on the 4th, and the Galway and Ballaghadareen detachments rejoined headquarters at Athlone on the 25th April and 22nd May respectively.

Under War Office instructions, it was directed that from the 1st April, the establishment of the 1st Battalion was to consist of 27 officers, 42 sergeants, 16 drummers, and 520 rank and file.

On the 9th July, the headquarters and 7 companies, 1st Battalion, proceeded by rail from Athlone to the Curragh Camp, the depot, 2nd Battalion, being left behind to do duty at that station, and, on the same date, the remaining 3 companies of the battalion proceeded from Castlebar, Tuam, and Newport to join headquarters at the Curragh Camp.

On the 9th August, a serious conflict took place at the Curragh between the North Cork Rifles and the Queen's County Militia Regiments, both of which were occupying the same lines. The 12th, 27th, and 57th Regiments were called on to suppress the disturbance, and disarm the combatants; which elicited from the General Officer Commanding, a Divisional Order, expressing his entire satisfaction at the promptitude with which the latter had assembled in the militia square, adding that he wished "for no greater proof of good discipline and soldierlike conduct."

1874

On the 28th February, the 2nd Battalion was inspected at Ferozepore, by Major-General Sir C. Reid, K.C.B., who expressed himself highly satisfied with its appearance and drill, and with all he had seen. To this, Lieut.-Colonel McKay was pleased to add, in Regimental Orders, a tribute to the strong *esprit de corps* in the battalion, on which Sir C. Reid had also commented most favourably.

On the 18th April, the 1st Battalion, with the depots of the 2nd Battalion and 86th Regiment attached, proceeded, under Colonel Foster, to Dublin, to take part in the procession, on the occasion of the public entry into Dublin of His Grace the Duke of Abercorn, as Lord-Lieutenant of Ireland, and returned the same day to the Curragh.

The establishment of the 1st Battalion was this year re-constructed, from the 1st April, to consist of 31 officers, 48 sergeants, 18 drummers, and 600 rank and file.

At an inspection of the 1st Battalion, on the 8th July, at the Curragh Camp, by Major-General Wardlaw, C.B., he was pleased to remark on the soldierlike and cleanly appearance of the battalion on parade, and on its steadiness, and general efficiency in drill.

On the 13th August, the battalion, and depot, 2nd Battalion, proceeded by rail, in two divisions, under Colonel Foster, from the Curragh Camp, as follows:—headquarters and 4 companies to Kinsale; 5 to Camden Fort; 1 to Bantry, and the depot, 2nd Battalion, to Charles Fort.

On the regiment leaving for Kinsale, the following appeared in *The Irish Times* :—

> "The 12th have concluded a lengthened sojourn at the camp, to the regret of every resident at or near the Curragh. Officers and men, from their genial Colonel downwards, had endeared themselves to all, and by their departure have left a wide gap in the society at the Curragh. From the day of their march into camp, until they left it for Kinsale and Camden Fort, the conduct of the men has been most exemplary. Their Colonel has been frequently congratulated, by the General in command, on the excellent discipline and soldierlike bearing of the men."

The remarks of the Commander-in-Chief, on the latest inspection of the 2nd Battalion, at Ferozepore, by Major-General Sir C. Reid, were conveyed to it, on the 8th October, as follows :—

> "The report on the 2nd Battalion 12th Regiment is highly creditable to all concerned, and His Royal Highness desires that his satisfaction and commendations should be made known to all."

1875

In the month of September, a very sudden and severe outbreak of typhoid fever occurred amongst the headquarters of the 1st Battalion, at Kinsale, and, on the 19th, they marched from Kinsale Barracks, mustering 9 officers and 215 of other ranks, to encamp on Charles Fort Green.

The company at Bantry proceeded on the 30th to Carlisle Fort, Cork Harbour, on relief by a detachment from the headquarter encampment at Kinsale, and, on the 1st October, one company proceeded from Camden to Carlisle Fort, the headquarters furnishing, on the 7th, further detachments to Bandon and Youghal, whilst the Carlisle Fort detachment returned to Camden.

On the 29th October, the headquarters, 1st Battalion, under Colonel Foster, moved to Cork, where the whole battalion was concentrated by the 30th of the following month.

On the 2nd November, General John Patton was appointed Colonel of the 12th Regiment, on transfer from the 47th Regiment, in succession to General Henry Colvile, deceased. (See Plate 32.)

On the 13th December, the 2nd Battalion, on relief by the 34th Regiment, left Ferozepore by route march to Delhi, to take part in the camp of exercise, in honour of the arrival in India of His Royal Highness the Prince of Wales. Arriving on the 22nd, the battalion joined the 2nd Brigade, 3rd Division, and was brigaded with the 15th and 45th Sikhs, under command of Colonel Appleyard, C.B.

1876

On the 11th January, the 2nd Battalion took part in lining the roads of Delhi, on the occasion of His Royal Highness's arrival, and in the review on the following day.

An Adjutant-General's letter, received this month, intimated that the Duke of Cambridge had received a most satisfactory report on the 2nd Battalion, 12th Regiment, and on its commanding officer, Lieut.-Colonel McKay.

On the 27th January, the headquarters and 4 companies, 2nd Battalion, under Lieut.-Colonel McKay, proceeded to Calcutta, arriving on the 31st, and were followed the next day by the remaining 4 companies, who arrived at Fort William on the 1st February.

The establishment of the 1st Battalion, from the 1st April, was directed to be 31 officers, 48 sergeants, 18 drummers, and 820 rank and file.

On the 3rd August, orders were received for the 1st Battalion to be held in readiness to embark for India, via the Suez Canal, on or about the 22nd September, and for the formation of the depot. Two depot companies were accordingly formed on the 1st September, mustering 4 officers, 6 sergeants, 2 drummers, and 100 rank and file, when it was found that the service companies became far short of their establishment, whereupon volunteers were called for from other corps, when 3 sergeants and 142 privates were transferred to the battalion, this number having been made up by volunteers from the 2nd Battalions 6th, 10th, 20th, and 25th, and from the 27th, 64th, 77th, and 108th Regiments.

As it was anticipated that the battalion could not possibly embark at a strength of 820 rank and file, as had been directed, orders were received for 2 captains and 2 subalterns to be left behind, to proceed to India early in the following year, to reinforce the service companies.

On the 21st September, the service companies, 1st Battalion, under Colonel E. H. Foster, consisting of 21 officers, 33 sergeants, 16 drummers, 614 rank and file, 74 women, and 112 children, proceeded from Cork to Queenstown by river steamers, and embarked on the Indian Troopship "Crocodile," sailing on the following morning for Bombay. The depot companies, with 4 officers, which had now been made up to 13 sergeants, 3 drummers, and 293 rank and file, remained at Cork, attached to the 77th Regiment, with the depot of the 2nd Battalion, both depots being under orders to proceed to England. They accordingly embarked at Queenstown on the 2nd October, the depot, 1st Battalion, proceeding to Gravesend, and that of the 2nd Battalion to Tilbury Fort.

The 1st Battalion arrived at Bombay on the 24th October, and disembarking the same day, proceeded by rail in two detachments to Umballa, where the headquarters and 6 companies arrived on the 31st, and the remaining 2 companies on the 4th November.

By Royal Warrant of the 30th October, the rank of 2nd Lieutenant in the Army was substituted for that of Sub-Lieutenant.

1877

On the 1st January, Her Majesty Queen Victoria was proclaimed "Empress of India."

The 2nd Battalion at Fort William, Calcutta, paraded with all the troops in garrison, and also those from Dum Dum and Barrackpore, to hear the Proclamation read, and to celebrate the occasion. A Royal Salute of 101 guns and a feu-de-joie were fired, and the parade terminated with three cheers for Her Majesty the Queen and Empress of India. To mark the event, the Government presented a day's pay to every soldier, British and native, in India. A silver commemorative medal was also given to every regiment, to be presented to a selected soldier.

On the *obverse* side, the medal bore a bust of Her Majesty, with the words "Victoria, 1st January, 1877," and on the *reverse* were inscribed the words "Empress of India," and the same in Persian and Denanagra characters. It was worn round the neck, attached to a maroon-coloured ribbon, with a yellow border.

The medal given to the battalion was presented to Colour-Sergeant Joseph Parker, orderly-room clerk.

On the 2nd January, the depot, 2nd Battalion, moved from Tilbury Fort to Milton Barracks, Gravesend, and, on the 17th March, the depot, 1st Battalion, marched from Gravesend to Fort Wallington, Fareham, detaching 1 officer and 50 men to Fort Nelson, where the 2nd Battalion depot arrived a month later.

On the 4th April, three companies, 1st Battalion, marched to Solan, there to be stationed during the hot season, and rejoined headquarters at Umballa on the 7th November.

The 2nd Battalion having received orders to return to England, volunteers to other regiments were now called for, to the number of 180, when 99 men transferred their services to the 1st Battalion, and 81 to various other corps.

On the 4th December, on relief by the 54th Regiment, the battalion proceeded by rail to Bombay, in half battalions, and embarked in the troopship "Jumna" for England on the 22nd, sailing the same day.

During the Indian tour of the 2nd Battalion, 1864–77, there died 14 officers, 24 sergeants, 15 corporals, 7 drummers, and 200 privates, to whose memory a monument is erected at Bury St. Edmunds.

1878

The 2nd Battalion disembarked at Portsmouth on the 26th January, and was quartered in the New Barracks, Gosport.

On the 4th February, the battalion was inspected by General Sir J. Garvock, G.C.B., who, in expressing his great satisfaction at its appearance, was pleased to add that he "had never seen a regiment disembark in such a creditable manner, and that he would communicate his high opinion to the Commander-in-Chief." The report was in due course acknowledged by His Royal Highness, who considered it "highly satisfactory, and most creditable to the commanding officer."

On the 20th February, volunteers, to join the new 55th and 59th Brigades, were called for, when 33 men of the 2nd Battalion, volunteering to the former, proceeded to join the 26th Regiment at Aldershot, and one man to the latter, left to join the 78th Highlanders.

On the 28th February, more volunteers were called for, to join, on this occasion, the 56th Brigade Depot.

On the 6th April, 3 companies, 1st Battalion, again marched to Solan for the hot season.

Towards the completion of the Brigade Depots, which were first formed on the 1st April, 1873, the 2nd Battalion, in April this year, received an order to detail 5 officers, 4 sergeants, 5 corporals, and 20 privates, to proceed to Bury St. Edmunds, under arrangements by the Quarter-Master-General, with a view to their being incorporated into the 32nd Brigade Depot.

On the 3rd April, Brevet Lieut.-Colonel Walker succeeded to the command of the 1st Battalion, *vice* Colonel Foster to half-pay, and, on the 10th, Brevet Lieut.-Colonel F. Bagnell was promoted to command the 2nd Battalion, succeeding Colonel McKay, who, retiring on half-pay, was immediately brought in to command the 32nd Brigade Depot.

The establishment of the 2nd Battalion was, on the 6th May, increased to 28 officers, 51 sergeants, 16 drummers, and 1000 rank and file—total 1095 of all ranks.

The depots of the 1st and 2nd Battalions proceeded, on the 18th June, to Bury St. Edmunds, to join the 32nd Brigade Depot.

The 2nd Battalion was inspected, in Division, on Southsea Common, on the 19th July, by His Royal Highness the Field-Marshal Commanding-in-Chief, when the Duke of Cambridge is reported to have expressed his "pleasure and unqualified approval of its appearance and smartness," and to have remarked that it was "a great credit to the commanding officer, and a splendid regiment."

On the 30th, the Cambridge and West Suffolk Militia Battalions, which had been attached to the 2nd Battalion 12th, left it, and proceeded to Ely and Bury St. Edmunds respectively.

A Special Army Circular, dated 1st September, directed a reduction to be made in the 2nd Battalion (with effect from the 1st August) of 4 officers, 9 sergeants, and 400 privates, making the total of all ranks 682.

The contingency of war with Afghanistan had for some time been foreseen, and the probability of such an event became changed to almost a certainty when, on the 31st September, a mission under General Sir Neville Chamberlain, as envoy from Lord Lytton, Viceroy of India, to Shere Ali, Amir of Kabul, was refused permission to proceed beyond Ali Masjid.

An ultimatum was thereupon sent to the Amir, by direction of the Home Government, embodying an explicit declaration, that unless a satisfactory reply was received by the 20th November, the English troops would cross the frontier.

This ultimatum was put into the hands of Faiz Mahomed Khan, Commanding Officer at Ali Masjid, on the 2nd November, and a copy was, at the same time, posted in the Amir's post-office at Peshawur.

When the specified date arrived, no answer had come; accordingly, on the 20th November, the order was given for the British troops to advance into Afghanistan.

In the meantime, the 1st Battalion, 12th Regiment, at Umballa commenced to move up country, and, on the 7th October, the headquarters and 5 companies proceeded by rail to Jhelum, where, on arrival, they encamped, awaiting the return of the detachment at Solan, which rejoined on the 12th.

On the 15th, the battalion, under Lieut.-Colonel Sillery, marched to Rawal Pindi, where it arrived on the 20th; the married families and sick of the battalion had been left at Umballa.

On the 29th October and 19th November, two detachments of the 1st Battalion marched to Fort Attock and Murree respectively, to be quartered; Lieut.-Colonel Walker arrived at Rawal Pindi, from England, on the 31st October, to assume command on promotion from the 2nd Battalion.

Between the 2nd and 20th November (the dates referring to the ultimatum) the Indian Government had been actively engaged in massing troops at various points on the north-west frontier, and, in reinforcing the garrison of Quetta, so that, by the 20th November, three advanced columns had been formed at Peshawur, Thull, and Quetta. The first of these was known as the Peshawur Valley Army; the second, the Kuram Column; and the third, the Quetta Army. The number of troops destined to operate from the above places was estimated in round numbers as follows :—

Peshawur Valley Army, 10,000 men, with 30 guns, Lieut.-General Sir S. Browne.
Kuram Column, . . 5,550 men, with 24 guns, Major-General Roberts.
Quetta Army, . . 6,250 men, with 18 guns, Major-General Biddulph.

On the arrival, shortly after, of a strong reinforcement to the Quetta Army from Mooltan, the command of it passed to General Stewart.

In addition to the above, two Reserve Divisions had been ordered to assemble; one, under General Maude, was to concentrate at Hassan Abdul, and the other, under General Primrose, at Sukkur, each consisting of about 6000 men.

General Maude's Reserve was known as the 2nd Division of the Peshawur Valley Army, and it was this force that the 1st Battalion 12th joined in April of the following year.

On the 2nd December, the 2nd Battalion crossed from Gosport (446 all ranks), and occupied the Anglesea Barracks, Portsmouth, in relief of the 107th Regiment.

CHAPTER XI

England, East Indies, Afghanistan, East Indies, Egypt, East Indies, England 1879–1898

1879

In compliance with an Army Circular, taking effect from the 1st January, a reduction took place of 100 privates in the establishment of the 2nd Battalion, making the total of all ranks 582, which was followed, on the 1st April, by a further reduction of 20 privates.

On the 28th January, the 1st Battalion continued its progress up country, when the headquarters and 7 companies, under Lieut.-Colonel Walker, marched from Rawal Pindi to Nowshera, arriving on the 2nd February, and, on the 11th, furnished a detachment to Fort Attock in relief of the one already there, which rejoined, the Murree detachment also rejoining headquarters two days later.

Brigadier-General Ross, C.B., commanding the Peshawur District, inspected the battalion on the 21st and 22nd, accompanied by Brevet Major W. J. Boyes, half-pay, 12th Regiment, as Deputy-Assistant Quartermaster-General. The latter writes in his "Reminiscences":—

> "I had the pleasure of attending the inspection, and General Ross was particularly well pleased with the smart and serviceable appearance of the battalion, as also with its general good tone."

Orders having been received for the 1st Battalion to join the 3rd Brigade, 2nd Division, Peshawur Valley Field Force, under the orders of Lieut.-General F. F. Maude, V.C., the headquarters and 7 companies, under Lieut.-Colonel Walker, marched on the 12th April towards Peshawur *en route* to Lundi Kotal, where they arrived on the 16th.

The following officers crossed the frontier to Ali Masjid on the 15th:—

Lieut.-Colonel G. F. Walker (commanding)
Brevet Lieut.-Colonel C. J. Sillery
„ W. T. Baker
Brevet-Major H. M. Lowry
Captain H. Magee
„ R. O. S. Brooke
„ O. Williams
Lieutenant C. D. Cave
„ J. B. McDonell
„ F. W. Scudamore
„ R. C. Onslow
Lieutenant W. F. Percival
„ J. M. Carpendale
2nd Lieutenant F. P. Hutchinson
„ A. M. Brabazon
„ E. J. Medley
„ F. Graham
„ W. Giles
Lieutenant and Adjutant C. R. Townley
Quarter-Master W. Cox
Surgeon-Major J. Wallace } attached
„ G. Andrew } attached

The battalion mustered a total of 751 of all ranks, which, three days later, on return of the detachment from Attock, was increased to 850.

On the 18th April, the 1st Battalion having arrived at Lundi Kotal, was inspected by Lieut.-General Maude, who expressed himself highly satisfied with its general state of efficiency.

On the 21st, owing to rumours of a gathering of Mohmunds at Palosi, on the Cabul River, General Maude despatched Colonel Norman, 24th Native Infantry, to the Shuliman Valley, and the neighbouring district, with 4 British and 4 native companies, and 2 mountain guns, the troops taking three days' cooked rations. To this force, the 12th contributed 2 companies, mustering 130 of all ranks.

The maliks (head men) of Loi Shuliman proved friendly, but the Mohmunds were threatening Kum Dakka, the inhabitants of which applied for assistance to Dakka. Captain O'Moore Creagh, with 2 companies (150 men) of the Mhairwarra Battalion, was sent in answer to this appeal, but was surrounded by an overwhelming force, and, for some hours, was in a very critical position. His small force had to retire on a cemetery, which was made defensible. Here they were attacked all day, up to 3 o'clock, and, when relief came, had exhausted nearly all their ammunition. Two more companies (24th Native Infantry) under Major Barnes, were sent to reinforce, while 3 companies 1st Battalion 12th, and 2 mountain guns were sent to Dakka under Colonel Sillery. On the 22nd, a portion of this last force, consisting of 1 company 12th, and 2 mountain guns, reached Kum Dakka, and heavy fighting ensued, when the repulse of the enemy was wholly due to the bayonet; a charge, by the 10th Bengal Lancers, under Captain Strong, finally routed them, many being driven into the river.

It having been found impracticable to hold the position during the night, with no supplies for men or mules, Major Dyce, in charge of the guns, ordered, at about 5 P.M., a retirement on Dakka, which was successfully carried out, notwithstanding the difficult nature of the ground. The men worked admirably, and reached Dakka at 10 P.M., having marched all day without food. The total loss on our side was 5 killed and 24 wounded, of which the 12th had 1 sergeant and 1 private killed, and 4 privates wounded, during the retirement. That of the enemy was considerable, and estimated at about 14 killed and 60 wounded.

General Sir Frederick Haines, Commander-in-Chief, was of opinion that, but for the gallantry of Captain Creagh, the force that accompanied him must have perished. He was later awarded the Victoria Cross.[1]

Preparations were made to renew the fight at Kum Dakka on the 23rd, when it was hoped that Colonel Norman would be in time to co-operate, if necessary, but nothing of importance occurred.

[1] At the date of concluding this history (1913) this officer (now General Sir O'Moore Creagh, V.C., G.C.B., G.C.S.I.) holds the high position of Commander-in-Chief in the East Indies.

On the 25th April, telegraphic communication with Jellalabad was broken, but soon restored, and, on the 5th May, Azmet Alla Khan, with 100 followers, came and tendered submission.

By this time, the indefinite rumours about the intentions of the Amir, Yakoob Khan, which had been flying about for some time, began to assume shape, and the beginning of the end was evidently at hand.

On the 8th May, Yakoob Khan, with his Commander-in-Chief and a small infantry escort, arrived at Safed Sung (the advanced post of the 1st Division) and was received with Royal honours, the British force lining both sides of the road for two miles out of camp, and a guard of honour of 3 officers and 100 men was drawn up at his tent.

An official list of the distribution of the Peshawur Field Force, dated 16th May, shows 7 companies, 1st Battalion 12th, forming part of a force distributed between Lundi Khana and Lundi Kotal, and 1 company at Dakka.

Cholera now began to show itself on the line of communications, and, when evacuating Afghan territory, many of the troops suffered considerably from it. The first death from cholera in the battalion occurred on the 19th May, and, on the 24th, four more men were admitted to hospital with it.

On the 26th May, a treaty between the Government of India and Yakoob Khan, Amir of Afghanistan, styled the "Treaty of Gundamuk," was made and signed, and the 1st Battalion 12th formed part of the Khyber Brigade at Lundi Kotal (the new frontier as fixed by the treaty), where the regiment remained until the commencement of the second phase of the Afghan campaign.

Between the 21st and 29th May, the 2nd Battalion, at Portsmouth, was called upon to furnish volunteers to the 11th and 19th Brigades, for active service in Natal, and, in that interval, 2 sergeants and 90 rank and file from the battalion had been passed by the Inspecting Officer.

On the 2nd August, new Colours were presented to the 2nd Battalion on Southsea Common, by Her Royal Highness the Duchess of Connaught, in the presence of a large number of military and civilian spectators, when the whole garrison paraded under the command of General His Serene Highness Prince Edward of Saxe-Weimar. The battalion, under Colonel Bagnell, was in line, with backs to the sea, and was composed of 6 companies of 20 files (385 of all ranks), the remainder of the troops being formed in line of quarter columns, at right angles to the flanks of the 12th, forming three sides of a hollow square. The new Colours were consecrated by the Chaplain-General (Bishop Claughton), and, on presentation, were received by Lieutenants Dowse and Pike.

Her Royal Highness assured Colonel Bagnell of the great pleasure she felt in presenting Colours to so distinguished a regiment, and expressed her conviction that they would be zealously guarded and defended. Colonel Bagnell made a suitable reply, particularly referring to the circumstance that these were the first Colours Her Royal Highness had presented since joining the Royal Family.

The new Colours were then saluted; the battalion, with the old Colours in the rear, went past the saluting flag, and the parade closed with a march past of all the troops on the ground.

The laying up of the old Colours of the 2nd Battalion[1] took place at St. Mary's Church, Bury St. Edmunds, on the 22nd August. They were escorted by Lieutenants Blunt and Kennedy and 4 sergeants, and, on arrival of the party, a guard of honour, under Major Marcon, of all available men from the 32nd Brigade Depot, was drawn up at the railway station, and escorted them to the church; here another guard of the Suffolk Volunteers awaited their arrival, when a procession was formed, headed by several clergy and the Mayor, which proceeded slowly up the aisle, when the service commenced, conducted by the Rev. Snape, vicar of the parish, and terminated with a hymn.

At an inspection of the 2nd Battalion, on the 6th September, His Serene Highness Prince Edward of Saxe-Weimar expressed himself highly pleased at its appearance and steadiness, remarking also that the general conduct had been very good since he had taken over command of the district.

The incidents that occurred between the 20th November, 1878 (commencement of the war) and the arrival of the 1st Battalion 12th in the Khyber on the 16th April, 1879, have not been herein recorded, being matters of general history.

Amongst the articles in the "Treaty of Gundamuk," it had been agreed "that a British representative should reside at Cabul, with a suitable escort, in a place of residence appropriate to his rank and dignity," whereupon the appointed Envoy, Sir Louis Cavagnari, with his escort, made their entry into Cabul on the 24th July, 1879, where they were honourably received.

In the beginning of August, several regiments of Afghan troops arrived in Cabul from Herat, and, on the 3rd September, aided by the populace, they besieged the British residents, who, after a brave resistance, were massacred.

The Afghan country was accordingly re-occupied by British and native troops; the 1st Battalion 12th was one of the regiments selected to take part in the forward movement with the Peshawur Valley Field Force, and, on the 6th September, General Roberts left Simla *en route* to Cabul.

The regiment, during its stay in camp at Lundi Kotal, throughout the hottest period of the year, had suffered much from sickness, including the deaths of 2 officers and 32 men from cholera, and until its removal from Lundi Kotal, it was chiefly employed in convoy and escort duties, 2 companies marching on the 29th September with the Guides to Dakka, which was entered without opposition, the companies returning on the following day,

[1] Issued at Glasgow in 1858.

and, on the 4th October, the same detachment marched to Lundi Khana as an escort to C Battery 3rd Brigade, Royal Artillery, continuing their march on the 7th to Basawal to be quartered.

On the entry of General Roberts' force into Cabul, the Amir Yakoob Khan was deposed, and, on the 5th December, 2 companies of the 1st Battalion, marching to Dakka, returned to Lundi Kotal on the 7th, in charge of the ex-Amir. On the 13th, 3 companies marched to Jellalabad, followed, on the 18th, by the headquarters, who arrived on the 21st.

By the 31st, the right half battalion had moved to Gundamuk, where the Basawal detachment joined on the same date. This last party had formed part of an expedition, under Colonel Mackenzie, 3rd Bengal Cavalry, against the Shinwarri villages.

1880

On the 12th January, 1 company, 1st Battalion, which had marched to Ali Boghan, to reinforce that post, was attacked during the night by about 50 of the enemy, who were driven back without any casualties on our side, and, on the following day, the headquarters and 3 companies, under Colonel Walker, marched there to surprise the Mohmunds, some 500 of whom, it was believed, had collected on the hills between Ali Boghan and Girdi Kus. Colonel Walker's force, which went out at 2.30 A.M., moved up the left side of the Girdi Kus heights, along the river side, and was fired at by the enemy, who were concealed in long grass on an island in the centre of the river, and kept up a brisk fire, but inflicted no loss. Several of them were, however, killed by the fire from the 12th Foot and the 27th Punjab Infantry, at long ranges.

Colonel Walker, passing round Girdi Kus, found that rafts were in readiness on the opposite side of the river, and then moved down the east side of the heights towards Charagee Chowki, on the Ali Boghan side, with a view to ascertaining if any of the enemy had crossed over; and, advancing again towards Girdi Kus, he came through the hills to Ali Boghan. A heliogram received about 1.30 P.M. informed him that the enemy were collecting in a fort opposite Ali Boghan, intending evidently to cross the river. General Bright accordingly moved out from Jellalabad with 2 guns, Royal Horse Artillery, and a troop of the Carabineers, under Colonel Fryer (Carabineers), and this force, rapidly advancing, found that several bodies of the enemy, with standards and mounted leaders, about 1500 strong, were about to cross the river, some of them being on an island in the centre. The guns opened fire with good effect at 1500 yards. The enemy retreated in disorder; and were shelled until out of range. They were taken completely by surprise at the rapidity with which the guns opened fire on them from the positions selected. One hour elapsed between the receipt of the order and the guns coming into action, the distance travelled being about 7 miles from Jellalabad, where they returned in the evening.

There were no casualties on our side ; those of the enemy were believed to have been 8 killed, including two standard-bearers, and 12 wounded. The troops that had moved out from Ali Boghan (including 100 sabres, 3rd Bengal Cavalry), returned there to encamp for the night, and, on the following day, the headquarters of the battalion returned to Jellalabad.

On the 27th January, a force, consisting of 2 guns, C/3 Royal Artillery, on elephants, 2 guns 11/9 Royal Artillery, 2 guns Hazara Mountain Battery, 2 squadrons Carabineers, one squadron 17th Bengal Cavalry, 4 companies each of the 12th and 25th Regiments, 2 companies Madras Sappers, 300 of the 27th Punjab Infantry, and the 30th Punjab Infantry, marched from Jellalabad to explore the Lughman Valley, of which, up to this date, but little was known. The brigade was commanded by Colonel Walker, 12th Regiment, and the whole movement superintended by General Bright, who accompanied the expedition with his Divisional Staff.

The first march was to Daranta, 7 miles, where it was decided that the Sappers should be sent to make a path over the Siah Koh, fit, if possible, for baggage animals. With them went, in the first instance, 150 each of the 12th Regiment and 27th Punjab Native Infantry, reinforced later by 200 of the 25th Regiment, aided by a contingent of Hazara coolies. The distance over the hills was not less than 1½ miles, over little else but rocks, and, though the working parties had only been out four days, when the path was inspected, it was certain that mountain batteries could use it next morning.

Early on the 28th, the passage of the Cabul River began. Infantry were sent across on elephants, and crowned the hills on the far bank. Camels, with their loads, crossed at the Daranta Ford ; also mules, whose loads had been deposited for transfer on rafts ; then the doolies of the artillery and cavalry, commissariat cattle, and other animals. In the meantime, some of the infantry were being rapidly ferried across on rafts, the baggage of each following it. At about 11 A.M. the Mountain Battery and 30th Punjab Native Infantry began the ascent of the Lakrai Pass, and after a 2½ miles' journey, along a rocky and somewhat narrow path, descended into the Lughman Valley. The infantry who had crossed by the ferry remained there for the most part to guard the baggage, whilst some remained at the ferry on the right bank, to protect stores not yet embarked, it having been quite impossible to get everything across that day. Ten days' supplies accompanied the force, and this, for 2000 men, and a certain complement of camp followers, involved a considerable amount of transport. A post of 150 or 200 men was established near the ferry, to hold the pass during the stay of the force in this valley. Once over the ford, and round the hill, Futteh, Mahomed Khan's fort, and a fort belonging to Asmatullah Khan came into view, the former being occupied by the 25th Regiment and the latter by the 30th Punjab Native Infantry with a company of Madras Sappers, both forts having been found deserted.

The Lughman Valley is wide and fertile. Many of the maliks and leading men promptly came in to pay their respects to General Bright, and, so far, not a vestige of opposition had been offered, nor was it even anticipated that a shot would be fired, though supplies did not come in very freely. A telegraph station was erected at the Daranta Post, and reconnaissances were pushed 12 miles up the valley, and a fairly good road found.

On the 29th and 30th a halt was made, to enable the baggage and remainder of the infantry to come up, and by the evening of the 29th everything had arrived. A new encamping ground having been selected close to Asmatullah's Fort, the brigade moved there on the 31st. A reconnaissance sent up the valley on this date was fired at, and obliged to return, when it had nearly reached the village of Soodoo Nussur, 8½ miles from the camp. A volley was fired into the party of Carabineers (about 15 and an officer), but fortunately neither man nor horse was hit, and it was supposed to be the work of some prowling robbers from the left bank of the river. On the following morning (1st February) a force of about 500 men with 2 guns, under Colonel Boisragon, 30th Punjab Native Infantry, left for Soodoo Nussur, accompanied by a survey party, to map the country as far as Dabeli, and then to return.

On the 14th, a portion of the force, consisting of 300 12th Regiment, and 100 30th Punjab Native Infantry, with the commissariat stores, crossed the Cabul River on rafts, and encamped on the right bank. The rest of the troops, consisting of 2 guns C/3 Royal Artillery on elephants, 2 guns Hazara Mountain Battery, a company Madras Sappers, and 400 more of the 30th Punjab Native Infantry, joined them next morning, and the whole marched at noon to Mandrawa (7 miles), one of the few important villages in the district, where the cavalry (1 troop each of the Carabineers and 17th Bengal Cavalry) had previously arrived. Colonel Walker, 12th Regiment, was in command, and Major-General Bright, with the Divisional staff, accompanied the force. At Mandrawa, crowds of the villagers turned out, and seemed astonished at the sight of the elephants carrying the guns of C/3, Royal Artillery. On the 16th, the troops halted, and, marching the following morning up the right bank of the Lughman River, pitched camp nearly opposite the village of Tigri.

On account of rumours of an armed body, with leaders of note, combining against us, it was determined to reconnoitre with a portion of our force to Buddeeabad (a village, which, for a time, had given shelter to the British captives of 1842), so, at 7 A.M. on the 19th, 2 guns, C/3 Royal Artillery, on elephants, 2 mountain guns, 150 of the 12th Regiment, half a company of Madras Sappers and Miners, and 250 of the 30th Punjab Native Infantry, the whole under Lieut.-Colonel Campbell, 30th Punjab Native Infantry, marched for Buddeeabad; 40 sabres of the Carabineers, and as many of the 17th Bengal Cavalry, leaving camp at 9 A.M., with whom went the General and Staff.

As there was no opposition of any kind, the force returned to camp about 4 P.M. This reconnaissance had very nearly brought the troops to the limits of the Lughman Valley, as a few miles beyond is the Koond Range, which separates Afghanistan from Kafiristan.

The next day they returned to the village of Gundi, marching over a fine plateau, where an army of 100,000 men could be encamped and manœuvred, but for the want of water. Here the troops encamped on the left bank of the Cabul River, close to a ferry which had been constructed for their crossing, and, halting on the 21st, returned on the following day to the old ground near Asmutullah's Fort, which had been held by a force during their absence. Here they halted until the 14th March.

The prolonged stay of the expedition in this valley had been beneficial in every way, whilst the political effect of the move had been excellent, over 70 maliks having come into our camp, representing nearly every head man of importance in the valley, and two of the leading chiefs had written to General Bright, expressing gratitude at their forts and property not having been destroyed, they, apparently distrustful of our intentions, having fled before the arrival of the force at Buddeeabad.

On the 14th March, orders were received to evacuate the Lughman Valley, and, on the 16th, the 1st Battalion 12th marched towards Gundamuk, where they arrived the next day.

The battalion furnished a detachment of 104 of all ranks, on the 23rd, to Pezwan, to be stationed, which rejoined headquarters on the 27th April.

Fines inflicted on the Kharbun Khugianis, of Rs.5000, had been paid up in the stipulated time, but the Waziri section having failed to pay their share, movable columns from Gundamuk and Jellalabad marched, on the 4th April, to Kailagu, and blew up the towers of the villages implicated in the Fort Battye raid on the 26th March. The Gundamuk Column, under Brigadier-General Arbuthnot, comprised 4 guns, I./A., Royal Horse Artillery, 1 squadron Carabineers, 12th Foot (headquarters, mustering 361 of all ranks), 31st Punjab Native Infantry, 1st Gurkhas, and a company of Madras Sappers. The Waziri section of the Khugianis, after suffering considerable loss in the demolition of their towers, and damage to crops, on which the transport animals were turned out to graze, paid up their portion of the fine, and, on the 6th, the troops withdrew. The next day, the battalion joined the Jellalabad Column, and returned to Rozabad, arriving back at Gundamuk on the 14th, one company, of 103 of all ranks, having marched in the meantime to Jellalabad for garrison duty.

On the 22nd April, the 1st Battalion 12th was inspected by a Board of Medical Officers, in order that its state of health might be reported on, the result being that a Divisional Order, issued on the 23rd, announced, that the battalion was to proceed to India, on relief by the 1st Battalion 5th Fusiliers, its destination "to be hereafter notified." The headquarters of the battalion (mustering 449 of all ranks) marched from Gundamuk, *en route* to India, on the 15th May, and, on arrival at Rozabad, the following day, received orders to detail 2 officers and 60 men to remain there, to escort 10/11 Royal Artillery and F/A Royal Horse Artillery to India, this detachment rejoining headquarters two days later at Jellalabad.

Referring to the 12th Regiment being detained at Jellalabad on the 18th May, owing to disturbances on the left bank of the Cabul River, a Lahore newspaper wrote :—

> "The battalion, which has throughout behaved with soldierly devotion, was lucky enough to come in for a nice little fight on the way down at Jellalabad."

On the 19th May, Brigadier-General Doran, with 600 infantry, made up from the 5th Fusiliers, a half battalion 12th Foot, 2 mountain battery guns, and a squadron of the Central India Horse, was ordered to cross the Cabul River, and disperse a gathering of about 2000 Afghans at a place called Besud. The brigadier made arrangements for an ambush, and concealed his troops behind a low hill, when the Afghans, seeing no force, advanced boldly, with standards flying, into an open plain. While the infantry attacked, the cavalry cut off the retreat to Tangi. The result was a complete success. The enemy broke and fled, many escaping along the right bank of the River Kunar, leaving 50 bodies, and carrying off many other dead and wounded. Seventeen Ghazis shut themselves up in a bastion of an old fort, which was carried in splendid style. Captain Kilgour, 5th Fusiliers, and Private Longworth, 12th Regiment, were the first to enter it. Longworth got badly wounded, and a few men from both regiments, rushing in, killed every Ghazi with the bayonet, after a desperate resistance. Our loss was 7 wounded, viz. :—Colonel Rowland, 5th, slightly; 2 men of the 5th and 1 of the 12th severely, and 1 native officer and 2 sowars, Central India Horse. All the wounds on our side except one were sword cuts.

On the 20th May, the headquarters of the 1st Battalion continued the march to India, escorting I./A. Royal Horse Artillery, 10/11 Royal Artillery, the 40-pounder battery, and captured guns. A correspondent to the *Civil and Military Gazette*, writing at Basawal on the 21st, says :—"the captured guns to the number of 160 were then arriving at that station."

After the action at Besud, the march of the battalion remained uninterrupted until its arrival at Peshawur on the 27th May, where a halt was made on the 28th, the regiment marching next day *en route* to Cherat, which was reached on the 1st June, and, on the 8th, the service depot rejoined from Nowshera.

The following is an extract from Brigade Orders, issued by Brigadier-General J. Doran, C.B., Commanding 2nd Section, Khyber Line Force, dated, Jellalabad, 30th May, 1880 :—

> "Brigadier-General Doran desires to convey his warmest thanks to the troops employed during the recent operations in Besud, including . . . 200 men, 1st Battalion 12th Foot, under Lieut.-Colonel Sillery. In those few days, every attribute of good soldiers was called for and displayed. The endurance of the troops was tested by severe marching and exposure under the fiercest sun. Their discipline was perfect; their steadiness was proved by the ascertained effect of their fire, whilst their gallantry, where opportunity offered, was conspicuous and seen by all.
>
> The passage of the Cabul River, with the very scanty appliances available, was, in itself, a feat of which the troops may well be proud. The whole of the horses, and a large proportion of men, of three troops of the Central India Horse, and the whole of the mules of a division of No. 1 (Kohat) Mountain Battery, with some of their drivers, swam this formidable river, at a place 400 yards in width, the current having a velocity estimated at between six and seven miles an hour. This was achieved with the loss only of one driver and one cavalry horse. The Brigadier believes that no record of such a feat can be found in our military history.
>
> The troops will hereafter learn that Brigadier-General Doran has, in his despatches, borne witness to their excellent services, in no measured terms. He is confident that not a page in the history of this campaign will bear higher testimony to the sterling value of our troops than that which tells of the operations in Besud, in May, 1880.
>
> He desires that this order may be entered in the records of every corps concerned."

The 1st Battalion, which mustered 850 of all ranks when concentrated at Lundi Kotal on the 19th April 1879, numbered, at Jellalabad, on the 17th May 1880, 550, including some 24 men employed on signalling and other duties along the Khyber Line. The return to England, during the above period, of 67 men on discharge, had not materially affected the above figures, it having been met by a draft of 72, who had joined on field service. When the 12th were at Lundi Kotal, cholera, fevers, and dysentery crept into their ranks, and, for a time, laid the regiment so low as to almost render them temporarily unfit for field service, although they advanced afterwards as far as Gundamuk.

The total casualties in Afghanistan amounted to 2 officers (Captain Reed and Surgeon-Major Wallace), and 71 of other ranks, who died from disease (the deaths of 2 officers and 32 men being from cholera), whilst 1 officer (Captain Magee), and 18 men invalided, had, during the same interval, died in India, from disease mostly contracted from service in Afghanistan.

On the 7th August, Colonel G. F. Walker, commanding 1st Battalion, was posted to the command of an infantry brigade, proceeding from Bengal to Candahar, composed of the 2nd (Queen's), 63rd Regiment, and 3rd and 4th Regiments Bengal Native Infantry, the command of the battalion devolving on Lieut.-Colonel Sillery.

On the breaking up of the force employed in Northern Afghanistan, General Sir D. Stewart, commanding, issued a General Order, dated Safed Sung, 10th August, 1880, which applied to the whole of the troops employed on field service, without mentioning any in particular, thanking all concerned for their cordial support, and commenting on the discipline and conduct of the soldiers, European and Native, who had served in the various field forces throughout Afghanistan, from first to last, as being beyond praise.

The above was conveyed to the Khyber Line Force, under a covering order from Lieut.-General Bright, commanding, who, in expressing his great satisfaction in publishing it, was pleased to add his praise to all employed on the line of communication, commenting on the good conduct of the troops, their high state of discipline reflecting the greatest credit on all ranks, their cheerfulness during duties of an extremely severe nature, and thanking them for their hearty co-operation.

On the 24th November, the 2nd Battalion embarked, for conveyance to Jersey, in the troopship "Assistance," which did not leave Portsmouth until the 27th, arriving next morning, when the headquarters, on disembarking, proceeded to Fort Regent, and a detachment of 2 companies to St. Peter's.

Major-General Nicholson, C.B., R.E., expressed himself highly pleased with the parade of the battalion, in marching order, at his inspection, at Fort Regent, on the 3rd December.

The 1st Battalion, mustering 635 of all ranks, marched on the 10th, from Cherat to Rawal Pindi, where they arrived on the 20th, and, next day, the right half battalion, under Lieut.-Colonel Sillery, proceeded by special train to Fyzabad, to be quartered, arriving on the 26th, the married families from Umballa, and a draft of 208 men from England, having previously arrived there, awaiting the arrival of the headquarters.

In the despatch of General Phayre, Commanding 2nd Division, Southern Afghanistan Field Force, describing his march for the relief of Candahar, dated Candahar, 16th December, 1880, Brigadier-General G. F. Walker, 12th Foot, Commanding 3rd Infantry Brigade, was honourably mentioned for his "valuable aid."

The system of the supply of army clothing by contract was continued up to this year, when, after a careful inquiry by a committee, it was decided that the clothing should in future be supplied from the Royal Army Clothing Department (with its factory at Pimlico), and that the allowance to commanding officers be withdrawn.

1881

The left half of the 1st Battalion, under Lieut.-Colonel Baker, which left Rawal Pindi for Seetapore, on the 23rd December, arrived at its destination on the 2nd January, and, on the 8th May, 2 companies of this half-battalion rejoined headquarters at Fyzabad.

On the 1st July, Brevet Lieut.-Colonel Baker was promoted to command the 2nd Battalion, in succession to Colonel Bagnell, retired. On the same date also many important changes were made in the organisation of the army, and in the titles and uniform of regiments of Infantry of the Line and Militia.

The 32nd Brigade (with all other infantry brigade depots) was abolished, and the corps formed into a territorial regiment, as follows :—

Territorial Regiment.		Composition.	Head-quarters of Regtl. District.	Uniform.		
Precedence.	Title.			Colour.	Facings.	Lace.
12th	The Suffolk Regiment	1st Battn. 12th Foot 2nd ,, ,, 3rd ,, West Suffolk Militia 4th ,, Cambridge ,,	Bury St. Edmunds	scarlet	white	rose

All correspondence, returns, &c., were ordered to be addressed : " Suffolk Regiment."

The establishment of officers of the regiment was also altered, and all officers with the rank of 2nd Lieutenant were to be styled Lieutenants, the former rank being abolished.

The following non-commissioned officers were also promoted to warrant grade, viz. :—sergeant-major, bandmaster, and schoolmaster, if of 12 years' service, and re-engaged. In September the ranks of trumpet-major, drum-major, bugle-major, and pipe-major were abolished and those of sergeant-trumpeter, sergeant-drummer, sergeant-bugler, and sergeant-piper substituted, these N.C.O.'s being accounted for in Returns, in the columns for sergeants, and not in those for drummers.

The re-engagement of private soldiers to complete 21 years' service was ordered to be discontinued, except by special authority from Army Head-quarters, and then only granted in special cases.

The total strength of all ranks of Territorial Line Battalions was to be :—India, 913 ; Home, 562 ; Depot, 69.

On the 1st December, the remaining 2 companies of the 1st Battalion at Seetapore, rejoined headquarters at Fyzabad.

The following rates of pay in the ranks (exclusive of deferred pay) were also this year established, viz :—sergeant-major, 5*s.* ; quarter-master sergeant, 4*s.* ; colour-sergeant, 3*s.* ; corporal, 1*s* 8*d.* ; drummer, 1*s.* 1*d.* ; private, 1*s.*

1882

On the 23rd and 27th January, 91 men, volunteers to nine regiments from the 2nd Battalion, proceeded to their respective destinations.

On the 1st April, Lieut.-Colonel W. O'Shaughnessy was appointed to command the 2nd Battalion.

At a review order parade of the 1st Battalion at Fyzabad, on the 15th April, the Afghan War Medal, 1878–80, was presented by Colonel J. Burn, Bengal Staff Corps, who made a speech suitable to the occasion.

An Army Circular, dated 1st May, directed the establishment of the 2nd Battalion to be increased from 480 to 500 rank and file.

On the 21st August, the 2nd Battalion, consisting of 18 officers and 487 of other ranks, embarked on the troopship "Assistance" for conveyance to Ireland, and, on landing next day at Kingstown, proceeded to the Curragh Camp. Before leaving Jersey, 99 men, from the 1st Class Army Reserve, joined the battalion on board ship.

On the 21st September, the headquarters and 4 companies 2nd Battalion left the Curragh for Galway, and detached the remaining half battalion to Oughterard, Gort, Tuam, and Athenry.

The observations of His Royal Highness the Field-Marshal Commanding-in-Chief on the last Inspection Report on the 2nd Battalion (notified on the 17th October) were as follows :—"This is a highly satisfactory report, and creditable to all concerned."

On the 7th November, the 1st Battalion, mustering 17 officers and 606 of other ranks, under command of Major Lowry, marched from Fyzabad to Lucknow, to form part of the escort to His Excellency the Viceroy of India, the Marquis of Ripon, K.G., arriving at Lucknow on the 14th. On the termination of escort duties, the battalion was detained at Lucknow over seven weeks, brigaded with the 2nd Battalion Derbyshire Regiment, and the 2nd Bengal Light Infantry, for drill and manœuvres, under the supreme command of Lieut.-General C. Cureton, C.B., Commanding Oudh Division.

1883

The return march of the 1st Battalion to Fyzabad commenced on the 22nd February, the battalion arriving at its destination on the 29th.

On the 5th April, Lieut.-Colonel H. P. Pearson assumed command of it, in succession to Colonel G. F. Walker, appointed to command the Agra Brigade.

By Special Army Circular, dated 23rd May, the establishment of the 2nd Battalion was increased by 20 privates, and a reduction of 1 sergeant, owing to the abolition of the title of Sergeant-Instructor of Musketry, the revised establishment now being 20 officers, 2 warrant officers, 38 sergeants, 16 drummers, and 520 rank and file, total of all ranks 600. The above was

FROM ENSIGN TO COLONEL OF THE REGIMENT.

to date from the 1st April, on which date the rank of Instructor of Musketry was also abolished, except in militia battalions, in which both the Instructor and Sergeant-Instructors were retained.

On the 20th November, the left half of the 1st Battalion left Fyzabad, by train, for Delhi, and arrived on the 23rd in relief of a wing of the 2nd Battalion Dorsetshire Regiment. The headquarters left, by train, next day for Meerut, and thence by march to Roorki, arriving on the 30th, in relief of the headquarters wing of the 2nd Battalion Dorsetshire Regiment. At the All Ireland Rifle meeting this year the 2nd Suffolks were second for the Castle Bellingham Cup, open to teams from all Corps in Ireland, when 15 regiments were represented. The battalion lost the cup by the last man getting a bull's-eye on a wrong target.

1884

On account of insufficient accommodation in the barracks at Roorki, a detachment of 3 officers and 150 men, 1st Battalion, marched, on the 10th March, for Chakrata.

From the 1st April, the appointment of Sergeant-Instructor of Musketry, abolished in the previous year, was again instituted.

On the 2nd Battalion leaving Galway, a resolution was unanimously passed on the 7th August, by the Galway Town Commissioners, praising the conduct of the men, and recording the good feeling that had existed between them and the citizens, during the two years' stay of the battalion in the garrison.

The headquarters 2nd Battalion moved from Galway to Cork on the 26th September, under Lieut.-Colonel O'Shaughnessy, and the detachments at Gort and Oughterard were withdrawn.

On the 1st October, a communication was received from the Deputy Adjutant-General in Ireland, referring to the resolution of the Galway Town Committee (dated 7th August), which stated :—"The Commander of the Forces is pleased to find that the good conduct of the men of the Suffolk Regiment, during the period they have been stationed at Galway, has been so much appreciated by the inhabitants of that town."

The 2nd Battalion, in October, furnished a detachment to Youghal.

On the 4th November, the headquarters and 4 companies 1st Battalion, under Lieut.-Colonel Pearson, marched from Roorki to Meerut, as a guard of honour to His Excellency the Commander-in-Chief, and arrived on the 10th. Returning, on completion of this duty, the detachment reached Roorki on the 23rd.

1885

The 2nd Battalion furnished a detachment of 2 officers and 40 men, on the 28th September, to Killarney.

In the observations by the Commander-in-Chief, dated 7th November, on the Confidential Report, by Major-General Young, at the last inspection of the 2nd Battalion, His Royal Highness the Duke of Cambridge remarked most favourably, and " considers its condition as most satisfactory."

A Special Army Circular, issued on the 10th November, authorised an increase of the 2nd Battalion to a total of 940 of all ranks, taking effect from the 1st April. In consequence, each company was increased by 1 sergeant and 25 privates. The officers were now to number 2 lieut.-colonels, 3 majors, 5 captains, 12 lieutenants, 1 adjutant, 1 quarter-master.

The 2nd Battalion were 9th this year, in signalling efficiency, of all units, Cavalry, Household Troops, and Infantry, in the United Kingdom.

1886

On the 4th March, the headquarters and 4 companies, 1st Battalion, left Roorki by rail, and arrived at Rawal Pindi on the 6th, where they encamped on the race-course, the wing that was stationed at Delhi rejoining there the same day.

On the 24th, the battalion was warned to be held in readiness for active service against the Bonerwals, and was medically inspected for that purpose, but the murderers of Major Hutchinson (Guides) having been given up, the expedition did not go.

From the 1st April, the establishment of the 2nd Battalion was reduced by 8 sergeants and 100 men.

On the 13th, Major W. Keough, promoted lieut.-colonel, was appointed to command the 2nd Battalion.

From the 22nd to the 30th April the 1st Battalion marched, by two-company detachments, to the Murree Hills, the headquarters being quartered at Kuldunna, and 2 companies at Camp Thobba.

At an inspection, at Cork, of the 2nd Battalion, on the 5th July, by Major-General Stevenson, the inspecting officer remarked on its efficient condition, whilst the steady bearing of the men under arms called forth his special commendation.

On the 26th August, the 2nd Battalion, under Lieut.-Colonel Keough, proceeded to the Curragh Camp, and arrived the next day, the detachment from Youghal joining the headquarters at Queenstown. The strength was 24 officers and 804 of other ranks.

On the 7th September, Major C. H. Gardiner was promoted Lieut.-Colonel in succession to Colonel Pearson, retiring on expiry of his four years in command, and, on the same date, Colonel R. H. O'Grady Haly was appointed to command the 1st Battalion.

Between the 5th and 16th October, the 1st Battalion marched from the Murree Hills to Rawal Pindi, by two-company detachments, at intervals, the last of these arriving on the 18th. The regiment was encamped on the race-course, and, on the 1st November, furnished a detachment of 2 officers and 88 men to Campbellpore.

The Rawal Pindi Camp of Exercise commenced on the 6th December, when the 1st Battalion was brigaded with the 14th and 45th Sikhs.

The rank of 2nd Lieutenant was this year revived by the Royal Warrant of the 31st December.

1887

On the 19th April, a detachment from the 1st Battalion of 1 officer, 1 sergeant, 1 drummer, and 43 rank and file was attached to the 1st Dragoon Guards, at Rawal Pindi, for instruction as "Mounted Infantry," the horses being supplied by that regiment.

Between the 22nd April and the 3rd May, the 1st Battalion marched by detachments to Murree, where the headquarters and 2 companies were again quartered at Kuldunna, and the remaining companies at Thobba Camp, the Mounted Infantry class rejoining headquarters on the 3rd June.

On the 4th July, a party of 4 officers and 150 men, 2nd Battalion, proceeded from the Curragh on eviction duty to a place called Coolgreany in County Wexford, and, after having discharged the duties allotted to them without any incident of special note, they returned on the 20th.

Shortly after, it was notified from Dublin Castle, that the Divisional Magistrate, in his report on the late evictions, had "specially remarked on the great assistance afforded to him by Major Dowse and officers, 2nd Battalion Suffolk Regiment, and the admirable behaviour of the men," whereupon His Excellency the Lord-Lieutenant had great pleasure in bringing the report to the notice of His Serene Highness, the Commander of the Forces in Ireland.

The annual Inspection, on the 4th and 5th August, of the 2nd Battalion, by Major-General the Hon. Thesiger, commanding Curragh Brigade, drew forth high encomiums, in every respect, from that officer, who, as a mark of approbation at the satisfactory state of the battalion, directed that all defaulters should be pardoned, and a general holiday granted to the battalion.

On the 6th August, the Commander-in-Chief's remarks on the Inspection Report of the 2nd Battalion, for the previous year, were published, in which His Royal Highness commented "on its being (notwithstanding the large number of young soldiers in its ranks) in all respects in an admirable state of discipline and efficiency, and desired that his high approbation be conveyed to the officers and men of this excellent battalion."

Lieut.-Colonel A. Tower joined the 1st Battalion on the 9th August, on exchange from the Derbyshire Regiment with Lieut.-Colonel C. H. Gardiner.

Between the 27th October and 1st November the 1st Battalion had returned from the Murree Hills by detachments to Rawal Pindi, and, on arrival, was encamped, as before, on the race-course.

The remarks of His Royal Highness, the Commander-in-Chief, on the

Inspection Report of the 2nd Battalion for this year were published on the 31st December, as follows :—" The report on this battalion is most creditable in all respects."

1888

Lieut.-Colonel A. Tower proceeded, on the 22nd February, to take up the appointment of Commandant of the Kasauli Depot. On the 28th February, General J. M. Perceval, C.B., was appointed Colonel of the regiment, in succession to General J. Patton, deceased. (See Plate 32.)

The 2nd Battalion, under Lieut.-Colonel Keough, left the Curragh on the 11th April, and embarked at Dublin in the troopship " Assistance " for Portsmouth, where, on disembarking, the battalion proceeded on the 14th to Aldershot, arriving the same day, and took over quarters in the South Camp.

Major-General the Hon. Thesiger, C.B., commanding Curragh Brigade, expressed, in a farewell order, his regret at parting with " so good a regiment, in which the behaviour of the men had been especially good, the Return of Crime for the Army, for that year, showing it honourably mentioned as a corps in which desertion was happily almost unknown, and as having the smallest number of courts-martial and minor punishments. Its efficiency also in drill and discipline reflected the greatest credit on Colonel Keough, and all under his command."

Headquarters and 3 companies, 1st Battalion, marched from Rawal Pindi to Upper Tubba, Murree Hills, on the 20th April, for the hot season ; here another company joined in July, 3 companies having been left at Rawal Pindi, and one was detached to Campbellpore.

On the 7th May, the establishment of the 2nd Battalion was raised, showing a total of 24 officers and 986 of other ranks.

Colonel Wm. Keough, 2nd Battalion, retired on half-pay on the 1st August, and was succeeded in the command by Lieut.-Colonel J. E. Harris.

Early in the month, an outbreak of cholera occurred in the 1st Battalion, the majority of cases taking place at Rawal Pindi. One sergeant and eleven privates died from it, " B " company losing 5 men within 24 hours.

The following Battalion Order was published on the 13th August :—

> " Colonel O'Grady Haly has the greatest pleasure in publicly placing on record the very excellent behaviour of 'B' Company and its attached men, on the occasion of the recent serious outbreak of cholera at Rawal Pindi. This company experienced, during its march into camp, unusual difficulties and the greatest discomfort, having been exposed to a violent storm for the whole night, without any shelter whatever, and all ranks displayed such endurance, pluck, and good soldierlike spirit, as called forth the special commendation of the Major-General Commanding the Division. Very great credit is due to Lieutenant C. F. Lennock, and the officer, non-commissioned officers and soldiers under his command, who have so well upheld the character

of the 'Suffolk Regiment,' in most trying circumstances. As a mark of the commanding officer's appreciation, this order will be embodied in the official records of the battalion."

For some years, the Black Mountain country had been in a very disturbed state, notably the Hazara District, a wild and rugged tract on the Indus about 80 miles from Peshawur. The tribesmen were considerably in arrears with sums due from them for various offences they had committed, whilst the infliction of fines had had little effect in preventing them from raiding across the British border, looting villages, and killing peaceful British subjects, particularly in the Agror Valley.

On June 18th, 1888, a serious affair occurred on the Agror frontier, which resulted in the deaths of two British officers and 4 men of the 5th Gurkhas, who, as a surveying party, were ruthlessly attacked and murdered. It was to avenge this latest outrage that the Government of India again considered the question of punitive measures against the Black Mountain tribes, and, an expedition having been decided on, orders were issued on the 7th September, for the formation of a force to be styled the "Hazara Field Force," under the command of Major-General McQueen, C.B., A.D.C., for the purpose of punishing the Black Mountain tribes for their repeated raids and acts of aggression. The force was to consist of:—

2 mountain batteries, British.	4 battalions British Infantry.
1 „ battery, Native (6 guns).	9 „ Native „
1 company, Sappers and Miners.	including 1 of Pioneers.

to be organised in two brigades, under Brigadier-Generals Channer, V.C., and Galbraith, and each brigade subdivided into two columns. In addition, a field reserve, of 1 regiment of cavalry and 2 of infantry, was ordered to be formed; the Nowshera Brigade, at the same time, being held in readiness for field service.

Headquarters and the 1st, 2nd, and 3rd Columns were directed to concentrate at Ughi, in the Agror Valley, by the 1st October, and the 4th Column at Darband, on the Indus, by the same date.

Infantry battalions were to take the field 600 strong. For baggage, staff officers were allowed ½ and regimental officers ⅓ of a mule. British and native non-commissioned officers and men were allowed 16 lbs. of baggage, and followers 10 lbs. each, and no tents were to be taken. Seventy rounds of ammunition were to be carried in the pouch, and thirty on mules with corps. Five days' supplies, with two days' grain for all animals, were ordered to accompany the 1st, 2nd, and 3rd Columns, and seven days' to accompany the 4th.

The Government of India also accepted the services of 2 battalions with 2 guns, of Kashmir troops, which had been offered by the Maharajah, and, also, those of a detachment, 300 strong, of the Khyber Rifles, who had volunteered for the expedition.

The 1st Battalion having received orders to take part in the operations, the headquarters and 4 companies marched, on the 21st September, from Upper Tubba, via Abbottabad to Kaluka, to join the 2nd Column, Hazara Field Force, the detachments from Rawal Pindi and Campbellpore rejoining at Abbottabad and Kalakka respectively, the latter detachment losing one man, from cholera, *en route*.

The following officers of the 1st Battalion proceeded on field service:—

Colonel R. H. O'Grady Haly, appointed Brigadier-General, in command of the 2nd Column.

Major T. Baker (Commanding)
„ O. Williams
„ R. J. Pike
Captain A. C. Cubitt
„ W. R. Lloyd (Joined from leave in England, about the 15th October)
Lieutenant E. Montagu
„ C. F. Lennock
„ C. M. De Gruyther
„ W. B. Wallace
Lieutenant E. P. Prest
„ C. A. H. Brett
„ C. R. Fryer
„ W. H. N. Glossop
„ G. H. S. Browne
„ A. B. Morgan
2nd Lieutenant F. G. Davies
Lieutenant and Adjutant L. J. Shadwell (Joined from leave in England, October 5th)
Quarter-Master W. Norris

The 2nd Column was composed as follows:—

3/1 South Irish Division, R.A., 41 guns
1st Battalion Suffolk Regiment
34th Pioneers, a wing
40th Bengal Infantry
45th Sikhs
2 Gatlings

under Brigadier-General R. H. O'Grady Haly, Suffolk Regiment.

On the 4th October, the 2nd Column moved from camp, Kalakka, at 11 A.M., and occupied Barchar, the position assigned to it, without meeting with any opposition.

At 6.30 A.M. on the 5th, the column continued its advance up the Barchar Spur, and reached the crest, at Bampar Gali, without any casualties, the slight resistance offered having been easily overcome by 2 companies of the battalion, and a wing of the 45th Sikhs, covered by the guns and Gatlings. The column bivouacked on the crest, to the south of Bampar Gali; there was much sniping all day and night.

One man, 40th Bengal Infantry, was killed after dusk.

On the 6th, the 2nd Column moved along the crest to Nimal, and occupied, throughout the 7th and 8th, the bivouac vacated by the 3rd Column.

On the 9th, a column of troops from the 2nd and 3rd Columns, proceeded, under Brigadier-General Channer, to Seri, without meeting with any opposition. Some villages were destroyed *en route*. The Seri villages had already been burnt by the Khan Khels themselves, but a tower and a fort, which had been left standing, were blown up by the engineers. Two other large villages were shelled from the Seri plateau, with good effect, the inhabitants clearing out with their goods and cattle. The force returned to camp at 5.30 P.M. This day, a commissariat driver was shot and cut up, on the road from Nimal down to Sambalbut, within British territory.

An ultimatum was sent on this date to the Hassanzais, increasing their fine to Rs.7500, on account of the active hostility shown by them, the other terms being as originally fixed, and giving them till midday of the 15th to submit. Other tribes were also informed that it was proposed to destroy their villages on the following day, giving them the opportunity to submit in the meantime.

On the 9th, a force from the 2nd Column, under Colonel O'Grady Haly, consisting of 2 guns, 3/1 South Irish Division, Royal Artillery, 200 Suffolk Regiment, 100 34th Pioneers, 300 45th Sikhs, and 100 Khyber Rifles, left camp at 7, and reached Seri at 11 A.M.

From Seri, a small column of 420 rifles, under Colonel Waterfield, 45th Sikhs, descended into the Shal Nullah, and, after an arduous ascent, reached the Kund villages, which they partially burnt, and returned to bivouac at Seri. Some opposition was met with during the attack, and the enemy's loss was estimated at 13 killed and wounded. No casualties on our side. From the detached force of the 2nd Column at Seri, a reconnaissance was pushed out in the direction of Sabe, and returned to Seri in the evening.

On the 12th, Colonel O'Grady Haly returned to the 2nd Column at Nimal, and Brigadier-General Channer took command of the detached force from this column at Seri, which had then moved to Karun. The village of Merwata or Merabad was this day destroyed.

At noon, Brigadier-General Channer, with 230 rifles Suffolk Regiment, 40 rifles 34th Pioneers, and 50 of the Khyber Rifles, proceeded to the village of Betband, with the object of making reconnaissance towards, and effecting a junction with, the 4th Column on the river. The men carried greatcoats, 1 day's cooked rations, and 50 rounds ammunition. The force bivouacked for the night at Betband, and continuing the march next morning, communication with the 4th Column on the Indus was effected. [The march was a very severe one, involving a descent of about 5000 feet.]

From the 2nd Column, a detachment, under Colonel Waterfield, 45th Sikhs, proceeded from Karun on the 13th and burnt the villages of Maira and Sabe, and returned to Karun the same evening.

On the 15th, Brigadier-General Channer also returned there ; the party was fired at on their return journey, but had no casualties. Two of the enemy were killed.

On the 18th, Brigadier-General Channer, from Karun, crossed the Shal Nullah, and completed the destruction of Bar and Kuz Kand, blowing up the towers of the latter. Six prisoners, including a leading malik, were captured. This punishment had the desired effect in this quarter, and, on the return of the troops to Karun, they were followed by the Akazai jirga, who were sent on to headquarters the following day.

On the 19th, a mixed force of 250 men, under command of Major Pike, Suffolk Regiment, of the 2nd Column, proceeded from Nimal and burnt the Akazai village of Dare, the men of which were concerned in the attack on Major Battye's party. There was no opposition. The Akazai jirga, arriving at headquarters on the 19th, accepted unconditionally all the terms that were imposed, and were given seven days to collect the fine, leaving five hostages in our hands, and orders were issued that, during the period of grace allowed, no more Akazai villages were to be destroyed.

A Brigade Order, dated 19th October, by Brigadier-General Channer, announced the Brigadier's wish :—

> "to place on record his high sense of the manner in which the detachment, 1st Suffolk Regiment, under Major Baker, and that of the 34th Pioneers, under Captain Hogg, had accomplished the march, a few days previously, which had effected a junction with the 4th Column. All ranks had worked willingly and cheerfully, not a man falling out. The Major-General Commanding the troops had himself acknowledged this, and sanctioned the circumstance being noted in Brigade Orders."

On the 20th, the 2nd Column moved to Chittabut, in relief of No. 1 Column.

By this date also, the operations against the Hassanzais and Akazais had been brought to a satisfactory conclusion. The punishment of these tribes had been very severe, and it was estimated that their loss in burnt villages, fodder, corn, and honey, amounted to Rs.50,000 at least.

The second phase of the operations may be considered to have commenced on the 21st October.

A 5th Column was now formed, and after a series of operations, it was resolved to advance with a lightly equipped force to Thakot, under Brigadier-General Channer.

On the 24th, a force of 200 men, Suffolk Regiment, covered by the 45th Sikhs and 2 guns, under Colonel O'Grady Haly, proceeded to co-operate with the 1st Column, in destroying the villages of Kopra and Garhi.

On the 25th, His Excellency the Commander-in-Chief, Sir Frederick Roberts, made an inspection of the troops at Churnang and Maidan. On the 26th, with his Staff, he rode along the crest of the Black Mountain, inspecting the troops at Mana-ka-dana, Chittabut, and Nimal, returning to Ughi, and leaving the same afternoon for the 4th Column, via Abbottabad and Darband. The 1st Suffolks were at Chittabut.

On the 26th, the Hassanzias made full submission.

On the 28th, the Chittabut peak was evacuated, and the headquarters of the 2nd Column established at Churnang, in order to bring pressure to bear on the Parari Saiads, by taking grain, forage, and wood from their country. Two companies Suffolks were now posted to Churnang, the headquarters and 450 men, under Major Baker, occupying Dilbori.

On the 29th, the sword belonging to Major Battye, which had been carried off when he was killed, was recovered, and handed to the General Officer Commanding.

On the 30th, the whole force retired to Maidan. The retirement was conducted without opposition, due warnings having been given that if any shots were fired during the retirement, unhappy results would follow.

Orders were now given for an advance to the Allai country, to punish a Khan who had not tendered his submission, nor accepted the terms offered by the Government. Accordingly, a force composed of troops from the 1st and 5th Columns, including a detachment of 50 marksmen of the Suffolk Regiment, under Major Pike and Lieutenant Wallace, marched on the morning of the 31st towards the Gorapher Pass, which was reconnoitred, and found to be occupied by the enemy.

On the 1st November, the advance on the pass leading into Allai was continued by the 1st Column (to which Major Pike's party was attached) supported by the 5th Column. The first shots were fired in the thick of the forest, at the foot of the ascent, at 10.30 A.M., and, before noon, the position was taken. Our casualties were only one man killed, and one wounded, both of the Northumberland Fusiliers. A few dead bodies of the enemy were found on the crest, and also some blankets, food bags, &c., showing they had gone off in a hurry.

On the 2nd, the headquarters Suffolk Regiment from Dilbori rejoined the headquarters of the 2nd Column at Churnang.

On the 3rd November, at 8 A.M. the column under General Channer, V.C., started for Pokal, Arsalla Khan's village, which was reached at 11 A.M., the force consisting of 160 Northumberland Fusiliers, 160 Seaforth Highlanders, 40 Suffolk Regiment, 40 Royal Sussex, 200 5th Gurkhas, 50 24th Punjab Infantry, 150 Khyber Rifles, and 2 guns. On reaching the village, the detachments of Suffolk and Royal Sussex Regiments (all picked marksmen) fired volleys on the retreating bodies of the enemy at long ranges with great effect. The village of Pokal was entirely destroyed with the exception of the mosque; the tower was blown up, and the force commenced to retire at 1 P.M. The distanceto Pokal was seven miles, and the descent 4300 feet. The troops all returned to camp before dark, after a very heavy day's work. Our loss had been 1 killed, and 1 wounded 5th Gurkhas, and 3 wounded Khyber Rifles, that of the enemy having been estimated at from 80 to 100 killed. Late in the evening an Allai jirga came into camp, as at Thakat, in two batches. Arsalla Khan, it was said, had fled across the Indus. The cold was intense, three-fourths of an inch of ice.

On the 4th, the whole force withdrew from the Allai country to Mazrai, the Major-General and Staff returning to Maidan, whence, on the cessation of hostilities, the troops were gradually withdrawn to British territory.

On this date also, 200 of the Kashmir Contingent were brought up from Mazrai to the foot of the Ghoraper Pass, and were employed in improving the road.

Major Baker proceeded home on the 4th, to join the 2nd Battalion, on transfer as second in command.

On the 5th, the 2nd Column, including the Suffolk Regiment, evacuated Churnang leaving a small detachment, and proceeded to Kalakka, where Major Pike's party joined on the 7th.

The object of the expedition having been now achieved, orders were received, on the 11th, for the force in the Agror Valley to be broken up, and to march back across the British frontier, whereupon the following Battalion Order was published :—

"The Commanding Officer wishes to express to the officers, non-commissioned officers and men of the battalion, his appreciation of the admirable bearing of all ranks during the recent operations. The 1st Battalion Suffolk Regiment has well maintained the high reputation of the old 12th, and Colonel O'Grady Haly will ever feel proud that he has had the honour of commanding the regiment during the Black Mountain Expedition."

The following farewell order to the Hazara Field Force was published on the 12th November.

"On the approaching departure, for India, of the troops composing the Hazara Field Force, Major-General McQueen congratulates the officers, non-commissioned officers and men of all corps and departments of the force under his command, on the thorough manner in which they have carried out the work entrusted to them by His Excellency the Commander-in-Chief, under the orders of the Government of India.

Briefly the result of the expedition may be summarised as follows. The Hassanzais, Akazais, Parari Saiads, &c., have tendered their submission in full, in accordance with terms dictated to them by the British Government. The Machai Peak, and the principal villages of the Allai country have been visited by the force, and the construction of roads along the Sudas Valley to the Chagarzia border, and to Thakot from the Agror Valley, give the power of rapid movement to troops in the future, which will go far towards maintaining the security of our frontier, and our friendly relations with Cis-Indus tribes. The enemy against whom the force has had to contend, have, hitherto, considered themselves secure from attack, owing to the mountainous and supposed inaccessible nature of their country, but the Hazara Field Force of 1888 has proved to these tribes that their highest and most difficult passes, hitherto untrodden by Europeans, can be traversed by any well organised force. Much hard work has been entailed both in marching and road making, and the exposure at high elevations by Nos. 1, 2, and 3 Columns during a period of six weeks has been most trying."

The farewell order then expresses the Major-General's pleasure in being able to report on :—

"the cheerful manner in which all duties had been carried out, the exemplary conduct of the force, and the soldierlike spirit which had animated all ranks on every occasion, whilst the total absence of sickness amongst the troops, from first to last, showed that the fine condition of the men left nothing to be desired by any Commander, and, in bidding the force farewell, he thanked all ranks for the very efficient help rendered throughout the operations."

Prior to leaving the base at Kalukka, Colonel O'Grady Haly had given orders for the regimental native tailor at Rawal Pindi to join the regiment here, with ample supplies of khaki clothing, pipe-clay, and native tailors, so that 14 days later, the battalion marched back to Rawal Pindi with clothes mended or replaced, belts pipe-clayed, beards cut off, &c., and, in general, sufficiently smartened to give the appearance of never having left it, on a tour of active service—a marked contrast to other regiments returning.

On the 14th November, the 2nd Column left Kalukka for Rawal Pindi, and arrived on the 23rd, when it was broken up.

The following were mentioned in despatches :—Colonel R. H. O'Grady Haly ; Major T. Baker, Major R. J. Pike ; Major A. J. Watson, Brigade-Major, 1st Column ; Lieutenant and Quarter-Master W. Norris, and Corporal Manning, in charge of regimental signallers. Colonel O'Grady Haly was later awarded the Distinguished Service Order.

A camp of exercise was formed at Rawal Pindi on the 27th, in which the 1st Battalion took part, brigaded with the 2nd Native Infantry and the 14th and 15th Sikhs.

On the 13th December, the 1st Battalion, mustering 21 officers and 841 of other ranks, left Rawal Pindi, by rail, for Jhansi, where they arrived on the 18th, when a draft of 95 men, which had just arrived from England, brought the total of other ranks up to 936.

The battalion now furnished detachments of 1 company to Sipri, and 4 companies to Nowgong, which arrived at their destinations, by march route, on the 27th and 31st December respectively.

1889

On the 10th January, a most favourable report was received from Major Hutton, commanding Mounted Infantry Regiment, on the excellent manner in which the men in Lieutenant Leader's division, 2nd Battalion, had performed all their novel and arduous duties in the Mounted Infantry Regiment, whilst No. 4 Company, under Captain Hore, gained very high praise from Lieut.-General Sir H. E. Wood, V.C., at his inspection, on the 7th instant, for the extraordinary degree of efficiency they had attained in so short a time, a standard, which he was pleased to say " had fairly astonished him."

On the following day, the Officer Commanding the 2nd Battalion drew attention, in Battalion Orders, to the high position held by the regimental signallers, not only in the Aldershot Division, but in the army, the Major-General Commanding the Brigade having congratulated the battalion on its standing amongst the regiments at home.

His Royal Highness, the Commander-in-Chief, was pleased to consider the Inspection Report of the 2nd Battalion for the previous year, as " generally very satisfactory, and was glad to notice the small amount of crime."

On the 21st November, 2 companies, 1st Battalion, of the Nowgong detachment, rejoined headquarters at Jhansi.

On the 17th December, the 2nd Battalion, consisting of 25 officers and

780 of other ranks, embarked at Portsmouth, on the troopship "Serapis," for conveyance to Egypt, and, disembarking at Alexandria on the 30th, marched to Ras-el-Tin Barracks.

A party of 61 non-commissioned offices and men, under Captain Mackenzie, proceeded next day to join the depot at Bury St. Edmunds.

The total strength of the 1st Battalion, on the 31st December this year, not including officers, was 1161.

1890

On the 29th April, at a brigade parade, at Jhansi, the Indian Frontier War Medal, with a clasp "Hazara, 1888," was presented to officers and men of the 1st Battalion, who had taken part in the Black Mountain Expedition, between October 3rd and November 19th, 1888.

A reduction was authorised on the 7th June in the establishment of the 2nd Battalion, which was now to consist of 28 officers, 2 warrant officers, 45 sergeants, 16 drummers, and 800 rank and file; total of all ranks, 891.

Colonel R. H. O'Grady Haly, D.S.O., 1st Battalion, retired on half-pay on the 26th August, and was succeeded in the command by Colonel A. Tower.

On the 7th September, at Mandara (a village near the sea, about 11 miles from Alexandria), a convalescent camp was formed, of which Captain Graham, 2nd Battalion, was appointed Camp Commandant.

Colonel Arthur Tower, 1st Battalion, died at Jhansi, on the 8th December, and Colonel J. C. R. Glasgow was appointed to the command.

1891

A wing of the 1st Battalion, and the drums, left headquarters at Jhansi on the 4th February, and marched to Nowgong, arriving on the 10th.

On the 11th, the 2nd Battalion, mustering 25 officers and 678 of other ranks, embarked at Alexandria in the "Serapis," for conveyance to Bombay, and, arriving on the 27th, proceeded, by rail and march, to Wellington, Madras, detaching 3 companies, which proceeded by sea to Calicut, Malapuram, and Carmanore respectively.

An alteration in the establishment of the 2nd Battalion (published in March) showed a reduction of 1 sergeant, and an increase of 1 officer and 121 rank and file; total of all ranks, 1021.

During the stay of the 2nd Battalion in Egypt, 24 non-commissioned officers and men died from enteric fever, and a tablet has since been erected to their memory in St. Mary's Church, Bury St. Edmunds.

A guard of honour, consisting of 3 officers, 4 sergeants, and 50 rank and file, with band and regimental colour, 2nd Battalion, under Captain V. Graham, marched, on the 5th July, to Ootacamund, to attend the presentation of the Victoria Cross to Major C. J. W. Grant, Indian Staff Corps, for conspicuous bravery in the defence of Thobal, Major Grant having commenced his career in this battalion.

On the following day, at 12.30 P.M., the presentation was made at Government House, by His Excellency the Governor and Commander-in-Chief of the Presidency, Lord Wenlock, whose band was drawn up to the right of the platform which had been erected.

In addition to being awarded the Victoria Cross, Major Grant (now Colonel, retired) had been promoted to the ranks of captain and brevet major for the following act of bravery.

On the occasion of an outbreak, or mutiny, at Manipur (a small native State, at the foot of the Eastern Himalayas) early in 1891, Major Grant, then a lieutenant of nine years' service, learning of the fate of those massacred there on the night of March 24–25, and of the danger the survivors, if any, were in, took with him 80 native soldiers, marched day and night through Northern Burma, and reached Thobal, near Manipur, where he defeated and almost destroyed the Manipur Army.

At the inspection of the 2nd Battalion, on the 18th December, by Major-General Rowlandson, he commented on the remarkable steadiness, smartness, and intelligence of the non-commissioned officers and men, and the general efficiency of all ranks.

By Army Order 187 of 1891, compound titles were introduced for officers of the Medical Staff, which consisted of the prefix "Surgeon" to the ordinary army ranks.

1892–94

In view of the approaching departure of the 1st Battalion to England, a party of 3 officers left Jhansi on the 4th January, 1892, for Secunderabad, with 276 men, transferred to the 2nd Battalion, who waited its arrival at that station from Wellington.

The 2nd Battalion marched to Secunderabad on the 21st January, 1892. On the 13th and 15th February, the companies at Nowgong and Sipri, 1st Battalion, joined headquarters at Jhansi, and, on the 19th, the battalion left Jhansi by rail for Bombay, and, embarking on the troopship "Euphrates," sailed the same day for Portsmouth, where the regiment arrived on the 2nd March, and proceeded to Colchester.

On the 1st April, 1892, Major Owen Williams was promoted Lieut.-Colonel to command the 2nd Battalion, in succession to Colonel Harris, retired on half-pay. On the 19th September, the 1st Battalion furnished a detachment of 1 company to Weedon. The 2nd Battalion formed part of the Secunderabad force at the camp of exercise held between Secunderabad and Beder, from the 30th November to the 18th December 1892.

On the 12th July, 1893, the 1st Battalion was inspected by Major-General Carr-Glyn, whose report was later considered by His Royal Highness, the Commander-in-Chief, to be "very satisfactory."

In the remarks of the Commander-in-Chief in India, on the latest confidential report on the 2nd Battalion, His Excellency was pleased to comment on the excellent conduct of the battalion, in which there had been only ten courts-martial and 260 minor entries, less than a quarter of the average of regiments in India.

In consequence of a lock-out, and strike of miners, and the apprehension of disturbances, the 1st Battalion, on the 8th September, furnished detachments in aid of the civil power to Garforth (near Leeds), Nottingham, Pontefract, and Barnsley. The Pontefract detachment was also broken up into small parties, which were mostly quartered in various collieries, a few men only being retained in barracks. None of the detachments were called out to quell disturbances, and, by the 21st October, all had rejoined headquarters on completion of the duty.

The 1st Battalion was inspected at Colchester on the 16th July, 1894, by Major-General Carr-Glyn, and an Adjutant-General's letter, received later, communicated the opinion of His Royal Highness, the Commander-in-Chief, that the battalion "maintained its high standard of conduct and general efficiency."

On the 21st August, the detachment of the 1st Battalion at Weedon rejoined headquarters, and, on the 27th September, the whole battalion moved by rail from Colchester to Warley.

Lieut.-Colonel Dowse was appointed to command the 1st Battalion from the 9th December, in succession to Lieut.-Colonel Glasgow retired on half-pay.

1895

The remarks of the Lieut.-General Commanding Secunderabad District, on his previous annual inspection of the 2nd Battalion, were published on the 19th March, commenting on its efficient state, the good conduct of the men, and the general high tone existing amongst all ranks.

On the 27th April, a detachment of 11 officers and 500 men of the 1st Battalion, with band and colours, proceeded to the Tower of London for temporary duty, in relief of a battalion of the Guards, ordered to Pirbright; and, during their stay, furnished the west end guards, including the Queen's Guard at St. James's Palace. They also took part in lining the streets at the state reception of His Highness Nasar Eli Khan, son of the Amir of Afghanistan, on the occasion of his visit to the City of London.

At a general parade of the 2nd Battalion, at Secunderabad, on the 3rd July, Lieut.-Colonel Williams presented parchment certificates, awarded by the Royal Humane Society, to Privates Harry Hodson and Stephen Parker, for having, at the risk of their lives, gallantly rescued Corporal D. Hobart, a comrade, from drowning in a lake, at Kundy, whilst bathing.

Since the arrival of the 2nd Battalion at Secunderabad, the Lee-Metford Magazine Rifle had been supplied to the regiment in place of the Martini.

The Tower detachment, 1st Battalion, rejoined headquarters at Warley on the 6th September, and, on the 10th, proceeded to Gravesend for the annual course of musketry (returning on the 23rd), which the remainder of the battalion, with recruits, had carried out at Landguard.

The 2nd Battalion, having received orders to move to Rangoon and Port Blair, Andaman Islands, the headquarters and 7 companies, proceeding to Madras, arrived at Rangoon on the 16th October, and detached 1 company to Port Blair.

1896

From the 1st April, Lieut.-Colonel Owen Williams was granted an extension of sixteen months in his period of command of the 2nd Battalion, and was given the brevet rank of Colonel.

In the *London Gazette*, dated 1st May, it was announced that the Queen had been pleased to confer the Distinguished Service Order on Captain K. J. R. Campbell, Suffolk Regiment, for services in the operations against the Chief of Benin, from August to October, 1894.

The 1st Battalion was inspected at Warley by Major-General C. J. Burnett, C.B., on the 30th and 31st July, when he expressed the greatest satisfaction at the general turn out and drill of the men.

The headquarters, consisting of 12 officers and 535 of other ranks, left Warley for the Tower of London on the 5th August, a detachment of 3 companies remaining at Warley; and, on the 14th September, both detachments left their respective stations, by rail, for Aldershot, and were quartered in the Blenheim Barracks, North Camp.

The 2nd Battalion, having now a detachment at Port Blair, the following extracts are of interest, from "A Retrospect of the Andaman Islands," by Captain C. C. R. Murphy, Indian Army, then serving as a subaltern in the Suffolk Regiment.

"Port Blair is a very large penal settlement, the number of convicts there in 1895 amounting to over twelve thousand. There were of course frequent attempts at escape, which were sometimes successful, and often of the most daring character.

It was a long time before we grew accustomed to our strange home. At first we used to wander about on the sands looking for footprints; and we had already begun to think seriously of discarding our clothes, when suddenly one day a ship arrived, bringing us letters from home. She often came after that. It was always the same ship, and whenever she came back it seemed like the dove returning to the ark with the olive-leaf. During her first few visits there were some amongst us who would have even changed their lot with the turtles she used to take away with her, but for the fact that there were khansamahs waiting for them in port to turn them into soup. But you can get accustomed to anything in time, and we gradually began to settle down; and long before we had been there a year, we thought that Port Blair was the throbbing centre of life, and that murderers were the only people worth knowing.

When we arrived, we found a filter in the Settlement Mess that had not been cleaned for eleven years; and the building itself, which was in reality a deserted fowl-house, was not far behind in the way of a curiosity. After we had been there for a few weeks, we asked for a new perch, or a coat of white-wash, or something equally trivial, for the mess-house. We received a reply to the effect that, as there were only 12,000 convicts on the books at the time, there was no labour available; and further, that the Settlement Officer had been there for twenty-five years, and that such a thing had never been found necessary before.

The head robber at the mess was one Abdul Guffoor. He was a bad man, and what was worse, he was a bad cook. His dinners were sad things. The menu used to be written on a piece of torn paper, and six days a week it read as follows :—Chicken soup—cocari fish—chicken cutlit. Fowl rost. Secan course. Sardin tost—Brown custad. Fruits.

The only eatable dish was the cocari, which the parson used to catch each morning before breakfast.

The barracks were known as Windsor Castle, on account of their battlemented design. The parade ground was far too small for anything but squad drill; and any serious soldiering had to be done on the other side of the harbour, facetiously called the mainland. We had a bathing parade every Tuesday morning, and now and then the men used to get hurt by mysterious things they called 'stingamarees,' which, though they could not be seen, produced very painful effects.

The Commandant at Port Blair, in our day, had no authority to convene a district court-martial. Now, as we have seen, there was only one ship which visited the Andamans; and that one wandered about between Madras, Calcutta, and Rangoon, in the intervals. The result was that when a man was remanded for court-martial, the period he spent in confinement awaiting trial depended on the vagaries of this good ship "Shahjehan." Whether a man does his sentence before or after his trial, is, of course, mere sentiment provided he is found guilty. But in Port Blair, by the time the court was convened and had assembled, most of the witnesses were dead; and the prisoner, with white hair and bowed limbs, was dragged out of his dungeon only to learn just before he died that he was acquitted.

While we were there, a party of seven Burmese convicts seized a dak-boat one evening, and put out to sea in the full burst of the south-west monsoon. The gale drove them along at a great rate, and in seven days they had reached land and all got safely back through Siam to their own country. No trace of them was ever discovered, except the boat in which they had escaped, and that was found bottom upwards on the beach somewhere near Tenasserim. Encouraged by the success of this escapade, a batch of Indian convicts soon afterwards made off on a raft, but it travelled so slowly that they were many days reaching land. When they did arrive, there were only two of them surviving, and they, being too exhausted to move, were easily recaptured.

The one event with which the Andaman Islands will always be associated in the minds of the general public is the murder of Lord Mayo in 1872, who, whilst Viceroy of India, was assassinated by a convict in the settlement. Some years after, another tragedy occurred close to the same place, which, though probably not so widely known, cast a great gloom over the settlement at the time, and became the event from which everything is dated to the present day. On that occasion one of those cyclones, which are so well known to the navigators of the Indian seas, swept over the islands with such violence that a ship called the 'Enterprise,' belonging to the Royal Indian Marine, was driven from her anchorage on to the rocks in front of the Female Gaol, where she foundered. A few of the crew had time to climb up the rigging, the top of which remained above water after the vessel had settled down, and managed to cling on until the morning broke, when the fury of the storm abated. Some female convicts, seeing their plight, improvised ropes with their clothing, and succeeded in rescuing them from their perilous position.

The aborigines of the Andamans are perhaps the most interesting race in existence. They are not in any sense the gollywogs they are often represented to be. They are black but comely. They do not use poisoned arrows; nor are they cannibals. They live peacefully in the jungle, and go about stark naked. If you trespass in their preserves they will attack you, but merely to protect themselves and their women. One of the settlement officers named Vaux, who had been sent into the interior, was killed in this way by a tribe known as the Jarawas; and any convicts who escaped into the jungle met the same fate. Their weapons consist of bows and arrows, the latter being made of bamboo sharpened to a fine point.

In the main islands there are virgin forests of padauk trees, the wood of which is a beautiful red, and takes a fine polish. Recent explorations show these forests to be of immense value, on account of their extensiveness and the excellence of the timber.

During our stay at Port Blair, the local guardship made three or four trips to Nicobar Islands, which lie off the high road as it were, and though they are close to a busy thoroughfare, no steamers ever call there except the Indian Marine ship from the Andamans; and for this reason comparatively few people ever get an opportunity of visiting them.

The only things in the way of curios that can be obtained in the Nicobars are shells and 'scare-devils.' The shells are of many kinds, and some of them are wonderfully pretty. The Nicobarese will not take money, and all bartering has to be done with rum or top-hats. With a few bottles of 'fixed bayonets' and some opera-hats, you might make quite a corner in shells. The 'scare-devils' are crude wooden effigies, made to keep away all evil spirits, except of course those that are intoxicating. Why any devil should be afraid of these harmless devices, and not of the Nicobarese themselves, who are positively hideous, I cannot imagine; but that is their business. If the military authorities ever think of bringing out a sealed pattern scarecrow for soldiers' gardens, they could not do better than take a Nicobarese as a model. They are not a pure race, like the Andamanese, but a mixture; and a very unpalatable mixture too. They seem to have made a hobby of trying to look ugly. They live in huts, like big beehives on piles, which are entered by means of a ladder leading up from the ground through a trap-door in the floor. At night time the father of the family calls the roll, and when everyone is present he pulls up the ladder to prevent anyone from breaking in or out of barracks.

There are some curious birds in these parts, notably hornbills with huge beaks, and megapodes, with society habits. The latter bird is a kind of small wild turkey. The hen hides her eggs in the sand and leaves them there to hatch themselves. It saves trouble and avoids scandal.

On October 24th, we embarked in the R.I.M.S. 'Clive' for Rangoon We sailed at 11 P.M. in bright moonlight, and arrived in Rangoon two days later, on relief by a detachment from headquarters of the regiment."

1897

Major C. R. Townley was promoted Lieut.-Colonel on the 24th February to command the 2nd Battalion, in succession to Colonel Owen Williams retired.

On the 10th April, the 1st Battalion, under Lieut.-Colonel Dowse, mustering 19 officers and 721 of other ranks, embarked at Southampton, in the P. & O. Steam Ship "Sumatra," for conveyance to Malta, one officer and 31 men having embarked on the previous day in another vessel.

On arrival at Malta on the 19th, the battalion marched to occupy the new barracks at Imtarfa (7 miles from Valetta) in relief of the King's Own Royal (Lancaster) Regiment.

Inspections of the 2nd Battalion in January and February, by the Brigadier-General Commanding Rangoon District, and by Lieut.-General Sir Mansfield Clarke, Commanding the Forces, Madras, resulted in most satisfactory reports being promulgated in April, in praise of the discipline of the battalion, its steadiness, general efficiency, and absence of crime.

In celebration of Queen Victoria's "Diamond Jubilee," the Government of India made a grant of Rs.450, to be expended for the use of each British battalion serving in the East Indies, whereupon the 21st and 22nd June were observed as general holidays, and were passed by the men of the 2nd Battalion in Garrison Sports, and in recreation. As an act of clemency on the occasion, and on account of the general good conduct of the battalion, the Commanding Officer directed the release of all cell prisoners, and remitted the unexpired punishments of defaulters at daybreak on the 20th June.

On the 9th November, this year, the Victoria Cross was conferred on Major and Brevet Lieut.-Colonel R. B. Adams, Indian Staff Corps (now Major-General Sir R. B. Adams, V.C., K.C.B., retired), who formerly served in the 12th Regiment, for the following act of bravery.

> "During the fight at Nawa Kila in Upper Swat, on the 17th August, Lieut-Colonel Adams, with Lieutenants H. L. Maclean and Viscount Fincastle, and five of the Guides, proceeded under a very heavy and close fire to the rescue of Lieutenant R. T. Greaves, Lancashire Fusiliers, who was lying disabled by a bullet wound, and surrounded by the enemy's swordsmen. In bringing him under cover, he (Lieutenant Greaves) was struck by a bullet and killed. Lieutenant Maclean was mortally wounded, while the horses of Lieut.-Colonel Adams and Lieutenant Viscount Fincastle were shot, as well as two troop horses." (See Plate 33.)

DISTINGUISHED OFFICERS WHO SERVED IN THE 12TH.

On the 3rd October, the bronze medal of the Royal Humane Society was presented on church parade to Lance-Sergeant E. Popple, for having assisted to save a comrade from drowning in the Quarantine Harbour.

The headquarters 1st Battalion, consisting of 18 officers and 770 of other ranks, left Imtarfa Barracks for the Isola Barracks on the 5th November.

News was received on the 16th of the death of Lieutenant G. D. Crooke, 2nd Battalion, who was killed in a rearguard action at Maidan, on the North-West Frontier, whilst attached to the 1st Battalion Dorsetshire Regiment.

1898

The 2nd Battalion was inspected at Rangoon, on the 7th and 8th January, by Brigadier-General J. Glasgow, formerly of the regiment, who was pleased to remark very favourably on the steadiness and drill, the cleanliness of the barracks, the good management of the institutions, and the excellent state of the interior economy of the battalion.

At a brigade parade of the 2nd Battalion on the 2nd February, the Brigadier-General presented the Royal Humane Society's bronze medal to Private Wm. Starling, for gallantry in saving a Eurasian from drowning on the 22nd July.

On the 17th February, the headquarters and 6 companies, 1st Battalion, consisting of 20 officers and 542 of other ranks, left Isola Barracks, and proceeded to Fort Manoel, leaving a detachment of 2 companies at the former.

In General Orders, dated Malta, May 17th, it was stated :—

> "His Excellency wishes to take this opportunity of recognising the humane conduct of Sergeant Burbridge, 1st Battalion Suffolk Regiment, in promptly jumping into the sea, to save a child's life, at Sliema, on the 15th instant."

By a Royal Warrant, dated, 23rd June, the title of "Royal Army Medical Corps" replaced that of "Medical Staff Corps," and compound titles were succeeded by the corresponding ranks of the combatant branches of the army.

On the 10th September, Major A. J. Watson was promoted to command the 1st Battalion, in succession to Lieut.-Colonel Dowse, removed on appointment to the command of the 12th Regimental District.

In General Orders, dated, Malta, November 19th, His Excellency the Commander-in-Chief was pleased to place on record the gallant conduct of Privates Crick and Goodwin, 1st Battalion, in saving the lives of two of their comrades (a third having been lost by drowning) in connection with the capsizing of a small boat on the 9th of that month. The boat having sprung a leak, and the five men being thrown into the sea, Privates Crick and Goodwin did their best to save their comrades, the former by endeavouring to support them until help arrived (in about half an hour), and the latter by swimming ashore for assistance, and returning to the boat, involving a swim of 900 yards. His Excellency was pleased to bring their gallant conduct to the notice of the Royal Humane Society.

Orders having been received, on the 27th November, for the 1st Battalion to return home, 96 men were, in December, transferred to the 3rd Battalion Warwickshire, Royal Fusiliers, and Lancashire Fusiliers.

CHAPTER XII

England, East Indies, Malta, South African War, England. 1899–1902

1899

The 1st Battalion, mustering 21 officers and 721 of other ranks, embarked at Malta on the 2nd January, on board the transport "Jelunga," and, arriving at Southampton on the 11th, proceeded to Dover.

On the 16th January, Major-General M. Protheroe, C.B., C.S.I., Commanding Burma District, presented the Indian Frontier Medal, with two clasps, and the Tirah Medal, 1897–8, to Captain C. H. Turner, of the 2nd Battalion.

The headquarters 2nd Battalion, under Lieut.-Colonel R. Townley, numbering 15 officers and 617 of other ranks, embarked at Rangoon on the 13th February for Karachi *en route* to Quetta. Port Blair was reached on the 15th, when the detachment of 2 officers and 140 men of the battalion came on board. On arrival at Karachi on the 28th, the battalion entrained for Quetta, and, arriving on the 2nd March, occupied the Right Infantry Barracks, vacated by the Border Regiment.

On the 16th February, the bronze medal of the Royal Humane Society, for saving life at Malta, was presented, at Dover, to Privates Crick and Goodwin, 1st Battalion, by Major-General Sir L. Rundle, K.C.B.

Between the 1st and 6th May, a most successful rifle meeting was held at Quetta, to celebrate the centenary of Seringapatam, in which nearly every man of the 2nd Battalion joined.

The 1st Battalion moved to Lydd on the 29th May, for the annual course of musketry, returning to Dover on the 26th June, and, on the 3rd July, proceeded to Salisbury Plain for manœuvres, forming part of the 2nd Brigade, 1st Division, under Major-General Sir L. Rundle, arriving back at Dover on the 29th. The Officer Commanding now had the satisfaction of publishing in Battalion Orders, a letter received from the Chief Staff Officer, S.E. District, expressing the Major-General's satisfaction at having had the battalion under his command at the recent manœuvres near Salisbury, and also his satisfaction at the state in which he found the battalion at his annual inspection.

The long-standing tension which had existed between the British and Transvaal Governments came to a climax in the autumn of this year.

On September 7th, at a meeting of the British Cabinet, the claim of the Transvaal to be considered a Sovereign International State was emphatically repudiated, and, on the 18th, the Boer official reply was received rejecting all British demands.

Late in September, 5000 British troops from India, together with some reinforcements from Europe, arrived at Natal.

On October 2nd, the President of the Orange Free State informed the Governor of Cape Colony that he deemed it necessary to mobilize his forces, and, on the following day, the mail train from the Transvaal to the Cape was stopped at Vereeniging, and about £500,000 worth of gold taken by the Boer Government. On October 9th, the reserves for the 1st Army Corps were called out in Great Britain, and Parliament was summoned. On the same date, an ultimatum was received from the Boer Government, giving 48 hours for a satisfactory reply. This ultimatum Her Majesty's Government refused to discuss.

The South African War.

The 48 hours having elapsed on the 11th, war was declared on the 12th October, 1899, by the South African Republic, and by the Orange Free State, against Great Britain, and, on the same date, the Boer forces crossed the frontier of Natal, both from the north and west.

On October 15th, patrols first came into collision, and, on the 20th, was fought the action of Talana Hill, followed by the retirement to Ladysmith of the Dundee Column, and the actions, on the 30th, of Lombard's Kop and Nicholson's Nek outside Ladysmith.

Further reinforcements now being required, the 1st Battalion Suffolk, with other regiments, received telegraphic instructions, on the 31st October, to mobilize for service in South Africa.

November 1st to 6th were the days allotted to mobilization, and, in that interval, 512 out of 514 reservists rejoined the 1st Battalion. Of the remaining two, one, in prison for debt, joined the regiment on field service, and the other, who was in India, joined at his own expense.

On completion of the mobilisation, the battalion was inspected on the 10th November, by Major-General H. Parr, C.B., in the presence of the Mayor and Corporation of Dover, and, on the following day, leaving Dover in two special trains, the regiment embarked at Southampton, on the Union Line Steamship "Scot," for Cape Town, arriving and disembarking on the 29th, at the South Arm Dock; strength, 22 officers and 1074 of other ranks.

The following officers embarked for field service :—

Lieut.-Colonel A. J. Watson, commanding.

Major A. C. Cubitt	Lieutenant S. J. Carey
„ E. A. Kemble	„ F. T. D. Wilson
„ V. Graham	„ C. A. White
Captain C. A. H. Brett	„ G. A. L. Thomson
„ W. Thomson	„ G. P. Newstead
„ H. F. H. Clifford	2nd Lieutenant A. H. W. Temple
„ S. E. M. Lloyd	„ A. L. Allen
„ J. A. S. Murray	„ F. W. Wood-Martin
„ A. W. Brown	Lieutenant and Adjutant F. A. P. Wilkins
Lieutenant A. S. Peebles	Lieutenant and Quartermaster A. Smith

Major W. J. Baker, Royal Army Medical Corps, in medical charge.

One hundred men, under 2nd Lieutenant Wood-Martin, having been left at Cape Town, the battalion entrained the same day, arrived at De Aar on the 30th, and detrained at Naauwpoort on December 1st, in the district commanded by Major-General French.

From the 1st to the 7th December, transport was allotted to the battalion, officers' swords returned, and there was practice in route marching, and in cyclist and machine gun drill.

From the 10th to the 26th, mounted troops continued to pass through to Arundel, the camp at Naauwpoort was strengthened with earthworks, rifle pits, &c., and a signal station connecting it with Arundel was established at Opperkop, north-east of Naauwpoort.

On the 26th, the battalion, on relief by the 1st Essex, moved to Arundel, where the new camp was 400 yards west of the railway, facing north.

The Cavalry Division in the camp consisted of O and R Batteries, Royal Horse Artillery, the Carabineers, Inniskilling Dragoons, 10th Hussars, Mounted Infantry, New South Wales Lancers, and New Zealand Mounted Infantry, while the 1st Suffolk, a half battalion Royal Berks, and 2 guns No. 14 Company Royal Garrison Artillery, formed the Divisional Troops under Lieut.-Colonel Watson, Suffolk Regiment.

By the 27th December, the Boer force under Schoeman, in the neighbourhood of Colesberg, had been reinforced to fully 4500 men.

On their evacuating, two days later, their position at Rensburg, it was occupied, on the 30th, by the 1st Cavalry Brigade, and a half battalion Royal Berks, and, on the 31st, the remaining troops of the Cavalry Division, and a half battalion Suffolks, moved there.

1900

On the 1st January, the half battalion Royal Berks occupied McCracken Hill, and the right half battalion Suffolks occupied Porters' Hill, near Colesberg, whilst the left half battalion Suffolks moved up to Rensburg.

This day, a supply train, standing at Rensburg Station, broke loose, and running down the incline towards Colesberg, derailed at Plewman's Siding, about midway between Rensburg and Colesberg Junction, and about four miles from each.

Next day, Major Kemble of the battalion, started with his company, at 1.15 P.M. by train, to try and bring back derailed waggons; but in the face of a very hot gun and rifle fire, the company could do nothing but protect themselves under such cover as was available, returning to camp after dark.

The waggons, set on fire by the Boer shells, burnt for two days and three nights. Private Chapman was dangerously wounded in two places.

Early on the 4th, the Boers, under Schoeman (upwards of 1000) dashed suddenly from their lines against our left. Eluding the cavalry picquets, posted on the outer flank of the Suffolks, the burghers galloped for a line of kopjes, which ran east and west, across the left rear of Kloof Camp. The enemy's artillery at once opened fiercely, from their main position, upon the entrenchments of the Suffolks, who, assailed from three directions, were, for some time, seriously threatened.[1] Major Graham's company, of the battalion, moved out to a kopje immediately facing the new Boer position, while Captain Clifford's company supported two other companies of the battalion, which were already occupying Brett and Thomson Hills, under Captains Brett and Thomson respectively. The Maxim of the battalion came into action under Lieutenant Peebles, while 2 guns Royal Horse Artillery, south of Kloof Camp, engaged the three Boer guns on Red Hill, and later, 4 more British guns opened fire on Schoeman's men. An advance by the 10th Hussars, a squadron 6th Dragoons, and 2 guns Royal Horse Artillery, on the right of the Boer attack, was now ordered, augmented about 1 P.M. by 200 Mounted Infantry under Captain de Lisle.

There was a short encounter, 21 prisoners falling into the hands of the Mounted Infantry. The rest of the Boers scattered in flight, and by 2 P.M. had disappeared. The Suffolks' casualties this day were :—1 sergeant and 1 private killed, and 6 privates wounded.

General French expressed himself as very pleased with the operations.

Some two miles north of Cole Kloof Camp is a position originally called Red or Grassy Hill, and now Suffolk Hill.

This appeared to General French and Lieut.-Colonel Watson, who together reconnoitred it from a neighbouring ridge, to be the key of the position of Colesberg. Lieut.-Colonel Watson had, from the first, expressed his belief that he could capture it with his battalion. On January 5th, his remaining half battalion had joined him, and, during an interview with Colonel Eustace (C.R.A., Cavalry Division) he asked him to obtain, from the General, leave to rush the position in the night with 4 companies. Colonel Eustace conveyed the request, and, in reply, a message was sent to Colonel Watson, authorising him to attack the hill, if he saw a favourable chance. He

[1] *Official History*, pp. 394–6.

was first, however, to inform the General, and all troops in the vicinity, of his intention.

Colonel Watson accordingly ordered, at 11.45 P.M. the same night (5th), that A, B, D and H Companies were to parade at 12.30 A.M., the object of the move being kept secret.

The remaining 4 companies of the battalion were on outlying picquet.

Major A. C. Cubitt was left in camp, with orders to warn the Royal Artillery and all picquets, and to be ready to send off the small arm ammunition and water carts on receipt of a signal from Thomson's Hill.

The column, numbering 305 non-commissioned officers and men (those who had them, wearing deck shoes), and carrying 200 rounds of ammunition per man, moved off noiselessly, headed by the colonel and adjutant, soon after 12.30 A.M., the night being very dark.

After half an hour a halt was made, when the colonel explained the objective of the march, stating that it was to attack the hill, seize it, and occupy it until the morning. He did not expect that any serious resistance would be offered, if indeed the hill was held at all. On no account were any rifles to be loaded, the assault being delivered with the bayonet alone, and absolute silence was to be maintained. The warning as to trusting to the bayonet he again repeated at the next halt, being careful to explain to the officers precisely what he intended each company to do. This was that A and B Companies would deliver the assault, A on the right and B on the left; D Company in support and H in reserve, the Commanding Officer to be with the reserve company. The march was then continued, and, at the last halt but one, the order of the companies was changed, and the last advance was made in quarter column in the following order,—H, D, A, B.

At about 3 A.M. the little column arrived without incident at the water-course at the base of the hill, and an officer of the regiment, describing the nature of the ground on Red Hill, says :—

> "At its base (about 500 yards from the crest), there was a deep narrow ditch about 2½ feet wide and 8 feet deep, which looked formidable in daylight, and must have been a serious obstacle that night to our men. The ground where the night attack was made was very difficult, being very broken at the top with deep ravines, down some of which were found dead and wounded men who must have charged right through the Boer position, and got killed in rear of it."

Three bayonet charges were made in the advance, which was led by H Company, supported on its right rear by D, and scarcely had the leading company approached the summit of the hill, when a single shot rang out from a Boer sentry, who was promptly bayoneted in the thigh, and this was followed by a heavy rifle fire from the Boer position on the crest of the hill. The first charge took place when H Company with a cheer dashed forward, but was driven back with very heavy loss, their captain being mortally wounded and their subaltern killed. The adjutant also was killed at this time, while Major Graham was wounded in two places.

The Colonel then gave the word "Retire," on which, while many made their way to the foot of the hill and back to camp, the remainder attempted to re-form under cover of an underfeature of the hill. After a short pause, the Colonel told Captain Brett to take A Company and support H if possible. The companies were much mixed, but Captain Brett, having got together half a company in the darkness, moved out to the right, with a view to coming up on the flank, and then, in compliance with a further order from the Colonel, the second charge took place, when Captain Brett advanced to occupy the crest in front. This was met by a very hot fire at very short range, supposed to have been only 30 or 40 yards; Lieutenant Carey was killed, Captain Brett and 2nd Lieutenant Butler wounded. Still, Colonel Watson, whose gallantry was a splendid example, strove to retrieve the situation.

It was now beginning to dawn (4 A.M.). The third charge was made when the Colonel personally brought up B Company on the extreme right, only to be met here, as elsewhere, by a murderous fire from a near but quite invisible foe. Then it was that the Colonel was killed, and 2nd Lieutenant Allen wounded, the Colonel's body being surrounded by the bodies of his men, 27 corpses lying within a radius of twenty yards.

The pouches of the dead were rifled for cartridges with which to continue the struggle, but no hope remained. The British shell fire, which now opened from Kloof Camp, inflicting several casualties on our men, became an added danger, which, combined with the increasingly accurate fire of the Boers, made the position untenable. A retirement under such circumstances would have led to a further useless loss of life, so that at 4.30 A.M., the survivors, 99 in number, of whom 29 were wounded, had to surrender.

One of the officers expressly states that the enemy shouted the word "retire," in the first instance, and that it was mistaken as coming from Colonel Watson.

The following were killed, or died of their wounds:—Lieut.-Colonel A. J. Watson, Captain A. W. Brown, Captain and Adjutant F. A. P. Wilkins, Lieutenant S. J. Carey, Lieutenant C. A. White, and 32 non-commissioned officers and men.

The following were wounded and taken prisoners:—Captain C. A. H. Brett, 2nd Lieutenants A. L. Allen and C. W. Butler, and 26 non-commissioned officers and men. Of those not wounded and taken prisoners, were Captain Thomson, 2nd Lieutenant Wood-Martin, and 68 non-commissioned officers and men. The wounded who returned to camp were Major V. Graham and 22 non-commissioned officers and men.

Total casualties, 11 officers and 148 non-commissioned officers and men.

The Boer losses were reported to have been 8 killed and 17 wounded.

With reference to this attack, the following letter has been received by an officer of the Suffolk Regiment, from Mr. P. A. du P. Naude (a Boer), P.O. Box 60, Heilbron, O.F.S., dated 2nd November, 1913. Whilst containing, apparently, some discrepancies with regard to numbers, &c., it is of particular interest, in showing the fierce and close nature of the fighting, and the Boer admiration for the gallantry of the Suffolk Regiment.

"Dear Sir,—With reference to your letter of the 13th September last, and the previous correspondence about the attack at night by 4 Companies of the Suffolk Regiment on Grassy Hill near Colesberg, on January 6th, 1900, I am now in a position to lay before you a few particulars which may be of interest, and which I have much pleasure in giving.

First of all, however, I must apologise for the delay in supplying these particulars. The fact is that Mr. Theron, M.P., from whom I thought much information might be obtained, was only back from Parliament a very short time when he commenced his parliamentary tour of the district, and as he lives a great distance from town, I have been unable to obtain a personal interview with him until just recently. I considered an interview important in preference to communicating with him by letter.

I shall now proceed to give the particulars as narrated by him (P. J. G. Theron), and according to questions put by you in your first letter :—

The Officer Commanding the Boer Forces on that particular night at Colesberg was *not* Naude but Commandant Stefanus Van Vuuren, and although aged, he is still alive, and is living at Kruisementfontein, P.O. Viljoensdrift Station, O.F.S. Besides the Commandant, there were in command Field Cornets P. J. G. Theron (now M.P.) and Johannes Viljoen. The former lives at Driefontein, P.O. Varkfontein, via Heilbron, and the latter in the district of Vredefort, O.F.S., but am not sure of the proper address.

The name of the Hill as known by the Boers is Suffolk Hill.

There is, of course, no certainty about your own killed and wounded, but the number quoted by you is about right, and 87 were captured by the Boer Forces that night.

There were only two distinct charges during the attack, and Colonel Watson was killed in the second.

It is not correct that some of your killed were found on the other side of the hill—they did not charge right through our lines. Some were, however, shot down 3 yards in front of our men! The British charged so close up that their rifle fire (from the muzzle) intermingled with that of our men. Mr. Theron thinks that during the whole war the attack on Suffolk Hill was the most determined he had witnessed and participated in. The Suffolks made a brilliant attack and fought very bravely.

It is incorrect that your movements were made known to the Boer Forces by signals from the flash of lamps and by a call like a bird. The first intimation of an attack was when our guard ran in and breathlessly announced the fact. This was at about 2.15 A.M., and almost immediately firing commenced. The attack that night was, however, anticipated. Your camp, as might be expected, was being watched continually, as well as your every movement outside camp, but mainly your scouting, and, by putting two and two together, it was not a very difficult matter for our officers in command to draw conclusions, and therefore, from your movements in and out of camp that day (5th) the fact was deduced that an attack was contemplated that very night, consequently our men were in readiness. It is an honest fact that no information of whatever description was conveyed to the Boer lines in any shape or form of the attack that night.

No reinforcements were received by us until the evening of the 6th January, when we were increased by 50 Free Staters, though 15 Johannesburg Police had arrived (at Colesberg) the previous evening—5th. The total number of Boers in position when the engagement began at 2.15 A.M. on January 6th, was 60, and this number was increased by the 15 Johannesburg Police, who joined them in the firing line at about 4.0 A.M.

The casualties on our side were nine dead and twenty wounded.

About the attack and defence much has been forgotten after this lapse of time. The night was a dark one, although fairly calm, and as the ground is pretty rough and rock-strewn all around Colesberg, it was almost impossible for an advancing Force to move up the side of the hill, or any of the hills, unnoticed, and it is thought that what drew the attention of the advance guard of the Boers was the grating of a boot-heel of the advancing party over a boulder. Our midnight guard was being relieved at about 2.0 A.M., and this relieving guard had hardly taken up their stand—in fact the guard going off duty had not left the spot—when this noise was heard by one of them, and the one who heard it thought that he had also seen, in the intense darkness, some moving figures. While this point—its correctness or otherwise—was still being debated by the men, a footfall was distinctly heard in the same direction, and almost immediately rifles flashed, and the attack had commenced.

From the outset the attack was most determined, and the British fought heroically, while the Boers in their turn defended the coveted position as bravely and heroically, never even attempting to retreat, but being shot as they lay.

I trust that the slight information which I have been able to lay before you may be of some use, and if I could be of any further service in this or any other matter, you need only drop me a line, and it will give me great pleasure.

Wishing you every success with your undertaking,

Believe me, Dear Sir,

Yours very truly,

(sd) P. A. DU P. NAUDE.

P.S. The war is over now—I hope for ever—between English and Dutch—even race war."

Lieut.-Colonel A. J. Watson, the gallant commander of the Suffolks, had, throughout a period of 27 years, repeatedly come to the front as a brave and able soldier, having, in the Bechuanaland and Hazara expeditions, already been mentioned in despatches for distinguished conduct, and selected, while with the Chitral Force, for an important staff appointment. Sir John French had the greatest confidence in his judgment and ability, and, after his death, is reported to have said :—" The Queen has lost one of her bravest soldiers, and the Army one of its best Infantry Commanders."

On the afternoon of January 6th, the battalion, under Major A. C. Cubitt, with Lieutenant F. T. D. Wilson as Adjutant, moved back to Macder's Farm, on relief by the 1st Essex, and, on the following day, marched to Rensburg.

On the 9th and 15th, the regiment entrained by half battalions to Arundel and Port Elizabeth respectively, to re-officer.

On the 25th January, General the Honourable Sir P. R. B. Fielding, K.C.B., was appointed Colonel of the regiment, in succession to General J. M. Perceval, C.B., deceased.

By the 10th February, the whole of the 1st Battalion had arrived at De Aar, where it had been ordered to concentrate, detaching one company, on the 5th, to Britstown.

On the 21st, the Suffolk Company of Mounted Infantry, under Lieutenants Thomson and Newstead (120 strong), left for Orange River.

Up to the 3rd March, the following officers had joined for duty, since the battalion took the field, on the dates mentioned :—

Lieutenant F. F. W. Hall, from West Africa, 12th January.

2nd Lieutenant R. M. Wilford, from the base, 19th February.

Lieut.-Colonel G. F. C. Mackenzie joined and assumed command, 21st February.

Major W. R. Lloyd, Captain W. Keates, Lieutenants Foster, F. A. White, and Frankland, 3rd March.

On the 6th, a memorial service was conducted by the Rev. C. Wilson over the regimental grave at Colesberg, a firing party being supplied by the 4th Battalion Argyll and Sutherland Highlanders (Militia), and the Colonial and Australian Contingents, and very many residents attended; also Major Cubitt and Captains Lloyd and Clifford. Three volleys were fired over the grave. There was a surpliced choir, the band of the 4th Argyll and Sutherland Highlanders (Militia) played dirges between the volleys, and the buglers sounded the "Last Post." The whole congregation joined in singing the National Anthem. Our burying party that had been sent out was received sympathetically by the Boers. They rendered assistance, also sang a hymn over the grave, and some of the leaders made impressive speeches, expressing abhorrence of the war, regretting the heavy losses on both sides, and declaring the hope that the war would soon be ended.

General Sir John French subsequently said :—

> "The Suffolk Regiment had reason to look back upon Colesberg with pride, as the circumstances attending the operation, on the 6th January, added lustre to the fame of that distinguished regiment, and glory to its splendid traditions. The idea, at one time, seemed to have got abroad that every single operation of war must always turn out exactly as it was wanted. They were to run no risks, and have no losses. War had not taught him that lesson. If they risked nothing, if they were not prepared to take the ordinary chances of war, they would do nothing."

Lieut.-Colonel Mackenzie, with the right half battalion, proceeded on the 7th March, without tents, to Britstown.

On March 13th, Lord Kitchener arrived there, bringing with him the remainder of the 1st Suffolk Regiment, 68th Battery Royal Field Artillery, Suffolk and Cheshire Imperial Yeomanry, one squadron Nesbitt's Horse, and one company South Australian Mounted Infantry, the troops already there being 1 company Warwick Mounted Infantry, Australian Mounted Infantry, 44th Battery Royal Field Artillery, 2 companies City Imperial Volunteers, and 4 companies 1st Suffolk Regiment. On Lord Kitchener leaving, on the 15th, for Prieska, 2 companies of the battalion, with the 68th Battery Royal Field Artillery and Yeomanry details, were left at Doonbergfontein. The remainder of the column followed on to Prieska.

In the meantime, the 1st Volunteer Company, Suffolk Regiment, under Captain Whitmore, had landed at Cape Town on the 7th March, and sailed to Beaufort West, Cape Colony, where they had an arduous time for some weeks. The officers with the 1st Volunteer Company were :—Captain G. E. Whitmore, 3rd (Cambs.) V.B.; Lieutenants H. S. Marriott, 2nd V.B.; and G. H. Mason, Camb. University. The 2nd Volunteer Company was commanded by Lieutenant C. L. Read.

Lord Kitchener had commented favourably on the marching of the battalion.

A draft of 200 men arrived on the 21st March at De Aar, and was joined there by 3 companies of the battalion, under Major Cubitt, on the 9th April, when the whole left for Piet River.

Headquarters and 5 companies of the battalion arrived at De Aar on the 22nd April, by route march from Prieska, and, on the 24th, entrained for Kaffir Bridge, detaching 1 company on the 27th to Kraalspruit, and 1 to Ferreira Siding.

On the 1st May, the battalion was further augmented by a draft of 100 non-commissioned officers and men under 2nd Lieutenants Nicholson and Pratt, and, on being concentrated at Bloemfontein, on the night of the 11th entrained for Vet River, where it was joined by the 1st Volunteer Company.

When the Suffolks first entered the Free State, they had been attached to the 23rd Brigade (Major-General W. G. Knox), which consisted of the 1st Suffolks, 1st and 3rd Royal Scots, 9th King's Royal Rifles, the 9th, 17th, and 39th Batteries Royal Field Artillery, and the Elswick Volunteer Battery, but the Suffolks having been ordered by telegram to join Lord Roberts, on his advance to Johannesburg and Pretoria, the battalion marched accordingly to Kroonstadt, the last 40 miles being covered in a day and night, with but short halts.

According to arrangements made on May 17th, for the safety of the railway, Kroonstadt was to be held by the 1st Suffolk Regiment, and by the 4th Argyll and Sutherland Highlanders, the battalion arriving there at noon on the 19th, and remaining until the 24th, when the march was continued to Rhenoster River, which was reached on the 26th.

Here the battalion awaited the arrival of a portion of the Siege Train, including four 6-inch howitzers, under command of Major Allen, and the two recently purchased 9·5-inch howitzers. These siege howitzers had been hurried up from Bloemfontein, in view of the possible necessity of bombarding the forts around Pretoria. Leaving on the 29th, with the howitzers, the battalion marched to Vredefort Road, on the 30th to Taibosch Spruit, and, on the 31st, after a very heavy march through deep sand, crossed the Vaal to Vereeniging.

On June 1st, the battalion entrained in two parties, with siege guns, on a supply train to Johannesburg, the transport going by road, and, arriving on the 2nd, bivouacked at the Park Station, where there was a refreshment room, which is here alluded to, because it is recorded that a waiter there had stated, apparently with truth, that he had been the sentry on Red Hill, on January 6th, who first alarmed the Boers, he himself having been bayoneted through the thigh immediately afterwards.

On the 3rd, the battalion marched north, and joined the main army under Lord Roberts, taking part next day in the action of Six Mile Spruit, when the naval guns shelled the fort and stations of Pretoria.

Lord Roberts had left Bloemfontein on the 3rd May, and, by a rapid and well-organised march, entered Pretoria on the 5th June, at the head of about 25,000 troops, including the 1st Suffolks, who entered the enemy's historic capital by the Artillery Barracks, and, marching through the town (none more proudly than the Suffolk Volunteer Company) bivouacked at its extreme eastern end, close to the Delagoa Bay Railway, 1 company with the siege guns being detached to Klapper Kop Fort.

From May 20th to June 7th the 1st Suffolks had been attached to the 14th Brigade. On the latter date the battalion formed part of a column under General Smith-Dorrien, which now consisted of the 3 remaining battalions of his (19th) Brigade, the 1st Suffolks, 81st Battery and 4 guns of the 74th, Ross's Mounted Infantry Corps, and a company of the City Imperial Volunteer Mounted Infantry. The column marched on the 7th to Irene, and on the 8th to Kaalfontein. Two companies and headquarters of the battalion were at Aanoluling, the junction of the Natal and Cape railways.

Two companies, with part of the Volunteer Company, held Zuurfontein Station; two others, with remainder of the Volunteer Company, held Meyerton, and the remaining two companies held Rietfontein. All these posts were well entrenched.

On the entry of the army into Pretoria, Captains Brett and Thomson, and Lieutenants Barnardiston and Wood-Martin of the battalion were amongst the released prisoners of war. Captain Thomson, however, died there on June 9th. Lieutenant Barnardiston had been taken prisoner at Stormberg on December 10th, in the previous year, when attached to the 2nd Royal Irish Rifles. General Smith-Dorrien's column was reinforced by 900 released prisoners (chiefly of the Northumberland Fusiliers and Royal Irish Rifles) who were armed with captured Martini-Henry rifles.

On the 8th July, the battalion concentrated at Irene, and, moving next day to Tigerpoort, had, by its timely arrival, set Colonel Pilcher free for more active operations, he having on the previous evening discovered an attempt by 400 Boers to pass between his outposts and those of the 11th Division to the north. The march to the Tigerpoort position was an arduous one, in which the waggons did not reach the bivouac until long after midnight. By an order received on the 11th, the Suffolks and the Royal Irish Fusiliers were to hold the ground from Tigerpoort Pass to Bronkhorst Spruit, a distance of twenty miles of ridge. On the 11th July, 4 companies of the battalion, including the Volunteer Company, under Major W. R. Lloyd, took part in the operations which resulted in our capture of the Witpoort and Leeuwpoort positions. Marching on the night of the 18th to Witpoort, the battalion joined General Hutton, who had under his command the 1st Mounted Infantry, 2nd Royal Irish Fusiliers, and Canadian Mounted Rifles.

General Hutton's column started on the 23rd, as the inner of 4 columns, which, under General French, were to make a wide sweep east and south-east on Middelburg, in conjunction with Lord Roberts' main advance along the Delagoa Bay Railway.

On the 25th, Hutton patrolled the line between Balmoral and Brug-Spruit, and bivouacked on the left of the cavalry. His two infantry battalions (1st Suffolks and 2nd Royal Irish Fusiliers) and his two 5-inch guns, dragged by oxen, had, on this date, covered 23 miles, and had shown great spirit in keeping up well with the mounted troops during the three days' march.

During the night of the 25th, a violent storm of wind and rain raged, causing much privation, and resulting in deaths from exposure of several men and many transport animals.

When entering Middelburg on the 27th, the British flag was hoisted by Pioneer-Sergeant Morrogh, 1st Suffolk Regiment, the battalion with the Volunteer Company holding the kopjes west of the town.

Seven officers (Captain Brett, Lieutenants Murphy, Hall, Wood-Martin, Taylor, Probert (3rd Suffolk) and Marriott (2nd Volunteer Battalion, Suffolk) and 175 non-commissioned officers and men joined the 1st Battalion on the 30th July, by direct road from Pretoria.

At a special parade held at Middelburg, on August 2nd, General French addressed the battalion as follows :—

"Major Cubitt, Officers and men of the 1st Suffolk, I wish first to say that I am very glad to have the pleasure of serving with you again, and I want to thank you all for the good work you have done in different parts of South Africa, since you were last under me. The work you did at Colesberg is fresh in my recollection ; there you were placed in a position of great responsibility on the left flank, the duties of which you performed thoroughly and well.

But what I want specially to recall is the sad event of the morning of the 6th January, and to express my sympathy with you. I wish to condole with you on the loss of your gallant leader, Colonel Watson, who showed splendid qualities as a noble and able officer, and of so many other brave officers and men.

Now it has lately come to my knowledge that there has been spread about an idea that that incident cast discredit of some sort on your gallant regiment, and I want you all, and especially the young soldiers I see amongst you, to, once for all, banish from your minds any such unworthy idea.

Some of you, to whom I am now speaking, took part in a night operation of extreme difficulty, in pitch darkness, and did all in your power to make it a success, so do not now let any false impression get into your heads ; think, rather, that what took place brings honour to your regiment, and add it to the long list of actions it has taken part in, in the past.

I want you to bear in mind that night operations can never be certain of success, and because such an attack sometimes fails, it does not necessarily bring discredit on those who attempted it.

If we were always waiting for an opportunity of certain success, we should never do anything ; in warfare, with a bold enemy, who is worthy of your steel, it is absolutely impossible to be always sure of success.

All that men can do is to try their very best, and you did that on the occasion I am speaking of, and I thank you for that, and all the good work you have done since, and remember, in conclusion, that no blame whatever attaches to your regiment as the result of that operation."

On the 4th August, the battalion marched to Parr, and, on the 14th to Wonderfontein with General Hutton, returning by rail to Middelburg on the 19th.

On the 2nd September, the battalion joined General Mahon's column, and again marched to Parr and Wonderfontein ; leaving the latter station on the 4th, Carolina was reached on the following day, where General Mahon joined General French's division, bringing with him the following troops : Imperial Light Horse, 3 companies New Zealand Mounted Rifles, Lumsden's Horse, the Queensland Mounted Infantry and Bushmen, the 3rd Mounted Infantry, 1st Suffolk Regiment, 2nd Shropshire Light Infantry, a 4·7-inch Naval gun, M Battery Royal Horse Artillery, 2 guns 66th Royal Field Artillery, and 2 Vickers-Maxims.

On the 9th September, General French struck eastwards from Carolina, making a wide turning movement on Barberton, which was brilliantly accomplished in the face of great difficulties. At 10 A.M. the Boers were encountered, 600 in number, with 3 guns, on what proved to be the first of three prepared positions. The first of these, attacked in front at close range by the guns of the 66th Royal Field Artillery, was won in two hours. The second, a much stronger line along the hills, took longer, and, after a two hours' bombardment, French deployed his two battalions of infantry for an assault. Before the determined advance of the Suffolk Regiment, the opposition gradually crumbled, the Boers being heavily punished by shrapnel, as they gave way in small bands.

The casualties in the battalion were :—1 sergeant, 1 drummer, and 2 privates all slightly wounded.

Referring to this action, Lord Roberts wired :—"I am delighted to hear of your successful advance, and of the gallant behaviour of the Suffolks ; please congratulate them from me."

On the 10th General French found that his opponents of the previous day had vanished, and pursued his march uneventfully.

On the 12th, the greater part of the troops moved up the very steep and difficult Nelshoogte Pass, and the battalion pressed on, after a convoy, which was retreating south, under cover of a long range gun and pompom fire. Six of the waggons were taken ; but the remainder of the convoy and guns got away. The Volunteer Company now occupied a hill, exceptionally difficult of access, and waterless, remaining there for the next two days. The first night, the company held the advanced kopje, two miles ahead, and their transport failing to reach them, they passed a trying night in the cold, at a high elevation, without rations or blankets, which was borne with their accustomed cheerfulness.

The 13th and 14th September were passed in getting the waggons and guns up the pass, which was done with the greatest difficulty, they having to be doubly and trebly spanned. The battalion, moving on the following day, was the first infantry regiment to reach Barberton on the 18th. Here, about 200 Boers surrendered, and, at the bank, a Cape cart was secured, full of specie, but largely containing government promissory notes. The railway station, abounding with supplies and stores, was also captured, with enough provisions, including flour, to ration the column and the forces of Generals Pole-Carew and Hutton for three weeks, and also 48 engines. Many bags of flour had been soaked with paraffin by the enemy.

The next move was by rail to Avoca on the 21st, where, assisted by large native working parties, the battalion improved the precipitous road up to Kaapsche Hoop. At Avoca, 52 more railway engines were secured, which were of great service later on.

On October 1st, Lance-Corporal F. Lane, of the battalion, was dangerously wounded, in defence of a train near Parr. On the 3rd, General French left Barberton with the 1st and 4th Cavalry Brigades, Mahon's Brigade, and 4 companies 1st Suffolk from Avoca, with the Volunteer Company, and marched *en route* for Machadadorp, which was reached on the 9th.

On the 10th, the brigades of the Cavalry Division were reorganised, when a half battalion of the Suffolks was posted to Major-General Dickson's

4th Brigade, which consisted of O Battery, Royal Horse Artillery, 7th Dragoon Guards, Lumsden's Horse, Imperial Guides, E Section Vickers-Maxims (1017, all ranks),[1] and ox transport, 150 waggons. At Machadadorp, on the same date, the Volunteer Company left, under orders for England, but did not get farther than the Pretoria-Elandsfontein section of railway, being detached to hold Olifantsfontein Post, which they left in December for Vereeniging on the Vaal River, where the Company remained until April, 1901.

On the 14th, the Division concentrated at Carolina, and, on the 16th, started in three columns (Gordon on the left, Dickson centre, and Mahon right) for Ermelo, which was reached on the 18th. Thence to Bethel, on the 20th, halting one day, and reaching Heidelburg on the 26th, where a four days' halt was made, owing to the exhausted state of the oxen. Throughout the march from Carolina to Heidelburg, the enemy hung round our flanks, but never pressed an attack home.

On the 30th October, the Division advanced to Springs, and, on the following day, was inspected by Lord Roberts.

The force started for Pretoria on the 1st November, arriving on the 3rd, when, after a period of two months' bivouacking, tents were again issued to the battalion, which, a few days later, was posted to General Smith-Dorrien's force, at the Komati River. On the 6th, he set out with a single column, viz., 250 5th Lancers, Royal Canadian Dragoons and Mounted Rifles ; 2 Canadian Horse Artillery guns, 84th Field Battery 4 guns, and 900 Infantry, Suffolks and Shropshires.

At 2 P.M. on the 7th, he detached the Royal Canadian Dragoons, with 2 guns, Royal Canadian Artillery, supported by two Vickers-Maxims, and 2 companies Suffolk Regiment, under Major W. R. Lloyd, to work round the Boers' left. At 4 P.M., the Suffolk companies, ably handled, obtained a lodgment on that flank, completely turning the enemy, who were confronting the Shropshire Light Infantry. The result was instantaneous ; the stronghold was hurriedly evacuated, and all the commandos fled across the Komati River. Our troops then took possession of the position, and advanced along the high ground to Lilliefontein, where they bivouacked at sunset.

On the 12th, 4 companies of the battalion, under Major W. R. Lloyd, rejoined headquarters from Belfast, where, with the 2nd Shropshires, they had been in General Smith-Dorrien's command.

The prisoners taken by the Boers at Red Hill, near Colesberg, on January 6th, had been released at Noitgedacht, and, after a short spell of rest, rejoined the battalion, when the following Battalion Order was published :—

> " In welcoming back to the battalion the non-commissioned officers and men who were unfortunately taken prisoners, on the morning of the 6th January last, the Commanding Officer wishes to place on

[1] " *Times* " *History of the War*, Vol. v, p. 48.

record his appreciation of the repeated gallant efforts made by them, and their deceased comrades, to drive the Boers from what proved to be, under the circumstances, a practically unassailable position. All honour to the men, who, in charging three times over rough ground, on a very dark night, under close musketry fire, made such a gallant effort to add fresh glory to the history of their regiment."

Under orders, received at noon, on the 27th November, the battalion entrained for Bethulie, 6 companies arriving on December 1st, 1 on the 2nd, and 1 on the 3rd.

At the half-yearly inspection of the 2nd Battalion at Quetta, on the 28th November, by Brigadier-General Sir R. Hart, V.C., Commanding District, the inspecting officer was pleased to report as follows :—

"The battalion is well officered, and will compare favourably with the best battalion in the service ; there is the nicest possible feeling between all ranks.

Non-commissioned officers, excellent ;

Men, keen, soldierlike, march well, and conduct exemplary.

Musketry, practical, and results very satisfactory.

Company training excellent, and has developed a spirit of independence and self-reliance in officers and non-commissioned officers.

Interior economy perfect, and workshops the best I know of.

Absolutely not the slightest imperfection to point out."

Referring again to the South African War, every effort was made, at the end of November, to frustrate De Wet's evident design of invading Cape Colony. To attain that end, he had gathered all available commandos on the banks of the Orange River, adding to his numbers by recruiting or impressing among the farmers of the district, who had either taken the oath of neutrality, or were British subjects.

The battalion was now with General Knox's force, and, on the 1st December, 3 companies 1st Suffolk, near Odendaal Stroom, with 300 of the 1st Mounted Infantry and 4 guns, under Lieut.-Colonel W. H. Williams, were part of the troops at hand to head back De Wet, whilst other columns were on the way south from Bloemfontein. On the 3rd, they came in contact, in difficult country near Silk Spruit, with a rear guard left by De Wet, he himself getting away over the Caledon River. The next day, Colonel Williams' force was at Carmel, on the above river ; on the 5th, under Colonel Long, the Suffolk detachment had moved to Bethulie, and, on the 6th, was railed south to Knapdaar to assist General Knox in his pursuit of De Wet.

On the 7th, the regiment proceeded by rail to Aliwal North, and, crossing the Orange River, moved to Beeste Kraal, twenty miles on the Rouxville Road.

The 14th December found the battalion returning by rail to Pretoria, which was reached on the 16th.

The short absence from Pretoria of less than three weeks, embracing as it did two long rail journeys, and some very hard marching in torrents of rain, with great transport difficulties, did a great deal of damage to the health of the battalion, several deaths, and later much sickness, being distinctly attributable to it.

On the 18th December, F Company, under Captain Massy Lloyd, was appointed bodyguard to Lord Kitchener, and, on the 31st, the following was the distribution of the battalion at Pretoria. A Company at Koodoo's Poort, B and G at Sunnyside, C at Arcadia, D at Proclamation Hill, E at East Fort, F with the Commander-in-Chief, and H at Johnson's Redoubt.

1901

On the 19th January, the battalion proceeded to Wonderfontein (120 miles) in open trucks, in heavy rain ; the 1st Essex, 1st Cameron Highlanders, and 2nd West York Regiments arriving a few days later. The whole column, including mounted troops, was under General Smith-Dorrien.

Her Majesty Queen Victoria died on the 22nd, and was succeeded by His Royal Highness the Prince of Wales, as King Edward VII.

The column, moving towards Carolina, on the 25th January, came upon the enemy in force, at Twyfelaar, on the Komati River, the battalion, as advanced guard, coming, at about 3 P.M., under a heavy rifle and Vickers-Maxim fire. After a heavy shelling, the enemy retired at dark, and was reported to have suffered heavily. The casualties in the battalion were :—Major W. R. Lloyd killed, and 6 rank and file wounded.

On the 27th, Major Lloyd was buried in Carolina graveyard, and, on the following day, the battalion started to return to Wonderfontein.

In a Battalion Order, dated 31st January, the Commanding Officer wished to record his sense of the very great loss suffered by the regiment, in the death, at Twyfelaar, of the late Major W. R. Lloyd, who, like his father, had been killed in action ; at the same time adding :—"A better soldier, or more kindly genial comrade, has rarely, if ever, worn the British uniform."

On the following day, the following extract from Force Orders, dated Wonderfontein, 1st February, 1901, was published :—

> "It has given the Major-General Commanding much pleasure to bring to the notice of the General Commanding-in-Chief the excellent and bold work done by the mounted troops, the Suffolks, West Yorks, and batteries in the recent Expeditions to Carolina, having also reported to him the extremely hard work, and severe weather, all the troops have endured with such cheerfulness."

On the 3rd February, a force composed of the 5th Lancers, 3rd and 4th Companies Mounted Infantry, 2nd Imperial Light Horse, and gun detachments from the 66th, 83rd, and 84th Batteries, 1st Suffolk, 2nd West Yorks, and 1st Cameron Highlanders marched to Twyfelaar.

On the 4th to Bosman's Spruit, south of Carolina, and, on the following day, to Bothwell Farm at the north end of Lake Chrissie. There was the usual rearguard action on the afternoon of the 5th, but nothing to indicate any unusual strength on the part of the enemy. The night posts on the south-west and west were held by the West Yorks, on the north by the Suffolks, and east by the Camerons ; the main body of the infantry being in rear of their own picquets, and the artillery and mounted troops in the centre. The night was dark and misty, and, covered by the darkness, 3 or 4 Boer commandos took up positions entirely surrounding our camp. At about 3.20 A.M. on the 6th, a very heavy Boer rifle fire broke out on the south-west and west of the camp, sweeping the ground occupied by the West York picquets, and the main body of the Suffolks, stampeding the horses of the 5th Lancers, and many transport animals. While the Suffolks were severely tested, the brunt of the attack was borne by the West Yorks. A mob of stampeding horses which had galloped into the enemy's lines was turned back by the Boers, and driven up to the outpost line. Rushing in under cover of the horses, a party of Boers overwhelmed and cut to pieces two of the West Yorks' picquets. The officer commanding their supporting company, and all the men, except one sergeant, were either killed or wounded. The Suffolks, meanwhile, though suffering from a galling fire at close quarters were receiving the enemy with steady volleys, whose sound spread confidence in the camp behind.[1] All the infantry having quickly deployed along their respective fronts, the Boers were unable to get into the camp, the commando on the south-west being the only one that made a serious attempt. In three quarters of an hour, the Boer fire had slackened, and at 4.30, the enemy were in full retreat. Their losses were credibly reported to have been 37 killed and 107 wounded. Those of the Suffolks were 1 private killed, and Captain Keates, 1 sergeant, and 7 privates wounded.

On the 9th February, the force continued the march to Umpilair, where 53 waggons and many head of cattle were captured.

On the 10th, to Warberton Farm, where 12,000 sheep had to be killed, which were delaying the column.

On the 11th, to Middle Drift, halting next day to enable a bridge to be made for waggon transport.

On the 13th, to the Isutu River, and, on the following day, to Litchfield Farm, where long range sniping took place at our bivouac, when 1 sergeant and 1 corporal of the battalion were wounded. Here, a halt was made until the 20th, when the force rejoined General Smith-Dorrien at Amsterdam. Biscuit and grocery rations were now reduced, but meat was most plentiful.

A halt was made at Amsterdam until the 24th, the continuance of wet weather, with greatly swollen rivers, rendering the rationing of all columns under General French most difficult.

On the 25th, a move was made to the Ingwempisi River, which was crossed next day by a temporary bridge, halting the night at Wolvekop.

On the 27th, to Derby Run, where a halt was made until the 15th March, the infantry being occupied in foraging, and the mounted troops in more distant raids, resulting in the capture of a few Boers, waggons, and cattle.

[1] "*Times*" *History of the War*, Vol. v, p. 168.

On the 15th March, entered Piet Retief, prior to occupying an outpost about 1½ miles south-west of the town.

On the 18th, 2 companies of the battalion, under Captain Murphy, moved to guard the bridge over the Assegai River, 4 miles south of the town.

The perimeter of defence being greatly reduced by the 20th, the battalion held the south fringe of the town, with a series of small fortified posts; this restriction of area, occupied by troops and animals, causing a good deal of fever.

On the 26th, a detachment of the 5th Lancers, 2 guns, and 6 companies Suffolks, under Lieut.-Colonel Mackenzie of the battalion, took a convoy of prisoners, women, &c., to Castrel's Nek, and there took charge of a full convoy, which had come from the line of communications, via Wakkerstroom.

On the 14th April, the battalion left Piet Retief, and moving by Chela River, reached Wonderfontein on the 27th. On this date also, the 2nd Volunteer Company, Suffolk Regiment, landed in South Africa, and proceeded to join the 1st Battalion. In this respect, both the Volunteer Companies were fortunate in having been attached for duty to the Line Battalion, many of the Volunteer Companies having been taken for duties on the line of communications.

Prior to marching to Middelburg on the 30th April, General Smith-Dorrien addressed the battalion in most complimentary terms, speaking most highly of its services rendered during the past three months, both on the march and in action.

In April, the 1st Volunteer Company, which had been left at the Vaal, railed down country to Cape Town, and embarked for England on the 4th May. The members of it had, throughout their tour of active service, taken their full share of all military duties, and whether on toilsome marches, or in squalid wet bivouac, had borne all privations cheerfully. An extract from a War Office official pamphlet, regarding them, states:—

> "The representatives of the Volunteers who marched and fought in South Africa, side by side with their comrades of the regular battalions, left a very favourable impression of their value on the minds of the generals under whom they served."

On the 2nd May, one company of the battalion was detached to Bronkhorst Spruit, and, on the same date, the 2nd Volunteer Company joined, under Lieutenant Read, 1st Volunteer Battalion.

The corrugated iron blockhouses, which afterwards became so general in South Africa, were now being manufactured and introduced, at Middelburg, by Major Rice, Royal Engineers. The battalion remained at Middelburg until July 3rd, when they left for Groot Oliphants River to form part of a column under Lieut.-Colonel Mackenzie, but before any work had been done, the destination was changed to Friederickstadt, Transvaal, which was reached on July 10th.

The column consisted of 2 guns, 83rd Field Battery 2 pompoms G.G., detachment 23rd Company Royal Engineers, and telegraph section, 1 company 11th Mounted Infantry, 1st Suffolks, and 2nd West Yorks; its object was to blockhouse the line of the Mooi River from Friederickstadt to its source, and then on to Naauwpoort W. This line, about 35 miles long, was completed and occupied by the 13th August.

This blockhouse line was afterwards extended to Breedts Nek on the north-east, and Oliphants Nek on the north-west, both these neks being on the Megalliesberg River. The battalion also held Grobler's Pass three miles east of Breedts Nek. The country south of the Megalliesberg was thereby effectually protected, and convoys proceeded with comparatively small escorts from Krugersdorp to Naauwpoort, and thence to Rustenburg.

The blockhouse line was built thus :—

Every ten miles there was a strong fortified post with a garrison, roughly of 100 men. The blockhouses were 800 yards apart, and were occupied by a non-commissioned officer and six men.

Around each house was a barbed wire entanglement, and between each a barbed wire fence of six strands of wire and a curtain each side of five strands. During the time of the battalion occuped this line a ditch was dug on either side of the fence five feet wide at top, four feet deep, and three feet wide at bottom.

The occupation of these blockhouses proved very tedious and trying for both officers and men, for, as each only contained one non-commissioned officer and six men, the look-out had to be continuous, and the sniping, which usually took place at night, kept all garrisons awake.

Also, many small intermediate works had to be built and garrisoned, in order to prevent crossings of the wire, and although the battalion guarded nearly 35 miles of blockhouse line, there was never a serious crossing effected, the total being only thirteen, and these were principally made by single men carrying despatches. Every crossing was known as the wire had to be cut; and each wire of the fence had an alarm at both ends; also batteries of rifles were fixed so as to fire down the wire, six rifles being fastened together, carefully laid and aimed before sunset.

On the 20th October, Captain H. B. Rowlands, of the battalion, joined whilst on leave from the Central African Regiment, and was appointed Signalling Officer to the column.

Brigadier-General Hart, V.C., inspected the 2nd Battalion, on the 19th November, at Quetta, when he was pleased to speak in the most eulogistic terms of the high state of efficiency of everything connected with the battalion, and remarked that it was "in every respect, as efficient for war as a regiment well could be."

In South Africa, at this period, every military district in our possession was busy in the erection of chains of blockhouses.

On the 12th November, the battalion, on relief by the King's Own Scottish Borderers, marched to Ventersdorp, to erect a blockhouse line from thence to Tafel Kop; on the 16th, 4 companies left, in conjunction with Lieut.-Colonel Hickie's column, which had been detailed to cover the erection of the blockhouses, and about the 20th, Tafel Kop had been denied to the enemy by the establishment on it of a fortified post of the Suffolk Regiment. The above 4 companies were then detailed to hold the hill until signalling communication had been opened with Magato, Oliphants Nek, Breedts Nek, and Ventersdorp.

Lieut.-Colonel Mackenzie then returned with Hickie's column to Ventersdorp for further action in that direction. The blockhouses on this line were to be not more than 1000 yards apart, while, from three miles north of Ventersdorp there was a stretch of 17 miles of absolutely waterless country, which entailed special arrangements for procuring water, whilst the blockhouses were building, and involved a large daily water convoy from both ends of the stretch during the remainder of the war.

1902

By the 29th January, the blockhouse line was completed and held from North to South, with posts at Waterjock and Baltfontein.

Major Graham and Lieutenant Jourdain, with a draft from England, arrived and joined the detached post of the battalion at Waterjock, on the 19th February.

On the 23rd, Colonel C. R. Townley retired, on completion of his extended period of command of the 2nd Battalion, and was succeeded by Lieut.-Colonel A. C. Cubitt, promoted from the 1st Battalion.

A draft of 150 men from the 1st Battalion, to join the 2nd in India, having left Ventersdorp, sailed on the 6th March, and, next day, a draft from India joined headquarters of the 1st Battalion at Ventersdorp.

On the 18th March, Captain C. R. Fryer joined the 1st Battalion, on vacating the adjutantcy of the 3rd Suffolk Regiment, on relief by Captain S. E. M. Lloyd, who had proceeded home in the previous month.

The 2nd Battalion left Quetta on the 2nd April, and, arriving at Karachi on the 4th, detached 3 companies to garrison Hyderabad, Scinde, which arrived there on the 3rd, and detrained the next day.

On the 7th, Lieutenant A. H. Temple, 1st Battalion, proceeded to join the Suffolk Company of the 28th Mounted Infantry *vice* Captain Rowlands, rejoining the King's African Rifles from leave, and a week later, 2nd Lieutenant C. H. Standbridge was taken on the strength of the battalion, which was further augmented, on the 16th May, by a draft of 114 non-commissioned officers and men from England.

On the 19th, the 2nd Volunteer Company, under Lieutenant C. L. Read, 1st Volunteer Battalion, left for Cape Town, *en route* to England, and Lieutenant Jourdain left on the same date, to join the 1st Mounted Infantry in Orange River Colony.

The war was now concluded by the "Treaty of Vereeniging," and, on the 1st June, a telegram was received from the C.S.O., Kleepsdorp:—"Peace signed last night, precautions not to be relaxed until further orders."

On the 3rd, Captain J. A. S. Murray and 11 non-commissioned officers and men, 1st Battalion, left for England, to represent the regiment at the coronation of King Edward VII.

The headquarters, 1st Battalion, were now at Waterzoek, three miles north of Ventersdorp, where 2nd Lieutenant W. M. Campbell joined, on appointment, on the 5th June.

On the cessation of hostilities, telegrams of congratulation were received by Lord Kitchener from His Majesty the King and His Majesty's Government, which were communicated to all the troops.

On the 17th June, 150 men, 1st Battalion, under Brevet Major Prest, were selected to proceed to England as part of Lord Kitchener's escort, with similar parties of the 1st Cameron and 1st Gordon Highlanders, 2nd Hampshire, and 1st Welsh Regiments, and 1st Royal Inniskilling Fusiliers.

The 2nd Volunteer Company, Suffolk Regiment, left Cape Town for England on the 25th June.

The battalion occupied the blockhouse line until the declaration of peace, when the line was dismantled, to the extent of collecting all wire and standards, and, having concentrated at Waterzoek at the end of June, marched on the 1st July to Potchefstroom, arriving on the 3rd.

Up to the 2nd August, 200 reservists had been sent home, and, on the 8th, the Suffolk Mounted Infantry Section rejoined headquarters, after an absence of nearly three years with No. 1 Battalion Mounted Infantry.

The 9th August was duly observed as a general holiday by the troops in South Africa, as the Coronation Day of King Edward VII, the event having been unavoidably postponed from the 25th June, on account of His Majesty's illness.

On the 15th and 16th, 2 Suffolk companies of Mounted Infantry joined headquarters of the battalion from the 28th and 15th Battalions of Mounted Infantry respectively.

The 1st Battalion entrained for Stellenbosch, Cape Colony, on the 1st September, arriving on the 5th. Private Barrett was accidentally killed on the way down country, in endeavouring to board a train while in motion.

Orders having been received to proceed to England, the battalion embarked at Cape Town, on the 9th September, on board the steamship "Canada," with the 1st Border and 2nd Hampshire Regiments; the Suffolks leaving at Stellenbosch a draft of 120 non-commissioned officers and men, under Major van Straubenzee, for the 2nd Battalion at Karachi.

The 1st Battalion arrived at Southampton, and entrained for Colchester on the 29th September; strength, 16 officers, and 515 of other ranks.

On arrival at Colchester, the battalion was met by Major-General Sir William Gatacre, K.C.B., and staff, and by a very large crowd, to welcome them home, and they occupied quarters in the Meeanee Barracks, Napier Lines.

On the 1st October, 373 non-commissioned officers and men, who had been attached to the 9th Provisional Battalion, joined the battalion, and practically the whole of the men who returned from South Africa went on furlough for two months.

On the same date, Colonel Dutley, commanding Royal Artillery, Eastern District (on behalf of the General Officer Commanding Eastern District) presented the battalion with the Queen's South African medals.

Throughout the South African War, the total numbers killed in action, and died of wounds and disease, were seven officers (Lieut.-Colonel A. J. Watson, Major W. R. Lloyd, Captains W. G. Thomson, A. W. Brown, Captain and Adjutant F. A. P. Wilkins, Lieutenants S. J. Carey, and C. A. White), and 144 non-commissioned officers and men. The following honours and promotions were awarded to officers and men. Lieut.-Colonel G. F. C. Mackenzie was appointed Companion of the Order of the Bath. Captain C. A. H. Brett, Lieutenants A. S. Peebles, F. A. White, S. J. B. Barnardiston, and A. P. Frankland were made Companions of the Distinguished Service Order, and Captains E. P. Prest and S. E. M. Lloyd received the brevet rank of Major.

All the above officers were mentioned in despatches (Colonel Mackenzie twice), and the following officers were also mentioned :—Major A. C. Cubitt, Major W. R. Lloyd (killed in action), Captains J. A. S. Murray (twice), G. F. Whitmore (1st Volunteer Battalion), and Quarter-Master and Hon. Lieutenant A. Smith. The Distinguished Conduct Medal was awarded to Colour-Sergeant B. Godbolt, Sergeants E. Ager, G. Ford, A. Wheaton, G. Claridge, Corporal A. Fuller, and Privates C. Childs, J. H. Darley, W. Hall, A. Oliver, and G. Risby. All the above non-commissioned officers and men were mentioned in despatches, also the following :—Colour-Sergeants J. Handscomb, H. A. Loader, Sergeant W. Blackwell, Corporals N. Bollingbrooke, C. Sharpe, and Private P. O'Connor.

The following officers, not already mentioned, were attached to, and served with the 1st Battalion from 1899 to 1902 :—Major W. F. Plummer, 5th Battalion, Royal Munster Fusiliers (Militia) ; Lieutenant P. Hudson, 3rd Volunteer Battalion, Suffolk Regiment, Lieutenant G. L. Mason, 4th Volunteer Battalion Suffolk Regiment, Lieutenant H. S. Marriott, 2nd Volunteer Battalion Suffolk Regiment, and Lieutenant C. A. R. Stowes, 3rd Battalion Suffolk Regiment (Militia), Lieutenants J. C. Bousall, 3rd Battalion, and Lieutenants Ashby and Black, 4th Battalion Suffolk Regiment, and Lieutenant F. S. Mason, 3rd Royal Scots Fusiliers (Militia), and Civil Surgeons J. Acland, W. S. Waring, and J. M. Caw.

The following medals were given for the South African War :—

The Queen's South African for 1899–1900, which carried with it the following clasps, viz :—"Transvaal," "Cape Colony," and "Orange Free State" ; (2) The King's Medal, 1901–02.

The honour to be borne on the Colours was "South Africa, 1899–1902."

SERVICES OF THE MOUNTED INFANTRY OF THE 1ST BATTALION DURING THE WAR.

A section under Lieutenant H. D'A. Smith, consisting of 1 sergeant and 35 of other ranks, mobilised as part of the Eastern Counties Company of the 1st Mounted Infantry, and sailed from England on the 11th October, 1899, taking part in the following engagements :—

Zoutpans Drift, 13th December, 1899
Sunnyside, 1st January, 1900
Relief of Kimberley, 14th February, 1900
Paardeburg, 18th to 28th February, 1900
Osfontein, Poplar Grove, Abraham's Kraal, Driefontein, 11th March, 1900
Bloemfontein, 13th March, 1900
Sanna's Post, 31st March, 1900
Relief of Wepener, 19th to 29th April, 1900
Bramfort, 3rd May, 1900
Vet River, Zand River, Kroonstadt, Johannesburg, 28th and 29th May, 1900
Pretoria, 4th June, 1900
Diamond Hill, 11th and 12th June, 1900
Bronkhorst Spruit and Middelburg, and other skirmishes to Kaapsche Hoop.

At the end of 1900 returned to Bethulie, and from that time to the end of the war, this section was perpetually engaged in "drives," which took place in Cape Colony, and in the east of the Orange River Colony.

On the 15th February, a Mounted Infantry Company was formed at De Aar, under Lieutenants G. A. L. Thomson, and G. P. Newstead, 1st Battalion, which moved on the 25th to Orange River Colony, where the command was taken over by Captain W. B. Hickie, Royal Fusiliers. The following is a diary of its movements.

From 10th March, to 24th April, employed hunting rebels in the north-west of Cape Colony, returning on the latter date to Orange River Station, and being joined, on the 4th May, by Lieutenant A. S. Peebles.

On May 14th, left for Bloemfontein, and, on the 21st, arrived at Kroonstadt, when the command fell to Lieutenant Peebles, on Captain Hickie taking command of the 8th Mounted Infantry. On the 22nd, started with the general advance under Lord Roberts, and, passing through Elandsfontein on the 29th, encamped overlooking the Johannesburg waterworks.

Johannesburg surrendered on the 31st, and, on the 2nd June, Lieutenant F. A. White joined the company.

When nearing Pretoria on the 4th June, the company came under heavy rifle and pompom fire, when Sergeant Housden was mortally, and several men slightly, wounded.

On surrender of Pretoria (5th) made straight for the race-course, being the first to arrive, to liberate some 80 to 100 prisoners who were in hospital, and, in the evening, marched back to Irene. On the 9th, entrained to Klip River, and from the 24th to 26th July, having returned to Pretoria, joined in an abortive march to Crocodile River and back, under Colonel Hickman.

On the 28th, entrained to Wolvehoek, and thence marched on August 5th to Parys, returning on August 21st to Kroonstadt.

On the 26th, marched from Kroonstadt, under Colonel Le Gallais, and started hunting De Wet. Heavy transport was done away with, and a system of Cape carts inaugurated, 1 cart for every 8 men and horses, to carry blankets, rations, &c., for four days. Arriving at Bloemfontein on the 2nd September, reached Ladybrand on the 5th.

On the 17th, joined the 21st Brigade, three miles north of Winburg; from here the 5th and 8th Mounted Infantry were the advanced guard, on the march to Ventersburg, and encountered a body of the enemy in force, with a large convoy at Rietkuil. Sergeant Burns was wounded, and the bivouac this night was without food or blankets.

The remainder of September and the whole of October were occupied in marching about in bad weather with no appreciable results.

Leaving Great Kom on the 5th November, the column came across fresh wheel-marks, showing that De Wet could only be a few miles in front. At 3 P.M., a reconnoitring party, of which the 8th Mounted Infantry formed a part, was sent out to follow the wheel-marks; about 1½ miles north of the river, the enemy opened fire with three guns, but gradually retired south of the river at Bothaville.

Colonel Ross, having with him the Suffolk and West Riding companies of Mounted Infantry, decided to cross it, about 500 yards east of the Bothaville, and take up a position on a kopje half a mile to the south. This was held by Maxim and rifle fire against three guns and a pompom until dark, when a retirement was made to the north bank of the river; but, at midnight, the 8th Mounted Infantry recrossed the river by Bothaville Drift and bivouacked on the south bank.

Advancing again at 4.30 A.M. on the 6th November, Major Lean, 1st Mounted Infantry, within two miles of the river, surprised and captured a Boer outpost, without any alarm being given, when it was found that the Boer laager was within 300 yards of this post. Three guns of V Battery came into action within 400 yards of the laager, and with 40 men of the 5th Mounted Infantry, held the front; the right flank was held by the 13th and 17th Companies, Imperial Yeomanry, while on the left was the 8th Mounted Infantry, 105 strong, with one gun, U Battery. Large numbers of mounted Boers made several attempts to break through the left flank, and get to the baggage.[1]

Early in the fight, both Colonels Le Gallais and Ross were dangerously wounded. At 9 A.M., the Suffolk Company commenced a wide turning movement on the left flank, in order to advance as close to the laager as possible, and got within 150 yards of it. Here, Lieutenant Peebles and several men were badly wounded, while, for half an hour, the shells from two guns of

[1] In a letter from Lieutenant Brooke, published in *The Oxfordshire Light Infantry in South Africa*, he wrote—"Here, on the left flank, we had a desperately hot fight. . . . . Two hundred of them got within 70 yards of one of our guns, and would have captured it, but for a magnificent man in the Suffolk Mounted Infantry who was escorting the gun with only six men. He held his ground, gave the order to fix bayonets, and then, looking round, saw a Maxim strapped on the back of a mule. He got up, calmly walked back, and brought the Maxim into action, driving off the Boers at once. Major Taylor, in his report, specially mentioned the Suffolk Mounted Infantry."

V Battery were bursting close to the company, the shrapnel producing the effect of a hailstorm.

At about 10.45 A.M., the enemy put up the white flag. Our captures were 114 Boers, of whom 17 were wounded:

One 15-pounder gun, captured from us at Colenso.

One 12-pounder gun, ,, ,, ,, Sanna's Post.

Three Krupp, 75 mm.; one pompom; one 37 mm. quickfirer, and 13 waggons of ammunition and supplies.

Lieutenant Peebles went into hospital at Kroonstadt, and was unable to rejoin the company again during the war.

On the 9th November, the company arrived at Kroonstadt, and, on the 19th, 3 sections, under Lieutenant F. A. White, moved to Honnig Spruit, and 1, under Lieutenant Thomson, to Roaderal, from which posts the company was employed on constant patrolling day and night.

From the 20th to the 29th December, the company having concentrated, formed part of a column under Colonel Hickie, operating in the neighbourhood of the railway.

From the 1st to 8th January, 1901, De Wet being in the neighbourhood, the Suffolk company, which now formed part of the 15th Mounted Infantry, was employed on constant patrol. Lieutenant White was appointed staff officer to Colonel Hickie, and Lieutenant Thomson, Adjutant to 15th Mounted Infantry. In May, the company, with the 15th Mounted Infantry, operating as part of a small column, under Pinecoffin, scoured the country, both east and west of the railway line.

From June to December, it was engaged in drives and rounding up operations.

On the 7th February, 1902, it left Bethlehem, to join in drives by columns towards the main line, Vereeniging to Kroonstadt, and caught up driving columns, ten miles from the main line, having marched 105 miles in 60 hours. The drive resulted in the capture of about 280 prisoners.

On the 12th February, starting six miles south of Kroonstadt, the company joined in a drive east of Lindley-Bethlehem blockhouse line.

February and March were occupied in perpetual driving.

On the 21st April, De Wet came into Frankfort under flag of truce, on his way to a conference at Reitz.

On May 1st, proceeded from Frankfort to Bezedenhoust Drift and back (90 miles) in 24 hours, capturing Commandant Manee Botha and staff.

On the 1st June, the telegram proclaiming peace was read out, and, on the 24th, all reservists from the company were sent back to the 1st Battalion. Orders were received on the 6th August for the 15th Mounted Infantry to be broken up, the Suffolk company leaving on the 14th, to join the 1st Battalion at Potchefstroom.

MEMORIAL, *at* IPSWICH, *to the County of Suffolk casualties of the South African War*, 1899-1902.

[*Elliott & Fry.*

LIEUT.-COLONEL A. J. WATSON.
Killed at Colesberg, South Africa, 6th January, 1900.

ERECTED BY THE COUNTY OF SUFFOLK IN MEMORY OF MEN OF SUFFOLK KILLED DURING THE SOUTH AFRICAN WAR, 1899-1902.
[*In the Market Place, Bury St. Edmunds.*]

CHAPTER XIII

England, East Indies, Malta, Egypt, England
1903–1913

1903

A communication was received on the 21st January from the General Officer Commanding Bombay Army, on the inspection of the 2nd Battn. in India, by Brigadier-General J. H. Craigie, in the previous year, which stated:—"A very satisfactory report; the battalion is well trained in every way, and is efficient and fit for service."

On the 15th March, the presentation took place, at Colchester, of King's South African Medals to the 1st Battn. by Major-General Sir Wm. Gatacre, K.C.B., commanding 10th Division, 4th Army Corps.

Captain H. E. Olivey and Lieutenant E. W. Bell, Suffolk Regiment, were killed in action at Gumburres, in Somaliland, on the 17th April, and, on the 17th June, Captain H. B. Rowlands of the regiment died at Bohotta, from wounds received in action at Daratolah, Somaliland.

On the 3rd July, Field Marshal the Earl Roberts, Commander-in-Chief, presented medals for distinguished conduct to five non-commissioned officers and men of the 1st Battn., at Colchester.

On the 5th, the battalion proceeded to London, and encamped at Regent's Park, on the occasion of the visit of the President of the French Republic to His Majesty the King.

The "London Gazette," dated, 7th August, announced that the King had been graciously pleased to confer the Victoria Cross on Captain W. G. Walker, Indian Army (now Colonel, 4th Gurkhas), formerly of the Suffolk Regiment, for conspicuous bravery in Somaliland, in conjunction with Captain Rolland, as stated against their names:—

> "During the return of Major Gough's column to Danop, on the 22nd April last, after the action at Daratolah, the rear-guard got considerably in rear of the column, owing to the thick bush, and to having to hold their ground while wounded men were being placed on camels.

At this time, Captain Bruce was shot through the body, from a distance of about 20 yards, and fell on the path, unable to move. Captains Walker and Rolland, two men of the King's African Rifles, one Sikh, and one Somali of the Camel Corps, were with him when he fell.

In the meantime, the column, being unaware of what had happened, were getting further away. Captain Rolland then ran back some 500 yards, and returned with assistance, to bring off Captain Bruce, while Captain Walker and the men remained with that officer, endeavouring to keep off the enemy, who were all around in the thick bush. This they succeeded in doing, though not before Captain Bruce was hit a second time, and the Sikh wounded. But for the gallant conduct of these officers and men, Captain Bruce must have fallen into the hands of the enemy." (See Plate 33.)

On the 24th September, the ceremony took place at St. Mary's Church, Bury St. Edmunds, by Major-General Sir W. Gatacre, K.C.B., of unveiling a memorial, erected by the regiment to their comrades who died in South Africa, 1899–1902.

Some most important changes took place this year in the administration of the Army.

On the 7th November, the War Office appointed a Reconstitution Committee to assemble for the reorganisation of the Army, presided over by Viscount Esher, with Admiral Sir John Fisher and Colonel Sir G. S. Clarke, members, when it was ordained that :—

The post of Commander-in-Chief should be abolished, and in lieu there should be appointed a council of senior general officers, specially selected, to be termed "The Army Council," in conjunction with the Secretary of State for War, and with the abolition of the Commander-in-Chief, a new office, viz., that of Inspector-General of the Forces, was created.

With a view to the introduction of entirely new ideas, officers holding the principal staff appointments, such as Adjutant-General, Quartermaster-General, Director-General of Mobilization, &c., at the War Office, were directed to vacate them.

Colonel Mackenzie, C.B., 10 officers, and 750 of other ranks proceeded to London, on the 19th November, to take part in the celebration in honour of the visit of the King and Queen of Italy.

1904

On the 7th January, Major Frank Graham was promoted Lieut.-Colonel 1st Battn., in succession to Colonel G. F. Mackenzie, C.B., retired on half pay.

On the 10th, Lieut.-General the Hon. B. M. Ward was appointed Colonel of the regiment in succession to General the Hon. Sir P. R. B. Fielding, K.C.B., deceased. The 2nd Battn. proceeded from Karachi and Hyderabad to a camp of exercise at Jungshahi on the 16th January, and remained until the

6th February, brigaded with the 129th and 130th Beluchis, and 2 field batteries, R.A.

Field-Marshal the Earl Roberts ceased to be Commander-in-Chief from the 11th February, the Army Council being installed from the following day.

The Finance Department of the army was also entirely reconstructed on a civil footing, with a view to abolishing the Army Pay Department, which, however, was reinstalled in 1909.

The Committee, moreover, were in favour of the restoration of the old regimental numbers to battalions as their first titles, when unlinked for purposes of the mutual supply of drafts, by which means "*the connection between the regiments and the finest pages in the history of the British Army would be re-established, and at the same time the great convenience of the numbers —in war time especially—would be regained.*"

This recommendation, however, was not acted on, though the regimental numbers have since been retained in the Army List.

The report of Colonel Cleary Hill, Inspector of Gymnasia in India, promulgated on the 15th March, commented on the 2nd Battn. as being in a very sound physical condition. "The companies at physical drill and free gymnastics were smart, healthy, active and strong. The young soldiers in the Gymnasium were well taught; the instructors very efficient."

On the 12th April, a captain's guard of honour, under Captain F. Wilson, 1st Battn. proceeded to Bury St. Edmunds, on the occasion of the visit of Her Royal Highness Princess Christian to open the Suffolk Regimental Cottage Homes, which had been erected by voluntary subscriptions, to accommodate disabled soldiers of the regiment.

The 1st Battn. moved by rail, on the 30th June, to Oxney Farm, near Bordon, Hants, for Brigade and Divisional training, returning, after the 1st Army Corps manœuvres on the Thames, to Colchester, on the 13th August, and took part in the Essex manœuvres between the 7th and 15th September, moving, on the 28th, from Colchester to the Cambridge Barracks, Woolwich.

The establishment of all ranks of the 2nd Battn. on the 1st October was 1203, with no 3 years' men, and, for the past two and a half years, the battalion had approximated a total establishment of 1200 of all ranks.

On the 19th October, the 1st Battn. lined the streets of Woolwich, on the occasion of the visit of His Majesty King Edward VII to the Royal Artillery.

A captain's guard of honour, with band and Colour, under Captain A. S. Peebles, D.S.O., proceeded to Bury St. Edmunds on the 10th November, on the occasion of the unveiling of the South African War Memorial to fallen Suffolk soldiers. (See Plate 34.)

On the 17th November, the 1st Battn. took part in lining the streets of London, on the occasion of the King of Portugal's visit to the city.

At the Bengal Presidency Rifle Association meeting, held at Meerut during November, Major W. B. Wallace, 2nd Battn., took first place and a silver medal, in the championship aggregate, winning the gold jewel, and first place for the field officers' cup ; also three revolver match prizes.

1905

During the week February 13th to 18th the 2nd Battn. was inspected by Major-General Smith-Dorrien, C.B., commanding 4th Quetta Division, under the scheme of Infantry tests, originated by Lord Kitchener, Commander-in-Chief in India. The battalion, at this time, was in a very hard and efficient condition, and unusually strong in officers, especially in company commanders and senior subalterns, and, during the six days' test, it was under arms on an average over twelve hours a day. On completion of the test Lieut.-Colonel Cubitt desired to place on record an acknowledgment of the very excellent, hard, and serviceable work, so cheerfully performed by all ranks, in executing the tests of the India Army Order, which gained the warm commendation of Major-General Smith-Dorrien, who carried out the inspection, adding :—" It must be very gratifying to everyone in the battalion to know that this very searching inspection has resulted in the battalion taking first place in the Quetta Division, from a military efficiency point of view, irrespective of health considerations."

The Major-General's Inspection Report commented most favourably on the " thoroughly good tone of this fine battalion, the good physique of the men, and the Colonel and officers were much to be congratulated on its very high state of efficiency."

On March 20th, the 2nd Battn., on relief by the 1st South Wales Borderers, marched from Karachi to Keamari, and embarked in the Indian Marine Troopship " Dufferin " for passage to Madras. The strength of the battalion on board was 25 officers and 1088 of other ranks.

The " Dufferin " anchored in Madras Harbour on the 27th March, and, on the two following days, the baggage was disembarked in boats. On the 28th, a half battalion entrained for Bellary, and, next day, the half battalion with the headquarters marched into Fort St. George, Madras. The battalion also furnished a detachment of 1 officer and 53 men to the sanatorium of Poonamallee.

On the 1st June, the 1st Battn., 460 strong, moved by rail from Woolwich to the Worthing Manœuvre Area for brigade and divisional training, returning on September 1st. The battalion took part in lining the streets of London on the 15th November, on the occasion of the visit of His Majesty the King of the Hellenes.

1906

During the visit of their Royal Highnesses the Prince and Princess of Wales to Madras from the 24th to 29th January, the 2nd Battn. furnished several guards of honour, and lined the streets on two occasions, and the officers were presented to His Royal Highness at a levée.

During January, the Adjutant-General for Musketry in India reported "the 2nd Battn. Suffolk Regiment well maintaining its reputation as a very well-trained good shooting regiment;" and in the Annual Signalling Report for 1905–06, on the infantry of the Southern Circle, the battalion, of five units inspected, was "distinguished."

On the 24th February, Major V. Graham was promoted Lieut.-Colonel, to command the 2nd Battalion, in succession to Colonel Cubitt, retired, on completion of four years.

Lieut.-General Sir C. Egerton inspected the 2nd Battn. on the 27th

Dungeons in which British Officers and Soldiers were chained up at Seringapatam.

April, under command of Major E. Montagu, and reported on it, as "in a high state of efficiency, and fit for active service."

In June, the Seringapatam gun, which was most kindly presented to the regiment by the Durbar of Mysore (as a trophy of the assault of the 4th May, 1799), arrived at Madras.

The acquisition of this trophy is due to Captain G. H. Walford. In visiting, in December 1905, the scene of the assault, and seeing a gun which had become buried muzzle downward in the dungeons where British officers and men used to be chained up as captives, he coveted this gun as a memento for the regiment, and suggested the idea in a letter to the Commanding Officer, Colonel Cubitt, who wrote to the Resident of Mysore, asking whether the gun referred to above, or one similar to it, could be given to the Regiment.

The Mysore Durbar, at the suggestion of the regiment, also kindly promised to undertake some small alterations to the memorial of the 12th Regiment, at their own expense.

On the 24th July, the 1st Battn., 720 strong, left Woolwich and proceeded by march route, via Gravesend and Tonbridge, to Brighton, for battalion, brigade, and divisional training, encamping at Brighton and Newhaven, and returned to Woolwich by the same route on the 8th September.

The 2nd Battn. was relieved at Fort St. George, Madras, by the 2nd Cheshire Regiment, and sailed from Madras on the 1st December, in the Steamship "Hardinge" for conveyance to Bombay, where 3 companies were disembarked for Deolali.

The headquarters and 5 companies transshipped to the troopship "Assaye," leaving Bombay, 7th December, and reached Aden on the 12th, relieving the 2nd King's Own Scottish Borderers. The headquarters and 3 companies were stationed at the Crater, and 2 companies at Steamer Point, five miles away. The strength of the 2nd Battn. on leaving Madras was 21 officers and 876 of other ranks, and before its departure, a brass tablet, inscribed as below, was fixed up in the Main Guard Room to record the two occasions on which the regiment had been quartered at Fort St. George, Madras.

"12TH REGIMENT
1797–1799,
2ND BATTALION SUFFOLK REGIMENT
1905–1906."

1907

Major-General De Brath, commanding Aden Brigade, inspected the 2nd Battalion on the 19th, 20th, and 21st January, and reported on it as "in a high state of efficiency, and quite fit for active service."

As the result of the signalling inspection for the year 1906–07, the battalion continued to be shown as "distinguished." The Commander-in-Chief's remarks on the inspection of the battalion for 1906–07 were "very satisfactory."

At Bury St. Edmunds, on Sunday, St. Patrick's Day, was witnessed a ceremony which called to remembrance one of the most glorious, though melancholy, episodes in British military history—being the unveiling, by the Marquis of Bristol, Lord-Lieutenant of Suffolk, and Honorary Colonel 3rd Battalion Suffolk Regiment, of a memorial erected by the officers past and present, non-commissioned officers, and men of the Suffolk Regiment to the memory of 55 men of the 12th (East Suffolk) Regiment who perished in the wreck of the "Birkenhead." (See Chapter IX.)

The unveiling ceremony was performed at the morning service, which was of a military character, and appropriate to the occasion. Previous to the service a muffled peal was rung upon the bells in the church tower. The Marquis of Bristol was accompanied by the Marchioness, and the Mayor and Corporation attended the service in state.

Among the officers present were Brigadier-General J. H. Campbell,

MURAL TABLET ERECTED TO THE MEMORY OF MEN OF THE 12TH REGIMENT WHO WERE DROWNED IN THE WRECK OF H.M.S. "BIRKENHEAD."

Unveiled in St. Mary's Church, Bury St. Edmunds, by the Marquis of Bristol, 17th March, 1907.

commanding the Grouped Regimental Districts of the Eastern Counties, and a great number of regimental officers.

There were also present the officers and men of the Depot, Suffolk Regiment, and the 3rd Battalion; also those of the 1st and 2nd Volunteer Battalions, the King Edward's School Cadet Corps, and a detachment of the Suffolk Imperial Yeomanry; the officers and men of the Bury St. Edmunds and the Westgate Fire Brigades, and the officers and lads of the Church Lads' Brigade.

A very interesting circumstance in connection with the ceremony was the attendance of two survivors of the "Birkenhead" disaster, ex-Corporal William Smith, of Banbury, Oxford (late of the Suffolk Regiment), and Private John Smith, of St. Ives, Hunts (late of the 2nd Queen's Royal Rifles).

As the Mayor and Corporation entered the Church, the National Anthem was played by the band of the 3rd Battalion, and the service commenced with the hymn "Onward Christian Soldiers," as the clergy and choir proceeded from the vestry to the chancel. The service, which was fully choral, was taken by the Rev. F. L. Astbury, and the preacher was the Vicar.

After a most impressive sermon, the hymn known as "St. Patrick's Breastplate," commencing :—

"I bind unto myself to-day,
The strong Name of the Trinity"

—was sung in observance of St. Patrick's Day, which had been selected for the unveiling of the memorial, because of the 55 men whose names are inscribed thereon, about 20 came from the North of Ireland.

Then came the ceremony of unveiling the memorial, which has been erected at the north-west end of the church, opposite to the memorial to officers, non-commissioned officers, and men of the Suffolk Regiment who fell in South Africa. The memorial was veiled by a large Union Jack. The Marquis of Bristol, who wore his uniform as Lord-Lieutenant, proceeded to the west end of the church, being accompanied by the clergy, the officers, and the two veteran survivors. His Lordship, after a speech most suitable to the occasion, pulled the cord attached to the Union Jack, and as this fell, revealing the tablet, he added :—"I unveil this memorial to the soldiers who were drowned in the 'Birkenhead,' in the name of God the Father, God the Son, and God the Holy Ghost. Amen."

The tablet, which is of handsome Sicilian marble, has, within the arch of the moulding, a representation of a foundering ship, and upon the lower moulding are the Regimental Arms, with the well-known inscription, "*Montis Insignia Calpe.*" (See Plate 35.)

On the 9th August, the 1st Battn., 770 strong, left Woolwich by rail for Ringwood, Hants, for brigade and divisional training, moving, on the 2nd September, to Salisbury Plain for command manœuvres, and returning to Woolwich, by rail, from Amesbury, on the 8th.

On November 20th, the 1st Battn., mustering 24 officers and 889 of other ranks, under Colonel F. Graham, left Woolwich for Southampton, and embarked on the s.s. "Dongola" for Malta, which was reached on the 27th, and, disembarking the same day, occupied Floriana Barracks and Fort Manoel, the headquarters and 4 companies to the former, and 4 companies to the latter.

The 2nd Battn., before leaving Aden, was re-armed with the Short Lee-Enfield rifle.

The battalion embarked on the troopship "Assaye," 3 companies at Bombay on the 5th December, and headquarters and 5 companies on the 11th, and, on landing at Southampton on the 27th, proceeded at once to Parkhurst, Isle of Wight, mustering 21 officers and 829 of other ranks.

On the departure of the 2nd Battn. Suffolk Regiment from Aden, Major-General De Brath, in a Brigade Order "desired to place on record his appreciation of the good discipline, and high state of efficiency of the battalion, which reflects great credit on Colonel V. Graham, and all ranks."

1908

Major E. Montagu was promoted Lieut.-Colonel, to command the 1st Battn. on the 7th January, in succession to Colonel F. Graham, retired on completion of the appointment.

In February, an Old Comrades Association was started in the regiment, with the object of promoting *esprit de corps*, mutual aid, and to benefit deserving old soldiers of the Suffolk Regiment who might be in needy circumstances, and who were subscribers while serving. The objects of the association were explained at the first regimental meeting that was held, which was largely attended by old soldiers, and it was shown that the association had no connection with the Regimental Cottage Homes. A list of rules was drawn up, and the officers of the Association were to consist of a president, ten vice-presidents, a treasurer, and two secretaries, who were appointed, with Colonel C. R. Townley as President.

Out of a total of 933 in the 1st Battn. in April this year, 893 had joined the Association.

The Territorial Force came into being on the 1st April, 1908. The existing Yeomanry and Volunteers were given until the 30th June to transfer to the new force.

In the calendar year 1908, 48,706 men transferred also from the Militia to what was termed the Special Reserve, it being arranged that Special Reservists of the Artillery, Engineers, and Infantry were to be drilled on enlistment for six months.

In the infantry the volunteer regiments now became, and were shown in the Army List as, extra battalions of the Territorial line regiments to which they were affiliated, the Militia only having hitherto been shown as extra battalions since the introduction of the Territorial system of nomenclature of line regiments in July, 1881. Another innovation was that the names of all line officers serving at regimental depots were now shown as in the Special Reserve.

In approving of the abolition of the Militia, King Edward VII, on the 21st February, thus thanked them for their services :—

> " I take the opportunity of expressing to the Force my keen appreciation of its services in the past. In peace and in war the Militia has never been asked in vain to make sacrifices for the good of the country."

It was also decided that the Territorial Force may carry Colours on the same lines as in the regular army, whilst a service decoration was approved for the officers and a medal for the men.

On the 30th April, the 1st Battn. proceeded on combined naval and military manœuvres in the neighbourhood of Mellieha, embarking with the 1st Line Transport on H.M.S. " Goliath," at the Grand Harbour, and disembarking in boats at Mellieha Bay, returning to barracks by march route on May 1st.

On the 19th, the battalion proceeded to Ghain Tufficha, for battalion training, returning to Floriana and Manoel on the 31st.

The 2nd Battn. went under canvas at Bulford Camp, Salisbury, on the 8th July, took part in the manœuvres and inspection by the Inspector-General of the Forces, and returned to Parkhurst on the 7th September.

The Evelyn Wood and Bowyers Prize Competition was held at Bulford on the 31st August, when 12 battalions competed, and the 2nd Suffolk, with 99 points, won the first prize.

The remarks on the annual inspection of the 2nd Battn. signallers, held at Bulford in August, were :—" A very satisfactory report ; the men appear to be all excellent operators, and to have been well instructed."

In his remarks on the Inspection Report of the 2nd Battn. for 1908, the General Commanding-in-Chief, Southern Command, observed :—" The 2nd Battn. Suffolk Regiment possesses an excellent system, and the material of which it is composed would be very hard to beat."

On the 12th November, the 1st Battn. moved from Floriana and Manoel to the Imtarfa Barracks, Malta.

1909

On the 15th March, the 1st Battn., at Malta, took part in a mobilisation and defence scheme of the island. The " Precautionary Stage " took place on the 15th and 16th, and the " War Stage " from the 17th to 19th, its strength on marching to the latter at 3.40 A.M. on the 17th being 23 officers and 830 of other ranks.

On the occasion of the visit of King Edward VII with the Queen and Dowager Empress of Russia to Malta, on the 14th April, and, during their stay, the 1st Battn. took part in lining the streets, and furnishing guards of honour, &c., and was highly complimented by His Majesty on its smart appearance and good marching at the review, which was held on the following day.

Major C. F. Lennock, 1st Battn., was presented, on the 14th April, with the Honorary Testimonial of the Royal Humane Society, inscribed on vellum, which was awarded for having, on the 11th August, 1908, saved a soldier from drowning at Malta.

At the annual inspection, this year, of the 2nd Battn. by Major-General Franklyn, the Inspecting Officer was pleased to remark on the "high order of its musketry training," the battalion at the same time distinguishing itself in signalling, a team of eight signallers obtaining 100 per cent. with the large flag, and 99·79 with the smaller one, being an average of about 16 words a minute through intermediate stations.

1910

On the 21st January, the following report by the Brigadier Commanding Infantry Brigade, Malta, on the annual musketry return of the 1st Battn., was received :—

> "This battalion carried out its musketry, as it does everything else, with great keenness and *esprit de corps*. The preliminary training of the men is particularly well looked after, and knowledge of theory by all ranks very good. The machine gun teams are on an excellent footing."

To this, General Sir L. Rundle, Commander-in-Chief at Malta, added :—"Most satisfactory ; I have formed the same opinion of this battalion as expressed by the Brigade Commander."

On the 24th February, Major C. H. C. van Straubenzee was promoted Lieut.-Colonel, to command the 2nd Battn. *vice* Colonel V. Graham, retired.

His Majesty King Edward VII died on the 6th May, and was succeeded by His Royal Highness the Prince of Wales, as King George V.

News was received on the 1st June of the death in action on the West Coast of Africa, of Captain Chapman, Suffolk Regiment, attached to the Southern Nigeria Regiment, West African Frontier Force, who, on the 15th May, was killed at Abedi, whilst gallantly leading his men to the attack.

In a complimentary Garrison Order, published at Malta, on the 1st June (with reference to some night operations, carried out on the 24th May), His Excellency the Governor was pleased to record that "the 1st Battn. Suffolk Regiment, which came under his personal observation, was a model of what night marching should be."

On the same date, the 1st Battn. won the Infantry Brigade Bayonet Fighting Competition, for teams of 5 non-commissioned officers and men, and also the Officers' Bayonet Fighting Competition for teams of 5 officers, open to the Infantry Brigade, Malta Command. The battalion machine gun team also took first place in the Infantry Brigade, in a series of tests, held under brigade arrangements, on the 14th and 15th June.

It was notified to the 1st Battn. on the 29th, that in the annual report by the General Commanding-in-Chief, Malta Command, he had commented on it as :—" A first-class battalion in every respect."

On the 8th August, the 2nd Battn. proceeded to Bulford, for battalion and brigade training, until the 5th September, when it left for Sturminster Newton, preparatory to divisional training and manœuvres, at the close of which it returned to Bulford, *en route* to its new station, Aldershot, which was reached on the 29th, when the battalion occupied Barosa Barracks, South Camp.

Referring to the departure of the 2nd Battn. from Newport (I. of W), a local journal observed that :—

> " The inhabitants of Newport, and of the Island generally, will part with the distinguished corps with very great regret. The exemplary conduct of the men has hardly been equalled, and certainly never excelled."

In the competition for the ***Douglas Shield***, held on the 2nd September, 1910, " A " Company of the Battalion distinguished itself by winning the first prize and shield from 13 other competing Battalions, with a score of 318, the 2nd and 3rd being the 1st Somersets, 312, and the 2nd Gloucesters, 305. The three leading places were thus all taken by the 9th Brigade to which the Battalion belonged.

On the 25th October, the 1st Battn. was inspected at Malta by General Sir Ian Hamilton, G.C.B., Commander-in-Chief in the Mediterranean, who expressed himself very pleased with the fine appearance of the men, and their steadiness on parade. The Signalling Report of the battalion this year, by the Brigadier Commanding Infantry Brigade, Malta, was also most favourable, and " the general system of training excellent."

In the annual Inspection Report on the 2nd Battn. for 1910, its sound training and high efficiency in signalling were specially commented on, whilst the combined remarks of three inspecting generals were that the battalion was " well trained in all respects, and well behaved. Marching powers and musketry very satisfactory. A very good battalion, both in the field and in barracks."

1911

On the 16th January, Major-General Egerton, C.B., Commanding Infantry Brigade, made his farewell inspection of the 1st Battn. at Malta, when, remarking on the regret of every one in the Colony at their departure, he expressed his belief that they were " a pattern corps to this or any other army in the world," and, referring to their sporting proclivities, commented on the remarkable part they had taken in all the Service competitions and athletic contests held at Malta.

After church parade, on the 22nd, His Excellency the Governor, Sir L. Rundle, K.C.B., in bidding farewell, spoke in equally high praise of the battalion, expressing confidence in their maintaining, in peace or war, the glorious traditions which they had so worthily upheld in their present brigade.

The 1st Battn. embarked at Malta, on the 25th January, strength 23 officers and 938 of other ranks, in the troopship "Dongola," for Alexandria, where they arrived on the 28th, in relief of the 1st Battn. Welsh Regiment.

On the 27th February, four Marindin range-finders were issued to the 2nd Battn., and its establishment was later increased by one subaltern for the machine gun.

In the Bronze Medal Tournament, the Officers' Fighting Competition was won by the team of the 2nd Battn., consisting of Captain Orford, Lieutenants Cutbill, Stubbs, and Attree, and Second Lieutenant Chalmers.

On the 21st June, 20 officers and 450 of other ranks of the 2nd Battn. proceeded to London, and remained in camp at Battersea Park till the 23rd, for duty in connection with the Coronation of His Majesty, King George V, and, on the 27th, 20 officers and 500 of other ranks proceeded to London for duty, on the occasion of the Royal Drive to the City and Borough, &c.

In Regimental Orders, dated 22nd July, of the 1st Battn., Colonel E. Montagu had the satisfaction of announcing that a year had elapsed since a court-martial had taken place in the battalion, and congratulated all ranks on the state of discipline and conduct, which had rendered it possible.

At the Aldershot Command Rifle Meeting, held on the 14th August and following days, the 2nd Battn. Suffolk Regiment were the second highest winners, with money prizes of the total value of £120 16*s.* 6*d.* (For Cups won, see Appendix V.)

From the 17th to the 21st August the 2nd Battn. were encamped at Victoria Park, Hackney, for duty in connection with what threatened to be a serious railway strike. Their stay, however, happily ended without incident.

The battalion proceeded, on the 9th September, to London, for guard duty at the Royal Palaces, &c., and was quartered at Chelsea Barracks until returning to Aldershot on the 22nd. During its stay, its smart and soldier-like appearance was a subject of general remark.

On the 25th, on the invitation of the County Council of Suffolk, it had been intended that the 2nd Battn. should, on conclusion of the manœuvres in East Anglia, visit Bury St. Edmunds and remain until the 30th September. Owing, however, to the cancelling of the manœuvres, the visit did not take place, but subsequently the battalion was presented by the Ladies of Suffolk with a pair of very handsomely embroidered Colour Belts, also one large and four small Camp Colours, in yellow silk, with the regimental crest and motto embroidered in crimson. (See Chapter XIV.)

The combined remarks of three Inspecting Generals in the Inspection Report of the 2nd Battn., for 1911, were :—

> "Signalling results good, and it is the best battalion in the Division at musketry. A very good battalion, well commanded, with a good reliable lot of officers, and excellent men ; likely to be very reliable in war, on account of their quality, being chiefly off the land, and of the farm labourer class. There is a good tone running through the unit, and it is quite satisfactory at manœuvres."

Some of the concluding remarks on this very creditable report, by Lieut.-General Sir H. Smith-Dorrien, Commanding-in-Chief, were :—

> "The discipline, interior economy, stamp of men, and tone of all ranks are excellent. Rather too large a proportion of old soldiers, which reacts on the influx of recruits, and size of the reserve. The battalion is in every respect fit for active service. I congratulate all ranks on having the highest figure of merit in the command."

The 2nd Battn. on the 18th December, was re-armed with the S.M.L.E. Mark III Rifle.

1912

On the 7th January, Major W. B. Wallace was promoted Lieut.-Colonel, to command the 1st Battn. in succession to Lieut.-Colonel E. Montagu, to half pay, on completion of term.

On the 23rd, the 1st Battn. (strength, 28 officers and 844 of other ranks), arrived at Cairo from Alexandria, and took over the Citadel Barracks, in relief of the 4th Battn. Rifle Brigade ; and furnished a detachment of 2 officers and 94 men to Cyprus.

At 8.30 P.M. on the 27th March, instructions were received for the 2nd Battn. to entrain at once, in connection with the great colliers' strike which was going on. The battalion was at the Troop Siding, and ready to move off at 10.15 P.M. The train left at 11.8 P.M., and proceeded to Chirk, where, on arrival, the battalion was encamped at Brynkinall Park. Fortunately its services were not called into active requisition, its relations with the miners, who commenced work shortly after its arrival, having been of a most cordial nature. The battalion remained at Chirk until the 10th April, when it moved at short notice to Wigan, and was accommodated in the drill hall. All being quiet at this station also, the battalion returned to Aldershot on the 17th April. On the 10th April, a team, composed of Captains Orford and Nicholson, Lieutenants Stubbs, Attree and Oakes, won the Officers' Team Bayonet Fighting Competition in the Bronze Medal Tournament, for the second year in succession, whilst the men's team took second place.

As the result of the inspection of the 1st Battn. at Cairo, on the 11th April, by Major-General Maxwell, C.B., the Inspecting Officer's remarks were :—

> " A highly trained efficient unit, and ready to go anywhere ; drill, manœuvres, and fire action, satisfactory. The officers and non-commissioned officers are a very good lot. The unit maintains the high reputation it brought from Malta."

In the report on the 2nd Battn. at Aldershot, for 1912, the Major-General Commanding 2nd Division, endorsed, in every respect, the very high praise bestowed by the Brigadier-General, adding :—" Lieut.-Colonel van Straubenzee may well be proud of his battalion." The General Officer Commanding-in-Chief, Aldershot Command, commenting on the very excellent report, attributed the high standard of the battalion as mainly due to the self-dependence of the company officers, so thoroughly inculcated by Lieut.-Colonel van Straubenzee, who " is to be congratulated on the success of this system of decentralisation."

On the 8th September, the 2nd Battalion entrained for Swaffham, to take part in the army manœuvres, at the conclusion of which it marched, on the 20th, to Bury St. Edmunds, to be entertained by the County of Suffolk, a visit which was to have been paid in 1911, when cancelling of the manœuvres led to its postponement for a year. On Saturday the 21st, the battalion reached the historic old town, which was quite *en fête,* and a great concourse of people assembled to accord the regiment a most enthusiastic welcome.

As they marched to the position they were to take up, their fine soldier-like bearing and athletic build were frequently commented on and admired, as was also the precision of their movements. At 3 P.M., the battalion, mustering over 600, was drawn up in square in front of the Abbey Gate, when the Lord-Lieutenant (Sir Courtenay Warner, Bart., M.P.) delivered his address of welcome, in which he paid an eloquent tribute to the great history of the regiment, this being followed by his reading an address, on behalf of the inhabitants of the county, extending to the officers, warrant and non-commissioned officers, and men of the battalion a warm and hearty welcome. It continued :—

> " Our greeting is all the more cordial because this is the first occasion on which a Service Battalion of the Regiment, since its formation in 1685, has visited the county whose name it bears.
>
> It is with much pride that we recall to mind on such an occasion as this, the valour and devotion to duty, even in the face of certain death, which has always characterised the Regiment, and especially would we remember such instances as occurred at the battle of Minden, and at the wreck of the troopship 'Birkenhead,' off the South African coast. We are proud to know that the majority of the Regiment are Suffolk men.
>
> We have every confidence and belief that whatever the danger, the Regiment will, whenever called upon, be ready, as in the past, to do and dare in the cause of King and Country, and so maintain the cherished traditions for bravery which have gathered round the name of 'Suffolks.'"

The handsome illuminated address was in a large oak frame, surmounted by the Royal Arms, and on either side are the names of countries where the regiment has served, and battles in which it has been engaged. These include, besides the historic defence of Gibraltar—Dettingen, Seringapatam, India, Afghanistan, 1878–80, South Africa, 1851–2–3, New Zealand, and South Africa, 1899–1902.

The Mayor (Major Vesey Davoren) then offered the Battalion a welcome on behalf of the city:—

> "He assured them that the inhabitants of that ancient borough, as well as the county, had looked forward to that visit, which they were disappointed in not receiving last year, as was originally arranged. That town was the nursery of their regiment, having been for many years the headquarters of the Depot of the Suffolk Regiment; and he assured them the relations between the civil and military authorities had always been most cordial. They were very proud to have the Depot of the Suffolk Regiment in the borough, and they trusted that the manœuvres the Battalion had recently taken part in would do good to recruiting, not only to the Line, but to the Reserve and Territorial Battalions. The people of Bury, as well as the rest of the county, would do all in their power to make their visit as pleasant as possible, and he hoped officers and men would be able to take away many pleasant recollections of their first visit to that town."

The Lord-Lieutenant then called for three cheers for the Battalion, which were accorded with great enthusiasm.

To this Lieut.-Colonel van Straubenzee made the following reply:—

> "In the name of all ranks which he had the honour to command, he thanked Sir Courtenay Warner, and all present, for the extremely kindly and hospitable invitation they had extended to the Battalion to visit the town and county which bore their name, and to which the vast majority of the men belonged. As Sir Courtenay had rightly observed, most of the men were recruited from Suffolk, and he believed he was stating a fact when he said the Battalion, if not *the* most territorial in the Army, was at least very near being so. That fact doubled the pleasure of their visit that day. The Lord-Lieutenant had alluded to historic episodes in which the regiment had borne a part, and he (Colonel van Straubenzee) might be pardoned if he said that so far as possible in times of peace, the regiment—and he meant all the Battalions of the Regiment—was doing its very best to uphold the traditions that had gathered round its name in the old days. He repeated his thanks for the enthusiastic reception the Battalion had received that day, and said he desired to include the Mayor and Corporation for the part they had played in the reception."

Three resounding cheers were given for the county by the Battalion, on the call of Colonel van Straubenzee, and they were repeated for Sir Courtenay Warner.

Historic Church Parade.

Sunday the 22nd will long be remembered in Bury St. Edmunds, for, amid dense crowds, all units of the Suffolk Regiment assembled for a church parade, and the scene is not likely to be forgotten by those who had the privilege of witnessing it. Glorious weather prevailed. In the barrack field, at nine o'clock, the units were organised, and the spectators secured their initial glimpse of the fine military display, which was to fascinate them during the morning. The soldiers marched to the old church of St. Mary's. Colonel van Straubenzee was in command, and the band of the 2nd Battalion led the way. Lieut.-Colonel Massy Lloyd was with the 3rd Battalion, who had with them their band. Following was Major Royce-Tomkin, in command of a detachment of the Suffolk Yeomanry, Captain Cockburn (Adjutant) was with a detachment, numbering about 75, of the 4th Battalion (Ipswich), and next came Captain Kenneth Greene, in command of the Bury St. Edmund's Companies of the 5th Battalion. Then followed the 6th (cyclist) Battalion, Suffolk Regiment, and there was a good muster of the National Reservists. The scene in the church was made more striking by the Colours of the battalions, which had been borne in the procession, having been placed in the chancel during the service, which was most impressive throughout.

During their stay in Bury, the battalion was most hospitably entertained by the inhabitants, among the principal functions organised for its benefit being regimental sports, a smoking concert at the Corn Exchange to the non-commissioned officers and men, a ball to the officers, and a dinner to the non-commissioned officers and bandsmen. During their stay also, the Colours, Colour Belts, and Camp Colours, which were a present from the county last year, and the whole of the plate and trophies of the Officers' and Sergeants' Messes of the 2nd Battalion were displayed to the public in the Town Hall, and proved a source of unending interest.

The Battalion left Bury St. Edmunds for Aldershot on Thursday, amid scenes of the greatest enthusiasm, the visit having been thoroughly enjoyed by hosts and guests.

All prosperity then to the County of Suffolk, for having set such a noble example to the remaining counties of the United Kingdom in so endearing itself to its regiment, an example which, it is to be feared, will be but tardily followed.

1913

On the 30th January, an intimation was issued from the War Office that His Majesty the King had been graciously pleased to approve of the 3rd (Auckland) Regiment (Countess of Ranfurly's Own), New Zealand, being shown in the War Office Army List as allied to the Suffolk Regiment.

On the 9th April, the Bayonet Fighting Team of the 2nd Battn. won the Bronze Medal Tournament of the Aldershot Command, the officers thus being the winners for the third year in succession.

On the occasion of the visit of His Majesty the King to Aldershot, on the 10th May, the battalion band had the honour of playing during dinner at the Royal Pavilion, and the battalion furnished the guard there on the following day.

In the annual report for 1913, on the 1st Batt., at Cairo, by Major-General the Honourable J. H. Byng, C.B., M.V.O., commanding in Egypt, the battalion received high praise in the following particulars :—

> "Manœuvre with fire direction and discipline very thorough, and intelligently taught ; the battalion's facilities for becoming thoroughly efficient for war being made the most of ; the work in the field of rather an exceptional number of very good young officers had rendered them conspicuous, the N. C. O.'s, and men having been reported on as a fine stamp of country men, very free from crime and disease, and well trained and disciplined. In all respects, in a very satisfactory condition."

On the 13th September, the 2nd Battn. left Aldershot, and marching to Haines Hill, near Twyford, took part during the ensuing two weeks in the Inter-Divisional and Command Manœuvres, and the Army Exercise. At the close of the latter, on the 27th, a twenty-mile march into Rugby was followed by entraining for Ireland, at 12.30 A.M. on the 28th. The battalion arrived at the Curragh at about 3 P.M. same date, and was quartered in the Gough Barracks.

It is worthy of note that the Regimental Journal published monthly (known as "The Suffolk Regimental Gazette,") is said to have been the first established among Line Regiments, having begun its existence under the title of the "East Suffolk Gazette," at Dublin, on the 10th December, 1863, and was continued until May, 1868. From then to 1889, it ceased to exist, but has been published continuously from the latter date up till now. A particular feature in connection with it is that it has always been printed in the regiment.

A Special Army Order, dated 16th September, directed a four-company organisation to be adopted in battalions of the Foot Guards, and in all Regular battalions of the Line, serving at Home, or in the Colonies, with effect from October 1st, 1913.

The establishments of the 1st and 2nd Battalions (Suffolk Regiment) on the 31st December, 1913, were as follows :—

1st Battalion (Colonies) : 29 officers, 2 warrant officers, 46 sergeants, 16 drummers, 40 corporals, 800 privates ; total rank and file 840 ; all ranks, 933.

2nd Battalion (Home) : 25 officers, 2 warrant officers, 39 sergeants, 16 drummers, 40 corporals, 680 privates ; total rank and file, 720 ; all ranks, 802.

On concluding this history on the 31st December, 1913, we find the 1st Battalion under orders to leave Cairo for Khartoum, in the following month, detaching a half company to Cyprus, and a half company to Alexandria ; and the 2nd Battalion quartered at Gough Barracks, The Curragh, Ireland.

CHAPTER XIV

NOTES ON THE UNIFORM, EQUIPMENT, AND THE COLOURS

1685–1701

THE present standing army of Great Britain, which may be said to date from 1661 (in the reign of King Charles II) began, from its formation, to develop steadily in organisation, dress, and equipment.

In July, 1685, after the regiment was raised, each company of infantry (excepting fusiliers and grenadiers) consisted of 14 pikemen, and 46 musketeers; the captains carried pikes; lieutenants, partisans; ensigns, half-pikes, and sergeants, halberts.

The *Pike* was in the form of a spear, the head being flat and pointed, and on a staff from 13 to 18 feet long, shod with a pointed iron foot, and was used by infantry when fighting at close quarters before the introduction of the bayonet.

The *Spontoon* (*esponton*), carried by officers up to 1796, was but another name for a sort of half-pike.

The *Partisan* was another term applied to a halbert or pike, and to a marshal's staff, being more in the form of a baton.

The *Halbert* was carried by sergeants of foot and artillery, a sort of spear, about 5 feet long, made of ash or other wood, its head armed with a steel point, edged on both sides. Sometimes it also assumed the form of a light ornamental kind of battle axe.

The *Bayonet* was not at first a military weapon, but simply a hunting knife, the handle of which could be fitted on the muzzle of the fowling piece, and then used as a spear against an animal at bay. Bayonets or short daggers were used by infantry as early as 1663, but the earliest issue at home was 1673, when every foot soldier carried a bayonet. By the invention of the ring method of attaching, it was possible to fire the musket with the bayonet fixed.

On the 4th March, 1687, a Royal Warrant was issued, forbidding officers and soldiers to carry a dagger or bayonet, at any time, except when on duty under arms.

The first fire-arm in use in our army was the *Matchlock*, a musquet, fired by means of a piece of slow match, and musqueteers used to carry bandoliers.

The *Snaphans*, firelock, flintlock, or fusil gradually superseded the matchlock, the principle of the fusil being the use of a flint and steel, in lieu of the match.

The *Arquebus*, as well as the hand gun, was at first, fired with a match, and does not appear to have been of any particular length or bore. It was a gun with a hook, which was the forked rest on which it was supported.

The *Petronel* was a kind of carbine, a compromise between the arquebus and the pistol.

The armament of a grenadier consisted of fusil with sling, cartridge-box with girdle, grenade pouch, bayonet, hatchet and girdle, and match-box.

The *Hand Grenade* or *Granado*[1] was a small globular shell of iron, from 3 to 4 inches in diameter, filled with powder, and having a touch-hole, into which was inserted a wooden tube, filled with a fuse, compounded of fine powder, tempered with charcoal dust. The grenadier, having quickened the fuse from his lighted match, threw the grenade with the hand, such missiles causing not only wounds, but possibly confusion which might be turned to advantage. His supply of grenades was carried in a brown leather pouch, slung on a shoulder-belt.

The *Hatchet* was common to dragoons and grenadiers for the purpose of clearing the line of march, or removing obstacles from a breach.

The *Match-box* was a thin tube, pierced with small holes to admit the air. Its use was to conceal the burning match when on sentry, or any enterprise at night, and was attached to the belt of the grenade pouch.

According to Cannon, the dress of the regiment, in June, 1686, was :—Hats, broad brimmed, turned up on one side and ornamented with white ribbons, red coats lined with white, blue breeches, blue stockings, and high shoes with square toes, and the pikemen wore white sashes round their waists.

The origin of facings was the coat lining and sleeves turned back.

Sword-belts were supposed to be of buff leather, but, in reality, as buff was an expensive material, belts (except perhaps in such regiments as the Household Cavalry) were of some other and cheaper leather and of a browner colour. The head-pieces in use during the half-century were of two kinds : the basinet or pott, and the skull cap. The pott was a low-crowned helmet with a brim ; it was sometimes bright and sometimes painted black, and, as a distinction for officers, was ornamented with plumes. Both pikemen and horse soldiers wore, besides the pott, a cuirass, or back and breast-piece, the latter, like the pott, being sometimes black and sometimes bright. By officers of pikemen a sash was worn over the right shoulder.

[1] "Granado," the Spanish for pomegranate, owing to its resemblance.

An Indent from Colonel Wharton's (12th) Regiment, on the] Ordnance Department, to complete requirements for the years 1688–89, was as follows :—

Item	Number	
Musquets	273	For the use of the regiment.
Pikes	96	
Halberts	10	
Drums	11	
Bayonets	290	
Pouches	17	For the grenadiers.
Cartridge Boxes	13	
Hatchets	43	

On the 9th November, 1688, the regiment received an order to exchange the matchlock musquets for the snaphans.[1]

1702–14.

In the reign of Queen Anne, the following was the dress of an infantry officer :—Hat, ornamented with feathers, broad brim, two sides of which were turned up, full flowing wig ; square-cut coat and long flapped waistcoat, with large pockets to both ; breeches, tied below the knee, with stockings drawn over up to the middle of the thigh ; shoes ; sword slung over the right shoulder by means of a rich shoulder-belt or baldrick.

Before the advent of the first King George, the crimson silk sash, denoting the rank of commissioned officer, was worn over the right shoulder, and the sword hung in the frog of a leather waist-belt, sometimes placed over and sometimes under the waistcoat. Armour for infantry being now completely thrown aside, the men wore an easy red coat with facings, a cocked hat, breeches with stockings, and a strap below the knee to keep them up.

The Secretary of State issued an order on June 12th, 1706, that every man was to be armed with a musket and bayonet, both of improved patterns. The musket carried a bullet weighing 16 to the pound, which was deemed an advantage over the French musket, made for bullets 24 to the pound.

1725

In March this year, the King decided that all non-commissioned officers and private men were to wear swords.

1729

By a Royal Warrant, dated November 20th (a re-issue of one issued in 1706), the following were the prescribed " Necessaries for a Foot Soldier," viz. :—A good full bodied coat, well lined, which may serve for the waistcoat the second year, a waistcoat, a pair of good kersey breeches, a pair of good strong stockings, a pair of good strong shoes, two good shirts, two good neckcloths, and a good strong hat well laced.

[1] *MSS. Records*, R.U.S.I.

It is certain that very little information is obtainable as regards the peculiarities which distinguished one regiment from another prior to the year 1742.

1742

In the British Museum is a work entitled "A Representation of the Clothing of His Majesty's Household, &c., in 1742," which depicts a private of the regiment at this period. The coat, very full and roomy, is similar in many respects to that worn in 1714–15, except that the skirts are hooked back, showing the colour of the regimental facing, which was yellow.

The edgings of the cuffs, lapels, pocket flaps and red waistcoat are trimmed with a distinctive regimental pattern of white lace, with a yellow line through it. As the choice of the lace was left to the colonels of regiments, who provided the clothing, it was not surprising if the pattern changed whenever a new colonel was appointed. A white neckcloth was worn, and the hat was three-cornered, trimmed with white lace. The breeches were red, and white gaiters were worn high above the knee, fastened with dark-coloured garters. The ammunition pouch was supported by a broad leather belt over the left shoulder. There is, unfortunately, no direct evidence of an officer's uniform at this period. (See Plate 7.)

1743

The "King's Regulations," dated September 14th, directed that the sashes of all infantry officers were to be of crimson silk and worn over the right shoulder, their sword-knots to be of crimson and gold in stripes, and their gorgets to be either of gilt or silver according to whether the laces of uniforms were gold or silver.

Sergeants were to wear, round their waists, worsted sashes, which were usually red, striped with the colours of the facings and linings of their regiments.

1751

On the 1st July, a Royal Warrant assigned regimental numbers to infantry regiments, and directed the uniform of the 12th to be scarlet, faced and lined with yellow.

In Windsor Castle is to be seen an oil painting, one of a series, which represents a grenadier of the regiment, with his musket slung over the right shoulder, the regulation way of standing on parade being with the legs 14 inches apart.

The men of the battalion companies wore three-cornered cocked hats, and in other respects the dress varied little from that of 1742, the coat being just as voluminous, but fastening higher, with the addition of a small pouch in front of waist-belt. (See Plate 7.) The imposing mitre cap, with its picturesque embroidery, was made of cloth, the front being of the same colour as the facing of the regiment, with the King's cipher surmounted by a crown. The flap was red, with the white horse and motto over it "Nec Aspera Terrent," a badge that was universally worn by grenadier companies of all infantry regiments. The tuft at top was yellow and white, the back part red, with turn-up the same colour as the front, and number of regiment in the centre at back of cap. The lace of the grenadier was white, with the yellow line (or worm) still retained. In marching order, a knapsack was carried in the form of a bag made of hairy goat-skin, and worn over the right shoulder. The uniform of the officers was made up in the same manner as that of the men, laced and lapelled with the colour of the facing, the buttons being set in the same manner as on the men's coats, and waistcoats and breeches of the same colour.

The coat could apparently be fastened in three different ways, either buttoned completely across, or partially buttoned, as in the picture at Windsor Castle, or with the lapels buttoned back. The musket of the period was the Brown Bess, and the brass match-box on the shoulder-belt was probably retained as a distinctive badge only, its use for carrying the fuse having ceased in the reign of Queen Anne.

1756

The Inspection Report, this year, shows the uniform of the officers as :—"Red, lapelled, faced and lined with yellow, bound and looped with a mixed binding of white and yellow."

1758

In the Inspection Report, the sergeants are shown as wearing gold laced hats and sashes.

1759

The Wearing of Roses.

All battalions of the Suffolk Regiment, on "Minden Day" (August 1st) wear roses in their head-dresses, and in the event of a parade, the Colours and drums are similarly decked in honour of that memorable victory.

As regards the selection of the rose, the accepted story is that when the regiment was following up the retreating French troops, they passed through a rose garden, and each man plucked a rose, which he fastened in his head-dress. Roses are also worn by the regiment on the Sovereign's birthday, in accordance with long-established custom.

1767

On the 21st September, a warrant was issued for the numbers of regiments to appear on the regimental buttons, which, up to this date, had been quite plain. Pewter buttons were worn by the men, and officers wore gilt or silver, in accordance with their regimental lace being gold or silver.

1768

In the Inspection Report on the 11th May, the officers are shown as wearing gold embroidered button-holes, white waistcoats and plain breeches, epaulets and laced hats, and the grenadiers and drummers wearing fur caps, with yellow pleated fronts.

By a Royal Warrant, dated December 19th this year, the style and cut of uniform underwent a considerable change. The grenadier's cloth mitre cap was replaced by one of bearskin similar in shape, on the front of which was a badge of the King's crest, with the motto "Nec Aspera Terrent," the whole in white metal on a black metal ground. Instead of the white neckerchief, a turned-down collar and black tie were now worn, and the coat was scantier, and cut away to show the figure. The waistcoat was shorter, and was changed from red to white. The officers' coats were lapelled to the waist, with the facing of the regiment, and to have cross pockets, and sleeves with round cuffs. The lapels and cuffs to be same breadth as the men's. The patterns of all the regimental laces were altered, that of the 12th being white with yellow, crimson, and black stripes. The lapels of the officers' coats, three inches wide, were fastened back by gold buttons (pewter for the rank and file) at equal distances, and the loops of lace on the lapels, cuffs and pockets were curved, similar to a pattern adopted by a few other regiments, what was known usually as the "bastion loop," which also had the terms applied it of "frogged" "flowered," "flower-pot," and "crooked." This shape of loop continued to be worn in the 12th Foot and many other regiments as long as loops of lace were worn by the men (up to 1855 on the breast of coat, and up to 1869 on the sleeve). (See Plate 7.)

The Bastion Loop.

Swords having been discontinued by the rank and file, were now worn only by the sergeants and grenadiers. In other respects the accoutrements remained the same. Officers of grenadiers wore an epaulet of gold lace and fringe on both shoulders, and battalion officers on the right shoulder only. Officers of the grenadier company carried fusils, and wore white shoulder-belts and pouches; the other officers carried the esponton, a light steel-headed pike, about seven feet in length, with a small crossbar below the blade. Sergeants carried swords and halberds, a light ornamental kind of battle-axe, with a long hand or shaft; and wore hats laced with gold, and crimson worsted sashes with a yellow stripe in the centre. The grenadiers' coats had wings of red cloth on the point of the shoulder, with six loops of regimental lace and a border of the same round the bottom.

Corporals' and privates' coats were laced with the regimental lace, as before described, corporals being distinguished by a silk epaulet on the right shoulder. Drummers and fifers wore yellow coats lapelled with red and similarly made to those of privates; also bearskin caps with a plate in front, and a short sword with scimitar blade. Each of the pioneers carried an axe and a sword, and wore an apron and a cap with a leather crown and a black bearskin front, on which was displayed the King's crest, with a saw and an axe in white on a red ground. At the Prince Consort's Library, Aldershot, can be seen an old book, dated 1768, showing the privates' uniforms of every British infantry regiment at this period.

In the Royal Warrant of December 19th, the following regiments are shown as wearing gold lace, viz. :—the 1st, 7th, 8th, 11th, 12th, 18th, 19th, 21st, 22nd, 23rd, 25th, 27th, 32nd, 36th, 39th, 40th, 48th, 49th, 51st, 53rd, 55th, 57th, 58th, 64th, 66th, 69th, and 70th.

The following are shown as wearing silver lace, viz. :—The 2nd, 3rd, 4th, 5th, 6th, 9th, 10th, 13th, 14th, 15th, 16th, 17th, 20th, 24th, 26th, 28th, 29th, 30th, 31st, 33rd, 34th, 35th, 37th, 38th, 41st, 42nd, 43rd, 44th, 45th, 46th, 47th, 50th, 52nd, 54th, 56th, 59th, 60th, 61st, 62nd, 63rd, 65th, 67th, and 68th.

1769

By the new Clothing Regulations, sergeants of infantry regiments were to carry fusils and pouches, and the annual allowance of powder for each regiment was increased from eight to ten barrels.

1770

Soldiers of the light company, which was added in December this year, wore short jackets, red waistcoats, and a leather cap, almost as small as a skull cap, with a large round peak, straight up in front ; their waistbelt had two frogs, one for the bayonet, the other for the hatchet, when the soldier was on duty ; at other times, on the march, the hatchet was to be tied upon the knapsack, and the light company gaiter was to be " as high as the calf of the leg and no higher."

1776

By the year 1776, it became the universal custom for officers to wear the white leather shoulder-belt for the sword outside, instead of inside the coat, when regiments, disliking the plain buckle and tip, which had hitherto been conspicuous on it, began to adopt something more ornamental in the shape of a small gilt or silver plate.

1784

An Adjutant-General's Order, dated 20th March, directed that belts of infantry soldiers were to be worn over the right shoulder as cross-belts to support the bayonet, instead of round the waist as hitherto.

A Royal Warrant, dated 21st July, laid down that :—The whole quantity of ammunition carried by each soldier was to be 56 rounds, 32 of which were to be carried in a pouch on his right side, and 24 in a cartridge box by way of a magazine. The pouch to consist of a tin box, with five divisions in it, each containing 4 cartridges placed upright, and another tin box beneath, to hold the remaining 12 cartridges, placed horizontally, with divisions in it to fit the length of the cartridges. The cartridge box, by way of magazine was fixed to the bayonet belt, so as to be easily removable, and was only intended to be worn on the march, and on field service.[1]

1786

In accordance with a General Order, dated 3rd April, infantry officers were to lay aside the esponton, and to provide themselves with a strong, substantial uniform sword, with straight cut-and-thrust blade, an inch broad at the shoulder and thirty-two inches in length. The hilt, if not of steel, was to be either gilt or silver, according to the buttons on the uniforms. The sword-knot was to be of crimson and gold, in stripes.

A Horse Guards Letter, dated 17th November, approved of white hats, instead of black, being worn by troops in India. They were to be cocked, and ornamented in accordance with the pattern rendered for the King's inspection, except that there was to be no fur on the brim, and the number of the regiment was to be on the button.[2]

A General Order, dated 22nd July, directed that officers wearing sashes on duty were to wear their swords slung upon the right shoulder over their uniforms, and when off duty and not wearing sashes, to wear their swords slung over their waistcoats. The sergeants to wear their swords in the same manner as the privates carried their bayonets.[3]

1791

On the 6th October, it was directed that all field officers of cavalry and infantry were to be distinguished by wearing two epaulets. Grenadier and light company officers, who had worn two for some time, were now to wear a grenade and bugle respectively on the epaulet, and, on the same date, halberts were replaced for sergeants by pikes.

1795

An order, dated 19th July, directed that the use of hair powder was to be discontinued by non-commissioned officers and men, but not by officers. Officers' silver gorgets were this year replaced by those of gilt metal, which were of universal pattern and worn up to 1830.

By a Royal Warrant, dated 1st September, the following was the schedule of necessaries for each soldier of infantry, viz. :—three shirts, two pairs shoes, two pairs stockings or socks, one pair of long gaiters, one forage cap, one pack, one stock, one black ball, and two brushes.

[1] *W. O.* 3, Book 26, pp. 154 and 162. [2] *W. O.* 3, Book 27, p. 21. [3] *W. O.* 3, Book 29, p. 34.

1796

Plate 11 shows the uniform of a light company officer and men of the regiment. The officer's uniform, briefly described by Captain Elers (who had been posted to it, on joining the regiment, as a subaltern, in the spring of this year) was as follows :—Having to supply himself with a sabre (scimitar) he adds :—" We (the light company officers) wore wings instead of epaulettes, blue pantaloons, edged with scarlet, a scarlet waistcoat, ornamented with narrow gold lace, and hats covered with the finest black ostrich feathers, with a stand up feather, composed of red and black."

The hat was similar to that worn at the time by light companies of the Foot Guards, and, according to the Regulations, this type of uniform was only authorised up to the 1st May this year, when a Warrant was published, which laid down that the coats for all ranks were to be fastened down to the waist, by which the sleeved waistcoat (which afterwards became an undress garment for the rank and file) was completely hidden. The lapels were continued down to the waist in such a way as to make a double-breasted coat, with a high stand-up collar to admit of the large neckcloth then worn. The jacket for the rank and file was single-breasted, with ten buttons, and loops of regimental lace across the chest.

1797

A War Office Circular, dated 13th September, signified His Majesty's pleasure that the feather to be worn in future in the hats of officers and men was to be red and white, red beneath, the same as the cavalry ; that for grenadiers to be plain white, and that for light infantry, green.

A Horse Guards Letter, dated 28th October, authorised that the wearing of lapels was, in future, to be discontinued in the ranks of dragoon guards, and infantry of the Line, but officers retained them in full dress.[1]

1799

An order was issued this year for officers and men of infantry regiments (except those of flank companies) to wear their hair queued ten inches long, including one inch of hair to appear below the binding and to be tied a little below the upper part of the coat collar.

By the year 1799, officers' shoulder-belts had become wider, and in many cases the ornaments, instead of being engraved, were raised upon the plate, much enhancing the effect. (See Plate 14.)

This is the earliest breast-plate that can be traced of the 12th Regiment :—A gilt oval, $3\frac{5}{8}$ inches long, with garter, crown, and " Gibraltar " in silver, and numeral gilt.

[1] *H. O.* 50, Book 6 P.R.O. London.

1800

By a General Order, dated 4th February, the cocked hat worn by the men was discontinued, and a cylindrical shako with a peak was introduced. It was of lacquered felt and ornamented with a large brass oblong plate of universal pattern, bearing the King's crest, with a red and white tuft in front rising from a small black cockade, and the number of the regiment on the plate, each side of the Lion. (See Plate 16.)

Two types of the shako known as the "Stove-pipe" pattern.

1801

A Warrant was issued in April directing that all men of the Foot Guards and infantry should be provided with a serviceable greatcoat.

1802

A Horse Guards' Letter, dated 1st July, directed epaulets and shoulder knots to be discontinued by non-commissioned officers of the Foot Guards and Infantry, and, in lieu, chevrons, made of lace used in their regiments (sergeant-majors, four; sergeants, three, and corporals two) to be worn upon the right arm.

1806

In October, a black felt shako was substituted for that of lacquered felt, the brass plate of which was smaller in front and more oval than that of 1800; it was surmounted by a crown, and the King's cipher, with number of regiment immediately below it. It had a red and white worsted tuft and cockade at the side, and suspended across the front was a crimson and gold twisted cord with tassels, the same cord in white for the men of battalion companies, and green for those of the light company. This shako was worn by most infantry regiments throughout the later Peninsular and Waterloo campaigns. (See Plates 16 & 17.) The material of which this shako and its predecessor was made was of such a flexible nature that no chain or chin strap was found necessary to be worn with either.

1808

A General Order, dated 20th July, directed that the use of queues be dispensed with in the army, care being taken that the men's hair was cut close to their necks in the neatest and most uniform manner.

1811

Infantry officers were authorised to discontinue wearing cocked hats and to wear a head-dress similar to that of the men, a double-breasted red jacket with very short skirts, and a grey overcoat and grey trousers, the same as the men.

The dress cap shown in Plate 17 does not seem to have come much into use until about 1811–12, although approved some years before.

1815

In the Peninsular epoch also, the jacket, or coat with abbreviated tails, spread in use from the light company to officers of all companies. It was discontinued for all except the light company in 1820, and for that in 1826. Immediately after the close of the great war, the comparatively small and neat shako was superseded by the broad-topped continental shako, possibly an effect of the Paris reviews in 1815. Several British officers are found, in diaries, &c., complaining of the poor appearance of the English infantry, owing to their small head-dress.

A General Order, dated 2nd August, authorised the introduction of a bell-shaped shako, eleven inches in diameter on the top, and seven and a half inches deep, with gilt chain scales, which were allowed to be fastened up in front, below the black cockade ; upright white feather twelve inches long, shako-plate silver (See Plate 19), star oblong, 4 by $3\frac{1}{8}$ inches ; gold lace ($2\frac{1}{2}$ inches), round top, and half-inch lace at bottom. Grenadiers retained their bearskins, with gold tassels for the officers and white for the men. Green feather and bugle badge for light company.

Officers' Gilt Breastplate, 1816–25.
(Reduced size.)

From 1815, the officers of the 12th wore blue grey cloth undress trousers with a double white stripe down the outer seams.

1816

This year, they adopted, for the first time, an oblong-shaped breastplate (reduced pattern here shown) to replace the oval one taken into use in 1799 (Plate 14.) The new one measured $3\frac{1}{4} \times 2\frac{11}{16}$ inches, and consisted of a gilt burnished plate, with corners slightly rounded, dead gilt crown, pierced garter, and gilt matted "XII."

For the next 14 years, the infantry uniform was at its highest splendour. (See Plate 19.) The epaulet had increased in size like the shako, and as company officers still wore one only, the effect was rather lopsided. About this time, the 12th officers wore white dress trousers with gold lace down the sides, and apparently four kinds of trousers had to be kept, blue grey, for winter, white for summer, and, in each case, laced for full dress, and plain for undress. Skirt ornaments, prior to 1820, are shown in a lace maker's book as: "Crow's feet (gold) on yellow, filled scarlet."

1820

This handsome skirt ornament consisted of a gold embroidered star, richly spangled, on scarlet cloth, gold "12" raised on scarlet, and silver "Gibraltar," on a blue garter.

Attention was directed this year to the fact that the gorget formed part of an officer's equipment, and must be worn when on duty. The facings at this period were pale yellow.

Officers' Skirt Ornament, 1820–40. (Reduced to half size.)

1823

The sword-belt now worn, by captains and subalterns only, was of black patent leather, $1\frac{1}{8}$ inches wide, with dead gilt clasp plate, and "XII" in dull silver. (See Plate 37.)

1825

A new breastplate, measuring $3\frac{3}{4} \times 3\frac{1}{8}$ inches, was now taken into wear by the officers, consisting of a dead gilt plate, with handsome silver star, surmounted by a wreath and crown encircling a garter containing the words: "Minden," "Gibraltar," and "Seringapatam," all in gilt. (Plate 17.)

The officers' gold trouser lace was $1\frac{3}{4}$ inches wide, of gold basket check and vellum.

The pattern of sergeants' sash was, this year, authorised to be three streaks of yellow with red alternately, and its weight, $7\frac{1}{2}$ ounces. (See Plate 22.) All supplies were to be of the same quality and make.

White linen trousers, for summer wear, were worn between the 1st May and 14th October inclusive.

1826

The private soldier's coat was altered in cut, the lace removed from the collar and a single loop placed on each side, the loop across the chest made broad at the top, tapering narrower towards the bottom, and the lace removed from the coat skirts, except the loops on the slashed pockets. Officers' rank was distinguished by the epaulet, according to the instructions in the General Order of February, 1810. Field officers wore two epaulets, a colonel having a silver star and crown embroidered on the strap; lieut.-colonel, a silver crown; major, a silver star; whilst captains and subalterns wore a single gold epaulet on the right shoulder.

1828–29

On the 22nd December, a Horse Guards Circular authorised another change in the shako, a description of which first appeared in the "Dress Regulations" of 1831.

Whilst it remained bell-shaped, the lace was to be stripped from it, its height reduced to six inches, and a large gilt star plate worn in front. (See Plate 22.) It had gilt scales to fasten under the chin, and with it was ordered a rich gold festoon, with cap lines and tassels of gold. The same was made of worsted for the men, white for battalion companies and green for light companies, and with the shako was worn a plume of white feathers twelve inches high.

Plates 20 and 21 show the large gilt star worn by officers of the battalion and light companies of the 12th, respectively, the regiment being one of those that were specially select in having a light company shako star for this head-dress of a particular pattern, the grenadiers wearing bearskins.

By a circular letter, dated 21st March, 1829, a forage cap for officers was authorised for the first time, of blue cloth, with a large flat stiffened top and peak, having only a band of cloth round it of the colour of the regimental facings, and no number.

The coatee now introduced was scarlet, with collar and cuffs of the regimental facings, Prussian collar with two loops, and uniform buttons at each end. It was double-breasted and the buttons (for the 12th) placed in pairs, with intervals between them. Yellow cuffs, with a scarlet cuff slash, and on it four buttons and laced loops, cuffs 2¾ inches, slashed flap on skirts with four loops and large buttons, white kerseymere turnbacks and skirt lining, regimental skirt ornaments, those of the rank and file consisting of a regimental button for the battalion companies, and of a metal grenade and bugle-horn respectively for the flank companies. All ranks of officers now wore two epaulets, a universal pattern having been adopted for the whole of the infantry. An example of the single epaulet worn prior to this date is shown in Plate 17.

A universal pattern of scarlet shell jacket for officers was first authorised this year. Previously, the 12th had adopted a scarlet shell jacket with yellow collar and pointed cuffs, at first with close studs up the front, but, just before 1829, with 14 small buttons, set in pairs, instead of the studs.

By this time also, the regulation blue frock coat had come into use, as shown in Plate 19. Gold shoulder cords were added this year. The coatee was to be worn at all parades, regimental as well as general, whether with or without arms, and on all garrison duties with the white shoulder-belt and sash. It was always to be worn at mess and at evening parties with a black leather belt beneath the coat, and a sword, but no sash.

Silver Ornament

(Reduced to half size), found amongst Colonel Forssteen's effects, in 1828, probably an early pattern cap-plate.

GILT GORGET
(exact size)

Worn by Major Thomas Craigie,
12th Regiment

Ensign to Major, 1780–99.

By the courtesy of Mrs. Craigie-Sandbach.

1830

The Warrant of August this year made several changes. Trousers of "Oxford mixture" were substituted for those of bluish grey; shako lines and tassels were abolished, plumes shortened to eight inches, and a green ball-tuft introduced for light infantry and light companies. (See Plate 22.) Sergeants of infantry were armed with fusils instead of pikes. A red fatigue jacket succeeded the white one hitherto worn by the rank and file, the colour of the collar and cuffs (as well as the chevrons for sergeants) to correspond with the facings of the regiment, and finally the gorget was abolished.

The dress of bandsmen, which had hitherto been arranged regimentally, was so far recognised by the authorities, that infantry bandsmen were this year authorised to wear white coatees.

1831

On the 18th May, field officers of infantry were ordered to discontinue wearing the shoulder-belt with slings, adopting instead a white leather waistbelt with a gilt plate in front, as taken into use in 1823. (See Plate 37.) Instead of the leather scabbard, they were also directed to wear one of brass; adjutants, at the same time, being directed to wear steel scabbards. Spurs for mounted officers were to be of yellow metal with necks $2\frac{1}{2}$ inches long, including rowels.

Officer's Gilt Tunic Button.

Epaulet Button.

The same pattern in pewter was worn by the rank and file to 31st March, 1855.

1833

In January, the narrow welt of red cloth down the outer seams of infantry trousers, as at present worn, was authorised for the first time.

1834

A new officer's forage cap was introduced of blue cloth, with a peak, a black silk oak-leaf band and the regimental number, one and a half inches deep, embroidered in gold on the front.

The undress single-breasted blue frock coat now presented a handsome appearance, being ornamented with scales, which consisted of blue cloth shoulder-straps, laced with broad regimental gold lace, and terminating with gilt metal crescents.

The white plumes in the shakos were replaced by white worsted ball-tufts, the light company wearing green tufts.

1836

By General Order this year, the coloured regimental lace, so long worn by the rank and file, was abolished, and replaced by plain white tape lace, but the mode of wearing the loops across the chest was retained. The sergeants were directed to wear double-breasted coats without any lace on the chest, with white epaulets and wings for those of flank companies. The coloured lace, however, of the drummers, of regimental pattern which had been worn since about the year 1820 was continued.

1839

The brass shako plate, with the regimental number, of universal pattern, worn by the rank and file, was this year replaced by a round brass plate three inches in diameter, with a crown above it and the number raised in the centre.

1840–41

The new percussion muskets were generally introduced into the army.

In accordance with the fashion at this period, gilt buttons were largely worn in plain clothes, noblemen and gentry wearing them of various patterns, whilst those holding official appointments wore them of a special design. Military officers accordingly adopted buttons of a special regimental pattern, of which this was the design chosen by the officers of the 12th Regiment, and it was from this that a smaller button came later into use for the officers' mess waistcoat. (See page 404.)

Officer's Gilt Button, 1840.

The officers of the battalion companies of the 12th adopted this year a new pattern of skirt ornament, those of the grenadier and light companies remaining as before, consisting of a gold embroidered grenade and bugle respectively. In the battalion company officer's skirt ornament here shown, the wreath and "12" were in gold, on a scarlet ground; the garter in blue silk, edged gold; the remaining scrolls in yellow silk, and all the lettering in silver.

Officer's Skirt Ornament, 1840–55.
(Reduced to ½ the size.)

These skirt ornaments were in use until the abolition of the coatee in March, 1855.

As far back as 1841, a crown was worn on the forage cap by the rank and file of the battalion companies, presumably under regimental arrangements, to fill a vacant space above the numeral "12" (in keeping with the grenadier and light companies, who wore their respective badges), as there is nothing in the War Office books at this, or any other period, to show that a crown, as a badge, was sanctioned by the authorities as a special distinction.

1843

A new breastplate measuring 4 by 3¼ inches was this year taken into use by the officers. It was of bright burnished gilt, with crown dead gilt, the remainder all silver mountings, and was worn up to March, 1855, when breastplates were abolished. (See Plate 14.)

1845

Bearskins were abolished for the grenadier companies of infantry regiments, and a new shako was introduced of the "Albert" pattern, 6¾ inches high, with a patent leather peak in front, and a narrower one behind. Instead of taking into wear a gilt star plate of one pattern for all the officers (as was the case with nearly all line regiments) the officers of the 12th adopted, for this head-dress, three shako plates of different patterns, which were worn by the officers of their battalion, grenadier, and light companies respectively. A gilt chain was worn with it, fastening at the sides with a rose pattern ornament, and a ball-tuft completed it, two-thirds white at top and one-third red, for battalion and field officers, all white for grenadiers, green for light companies. The men retained the brass plate they had worn with the large-topped shako. (See Plates 23, 24, and 25.)

A new waistbelt clasp was also introduced this year, of bright burnished gilt with silver mountings, and was worn by field ranks only. (See Plate 37.) Also the red and yellow striped sashes worn round the waist by the sergeants of the regiment were discontinued, and a crimson one substituted, with one yellow stripe through the centre. Plain crimson sashes, 2½ inches wide, came into use before the Crimean War.

1847

The officers adopted a new forage cap badge; curiously enough, the key of the castle is shown in the centre of the gate. (See Plate 30.)

1848

By a General Order, dated 30th June, the laced loops and buttons on the skirts of the officers' coatee were abolished, leaving only the skirt ornaments. The blue frock coat with shoulder scales was also discontinued, and a plain shell jacket, with collar and cuffs of the regimental facings, was introduced. A black patent leather sling sword-belt with snake clasp was worn with it, and a grey greatcoat was also taken into use by officers, in lieu of the blue cloak.

1850

A plain shoulder belt, without breastplate, to carry the pouch, was authorised to be worn in the ranks, the bayonet being hung on a frog from a waist belt.

1855

On the 1st April, the coatee had given place to a long double-breasted tunic, with lapels (for officers), which were made to fold down at the top and show the yellow lining, but the lapels had to be buttoned over on parade or duty. This pattern tunic did not find favour long, and was succeeded, two years later, by the single-breasted tunic of to-day. Slashes of lace, with buttons, were worn on the cuffs, and similarly on the skirts. The crimson sash was worn over the left shoulder, retained by a twisted cord of crimson silk, and the same, with the blue frock coat, in undress. There were then no collar ornaments, beyond an officer's rank, which was designated on the collars of the tunics only, except field officers, who also wore it on the collars of their frock coats.

The respective ranks were shown as follows :—

Ensign, a silver embroidered star,
Lieutenant, a silver embroidered crown.
Captain, a silver embroidered crown and star.
Major, a gold embroidered star.
Lieut.-Colonel, a gold embroidered crown.
Colonel, a gold embroidered crown and star.

Trousers, with the scarlet welt, were of the same pattern as the present day, dark-blue serge being worn from 1st May to 14th October, and Oxford mixture cloth from 15th October to 30th April.

Officer's Gilt Button, 1855–57.

The shako was of the "Albert" pattern, with peaks front and back, and lighter in make than its predecessor, the first double-peaked shako, introduced in 1845. Two rows of regimental lace were worn round the top of it by lieut.-colonels, and one row by majors only.

The shako plate, of this date, was an eight-pointed star ($3\frac{5}{8}$ inches in extreme diameter), surmounted by a crown, having the regimental number in bright gilt on a black leather ground, inside a garter proper. (See Plate 30.)

A white enamelled leather belt with new clasp (See Plate 37) was worn outside the tunic and the frock coat, and sword scabbards were of black leather, with gilt mountings, except brass for field officers and steel for adjutants, and at a later period, when regimental instructors of musketry were appointed, they also wore steel scabbards. The sword knot was of gold embroidery, with crimson stripes.

The band wore white tunics, double-breasted, with yellow facings, and a red piping up the back and sleeves. A double-breasted blue frock, for undress, had been authorised for officers, and a new blue cloth forage cap with a straight peak showing the regimental number embroidered in front. The white tape lace in the ranks now disappeared, lace of regimental pattern being retained by the drummers, whilst white piping was introduced for the first time, and brass buttons, for the rank and file, replaced the old pewter ones.

1857–8

With the new issue of clothing this year, single-breasted tunics were introduced. Together with white shell jackets, they continued to be worn in white cloth by bandsmen, showing the regimental facings. (See Plates 24 and 29.) The large size tunic button now taken into use by officers was larger than its predecessor, with a rim to it, and of this pattern.

Officer's Gilt Button, 1857–72.

1858

On the abolition of flank companies in 1858, the white and green shako ball-tufts, specially worn by them, disappeared, the whole regiment now wearing these alike, viz.: upper part $\frac{2}{3}$ white, and lower $\frac{1}{3}$ red.

1859

The coatee, succeeded by the tunic, had, up to now, been the authorised mess dress for infantry officers, but for many years it had been customary to wear the shell jacket open with a waistcoat, the pattern and material of which were decided solely by officers commanding regiments. The wearing of the shell jacket at mess was authorised by Circular Memorandum, dated 9th June, this year.

Bandsmen of the 2nd Battalion at this period wore round forage-caps of scarlet cloth. (See Plate 24.)

Of the three buttons here shown, No. 1 was the first worn with

1. 1858.

2. 1861.

3. 1881 to present date

the mess waistcoat, its pattern being taken from the button shown at page 401. It was worn by the 2nd Battalion for some years with a blue mess waistcoat.

1861

This year saw the introduction of a new shako of blue ribbed cloth, single-peaked, and smaller and lighter. The plate was a gilt star of eight points surmounted by a crown, and having the number of the regiment cut out, within a garter. (See Plates 29 and 30.)

1866

Officers' black leather scabbards were replaced by steel ones, field officers retaining theirs of brass.

1867

An officer's blue patrol jacket replaced the blue frock coat, and pioneers' white leather aprons were discontinued.

It was about this time that white cork helmets for tropical wear were introduced into the Army, and they were of universal pattern.

It was also about this period that the tropical scarlet patrol jacket appeared in the regiment, edged with narrow gold braid, which also decorated the collar, cuffs, and outside breastpocket, the men wearing scarlet jumpers of the same shape. (See Plate 31.)

In October, the Snider breech-loading rifle was issued to the 1st Battalion on its return from New Zealand.

1868

The slashed cuff on the tunic was abolished, and pointed cuffs, with distinctions in the amount of lace for various ranks, were introduced. A levee dress was instituted, consisting of gold-laced trousers, and a crimson and gold sash, and sword-belt, all of universal pattern.

Bandsmen wore a crown over the "12" on the round forage-caps, and in the 1st Battalion these were white with a red band.

Glengarry Cap Badge, 1871.

1871

Early this year, a new shako, ornamented with narrow gold braid and a gilt chain, was introduced for officers, that for the ranks being without the braid. The old star plate was replaced by a garter with number inside, surrounded by a laurel wreath, the whole surmounted by a crown, and it was the last pattern of shako worn by British infantry. (See Plates 30 and 31.) The men's tunics were changed in colour from red-brick to scarlet, and a Glengarry cap, introduced as a forage cap in the previous year, became universally worn in the ranks of the infantry.

In some regiments, a scarlet Glengarry was also issued to bandsmen only, but it did not find favour very long, and was discontinued.

The 12th Regiment adopted the badge shown above as that worn by the rank and file, on the Glengarry cap ; a twelve-pointed star, surmounted by a unique castle, with three turrets, and key.

The *Suffolk Gazette*, dated 1st February, 1899, says :—

" The key which stands so prominently on the top of the centre turret, was on moulding the new pattern, placed in front of the door, for which the reason is given, that from its being so frail it was easily broken, but its actual place is shown in this print."

In September, white cloth clothing was abolished for bandsmen, who were now designated by a worsted badge of crossed trumpets, on the right arm of their scarlet tunics, this badge being later represented in metal. Scarlet serge frocks also replaced the red-brick coloured cloth shell jackets for the men, and, at the same time, distinctive regimental patterns of drummer's lace were finally abolished, having been gradually discontinued since 1866. Thus there disappeared the accompanying fringe, also of red, white, black and yellow (the same colours as the lace), which the drummers had worn for about fifty years. These were replaced by a smaller red and white fringe, and white lace with red crowns, both of universal pattern. November, this year, saw the introduction of a button in the ranks, of universal pattern, showing the Royal Arms, and without any regimental numbers.

The men's new serge frocks were made with an " eye " of narrow white braid on the cuff, and similar braid round base of collar, and their new tunic had a " crowsfoot " on cuff.

On the 1st September, the Snider breech-loading rifle, replacing the Enfield, was issued to the 2nd Battalion.

1872–81.

1872

Owing to the great variety of mess uniforms that now prevailed throughout the infantry, one of universal pattern was introduced, fastening at the neck, and both jacket and waistcoat edged with teat buttons.

A new pattern button was this year taken into wear by the officers, the wreath of which was more elaborate.

It has been suggested that some glory was attached to a laurel wreath on this button, but the War Office books do not show such a grant to the regiment, or of a laurel wreath having been conferred on any of the appointments. No regulation button design existed in the Army until the introduction of the Territorial System in July, 1881 ; it was therefore optional with commanding officers to choose their own designs. As a matter of fact, a great number of line regiments adopted a wreath on the buttons, and, for years, it had been no easy matter for tailors and button-makers to invent different designs. The above pattern of button was in use until the 1st July, 1881.

The officers, this year, also took into wear a new pattern waist-belt clasp, on which was introduced for the first time the Castle of Gibraltar, with its inscription. (See Plate 37.)

OFFICERS' WAISTBELT CLASPS.

NO. 2 WORN BY FIELD RANKS ONLY.

(*Exact sizes.*)

1873

The Castle and Key were authorised to be worn on officers' forage caps (*vide* "Dress Regulations, 1874") and a wreath was also added. (See Plate 31.) Collar badges of brass were also authorised to be worn in the ranks, the first pattern being as here shown.

1874

The white helmet now in use continued to be of the same shape until 1890, and the battalion serving abroad wore with it a Castle badge within a garter and Crown on yellow cloth, on the right side of the pugaree.

1877

On the 27th September, the 2nd Battalion was armed with the Martini-Henry rifle.

1878

A cork helmet, covered with blue cloth, with a gilt spike and chain, replaced the shako as a head-dress for the infantry, the plate being gilt, with number of regiment in the centre within a garter bearing the Royal motto, a laurel wreath round the garter and a crown above, and was taken into wear in the following year. (See Plates 30 and 36.) On the 26th May, men of the 2nd Battalion, furnishing the Double Gate Guard at Gosport, wore the new regulation helmet for the first time.

1880

Officers' badges of rank were removed from the collar and placed on the shoulder cords of the tunic and mess jacket; rank was also similarly shown on the cloth straps which had been added to the patrol jacket.

The officers in both battalions now wore red mess waistcoats.

1881

The introduction of the Territorial system, from the 1st July, and the abolition of regimental numbers, caused many sweeping changes in regimental distinctions, which, however, were less marked in the first twenty-five double battalions of the line than in the single battalions which followed, on account of the indiscriminate linking which the latter underwent.

The regiment (now "The Suffolk") had to adopt white facings, which, under the new system, had become the established colour for all English (except Royal) regiments.

A new officers' forage cap was introduced, with drooping gold-embroidered peak, the first Territorial badge consisting of the Castle and Key, within a laurel wreath, in gold embroidery, on a raised blue cloth ground; above the Castle, the Crown; below, a scroll in blue velvet, inscribed "Gibraltar."

The Glengarry cap was now authorised to be worn by officers on active service, and for peace manœuvres, with a silver badge on it, the oak-leaf wreath now introduced, being for no particular reason, but merely a design of the War Office.

The officers' gold tunic lace at this period also underwent a change, and was replaced by gold lace of "rose pattern," to be universally worn by all English regiments.

1882

Saddle-cloths, hitherto used in review order by infantry mounted officers, were discontinued. This year a plain round cuff replaced the pointed one on the men's tunics, which was revived in 1903.

1884–85

The new pattern (1882) valise equipment was issued to the 1st and 2nd Battalions of the regiment on the 8th February, 1884, and 10th April, 1885, respectively.

1887

Brown gloves first came into wear, and, except in review order, became universally worn from the year 1900.

The officers, 2nd Battalion, changed their mess waistcoats this year to white cloth, and in 1889 to a white washing material, which had been adopted some years before by the 1st Battalion.

1890

For about two years from now, the serge frock was issued to the men with the shoulder-straps, only, of the facings colour, and the officers' Indian pattern of scarlet patrol was somewhat altered, the collar only being now of the facings colour, and the cuffs ornamented with narrow gold braid, similar to the mess jacket cuffs.

A new scarlet undress patrol jacket was introduced for home service, with breast pockets and facings.

1893

In May, a field service cap of a special pattern came into use, and the 2nd Battalion at Secunderabad was armed with the Lee-Metford rifle, which involved a change in the type of bayonet from the old triangular shape to one of a straight two-edged pattern, this remaining in use until 1908.

1897

Under authority, dated 20th May, a mess jacket with yellow roll collar was approved for wear by the officers, with the addition of badges of rank on the shoulder straps, and white washing waistcoats were worn.

1898

The badge in the margin was approved this year by the War Office for the foreign service helmet and the field service cap. It was worn by both battalions, of silver for the officers, and made at first of brass for the rank and file, but afterwards altered to white metal.

Badge for the Foreign Service Helmet and Field Service Cap, 1898.

1899

A War Office letter, dated 30th October, restored the old yellow facings to the regiment; on tunics only for the men, and on other garments as well for the officers.

1900

The War Office having notified, under date 30th January, that the Castle of Gibraltar was represented by a different design in each regiment to which the distinction had been granted, a correct representation was now forwarded, as here shown, of the Castle as depicted upon the Seal of Gibraltar, granted in 1502, and subsequently upon its coinage. This design was to be followed in future, in regimental badges. It seems necessary to have drawn attention to the want of uniformity existing, if only on the ground that some regiments bearing the distinction, had, for many years, adopted as a pattern, a castle with two, instead of three turrets.

With regard to the correctness of the Arms of Gibraltar, the following extract from a letter to an officer of the regiment, in 1908, from the Garrison Librarian at Gibraltar, is quoted :—

> " As to the *Arms of Gibraltar*, the original warrant of Ferdinand and Isabella, the original of which is at San Roque, and of which I have secured a copy, runs as follows :' . . . . we give you as Arms an escutcheon on which two-thirds in the upper part shall have a white field, in the said field set a Red Castle, underneath the said Castle as the other third of the escutcheon which must be a red field in which

there must be a white line between the castle and the said red field, on this (evidently the red field) a golden key, which (" shall be " is to be understood here) on that with a chain from the said Castle as they are shown; and the said arms which we give you you may and shall place on the seal of that, the said city.'

The placing of the colours, and the bearings are therefore fixed. The shape of the castle and the direction in which the key is turned is practically optional. Those which the Deputy Librarian sent you are the usual and traditional shapes, but a variation in them would not appear to be vital.

The words 'Montis Insignia Calpe' which you have underneath the badge on your paper, are not a motto but simply mean 'The Arms of Gibraltar.' But if they are placed there it would seem that the Arms should be the correct ones. Without these descriptive words there would seem to be no reason why the War Office or a Regiment or anyone else should not manufacture a badge for certain uses out of any portion of the arms.

The Castle in Captain Drinkwater's History is very much more artistic, but it does not represent the Arms of Gibraltar, as the shield and Colours are not shown." [1]

1900

The sabretache for mounted officers, steel chain reins, and bearskin cover to wallets were abolished, also the levée dress.

1902

Throughout the South African War, infantry great coats were mostly of khaki-coloured cloth, and, at its conclusion, numerous changes in dress took place. Brass scabbards and brass spurs were replaced, for field ranks, by those of steel, and the great coat was to be carried on the saddle. The following articles were introduced: a double-breasted blue frock coat for officers; a new forage cap for all ranks, with a large flat stiffened top and sloping peak; a new pattern skirt to officers' and men's tunics, with a new sword belt, now worn beneath the tunic, and the sash worn round the waist; also a pointed cuff revived on the men's tunics.

From this date, the khaki service dress was universally worn as undress uniform on all parades and regimental duties, with putties and the Sam Browne belt, and levée dress was abolished for infantry officers.

1907

In November, the 2nd Battalion at Aden were armed with the Lee-Enfield rifle.

[1] Calpé, in the south of Spain, and Abyla, on the opposite Coast of Africa (about eighteen miles distant), were celebrated as the *Pillars of Hercules*; and according to heathen mythology, these two mountains were united, until that hero separated them, and made a communication between the Mediterranean and the Atlantic seas. Calpé received the present designation of Gibraltar from the Arabic "*Gib-el-Tarif*," or "*Mountain of Tarif*," being the spot where that Moorish Chieftain landed on his invasion of Spain in the year 711.

1908

This year, a new bayonet was introduced, 21½ inches in extreme length, the blade being 17 inches long; triangular in formation, of a sword-bayonet type, with a sword hilt, the double edge giving way to a rib along the upper, or thicker side of the blade, gradually vanishing towards the point.

1910

The web equipment of 1908 pattern was, in January, taken into wear by the 2nd Battalion.

1911

On March 13th, the 1st Battalion at Malta were ordered by the General commanding the Brigade to remove the yellow cloth "Castle" patch from their khaki helmets.

The officers of the 2nd Battalion received, on the 14th November, a very handsome and appreciated gift from the ladies of Suffolk, in the form of a pair of Colour Belts; also, one large, and four small Camp Colours, which were suitably acknowledged by Lieut.-Colonel C. van Straubenzee, commanding. They are made on a ground of morocco scarlet leather, with an inch and a half of rich gold lace in the centre, the edges being fringed with scarlet and gold, and fitted with massive gilt buckles, tip and side, with the Regimental crest in silver, standing in relief in the centre of the buckles. The battle honours are remarkably well executed in embroidery of silver thread, edged with the same work in gold, on a yellow ground. The edgings are of broad gold lace. An inscription is engraved on each of the lower buckle flaps as follows:—

Presented to the 2nd Battalion Suffolk Regiment
by the Ladies of Suffolk.
September, 1911.

The initial use of this very handsome gift to the battalion was made while on duty in London, when mounting Guard at the Royal residences, a most appropriate christening of an appreciated gift.

The four Camp Colours and Regimental flag are of rich yellow silk, handsomely embroidered with the regimental crest in scarlet silk twist; the work of the School of Art Needlework, and is of admirable design and workmanship. This set of Camp Colours is much valued, and will only be used on special occasions, to do honour to the Ladies of the County.

1913

In February, the yellow cloth "Castle" patches, which had been removed from the khaki helmets at Malta, were restored to the 1st Battalion, at Cairo, by Major-General the Honourable J. H. Byng, C.B., M.V.O.

Notes on the Colours.

1685–89

Since the early days of the formation of the British Army as a permanent force, Standards and Colours have been in use, mainly as rallying points in action, the former appertaining to Cavalry, and the latter to Infantry.

A cavalry Standard or Guidon was also known by the name of Cornette, an infantry Colour being termed an Ensigne, and, as far back as 1661, it was customary for each troop or company to have its own Standard or Ensign, bearing some device, or possibly a number.

All field officers of infantry were nominally captains of companies up to June 1803, and Plate 38 represents, in 1686, the colonel's, lieut.-colonel's, major's and first captain's Colours, copied from coloured drawings preserved in the Royal Library at Windsor Castle.

In the Duke of Norfolk's Regiment of Foot (afterwards 12th Regiment) the colonel's flag was red, having the Howard crest in the centre, on a cap of maintenance gules, turned up ermine, a lion statant with tail extended, or, ducally gorged argent; the lieut.-colonel's also red, with the St. George's Cross bordered white; the major's similar, with the addition of a white flame; eldest captain's flag as that of lieut.-colonel, with a cross crosslet (fitchée argent), a well-known charge on the Howard arms, in the centre. There is doubt as to whether other captains' Colours were distinguished by numerals or crosses crosslet.

Richard Cannon states that the regiment's Colours were white in 1686, but gives no authority for the assertion; the facings or coat linings were certainly described as white in the list of James the Second's army, at Hounslow Heath, July, 1686, but from evidences in the MS. lists at Windsor Castle, it by no means follows that the colour of the flags was the same as that of the coat lining or facings. Nor does it follow that the sketches or drawings were made during the Duke of Norfolk's term of command, he having resigned the coloneley in 1686; his flags may have been retained for a few months.[1]

1689–1743

During the reign of King William III (1686–1702) Standards and Colours were gradually reduced from troop or company to three for a regiment.

According to Cannon, regiments at this period were drawn up in three divisions, one colour being allotted to each.

In the reign of Queen Anne (1702–14) the Union with Scotland, which took place in 1707, had a marked effect on the appearance of Colours of the army. The combination which took place, known as the Union or Union flag, was brought about as follows :—

[1] *Standards and Colours of the Army*, S. M. Milne, pp. 50, 51.

"STABILIS" COLOURS.

Figure 1 shows the white saltire cross of St. Andrew on a blue field; Figure 2, the red cross of St. George, with a white edging or border, on a white field. By simply placing the red cross on the Scottish Ensign, the combination is as seen in Figure 3, which might be definitely termed the origin of the present "Union Jack."

A further development, later, showed the Rose and the Thistle, entwined

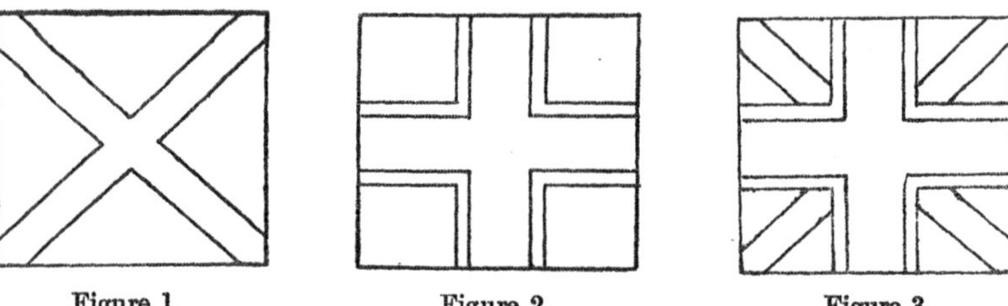

Figure 1. Figure 2. Figure 3.

in a wreath on the regimental Colour, encircling the number or title of a regiment.

No regulations concerning Colours had come into force until 1743, up to which date colonels of regiments had gone to great lengths in placing on them their own armorial bearings, crests, mottos, &c.

The upper Colour in Plate 39 shows, what was to all appearances, a private flag which had been made to Colonel Duroure's order in 1745, the year in which he was mortally wounded at Fontenoy.

On the edge is inscribed the following in silk: "Michel de Rynck fecit. Gand" (Ghent), "1745." It has been kindly lent by the owner, Colonel Sir Godfrey Thomas, C.B., D.S.O., late Royal Artillery (also a member of the Duroure family), for this photograph to be taken.

The Duroure Arms.

The Duroure arms, here represented, are sufficient to show the origin of "Stabilis" on a Colour of the regiment, whilst the flag above-mentioned also displays them, more completely. Its dimensions are 6 feet 6 inches by 5 feet 10 inches.

1743–48

The patterns of regimental Colours, and the devices on them had been optional with colonels of regiments until 1743, when, on September 14th, King George II caused the following orders to be issued regarding them :—

> "No Colonel to put his arms, crest, device, or livery, on any part of the appointments of his regiment."

The regulations affecting Colours were as follows :—

> "The first Colour of every marching regiment of Foot is to be the Great Union. The second Colour to be the Colour of the facing of the

regiment, with the Union in the upper canton, except those regiments which are faced with white or red, whose second colour is to be the red cross of St. George in a white field, with the Union in the upper canton.

In the centre of each Colour is to be painted, in gold Roman figures, the number of the rank of the regiment, within a wreath, except those regiments which have royal devices or ancient badges; the numbers of their rank are to be painted towards the upper canton. The length of the pike, and the colour itself, to be the same size as those of the royal regiments of Foot Guards, six feet six inches flying, and six feet deep on the pike; length of pole, spear and ferril included, being nine feet ten inches. The cords and tassels of all colours to be crimson and gold. The camp colours to be of the facing of the regiment, with the rank of the regiment thereon."

By this Warrant, the company Colours of colonels, lieut.-colonels, &c., in infantry regiments finally disappeared.

1749–1800

In 1749, attention had to be again drawn to the above General Order, which was republished on November 11th, with instructions to the Clothing Board, by the Adjutant-General, Colonel Robert Napier.[1]

Coloured drawings of the flags were also provided by him, showing the dimensions of them to all regiments, depicting their various regimental badges, and the manner in which they were to be emblazoned on the flag. In every case, two Colours were given: the first, the Union, and the second of the same colour as the facings of the regiment.

In the Inspection Return for May, 1768, the Colours were reported on as "bad." The old War Office books do not contain any such Returns for the two following years, but, in that for May, 1771, the Colours are shown as "good," from which it is clear that new Colours had been received in the interval, either in 1769 or 1770, but most probably prior to the regiment embarking for Gibraltar in May, 1769. There is no record of the date.[2]

The remnants of the lower Colour in Plate 39 came into the regiment's possession after having been exhibited at the Edinburgh Military Exhibition of 1889. In 1893, it was given to Lieut.-Colonel Glasgow, commanding the 1st Battalion, by Miss Craigie, whose ancestor (Major Craigie) served with the 12th Regiment at Seringapatam.

The fact that the wreath of this old King's Colour is practically identical with that in a Colour issued to the 9th Regiment, in 1772 (See p. 106, Mr. Milne's "Standards and Colours"), is strong presumptive evidence that the 12th Colour here shown, is one of those issued about 1769–70.

[1] Later appointed Colonel, 12th Regiment.

[2] Apparently it was this set of Colours which served throughout the Great Siege of Gibraltar, and at the storming of Seringapatam, as the next issue did not take place until after the Union with Ireland, in 1801.

The regiment must have still retained the word "Stabilis" at that period, but there is no evidence to show that it had been specially granted.

Mr. Milne states that there is no record of the word "Stabilis" appearing on any other regimental Colour.

With regard to the disposal of old Colours at this period, the Records of the 25th Regiment relate that on new Colours being issued to the regiment in 1763, at Newcastle-on-Tyne, the old ones, which had been in use since 1743, were buried with full military honours.

Up to 1768, no battle honours had appeared on regimental Standards and Colours, a commencement having been made that year by the 15th Light Dragoons becoming entitled to bear "Emsdorf" on their Standards and appointments.

In 1784, the King determined to mark his appreciation of the gallant services of the garrison of Gibraltar by a similar mark of Royal favour, and an Adjutant-General's letter, dated 14th April, intimated that:—

> "in commemoration of the glorious defence made by those regiments of infantry of the garrison of Gibraltar, during the late siege (of which the 12th was one), His Majesty had been graciously pleased to permit the word 'Gibraltar' to be placed on the grenadier and light companies' caps, and also on the accoutrements and drums, and below the number on the Regimental Colour."

Until about 1825, it was not customary in the army for Colours to be formally presented, but they were received and accounted for merely as ordinary stores.

As time advanced, the aspect of the ornamental part of Colours underwent a change. The central shield, with its often grotesque border, rococo style, became out of date, and regimental numbers were placed on a heart-shaped shield, whilst the Union wreath became much simpler.

Regiments of regulars received their Colours from the full Colonel as his private gift, and though it was customary to hand over disused Colours to him when he provided new ones, this was sometimes omitted.

County titles were conferred on infantry regiments in 1782, but they were never used on the Colours until 1816.[1]

1801–06

The Union of Great Britain and Ireland in 1801 caused a further alteration in the Colours of the Army. To the Union flag was now added the red saltire cross of St. Patrick, to the two crosses already displayed on it, and it was further ordered that the shamrock should be introduced into the Union wreath wherever that ornament or badge should be used.

[1] S. M. Milne, *Standards and Colours of the Army*, pp. 110, 115, 132, 223.

The King's authority was this year received to bear "Minden" on the regimental Colours and appointments in commemoration of the gallantry of the regiment in that action.

On the appointment, in 1806, of Mr. George Naylor, as Inspector of Regimental Colours, he requested every commanding officer to send him a sketch of the Colours then in use, and, with this note, he appears to have enclosed a drawing of two regimental Colours, with the centres of each left blank, to enable such devices or numbers as existed, to be inserted.

The following is a reply from the Officer Commanding 12th Regiment, dated, Cannanore, 20th February, 1808 :—

> "Enclosed, I return your sketches of the Colours, 12th Regiment, which have been received since the Union. The Colours were sent to the regiment about three years since by General Picton, who will furnish you with any further explanation you may require.
>
> (Signed) John Picton, Major."

1807–17

According to Major Picton's letter, the year in which a new set of Colours was received would have been about 1805, but there is no separate record of the date.

Centre of King's and Regimental Colours, 12th Foot, 1807.

Most of the replies from commanding officers to the Inspector of Colours were dated in the summer of 1807, and, according to the Return, the regiment was using devices as above.

The King's Colour with the Royal Cipher on an oval shield, with crown above it, and surrounded by the correct Union wreath, the regimental number being placed in the upper corner, near the spear head. The regimental Colour, with a similar oval, probably of red silk, containing XII REGT., above, a scroll with "Minden," and below, "Gibraltar," the whole combination extremely tasteful, though the "&c." seems an addition which might have been dispensed with.

There is every reason to believe that the representation following is that of the centre of an old Regimental Colour, issued about this time, which, in all probability, was an improved pattern of the previous one.

Its identity is further established by its having been found amongst the effects of the late Colonel W. H. Forssteèn, who died in 1828, a year after the regiment had received a new issue of Colours at Gibraltar, it having

been customary up to this period for disused Colours to become the property of the Colonel.

This fragment of the Colour has been kindly given by Miss Craig, grand-daughter of the late Colonel Forssteen, to the 2nd Battalion.

1818

In commemoration of the services rendered by the 1st Battalion throughout the siege of Seringapatam, the regiment was permitted (under authority dated 8th May, 1818) to bear on its Colours and appointments the word "Seringapatam."

1826

A Horse Guards' letter, dated 26th September, replies to an application from the Colonel of the Regiment, for the 12th to be permitted to bear on their colours and appointments the word "Mauritius," and for the flank companies to bear also the word "Bourbon," as follows :—

> "H.R.H. would really think it advisable that you should pause before you require a grant of this nature, for which he much regrets no precedent has been afforded, and it would appear to H.R.H. that the distinguished services of the 12th Regiment upon all occasions

do not require that that old corps should condescend to seek for distinctions, as arising in an obscure service."[1]

In both the Inspection Reports for the above year, the Colours had been reported on as "unserviceable."

1827

A new pair of Colours having been provided for the regiment, bearing the honorary distinctions, "Minden," "Gibraltar," and "Seringapatam," they were presented to the corps on the 28th June, at Gibraltar, by General Sir George Don, G.C.B., who addressed the regiment as follows :—

"It appears by the record of the 12th Regiment, to which I have the honour of presenting these Colours, that among the many valiant deeds of the corps, it achieved distinguished glory at the Battle of Minden. In 1799, I attended the renowned Duke of Brunswick on the spot where this battle was fought ; after His Serene Highness had shown me the position occupied by the British, he said 'it was here the conflict was most obstinate, and it was here that the British Infantry gained immortal glory.' This Rock and Seringapatam were afterwards among the scenes where the 12th Regiment distinguished itself, and which are immortalized in the history of our country. Being myself a soldier of fifty-seven years' standing, I am alive to every instance of meritorious conduct in my brother soldiers, and it is extremely gratifying to me to reflect that the 12th Regiment, which so early established its fame, has continued to augment it on every occasion ; and I am confident that whenever these Colours shall be displayed before an enemy, the regiment will, by its valiant conduct, add to the number of glorious records with which they are graced."

1832

A reply, in a Horse Guards' letter, dated February 4th, declined a request for the word "Tournay" to be borne on the Colours and appointments of the 12th Regiment, with reference to its services there in 1794, and, in the same letter, the word "India" was disallowed, in consideration of the regiment having already the word "Seringapatam."

1836

In 1836, application was again made for the words, "Tournay," "Bourbon," and "Isle of France" to be borne by the regiment, the reply being that

"The General Officer Commanding-in-Chief does not see any point on which to recommend that these distinctions may be granted, as the 12th Foot is not mentioned in the *Gazette Extraordinary* as having been engaged at the battle of Tournay on the 22nd May, 1794, nor is the regiment alluded to in the public despatch, which announced the capture

[1] *W. O.* 3, Book 76, p. 304.

of the Isles of France and Bourbon, as having performed any peculiar or prominent service on those occasions."

Horse Guards' letters, received on the dates mentioned, announced tha His Majesty the King had been graciously pleased to permit the 12th Regiment to bear the following on its colours and appointments :—

(*a*) *Under authority, dated 5th May*, 1836.

"In addition to the word 'Gibraltar,' the following distinctions, viz., The Castle and Key, and the Motto 'Montis Insignia Calpe,' in commemoration of the service of the Regiment during the memorable defence of Gibraltar, in the year 1782."

(*b*) *Under authority, dated 16th June*, 1836.

"In addition to any other badges or devices which have been already authorised, the word 'India,' in commemoration of its distinguished conduct during the regiment's services in the East Indies from the year 1797 to the year 1809."

1844

In January, 1844, new regulations were issued "forbidding any regimental record or device being placed on the Queen's Colour, other than the number of the regiment, in gold characters, surmounted by the Imperial Crown"; from which, it is clear, that the colours shown in Cannon's plate were intended to represent those of the latest regulation pattern up to 1848, the date of closing his Regimental Record.

1849

New Colours were presented to the 1st Battalion at Weedon on the 14th July by the Hon. Mrs. Arbuthnot. (See Plate 40 and Chapter IX.)

1858

This year a Royal Warrant was issued altering the colours; they were to be only 3 feet 9 inches flying, and 3 feet deep, ornamented with gold and crimson fringe for the Queen's, and white and gold for the regimental Colour; the poles to be surmounted with the crest of England instead of the ornamental spear-head; cords and tassels 3 feet long of crimson and gold.

1859

A new set of Colours, issued from the Ordnance Stores in 1858 to the 2nd Battalion, at Glasgow, was taken into use this year without any formal ceremony.

1870

A Horse Guards' letter, dated 4th June, stated that the Queen had been graciously pleased to approve of the words "New Zealand" being borne on the Colours of the 12th Regiment, in commemoration of its services in that country, during the years 1860–66. The New Zealand War was the last in which the Colours of the regiment were taken into action.

1879

New Colours were presented to the 2nd Battalion on the 2nd August on Southsea Common by Her Royal Highness, the Duchess of Connaught. (See Chapter XI.)

1881

With the introduction of the Territorial system, on the 1st July, there disappeared from the regimental Colour the small Union in the dexter canton, which had been authorised for 130 years, it having been first approved in the Royal Warrant of the 1st July, 1751. In lieu of it, the following was directed in the "Queen's Regulations," dated 1st July, 1881.

> "In regiments which are faced with white, the Second Colour is to be the red Cross of St. George, in a white field, with the Territorial designation and the title displayed within the Union wreath of roses, thistles, and shamrocks, and ensigned with the Imperial crown."

General Order No. 56, this year, announced that in commemoration of the services of the 1st Battalion, during the campaign in Afghanistan, the regiment was permitted to bear on its Colours the words: "Afghanistan, 1878–80." In compliance with an order received, the Colours of the regiment were deposited in the Fort at Peshawur, prior to crossing the frontier on field service.

1882

In commemoration of the services rendered by the 1st Battalion 12th Regiment, in South Africa, during the years 1851–2–3, the regiment was this year permitted to bear on its Colours the words "South Africa, 1851–2–3" (G. O. 252 of 1882).

Notification was received from the Horse Guards, on the 27th September, that the Queen had been graciously pleased to approve of the Suffolk Regiment being permitted to bear on its Colours "Dettingen," in commemoration of the battle fought there on the 27th June, 1743.

1909

In the Clothing Regulations for 1909, Part I., Appendix 11, para. 11, it was laid down that:—

> "In those regiments where the number of actions exceeds nine, laurel branches are to be introduced, and the scrolls bearing the names of the actions entwined thereon."

The height of the staves, with ornaments, by present regulation, is 8 feet 7½ inches, and the dimensions of Colours have not been altered since the Royal Warrant of 1859.

It is worthy of note also that honours are now borne on Colours in chronological sequence, the oldest action being sewn on next the pole, the next in date being on the opposite side of the Union wreath, nearest the fly. In the case of regiments where the number of actions does not entitle them to the distinction of a laurel wreath, their honours are in parallel lines across. Where the laurel is borne on the Colours, the honours are arranged round the wreath, but the same order of sequence is followed. There are, however, slight departures from the above, as, where there is an odd number of actions, the last in order is sometimes placed under the bow of the Union wreath.

In July Army Orders, the King having approved of the dates 1779–1783 being added to the honorary distinctions already awarded for the defence of Gibraltar, these dates were placed on the Colours of the 1st, 2nd, and 3rd Battalions.

1910

In March, an application to the Army Council for the following battle honours on the Colours was refused, viz. :—

1. Flanders (1708 and 1743) ; 2. Lille, or Lisle (1708) ; 3. Kirsch Denkern (1761) ; 4. Wilhelmstahl (1762).

By an Army Order, of the 1st October, it was approved that battle honours and honorary distinctions borne on the Colours and appointments of regular battalions, were in future to be borne in the same way by the Special Reserve battalions affiliated to them.

1912

An application having been made in December, 1911, to the Army Council, for the motto " Stabilis " to be adopted by the regiment on its Colours, a reply was received on the 13th November, that for want of sufficient evidence to establish the claim, it could not be acceded to.

CHAPTER XV

A Short Historical Record of The East and West Suffolk Militia; The latter becoming, in July 1881, The 3rd Battalion Suffolk Regiment.

1660. Shortly after the Restoration in 1660, an Act was passed for ordering the Forces in the several counties of this Kingdom.

By this, and succeeding Acts, the power was entrusted to Lords-Lieutenant of counties, to raise and arm Militia by the King's Commission, and to lead them to battle, in case of internal war.

It must not, however, be imagined that this was the earliest Act of the kind.

The rules for levying the force were as follows :—

Each person in the county whose estate was equivalent to £500 a year (or a personal estate in money and goods of the value of £6000, exclusive of house furniture) was liable to "a horse, horseman and arms."[1] Each person having an estate of the value of £50 a year (or personal estate of £600 value, exclusive of stock upon the soil) was bound to find a foot soldier and arms.[2] Persons holding estates of intermediate value were to pay a proportionate quota towards finding one horse soldier for every aggregate of such estates, amounting to the fixed standard ; and, in like manner, the lesser holders were conjointly to maintain foot soldiers, the liability of some being limited to as little as one tenth of a soldier.

Those who found the men were to furnish them with a specified quantity of ammunition, and were also to find pay for them at the daily rates of 2*s*. for a horse soldier, and 1*s*. for an infantry man ; there was no pro-

[1] The arms of the horse soldier were to include a sword, a case of pistols (whereof the barrels were to be at least 14 inches long) ; a back, breast, and pott (helmet), all pistol proof, besides horse furniture.

[2] The arms to be found for foot soldiers were :—

For a Pikeman, a pike of ash wood, 16 feet long, including head and foot ; a sword ; a back, breast, and head-piece. For a Musqueteer, a musket with barrel 3 feet long, and with a bore for 12 bullets to the pound ; a collar of bandoliers, and a sword.

hibition against serving in person, but not out of the militiaman's own county, except in case of invasion or civil war.

The disciplinary provisions empowered Lords-Lieutenant to punish ordinary delinquencies by fines not exceeding 5*s*., or by imprisonment not exceeding 20 days; but for desertion, the penalty was a fine of £20, or 3 months' imprisonment, and, for failing to appear at the appointed hour of muster, £1, or five days' imprisonment. Lords-Lieutenant also had the power of conferring commissions. In case of the militia being called out for active service, the payment of the men fell upon the Crown after the first month, and the Crown paid the officers at all times.

Oaths of allegiance and supremacy were to be administered to all who presented themselves to serve.

The county and city forces used to be divided into "Trained Bands," and "Auxiliaries," but, in 1663, all the former were treated under the common head of militia.

The London regiments of Trained Bands were of very ancient standing, and had been remodelled by Queen Elizabeth in 1588.

The Militia in other parts of the country were similarly organised into regiments, having their Colours and other appointments in the same way as the regiments of regulars. And their duties were not confined to mere parade and show.

1666. On the declaration of war with France, the militia of the Eastern coasts were summoned to guard the sea-board.[1] The East Riding of Yorkshire furnished 6 regiments, the North Riding, 3 regiments, each with their quota of cavalry, and it was computed that, in the maritime counties alone, above 60,000 well-equipped soldiers turned out on this occasion, and with little more than a day's warning.

In the same year, the militia were again called out all over the kingdom,[2] as well as in Ireland, and large camps were formed for instruction and practice.

1667. In July, the Suffolk Militia, under the Earl of Suffolk, were under fire in an attack upon Aldborough and Landguard Fort by the Dutch.[3] The enemy were in force, about 3000 strong, but the men of Suffolk beat them off with a loss of about 150, and to themselves of only some ten or twelve.

In Charles the Second's reign (1660–85), the Militia were under the Commander-in-Chief, who was authorised by his commission to call out and direct moves of such numbers as he might consider necessary.

1684. Blackheath and Putney Heath were now the favourite spots in England for encampments, and, on the 1st October, 1684, a review was held of all the troops then encamped on Putney Heath.

[1] *London Gazette*, 12th to 15th February, 1665–6.
[2] *Ibid.*, 19th and 26th July, 1666, and 5th and 29th July, 1667.
[3] *Ibid.*, 1st to 4th July, 1667.

1690. The Militia were called out in 1690, during the wars between William III and his exiled father-in-law, James II.

1691. In 1691, considerable doubt existed as to whether militia were subject to martial law, or whether they were to be regarded as a body quite separate from the regular troops, and therefore incapable of taking orders from any but their own officers, or such civil functionaries as Lords-Lieutenant. No decision of this important question can be traced.

In the 17th Century, the duties of modern yeomanry were undertaken by mounted militia.

1697. On the subject of uniform, it may be mentioned that, as early as 1697, militia officers wore crimson silk sashes, as worn by officers of the regulars, and the Suffolk Militia consisted of regiments dressed in red, white, blue, and yellow uniforms.[1]

1715. The Militia were called out this year, during the Jacobite Rebellion.

1743. The officers ceased to carry half-pikes, and were armed with spontoons instead.

1745. The Second Jacobite Rebellion, which took place this year, involved the Militia being again called out.

1752. The Militia of Great Britain and Ireland, had, by this time, practically ceased to exist.

1757. The necessity for obtaining troops from our Hanoverian dominions, at the beginning of the Seven Years' War in 1756, called attention to the general inefficiency of the English Militia, and an Act was passed in 1757, reorganising the force. This Act abolished the liability of individuals to provide men, and instead, men between 18 and 50 were chosen by ballot, under the superintendence of Lords-Lieutenant of counties, and had to serve three years or provide a substitute.

1758. The present system of voluntary enlistment was practically introduced by the Act of 1758, which authorised Lords-Lieutenant of counties to allow any parish to avoid the militia ballot by voluntarily furnishing the quota of men required. Militia officers were required to be qualified by the possession of landed property in the counties.

1759. A War Office letter, dated October 16th, 1759, to His Grace, the Duke of Grafton, announced the King's command for the militia of the County of Suffolk to be embodied in two battalions, the 1st or Western Battalion, under command of Colonel the Honourable Nassau, to march to Bury St. Edmunds, and the 2nd, or Eastern Battalion, commanded by Colonel Francis Vernon, to Ipswich.[2]

The following shows the establishment, and daily rates of pay to each battalion :—

[1] Colonel Clifford Walton, *History of the British Standing Army*, Vol. ii.
[2] *W. O.* 4, Book 757, p. 252.

	£	s.	d.
1 Colonel without a company	0	14	0
1 Lieut.-Colonel and Captain	0	17	0
1 Major and Captain	0	15	0
10 Captains at 10s.	5	0	0
12 Lieutenants at 4s. 8d.	2	16	0
12 Ensigns at 3s. 8d.	2	4	0
1 Adjutant	0	4	0
1 Quarter-Master	0	4	8
1 Surgeon	0	4	0
12 Sergeants at 1s.	0	12	0
12 Corporals at 8d.	0	8	0
12 Drummers at 8d.	0	8	0
468 Private men, each at 6d.	11	14	0
Allowance to the agent	0	6	0
Total	£26	6	8

In the first estimate of pay rendered, each Colonel was shown as in charge of a company, drawing £1 4 0 a day.

The Militia were to be paid only from the day of marching, and a grant of £7 10s. was extended to each battalion, to defray hospital expenses, when encamped, for such sick men as could not be properly taken care of in quarters.[1]

With reference to the precedence or seniority of militia regiments, it was directed that :—

> "The regiment of Militia which first arrives in camp, or quarters, shall have seniority there, the next which arrives being deemed junior, and so on, as they march in successively."[2]

This was altered in the following year.

1760-62. On the 26th June, 1760, His Majesty directed that "when any regiments or detachments of militia shall meet, they shall draw lots for taking post as to precedence while they remain together, this order not to disturb the precedence which may have been established by agreement between the several battalions of Militia belonging to the same county."

The following were the moves of the two battalions of the Suffolk Militia, to the date of their disembodiment, December 20th, 1762 :—

1st, or Western, Battalion.

1759. December, to Peterborough and Oundle.

1760. October 14th, to Bury St. Edmunds.
,, October 23rd, to Sudbury.

1761. May, to Hilsea, where the battalion suffered much from sickness.
,, October, to Sudbury, Ballingdon, and Lavenham.

1762. June, to Bury St. Edmunds, for disembodiment in December.

2nd, or Eastern, Battalion.

1759. November, to Peterborough, with powers to the commanding officer to detach companies as he might consider necessary.
,, December, to Leicester.

1760. October 14th, to Bury St. Edmunds.
,, October 23rd, to Beccles, and later, Woodbridge.

1761. May, to Ipswich, 5 companies to Landguard Fort in June.
,, October, to Woodbridge.

1762. January (now 12 companies), 7 to Ipswich, and 5 to Landguard Fort.
,, June, to Sandheath Camp, near Ripley (Lord Orwell commanding).
,, October, to Woodbridge.
,, December, to Ipswich, for disembodiment.

[1] *W. O.* 4, Book 757, p. 164.

[2] *Ibid.*, p. 181.

Letters to the Duke of Grafton and to Lord Orwell, dated 20th December, 1762, directed the disembodiment of the two Suffolk battalions, stating that:—

> "His Majesty expressed to the Militia the great satisfaction he had received from their seasonable and meritorious service, and, as a mark of his Royal approbation, was pleased to permit each non-commissioned officer and private man to keep his clothes and knapsack, which are in present wear, and also to allow them, respectively, 14 days' pay from the date of their being disembodied."

1778. A War Office letter, dated 26th March, 1778, to His Grace the Duke of Grafton, announced the King's Command for the Militia of the County of Suffolk to be embodied in two battalions, the 1st, or Western Battalion, under command of the Duke of Grafton, being encamped that summer at Coxheath Camp, and the 2nd, or Eastern Battalion, under Colonel Woolaston, at Warley Camp.

The establishments of each battalion, and rates of pay, were as prior to disembodiment in 1762.

On the 10th August, 1778, the 1st, or Western, Battalion was inspected by Lieut.-General Keppel, who expressed to the Duke of Grafton his approbation of the appearance and behaviour of the men, and, on the 3rd November, the whole of the forces assembled at Coxheath Camp were reviewed by King George III.

The Regimental order for dress at this parade was:—

> "The men to be particularly clean, clothes well brushed, hair neatly tied up and powdered, and accoutrements well put on; the officers to wear white cloth breeches, boots, and laced hats."

1779. On the 5th July, 1779, the Duke of Grafton issued orders to officers commanding companies of the 1st, or Western, Battalion to form their men into messes, "every soldier to be stopped *2d.* a day for meat and roots for the mess he belongs to." The Duke, being willing to give every encouragement to those who were most ready to enter into this regulation, undertook to give three pounds of good beef or mutton to each of the 40 messes who formed themselves quickest in a regular manner.

A War Office letter, dated 30th September, 1779, notified that: "His Majesty having approved of all Colonels of Militia being appointed Colonels in the Army, to rank as such, so long only as the Militia shall be embodied," the two Colonels of the Eastern and Western Battalions Suffolk Regiment were accordingly gazetted as such, the Duke of Grafton's commission bearing date, 2nd July, 1779.

1780. With reference to a surgeon's bill for inoculation, sent in by the Duke of Grafton, to the War Office, with other contingent expenses, it was ruled, on the 25th January, 1780, that it had been the rule of the War Office for many years, to allow a sum not exceeding one guinea for each man inoculated.

The Duke of Grafton resigned his commission as Colonel of the Western

Battalion Suffolk Regiment, on account of ill health, in February, 1780, and was succeeded by the Earl of Euston, and in the same year, Colonel Woolaston, Eastern Battalion, was succeeded in the command by Lieut.-Colonel Gibb.

1781. In a Regimental Order, dated 28th July, 1781, Lord Euston recommended all officers commanding companies to supply their companies with half a dozen wooden instruments with which to tie their men's hair both uniformly and expeditiously, and no man was required to powder his hair more than once a day unless particularly ordered.

The following were the moves of the two battalions to date of their disembodiment, 28th February, 1783 :—

1st, or Western, Battalion.

1778. May 24th, to Sudbury, Long Melford, and Lavenham.
,, June 2nd, to Coxheath Camp.
,, November 7th, to Bury St. Edmunds.

1779. April 3rd, 7 companies between Chelmsford, Springfield and Moulsham, and 1 between Bromfield, Woodford, and Writtle Green.
,, July 5th, to encamp on Warley Common.
,, November 26th, (now 9 companies) headquarters and 3 companies to Hitchin, 2 to Hatfield and Bell Bar, 1 between Welwyn, Codicote and Lemsford Mills, 2 to Stevenage, and 1 to Baldock for the winter.

1780. April 27th, to Newcastle.
,, May 10th, to Hull.
,, October 30th, 3 companies to Newcastle-on-Tyne, 3 to Alnwick, and 3 between Morpeth, Bedlington and Blythe (winter quarters).

1781. June 1st, to Warley Common.
,, October 30th (now 8 companies) headquarters and 4 companies to Newmarket, 1 to Mildenhall and Barton Mills, 2 to Stowmarket and Needham, and one between Botesdale and Ixworth (winter quarters).

1782. July 5th, to Danbury Common.
,, November 8th, 4 companies to Newmarket, All Saints (Halesworth), and Woodditton, 3 to Bury St. Edmunds, and 1 between Mildenhall, Barton Mills and Brandon.

2nd, or Eastern, Battalion.

1778. June 1st, to encamp on Warley Common.
,, November 4th (now 8 companies), 4 to Romford and Hare Street, and remainder between Woodford, Epping, Ongar, Dagenham, and Ilford.

1779. June 1st, to Coxheath Camp.
,, November 27th, to Ipswich, and then the regiment divided between Beccles, Bungay, Halesworth, Yoxford, Saxmundham, and Lowestoft.

1780. May 30th, to Chatham.
,, September 29th, the recruits to Faversham.

1781. October 20th, headquarters and 4 companies to Beccles, Lowestoft, and Bungay, 2 between Woodbridge, Wickham, Framlingham, and Aldborough, and 2 between Blythborough, Halesworth, Saxmundham, and Southwold.

1782. June 22nd, The battalion was ordered to march to Brentwood, and on the 8th November (now 7 companies), it was ordered to be distributed in 12 detachments in Suffolk and Essex.

1783. A Circular Letter to Officers Commanding Militia Regiments, dated 28th February, 1783, directed the disembodiment of the Eastern and Western Suffolk Militia Battalions, stating that :—

> " His Majesty expressed to the Militia the great satisfaction he had received from their seasonable and meritorious service, and, as a mark

of his Royal approbation was pleased to permit each non-commissioned officer and private man to keep his clothes and knapsack, which are in present wear, and also, to allow them respectively 14 days' pay from the date of their being disembodied."[1]

An order, dated 4th March, 1783, directed that these battalions were to march to such place or places as their commanding officers should judge most convenient, to carry into execution His Majesty's orders for their being disembodied.[2]

. . . .

1792. By an order, dated 4th December, 1792, the two battalions of Suffolk Militia were re-embodied, with effect from the 1st December, and their respective Colonels were informed accordingly, viz. :—Lieut.-Colonel the Earl of Euston, commanding the 1st, or Western, Battalion, headquarters, Bury St. Edmunds, and Lieut.-Colonel Goate, commanding the 2nd, or Eastern, Battalion, headquarters at Ipswich.

The authorised establishments were to date from the 25th December, and each battalion of 8 companies was to number 21 officers, 24 sergeants, 17 drummers for the Western, 18 for the Eastern, and 504 rank and file, the total annual cost estimated for each battalion being £9,221 6*s.* 9*d.* for the Western, and £9,250 2*s.* 10*d.* for the Eastern.

Indents on the Ordnance Department, dated 19th December, 1792, showed that 63 stand of arms were required for the light company of each battalion, with swords for the sergeants and drummers of the battalion (also drums for the latter), and a letter, dated 23rd August, 1793, applied for a new set of Colours for the Eastern Battalion.

The monthly allowance of £7 10*s.* for hospitals was continued as during the previous embodiment, and, under an Act of Parliament, volunteers numbering 3 sergeants and 59 rank and file were added to each battalion.

1793. On the 31st January, 1793, surgeons' mates were sanctioned to a regiment, but not to be considered part of the establishment. On the 30th March, on application from the Earl of Euston, the authorities sanctioned a chaplain for the Western Battalion, at 2*s.* 6*d.* a day, whilst at Hilsea.

1794. The West Suffolk Battalion, after encamping at Brighton, from May to October, 1794, were directed, when marching from Brighton, to leave their battalion guns, taking over those of the Somerset Militia on arriving at Margate, the same course being observed when leaving Margate in the following May (1795) again to encamp at Brighton, when they received their own guns again.

On the 9th May, 1794, certain allowances were authorised to Colonels of Militia, to provide "slop clothing" for such volunteers to their regiments

[1] This letter is identically worded with that of 1762.

[2] *W. O.* 5, Book 101.

as should be actually raised and added to the establishments, prior to the 24th June, following. The articles were to consist of :—1 jacket, 9*s.* 6*d.*; foraging cap, 1*s.* 6*d.*; pair breeches, 6*s.*; pair shoes, 6*s.*; pair gaiters, 4*s.*; shirt and stock, 6*s.* 6*d.*; pair stockings, 1*s.* 6*d.* Total £1 15*s.*

1795. In 1795, militiamen who enlisted into the Line received a bounty of 5 guineas.

Amongst the large number of Fencible Regiments raised between 1796 and 1802, was one termed "The Loyal Suffolk," under command of Colonel John Robinson, which was ordered for service in Ireland in 1799, and, on the 1st July, disembarked at the Pigeon House Pier, Dublin, strength, 27 officers and 525 of other ranks.[1]

1797. A Horse Guards' Circular, dated February 17th, 1797, directed that, for the ensuing summer, the militia forces should be formed into brigades, whereupon the West Suffolk was posted to the 2nd Brigade of the Division commanded by General Sir Wm. Howe, K.B., together with the East Norfolk, Warwick, and Cambridge Militia Regiments, whilst the East Suffolk was posted to the 4th Brigade of the same division, commanded by General Sir Chas. Grey, K.B., the other corps in this brigade being the East Hants, Bedford, Derby, and Yorks (West Riding).

1798. A Horse Guards' Circular, dated, January 31st, 1798, sanctioned the calling out and embodying a certain portion of the supplementary Militia, with a view to augment regiments of the Line.

Ipswich was to be the place of assembly of the supplementary battalions of the West and East Suffolk Militia, and Colchester and Ashford respectively their destinations.

An Order, dated September 1st, 1798, to Ireland, for the augmentation of the West Suffolk Battalion, directed it to be 10 companies, and 1125 of all ranks, whilst the East Suffolk Battalion of 8 companies was to muster 1073 of all ranks.

The West Suffolk Battalion embarked at Rockhouse Ferry for Ireland on the 10th September, 1798, and, on arrival at Dublin, proceeded to Dundalk, under Colonel the Earl of Euston. The battalion returned to England in June the following year, embarking on the 10th of that month, at Warrenpoint, in the ships "Amity," "Henrietta," "Elizabeth," "Sutton," "Nelly" and "Packet," strength, 20 officers, 656 of other ranks, 46 women, and 8 children. A detachment, which had been left in England in charge of the heavy baggage, had been ordered to Huntingdon, and to convey it thence to Norman Cross, being moved in April, 1799, to Chelmsford.

1799. An order, dated December 5th, 1799, directed the battalion's supernumerary detachment to march to Newmarket, or any other convenient place, to be disembodied, permission having been given in the previous month for a supernumerary captain to command the Colonel's Company.

[1] *Government Correspondence Book,* No. 141, Ireland, P.R.O. Dublin.

On the return of the West Suffolk from Ireland, it was reduced in the autumn of 1799 from 10 to 7 companies, and, on that account, an application from Lord Euston, for the battalion to have 2 flank companies, was refused by the War Office (letter dated December 17th, 1799), one flank company only having been allowed to a battalion of 7 companies.

1800. An Inspection Return of the East Suffolk Militia shows that the battalion was inspected by Major-General Lord Mulgrave, at Leeds, on the 17th September, 1800, mustering 32 officers and 459 of other ranks, when it was reported on as " a serviceable body of men, fit for any service, and went through the regulated manœuvres with tolerable accuracy."[1]

1801. The West Suffolk Battalion, commanded by Lieut.-Colonel Parker, was inspected at Colchester by Major-General Chas. Lewis, on April 15th, 1801, mustering 25 officers and 471 of other ranks, and was reported on as :—

> " extremely well dressed, and in very high order, steady under arms, and manœuvres done with great precision, except in lifting their feet rather too high when marching in ordinary time. The Officers and N.C.O.'s attentive ; the men strong, chiefly young and active, though not tall. The regiment in every respect fit for service, and does credit to the attention of the commanding and other officers."

1802. On the 6th April, 1802, both battalions were inspected at Ipswich by Major-General Grimfield, and received, on the whole, good reports, and by an order, dated 14th April, both battalions were disembodied, in compliance with the order for the general disembodiment of the Militia.

The following were the moves of the two battalions from December, 1792, to date of their disembodiment, April, 1802 :—

1st, or West, Suffolk.

1793. January 14th, to Hilsea.
,, June 19th, to Waterdown to encamp.
,, October 11th, 3 companies to Farnham, 3 to Alton, Chewton, and Farringdon, 2 to Alresford.
,, November, detachments to Bentley, Arundel and Aldershot.
1794. Companies detached in March and April to Petersfield, Odiham, Elvetham and Harslem.
,, May 12th, to Brighton to encamp.
,, October 29th, to Margate, with 10 different detachments for the winter.
1795. May 13th, to Brighton to encamp.
,, October 14th, 4 companies to High and West Wycombe, and 4 between Maidenhead, Marlow, and Bexham.
,, October 26th, headquarters to Marlow, with several detachments.
1796. Headquarters in January at Maidenhead with several detachments.

2nd, or East, Suffolk.

1793. February 7th, 6 companies to Yarmouth, and 2 to Lowestoft.
,, June 19th to Harwich to encamp
,, October 10th, to Newmarket, with 11 detachments.
1794. May 7th, to Caister, to encamp.
,, October 31st, 5 companies to Yarmouth, 1 to Lowestoft, and the other 2 split up into several small detachments.
1795. May 12th, to Warley Camp.
,, October 16th, 2 companies to Harwich, 4 to Landguard Fort, and 2 companies split up into several detachments.
1796. Harwich to Canterbury, April 19th.

[1] *W. O.* 27, Book 83.

1st, or West, Suffolk.

1796. April 29th, To Uxbridge with several detachments.
,, June 8th, to Warley Camp.
,, October, to Bury St. Edmunds for the winter.
1797. March 20th, at Ipswich.
,, In July, to Colchester.
1798. January 27th, to Ipswich, returning in March.
,, September 10th, to Ireland, and proceeded to Dundalk.
1799. In June to England. Proceeded to Chelmsford.
,, November 21st, to Cambridge.
,, December 29th, to Sudbury.
1800. May 4th, to Bury St. Edmunds.
1801. March 15th, to Colchester.
,, September 20th, to Ipswich.
1802. April 9th, to Bury St. Edmunds to be disembodied.

2nd, or East, Suffolk.

1796. May 17th, 2 companies to Tunbridge Wells, 2 to Penshurst, and 4 between Tonbridge, Hadlow, Mereworth, Teston, and Yalding. In the autumn to Ashford for the winter.
1797. March 20th, at Ashford.
,, In May, to Dover Castle.
1798. In May, to Ashford Barracks.
,, In the autumn, to Dungeness.
1799. April 16th, to Hull, the grenadier company at Canterbury joining at Sevenoaks.
1800. In September, at Leeds.
,, November 12th, on arrival at Sheffield, to march in 2 divisions to Stourbridge, Dudley, and adjacent places.
1801. January at Wolverhampton.
,, 2nd September to Chelmsford.
1802. April 12th, to Ipswich, to be disembodied.

. . . .

1803. By the King's command, an order, dated 12th May, directed that the Militia were to be re-embodied, the establishments of the West and East Suffolk battalions numbering 30 officers, 28 sergeants, 19 drummers[1] and 547 rank and file. The Earl of Euston again commanded the West Suffolk, while Colonel Goate again assumed command of the East Suffolk, the headquarters of the two battalions being Bury St. Edmunds and Ipswich respectively, as before.

Adjutants of militia were now placed on the same footing as those of the Line, receiving 8*s*. daily.

The West Suffolk Battalion, whilst encamped this summer at Thorington Heath, Essex, furnished picquets and patrols to watch the coast from Brightlingsea to Walton, some excitement now prevailing in England in consequence of Napoleon's threatened invasion.

On the 20th October the battalion was inspected by Major-General Hope, who specially praised the manner in which the extended order was performed.

1805. On the 13th September, the addition of a second major for each of the Suffolk battalions was authorised.

Viscount Brome had now been appointed Lieut.-Colonel of the East Suffolk.

On the 15th November, the establishment of each battalion was reduced to 8 companies, each with 4 field officers, and there were to be no supernumerary captains.

[1] This number included the drum-major, who was on the same rate of pay, no higher rate being allowed in the Line.

1807. The West Suffolk Battalion was inspected at Ipswich on the 8th April, by Lieut.-General Chas. Fitzroy, strength 588 rank and file, under Colonel the Earl of Euston, and was well reported on, the men being "active, quick in the field, and well set up."

Viscount Brome had now succeeded to the title of Marquis of Cornwallis, and the East Suffolk Battalion, under his command, was inspected at Aberdeen, on the 9th May, by Major-General Macdonell, and was well reported on, the Inspecting Officer observing :—

> "the exemplary conduct of the officers, and the orderly behaviour of the men has afforded me much pleasure, and has gained the approbation of the magistrates and inhabitants of this city."

On the 18th August, 1807, men of the West Suffolk Militia were invited to volunteer into the 10th, 48th, and 63rd Regiments. Option was given them either to enlist for the limited period of seven years, receiving a bounty of 10 guineas, or to enlist, without limitation of time, receiving a bounty of 14 guineas.

To men who might afterwards enter the militia, a bounty of 6 guineas would be allowed if they enlisted for seven years, and 10 guineas if they enlisted without limitation of time.

1808. Instructions, issued in February 1808, to the Colonels of both battalions, directed their establishments (with effect from the 25th December, 1807), to consist of 29 officers, 35 sergeants, 1 drum-major, 18 drummers, and 711 rank and file respectively, a second surgeon's mate being sanctioned in September to the West Suffolk Battalion at Tynemouth.

1809. On the 25th October, 1809, a jubilee was held to celebrate the commencement of the 50th year of King George III's reign, in which both battalions took part.

1810. The voluntary enlistment of militiamen by beat of drum was this year authorised.

1811. The West Suffolk Battalion was inspected at Liverpool on the 16th May, 1811, by Major-General Robinson, who expressed entire satisfaction at its state.

1812. In 1812, the battalion was inspected on the 27th October, by Major-General Acland, at Bradford, who found great fault with the state of the ammunition in the men's pouches, Lieut.-Colonel Wollaston commanding.

On November 1st, the following appropriation of the militiaman's pay was directed :—4*s*. 6*d*. to be expended weekly for the mess ; 1*s*. 6*d*. weekly to the soldier, subject to the usual deductions for washing, and small articles for cleaning purposes, and the remaining 1*s*. 6*d*. to be retained for necessaries, and accounted for monthly. Woollen mits were to be worn in future, instead of leather gloves, and false frills or dickies were forbidden, the frills to be sewn on to the shirts.

1813. On the 2nd March, 1813, grey trousers were taken into wear. On the 11th, the West Suffolk Battalion marched from Halifax and Bradford, in two divisions, to Berwick and Tweedmouth, and, on the 13th April, to Port Patrick, where the battalion embarked later for Donaghadee, in the ships "Earl of Stair," "Bachelor," and "Melville Castle," whilst six officers' horses only were conveyed in the packet "Hillsborough." The Battalion left a detachment at Berwick-on-Tweed, and the Disembarkation Return shows the strength as 20 officers, 26 sergeants, 19 drummers, and 390 rank and file, 44 women and 20 children. The battalion disembarked on the 2nd May, arrived at Armagh on the 3rd, and on the 14th December marched to Tullamore for the winter.

1814. An order, dated 22nd January, 1814, directed all men of the East Suffolk Battalion, extending their service in the militia, to proceed to Danbury Barracks, to join a Provisional Battalion.

On the 18th February, this battalion embarked at Gosport for conveyance to Ireland, in two ships, the "Essex" and "Mary and Margaret," and a Disembarkation Return shows it disembarking at Monkstown on the 19th; strength, 12 officers, 18 sergeants, 10 drummers, 157 rank and file, 46 women, and 47 children.

In July, the detachment of the West Suffolk Battalion at Berwick was ordered to march to Bury St. Edmunds, in compliance with the King's command, which had been given out on the 6th June, for the disembodiment of the Militia, His Majesty thanking the regiments for their ready and cheerful obedience.

Each subaltern and surgeon's mate (if any) was to receive, on disembodiment, an allowance equal to two months' pay, but the Adjutant, Paymaster, Surgeon, and Quarter-Master, being retained on duty, were not to receive this, their services being probably required with some of the Provisional Battalions which were being formed, of men who volunteered for extended service in Europe.

Each private man disembodied was to receive a bounty equal to 14 days' pay from date of disembodiment, but non-commissioned officers and drummers were excepted, the latter to wear their present clothing until Christmas, when a complete supply of new clothing would be issued to them, that for each rank to consist of a coat, a waistcoat, a cap, cockade, feather or tuft, a plate and cap case, a pair of brushes, and a pair of military shoes.

On the 30th August, 1814, the West Suffolk Battalion left Tullamore and marched to Dublin, where the regiment embarked on the 31st for Bristol, in the ships "William Luce," "Countess of Mexborough," "Regent," "Liberty Brig," and "Liberty Sloop," and the Embarkation Return shows the strength as 16 officers, 25 sergeants, 19 drummers, 361 rank and file, 52 women, 24 children, and 7 officers' horses.

Arriving at Bristol on the 5th September, the battalion was ordered to march, in two divisions, to Ipswich, to be disembodied; on the 15th, Brentford and Isleworth were substituted for Ipswich. The battalion, however, finally marched to Bury St. Edmunds, arriving on the 24th, and was disembodied on the 4th October, 1814.

The West Suffolk Militia, during the war of 1803–1814, supplied 33 officers and 1145 men as volunteers to the regular army, and in the same period the Militia Force supplied nearly 200,000 well-disciplined men to different regiments of the Line.

In 1814 the men of the West and East Suffolk Militia wore yellow facings with white lace, the West Suffolk lace showing a yellow worm in it, and that of the East Suffolk, a red worm.

1815. Notwithstanding the order for disembodiment, the East Suffolk Battalion remained in Ireland throughout the year 1815, and old War Office books show that on the 16th August and 25th November, that year, two strong detachments of recruits were ordered to march from Ipswich to Liverpool, "to await orders, pending conveyance to Dublin." The battalion, however, on arrival from Ireland, in two detachments, in January, 1816, was ordered to Ipswich, and was disembodied in February.

1816. The following were the moves of the two Suffolk battalions of Militia from May 1803 to date of disembodiment, February, 1816.

1st, or West, Suffolk.

1803. May 28th, Bury St. Edmunds to Ipswich.
,, August 3rd, encamped at Thorington Heath, Essex.
,, November, to Colchester.
1804. July 11th, at Ipswich.
,, August, to Lexden Camp and Colchester for the winter.
1805–06. February, 1805, to Ipswich, and remained throughout 1806.
1807. October 5th, to Chatham.
,, To Ipswich for the winter.
1808. To Tynemouth.
1809. Remained throughout the year.
1810. June, at Sunderland.
,, October 17th, to Liverpool.
1811. Remained throughout the year.
1812. August 20th, to Halifax and Bradford.
,, December, at Wakefield.
1813. March 11th, to Berwick and Tweedmouth.
,, April 13th, to Ireland, and proceeded to Tullamore.
1814. August 30th, to England, landing at Bristol.
,, September 24th, to Bury St. Edmunds to be disembodied.

2nd, or East, Suffolk.

1803. Remained at Ipswich.
1804. June 21st (10 companies), marched to Hull.
1805. July 4th, to Berwick and Tweedmouth.
,, In November at Aberdeen.
1806. Remained throughout the year.
1807. To Sheerness for the winter.
1808. To Maidstone.
1809. February 15th, to Sheerness.
,, June, at Ashford, Kent.
,, October 14th, to Winchelsea.
,, November 28th, to Hungerford and Newbury Speen.
1810. April 8th, to Brentford.
,, May 14th, to Acton, Ealing, and Hanwell.
,, June, a detachment at Tower of London.
,, August, at Portsmouth.
1811. October 11th, at Porchester.
1812–13. To Gosport, and remained throughout 1813.
1814. February, to Ireland, disembarked at Dublin.
1815. Remained throughout the year.
1816. January to Ipswich, and disembodied there in February.

. . . .

The Acts of 1816 and 1817 permitted the annual training of Militia to be suspended by Order in Council, and from 1815 to 1852, the Militia practically ceased to exist, but the permanent staffs of each regiment were maintained.

1829. In 1829, sergeants gave up the use of halberds. Since that year, an Act was for some time passed annually, suspending all proceedings for raising Militia by ballot, unless directed by Order in Council. In 1830, the Earl of Euston was appointed Honorary Colonel of the West Suffolk Militia (24th May) in succession to Colonel Sir Wm. Parker, Bart., deceased.

1831. On the 14th March, a letter was sent to the Colonel of the regiment from the aldermen, recorder, and magistrates of Bury St. Edmunds, expressing their

> "extreme satisfaction at the very orderly and good conduct of the regiment, during the whole time of embodiment in that town, which they feel, in conveying this testimony, is not more than an act of justice to the corps."

1833. In 1833, a circular was issued by Lords-Lieutenant of counties, announcing that lots had been drawn, in presence of King William IV, and several Lords-Lieutenant and colonels of Militia, to determine finally and permanently the precedence of militia regiments in Great Britain. By this method, the West Suffolk Militia became tenth in order of precedence.

In consequence of the threatening armaments of France, in 1852, an Act was passed to reorganise the Militia by voluntary enlistment, with power, in case of its insufficiency, to resort to the ballot. This was given effect to by Lord Derby's Ministry in 1853, and has since been carried on without recourse to the ballot.

. . . .

1853. The full training bounty of one guinea was paid to each volunteer for the year, excepting those who were enrolled after the 30th September, who were paid under different conditions.

The East Suffolk Battalion was this year converted into Militia Artillery.

The West Suffolk Militia assembled for training on the 19th September, at Bury St. Edmunds, under Colonel Lord Jermyn, and was inspected, on the 15th October, prior to dismissal of the men to their homes.

Officer's Skirt Ornament, 1853–55.

In the margin is shown (considerably reduced in size) a battalion company officer's skirt ornament worn at this period, and until the abolition of the coatee in March, 1855. Handsomely worked in silver embroidery, it bears a Castle on a black velvet ground, with the grass in front of it in green silk, the lettering being in gold, on a ground of light blue silk. The Castle, it will be noted, is not that of Gibraltar, but a badge of Suffolk.

1854. On December 14th, the West Suffolk Militia was permanently embodied, and billeted at Bury St. Edmunds; strength, 15 officers and about 670 of other ranks. The "Angel Hill" was named as an alarm post, and in case of the "Assembly" sounding at any time during the night, all officers and men were immediately to proceed there without arms, and to form up and await orders.

1855. On March 15th, the regiment had the "Albert" Shako issued, with black leather chin-strap, and peaks front and back, and, in May, a Regimental Bugle Call was established, and ordered to be sounded before all other bugle calls.

Colonel Earl Jermyn retired on the 3rd August, on being appointed Honorary Colonel of the regiment, and was succeeded in the command by Lieut.-Colonel Deare.

The regiment was inspected on October 15th, at Bury St. Edmunds, by Major-General Slade, who reported on the appearance of the men as "most satisfactory," and their "movements well and steadily performed."

1856. The battalion having, since its embodiment to 30th January this year, given 226 volunteers to the Regular Forces, the Colonel had, in the interval, become entitled to recommend, at intervals, three of his officers for commissions in the Line.

On the 19th February, the regiment proceeded to Colchester, and returning on the 6th June, was, with the remainder of the Militia Force, again disembodied.

The letter from the Lord Lieutenant of the county emphasised

> "the high sense entertained by Her Majesty the Queen, of the conduct of the Militia, and of the zeal and spirit which they had manifested since they were embodied."

The letter further recorded the Lord Lieutenant's best thanks

> "for the great credit attached to the county of Suffolk, in consequence of the excellent discipline, great military knowledge, and general good conduct, which has distinguished the West Suffolk Militia during the time it has been embodied."

. . . .

1858. The Permanent Staff of the battalion was still kept up at Bury St. Edmunds, and, on February 15th, orders were issued for it to be employed in recruiting for the regular army.

Up to October this year, the numeral "X," signifying the battalion's seniority, had been worn by all ranks on forage caps. It was now exchanged for the half moon scroll, bearing the name of the regiment "West Suffolk Militia," sergeants of the Permanent Staff only retaining the "X."

The magistrates of the Borough of Bury St. Edmunds again sent a letter to the Officer Commanding, on the expiration of the Militia services, expressing their great satisfaction at the general good conduct of the men.

1859. This year, the Militia of England, Ireland, Scotland, and Wales became one national force, liable to serve in all parts of the United Kingdom.

On December 31st, percussion muskets were returned into store at Harwich, and the Enfield Percussion Rifle was issued to the regiment.

Notwithstanding the official disembodiment in June, 1856, the following annual trainings of the West Suffolk Militia are recorded in its manuscript "Digest of Service," viz. :—

In 1860, assembled 30th April, for 27 days' training (801 present).
,, 1861, ,, 22nd ,, ,, 27 ,,
,, 1862, ,, 8th May, ,, 21 ,, (strength 1000 men).

For the first time, since enrolment in 1852, the regiment was completed to its full strength, viz. 1033 rank and file, and application was made to the Secretary of State for War, to enrol over the establishment.

In 1863, assembled 24th April, for 21 days' training (strength 1001).
,, 1864 ,, 21st ,, ,, 21 ,,
,, 1865 ,, 1st May ,, 27 ,,
,, 1866 ,, 30th April ,, 27 ,,
,, 1867 ,, 29th ,, ,, 27 ,,

During the training, there having been no hospital accommodation, sick men had to remain in their billets.

1867. This year, a Militia Reserve was formed, one-fourth of the regiment being liable to be called up for army service at any time if required, receiving £1 bounty in addition to the Militia bounty.

On the 24th December, the companies of the battalion were designated by letters instead of numbers, from A to K, excluding J.

1868–69. New pattern belts and pouches were received and issued to the men, and, on the 9th April, nipples were replaced in the rifles. Colour-sergeants' badges were changed from crossed swords and flag with crown above and double chevron below, to crossed flags, with crown above and three chevrons below.

In 1868, assembled 27th April for 27 days' training.
,, 1869 ,, 26th ,, ,, ,,

Sergeants' stripes were this year changed from tape to narrow silver lace.

1870. In March, Snider rifles were issued, and Enfield rifles returned into store. Lieut.-Colonel Deare, commanding, died on the 27th March, and was succeeded by Major F. M. Wilson, on the 24th April.

Assembled, 25th April, for 27 days' training.

1871. ,, 8th May, ,, ,,

The Militia were this year placed under direct control of the Crown, and re-organised in connection with the different Territorial Districts.

1872. Assembled, 6th May for 27 days' training.

On the 31st, the highly satisfactory inspection of the battalion by

Major-General Sir E. Greathead, C.B., elicited later the following remark from the Commander-in-Chief :—

> "His Royal Highness observes with much satisfaction, that the officers throughout, as well as the Permanent Staff, non-commissioned officers and men, are reported to be thoroughly efficient."

The Militia Reserve being now 30 below strength, volunteers were called for, when 170 of the battalion volunteered to it. On application being made, to accept men beyond the quota, it was ruled that the standard height for all in excess of it should be 5 feet 6 inches, and 33 inches chest measurement.

Mr. Cardwell was at this time Secretary of State for War.

1873. Infantry Brigade Depots having been formed this year, Bury St. Edmunds became the permanent depot of the 12th (East Suffolk) Regiment, and was numbered "32," it having for years been the head-quarters of the West Suffolk Militia.

Shell-jackets of red-brick cloth were this year replaced by scarlet kersey frocks, with no facing on the cuffs, but a white braiding instead, the depth of the facing, finishing in a peak on the arm, in a Prussian knot, and buttons were altered to those of the universal pattern.

On formation of the Brigade Depot, No. 32, vacancies in the Militia were to be filled from the 12th (East Suffolk) Regiment, and if none were available, they were to be filled from other corps.

Assembled, 28th April, for 27 days' training.

1874. In February, shakos were returned into store, and Glengarry caps were issued instead, one every second year for dress, and one every sixth year for fatigue.

Assembled, 28th April for 28 days' training.

1875. ,, 19th ,, ,, ,,

The Inspection Reports on the battalion for the two preceding years had brought forth great praise from the Commander-in-Chief, on account of its creditable state, in point of efficiency, steadiness under arms, cleanly appearance, and interior economy, His Royal Highness specially remarking on the general efficiency of the officers, and that the commanding officer, adjutant, and permanent staff, deserved great credit for the manner in which they had discharged their duties.

Similarly this year, the regiment was highly complimented by the Inspecting Officer (Colonel J. Nason) on its drill, general smartness, and good conduct in billets, during training.

Lieut.-Colonel Wilson died on the 4th September, and was succeeded in the command by Major W. J. Marshall.

1876. On March 13th, the old wooden shell side drums, and old pattern bugles were returned to the Harwich stores, and replaced by brass shell side drums, and bugles of small pattern.

On April 21st, Glengarry cap and collar badges, in white metal, were issued, showing:—Castle, with crown above, encircled with laurel leaf, and words "West Suffolk Militia" beneath, replacing the half moon scroll, for the Glengarry cap, the collar badges showing:—Castle, with laurel leaf border, and words as above beneath the badge, the dimensions being $\frac{3}{4}$ inches high, and $1\frac{3}{4}$ inches long.

"Small books" were issued this year to the men of all militia regiments.

Assembled, 1st May, for 27 days' training.

1877. ,, 23rd April, ,, ,,

On August 16th, drummers' swords, of scimitar pattern, were returned into store, and replaced by gladiator swords, of army pattern.

Orders were received, forbidding any man to be promoted to the rank of sergeant, who had not passed an examination, and obtained a certificate of efficiency, after undergoing a course of instruction for 61 days.

On October 25th, the militiaman's pay was altered from 1*s.* and 1*d.* beer money, daily, to 1*s.* per day, and free rations. Beer money was abolished, and the following rates of pay adopted for non-commissioned officers and privates daily:—

	s.	*d.*
Volunteer Sergeant, with a certificate of efficiency	1	8
Other Volunteer Sergeants	1	$6\frac{3}{4}$
Corporals	1	$2\frac{1}{4}$
Privates	1	0
Boys, under 16 years of age	0	8
PERMANENT STAFF.		
Sergeant-Major	2	11
Quarter-Master-Sergeant	2	5
Other Sergeants	1	9
Drummers, over 16 years of age	1	0
Drummers, under 16 years of age	0	9

Lodging allowance for the above ranks at 4d. per diem.

1878. In April, 144 Militia Reserve men were called up for duty, with the 2nd Battalion 12th Regiment, at Gosport, and returned to their homes on the 30th July, the Officer Commanding 2/12th having been pleased to express his satisfaction with their general good conduct, attention to drill, smartness on parade, and the soldierlike manner in which they bore themselves in and out of barracks.

The battalion assembled on the 6th May, for 27 days' training, and was inspected on the 31st July, in review and in marching order, by Colone McKay, resulting in His Royal Highness the Commander-in-Chief being pleased to observe that the inspection was very satisfactory. This was farther enhanced, later in the year, by His Royal Highness having had the opportunity (at a review, on Southsea Common, of the troops, Southern District) of personally bestowing praise on the drill and appearance of the battalion, nearly half the men in the ranks at the time being "Militia Reserve" men.

On November 18th, the county buildings, lately used as a Militia Barrack, were dismantled, and handed over to the county.

The enrolment of the Militia was altered, on November 25th, from five to six years, and separation allowance, at the rate of 6*d.* per diem for each woman, was granted to the wives of Militia Reserve men called out for army service, and 1½*d.* for each child.

1879. Assembled, 5th May, for 20 days' training.

1880. ,, 3rd ,, ,, 27 ,,

1881. ,, 25th April ,, 27 ,,

At this year's training, and those of the two previous years, the regiment had been highly commended by the Inspecting Officer (Colonel McKay) for cleanliness of arms and accoutrements, soldierlike appearance, and general good conduct in billets.

On the 1st July, the West Suffolk Militia became the 3rd Battalion Suffolk Regiment, to take precedence accordingly as No. 12. The facings of the battalion were changed from yellow to white, and the letter "M" was directed to be worn, in brass, on shoulder straps, by officers and men.

1882. In April, Martini-Henry rifles were issued, and Snider rifles returned into store.

Assembled, June 26th, for 27 days' training.

On an expedition being sent this year to Egypt, the West Suffolk Militia unanimously volunteered to place their services at the disposal of Government for foreign service, which was duly noted by the War Office.

On December 19th, orders were received that sergeants and others of the Permanent Staff were invariably to appear in uniform ; prior to this, the Permanent Staff, during non-training period, were allowed to wear civilian clothing.

1883. From April 1st, the battalion was reduced from 10 to 8 companies, and the numbers reduced from 1000 to 800 privates, with officers and non-commissioned officers in proportion.

1884. Assembled, 5th May, for 27 days' training.

On May 26th new Colours were presented to the battalion by the Marchioness of Bristol, the General Officer Commanding, and staff, Eastern District, being present, after which the old Colours (received in 1853) were marched to St. Mary's Church, Bury St. Edmunds, and deposited with the usual ceremony.

1885. Assembled, 4th May, for 27 days' training at Bury St. Edmunds.

1886. Assembled, 3rd May, for 27 days' training at Bury St. Edmunds, and proceeded by rail the same day to Colchester.

1887. New Valise equipment (1882 pattern) was issued, and the knapsack equipment returned into store.

The figure "3" was this year taken into use, to be worn above the letter "M."

Assembled, 16th May, for 27 days' training.

1888. Assembled, 28th May, for 27 days' training and proceeded to Colchester.

On the 11th June, the 3rd and 4th Battalions Suffolk, and 4th Bedford Regiments paraded for the inspection of the Commander-in-Chief, who expressed his astonishment at the steadiness, good appearance, and drill of the militia battalions, and highly complimented them. At the annual inspection also of the 3rd Suffolk Regiment, the Inspecting Officer was extremely gratified at the cleanliness and good drill of the men.

Two silver Colour-belt badges were this year presented by Lieutenant C. R. Ellice, on resigning.

1889. Assembled, 13th May, for 27 days' training, and proceeded by rail the same day to Aldershot, where the battalion encamped at Scrogg's Bottom; strength 22 officers and 777 of other ranks.

On the 25th May, the battalion, brigaded with the 4th Suffolk, paraded with the rest of the troops in camp, in honour of Her Majesty's birthday. The Militia Brigade was under the command of Colonel H. P. Pearson, C.B., and this was the first occasion on which the officers and men of the 3rd and 4th Suffolk Regiments wore roses on their Colours, and in their hats. On the three Sundays during the training, the 2nd, 3rd and 4th Battalions held an open air service at "Scrogg's Bottom," this being the first occasion, in the annals of the corps, of these three battalions being brigaded, with the unique addition of the brigade being under Colonel Pearson's command, who formerly commanded the 1st Battalion.

On the 31st May, the battalion took part in a Royal review held by Her Majesty Queen Victoria, when it was brigaded with the 4th Suffolk, and 3rd and 4th Bedfordshire Regiments, under Colonel Pearson. The local press, reporting the march past, stated :—

> "The Suffolk battalions went by wonderfully well, the 3rd being the better of the two. The physique of the men was all that could be desired, and they compared most favourably with the regulars; indeed, it is not going beyond the mark to say that the Suffolk battalions surpassed one, if not two, of the Line regiments, in marching past."

1890. Assembled, 12th May, for the annual training, and encamped on the barrack field, at Bury St. Edmunds.

This year, the letter "M" was discontinued on the shoulder straps.

1891. Assembled, 11th May, for the annual training and proceeded by special train to Colchester, where the battalion encamped on the Abbey Field.

At the celebration of the Queen's birthday, on the 30th May, this year, the 3rd and 4th Battalions Suffolk Regiments, in brigade, again wore roses.

1892. Assembled, 9th May, for the annual training, and proceeded to Colchester. The 1st Battalion Suffolk Regiment being now in garrison here, the 3rd Battalion was, on the Queen's birthday, paraded with it, when both battalions wore yellow roses in their head-dresses.

On the 30th May, the 3rd Battalion was inspected by the General

Commanding Eastern District, and highly complimented on its efficiency, steadiness on parade, good conduct, and discipline in camp.

1893. Assembled, 8th May, for the annual training at Bury St. Edmunds. On the 23rd, the right half battalion (the left being at Landguard for musketry) was inspected by the Honorary Colonel of the battalion, the Marquis of Bristol, who expressed himself thoroughly satisfied with all he saw, and remarked on the general conduct of the men during the training.

1894. On the 28th February the battalion was armed with the Lee-Metford (Magazine) rifle, and, assembling for the annual training on the 7th May, at Bury St. Edmunds, proceeded by rail to Landguard for musketry, moving on completion of it to Colchester, to finish the training, and returned to Bury St. Edmunds on the 2nd June.

1895. Assembled 6th May, for the annual training, and proceeded to Colchester, the companies leaving two at a time for musketry at Landguard. On the 9th, blue cloth helmets were issued, and taken into wear by the battalion. The General Officer Commanding Eastern District again expressed his entire satisfaction at the manner in which the training of the 3rd and 4th Battalions had been conducted, also at the cleanliness and regularity of the camps, and the good behaviour of the men.

1896. Assembled, 4th May for the annual training, and proceeded to Colchester.

1897. Assembled 10th May, for the annual training, and proceeded to Colchester.

At the inspections of the battalion, this year, by the Honorary Colonel, the Marquis of Bristol, and by the Officer Commanding 12th Regimental District, it was complimented on its excellent discipline, drill, and admirable conduct in quarters.

Her Majesty Queen Victoria, in commemoration of the sixtieth year of her reign, made, this year, amongst other honours, the following appointment to the most honourable order of the Bath (C. B. Civil) " Lieut.-Colonel and Honorary Colonel M. C. Browning, 3rd Battalion Suffolk Regiment."

1898. Assembled, 9th May, for the annual training, and proceeded to Colchester.

1899. Assembled, 9th May, for the annual training, and proceeded to Great Yarmouth, where it encamped on the North Denes.

On the 26th May, the battalion (in brigade with the Norfolk Artillery, and 3rd and 4th Norfolk Regiments) was inspected by His Royal Highness the Prince of Wales, who expressed his approval of the steadiness of the troops under arms, and their excellent marching.

On the 7th July, Private F. Leggatt, of the battalion, was presented, at Marlborough House, by His Royal Highness the Prince of Wales, with the bronze medal of the Order of St. John of Jerusalem, which had been awarded to him for an act of gallantry in saving life on land.

War having been declared, on the 12th October, by the South African

Republic, and by the Orange Free State, against Great Britain, the 3rd Battalion Suffolk Regiment was ordered to be embodied, and, assembling at Bury St. Edmunds on the 4th December, proceeded the same day to Dover under Colonel R. Norton, mustering 19 officers and 453 of other ranks, and was quartered in the South Front Citadel, and Fort Burgoyne Barracks.

On the 9th December, the battalion was inspected, at Dover, by Major-General Parr, C.B., in marching order, who expressed great satisfaction at its appearance.

On an enquiry from the War Office, as to whether the battalion would volunteer to serve out of the United Kingdom on the distinct understanding that they might be called upon to serve at any station, when required, about 80 per cent. of the battalion volunteered for active service.

1900. On the 5th January, the battalion left Dover by rail for Southampton, and embarked for Guernsey and Alderney, the headquarters and left half battalion (13 officers and 293 of other ranks) proceeding to Guernsey, and the right half battalion (8 officers, and 245 of other ranks) to Alderney. Colonel Norton was commanding. There were only 7 men who did not volunteer for service abroad, and they were promptly sent to their homes from Dover.

On the 10th January, the half battalion at Guernsey was inspected by Major-General Saward, Commanding Guernsey and Alderney District, who was pleased with the appearance of the men. On March 15th, a special grant of £1 was made by the Government to every non-commissioned officer and man of the militia, who embarked for service abroad, or to the Channel Islands.

On May 3rd, orders were received for a draft of 50 men from the Militia Reserve of the battalion, to proceed to Shorncliffe, to join a draft being prepared to reinforce the 1st Battalion in South Africa. No difficulty was found in obtaining the required number of volunteers, and the draft proceeded to Shorncliffe, under command of Captain R. M. Dowie, who, in the following year, died from wounds received in action.

The facings on the tunics were this year altered to yellow in place of white.

On June 16th, orders were received to prepare a draft of 55 men from the Militia Reserve of the battalion, for permanent service with the 1st and 2nd Battalions, Manchester Regiment, in South Africa. They were accordingly despatched to Aldershot, under Captain R. N. Darbishire, to join the draft being prepared by the 5th Battalion, Manchester Regiment.

On July 18th, about 50 men of the battalion availed themselves of the privilege of proceeding on furlough, to attend the harvest in Suffolk.

On August 17th, an exchange of quarters took place of the half battalions of the regiment, at Guernsey and Alderney respectively, six months being the usual time for which detachments are stationed at Alderney.

A week later, the battalion, supplying another draft to augment the 1st Battalion in South Africa, had now sent out a total of 147 men, and, in October, 84 men of the battalion joined the regular army, the vacancies being constantly replaced by recruits from the depot.

1901. On April 30th, the battalion embarked in two ships at Guernsey and Alderney, and proceeded to Colchester, where, on arrival, it was quartered in the Sobraon Barracks.

The Lieutenant-Governor of these islands, in bidding farewell to the 3rd Suffolk Regiment, congratulated the battalion, in a District Order "on its progress made in military requirements."

The following complimentary letter to the Officer Commanding Battalion, was received from the Chief Constable of Guernsey:—

> "We desire to convey to you the high sense which we entertain of the exemplary and orderly conduct of the 3rd Battalion Suffolk Regiment, whilst quartered here. No one having better opportunities than ourselves of witnessing the admirable discipline of the corps, we beg to express to you, and through you, to the officers, non-commissioned officers and men, under your command, the admiration which their conduct has called forth from all classes of the community.
>
> The gallant Suffolks carry with them the best wishes of the inhabitants for their future welfare and prosperity, in which none more cordially participate than
>
> Your obedient Servants."

(Signed by two Chief Constables of St. Peter's Port, Guernsey.)

On June 10th, the battalion supplied a party of 5 officers, 5 sergeants, 2 drummers, and 142 rank and file to undergo a course of training as mounted infantry, and the company was formed as "No. 4, the Suffolk Company of Mounted Infantry."

On the 14th, it was notified that a special gratuity of one half the rate of the war gratuity, would be granted to officers and men of Militia, including permanent staff, who had been employed in the Channel Islands, during the present emergency. The battalion accordingly became entitled to this gratuity, on disembodiment.

On the 30th June, the battalion was inspected by the General Officer Commanding Eastern District (General Sir W. F. Gatacre, K.C.B.), who afterwards inspected the barracks, and made the following remarks:—

> "The Battalion is in excellent order, one of the best I have seen; officers, non-commissioned officers and men do their duty satisfactorily."

On July 3rd, the battalion entrained for Bury St. Edmunds, where the men were paid, and dispersed to their homes. At the date of disembodiment, 123 N.C.O.'s and men of the battalion were still engaged on permanent service in South Africa.

On September 30th, Captain Dowie and Lieutenants Bailey and McDougall were seconded for service with Mounted Infantry in South

Africa, and, in October, Colonel R. Norton (commanding the battalion), Major Burnand, Captains Cautley and Halhed, and 2nd Lieutenants Skinner and Nunns offered their services with Provisional Battalions, Captain Castillain, at the same time, joining the South African Constabulary.

1902. On February 24th, the battalion assembled for embodiment, and, on proceeding to Colchester, was quartered in the Meeanee Barracks. For some days following, volunteers were called for to serve in South Africa, when 80 per cent. of the battalion volunteered.

On March 14th, the battalion proceeded to Dublin, via Holyhead, arriving at 10 A.M. the next morning, and was quartered in Wellington Barracks. On the 20th, the battalion was inspected in the Phœnix Park by His Royal Highness the Duke of Connaught, commanding the forces in Ireland, who expressed himself highly pleased with its appearance and steadiness on parade, and paid a similar compliment to the guard of honour, which was furnished to him by the battalion when he was leaving Ireland.

On the 1st July, peace with the Boers having been declared, the Militia Reservists in the United Kingdom (on additional rates of pay, owing to the war) were now discharged, eleven serving with the battalion being thus affected.

On August 25th, the battalion was inspected by Major-General Vetch, who, at the conclusion, spoke in high terms of its cleanliness, smartness and *esprit de corps*, remarking that he "had never seen so fine a militia battalion, and that it was a credit to the whole militia force." Medals for the South African Campaign were presented by him to Captain and Adjutant S. E. Massy Lloyd, Captain W. G. Probert, and 33 N.C.O.'s and men.

The battalion left Dublin on the 26th September, and arrived on the following morning at Bury St. Edmunds, where the men were quickly despatched to their homes.

Amongst the rewards to officers, in recognition of services in the South African War, the following was published in the *London Gazette*, to bear date from 22nd August, 1902 :—

> "The Suffolk Regiment—To be Brevet Major : Captain S. E. Massy Lloyd (Adjutant, 3rd Battalion)."

1903. Assembled, May 11th, for annual training at Bury St. Edmunds, and, on proceeding to Colchester, were encamped at Middlewick.

1904. Assembled, May 9th, for annual training, and encamped on the drill field at Bury St. Edmunds. The battalion proceeded to Landguard for musketry, returning 24th May, and, in June, was congratulated by its Honorary Colonel, the Marquis of Bristol, and by Colonel Townley, both of whom inspected it, on its general efficiency, and on its numerical strength.

A Special Army Order, dated 9th November, notified that His Majesty,

King Edward VII, had been pleased to approve of a medal for long service and good conduct being granted to non-commissioned officers and men of the militia, and stated the conditions on which it was given.

1905. Assembled, 8th May, for annual training at Bury St. Edmunds, and proceeded as usual to Landguard for musketry. The battalion was inspected on the 23rd May, by Colonel W. A. Ramsay, and reported on as "in excellent order, and an unusually smart Militia battalion."

1906. Assembled, 14th May, for annual training, at Bury St. Edmunds, and was inspected at field operations on the 30th, by General Lord Methuen, G.C.B., who expressed himself in terms of the highest praise at all he had seen. The remarks, also, on the annual inspection by Brigadier-General Townley, on the 6th June, showed the battalion in a most efficient state throughout, the Inspecting Officer alluding to Colonel Scudamore as "an excellent commanding officer," and to Brevet-Major Massy Lloyd, as "a first class adjutant."

1907. Assembled, 13th May, for 27 days' training, and proceeded to Landguard Camp. The battalion, this year, won the United Service Challenge Trophy, of the Militia Rifle Association, with a score of 748 points. The officers who competed in it received a small silver cup each, and the non-commissioned officers and men a money prize.

A Special Army Order, dated 23rd December, directed that the 3rd Battalion Suffolk Regiment was to become a "Special Reserve" battalion, losing its militia title and functions, and the 4th Battalion was one of 23 to be disbanded. The new peace establishment was to consist of 34 officers, 1 warrant officer, and 669 non-commissioned officers and men. The new functions of the battalion were laid down as :—

(1) A reserve to the two Line Battalions of the regiment.

(2) A training centre for the Territorial Force, about to be formed.

1908. The Territorial Force came into being on the 1st April, and the existing Yeomanry and Volunteers were given until the 30th June to transfer to it.

On the abolition of the Militia, the King's message was as follows :—

> "At this time, when the Militia is to be asked to undertake new duties, and fresh liabilities, I take the opportunity of expressing to the force my keen appreciation of its services in the past.
>
> In peace and in war the Militia has never been asked in vain to make sacrifices for the good of the country. The devotion to duty which has ever distinguished the Militia, will, I am convinced, continue to be shown by the officers and men of the force, whatever calls may be made upon them.
>
> I express my special thanks for past services to those battalions and other portions of the Militia to which, to my great regret, no place can be assigned in the new organisation.
>
> I desire that this message may be promulgated for the information of the whole Army."

The battalion assembled, on the 11th May, for the annual training, at Bury St. Edmunds, and, on its completion, was transferred as a unit of the Army Reserve. Such officers as assented to being transferred, were to be appointed to the "Special Reserve," retaining the rank and seniority they held in the militia from the dates stated.

The battalion again headed the list of competitors, this year, for the United Service Gazette Militia Challenge Trophy, with a total score of 732 points.

The General Officer Commanding-in-Chief, Eastern District, was pleased to remark on the annual inspection of the battalion for 1908, as follows :— "A smart battalion, well commanded."

1909. Assembled, 10th May, for 21 days' training at Bury St. Edmunds, six days of which included the musketry course, carried out at Great Yarmouth.

On the 20th, the battalion was inspected by Lieut.-General Sir A. H. Paget, K.C.B., who expressed himself in high terms on its general efficiency, and observed that "if the battalion went on service, he felt confident it would make history for itself."

1910. Assembled, 9th May, for 20 days' training, and proceeded to Landguard. Colonel Watts, inspecting the battalion this year, expressed himself very well pleased with all he saw, both in camp and at the outpost scheme.

On the 25th June, Major S. E. Massy Lloyd, Reserve of Officers, late Suffolk Regiment, was promoted Lieut.-Colonel to command the 3rd Battalion, in succession to Colonel Scudamore, retired.

By Army Order No. 251, of 1st October, 1910, His Majesty the King directed that Battle Honours and honorary distinctions borne on the colours and appointments of regular battalions, shall in future be borne in the same way by the Special Reserve battalions affiliated to them.

1911. Assembled, 15th May, for 27 days' training at Bury St. Edmunds, and proceeded to Landguard for musketry.

At the annual inspection, Colonel Watts expressed himself well pleased with everything, and was glad to be able to report very favourably on the battalion.

With reference to the Army Order of October, 1910, notifying that the King had approved of the honorary distinctions now borne on the Colours or appointments of the regular Battalions of each Regiment of the Infantry of the Line, being borne on the Colours or appointments of the Special Reserve Battalions, and that the battle honours borne by Special Reserve Battalions (awarded when Militia) should lapse ; it has been decided that if any particular Reserve Battalion wishes, as a temporary measure, to retain its existing Colours unaltered—*i.e.* bearing the old Militia honours without the full regimental honours, it shall be optional for that Battalion to do so. This arrangement to hold good until such time as it may become necessary to have a presentation of new Colours.

1912. Assembled, 13th May, for 27 days' training at Bury St. Edmunds.

On the 23rd, the battalion, under command of Lieut.-Colonel Massy Lloyd, paraded at 3 P.M. on the cricket ground at Bury St. Edmunds, for the presentation of new Colours by Her Royal Highness, Princess Louise, Duchess of Argyll.

The ceremony took place in the presence of a large number of military and civilian spectators, the day having been proclaimed a general holiday in the town.

Her Royal Highness was accompanied by the Duke of Argyll, K.G., and there were also present Colonel Sir Courtenay Warner, Bart., C.B., M.P. (Lord-Lieutenant of the County), Rear-Admiral the Marquis of Bristol, M.V.O., the Marchioness of Bristol, Earl Cadogan, K.G., and Countess Cadogan, Sir E. W. Greene, Bart. (Honorary Colonel 3rd Battalion), and a large number of the officers' personal friends.

The service of consecrating the Colours was performed by the Right Rev. Bishop Harrison, D.D., assisted by the Vicar of St. Mary's, the Rev. T. B. Waters, officiating Military Chaplain.

The pronouncement of the Benediction closed the Consecration Service, after which Major F. E. Allfrey handed the King's Colour to the Princess, from whom it was received by Lieutenant W. B. Squirl Dawson. Major W. O. Cautley in a like manner handed the Regimental Colour to Her Royal Highness, from whom it was received by Lieutenant G. B. Pollock-Hodsoll.

The Commanding Officer (Lieut.-Colonel S. E. Massy Lloyd), in thanking Princess Louise for giving the Colours to the Regiment said :—

> "Your Royal Highness,
>
> The ceremony to-day is one that will be for ever memorable in the annals of the 3rd Suffolk Regiment. The honour which your Royal Highness has conferred upon the Battalion in presenting it with new Colours is not only deeply appreciated by all ranks, but also future generations of the 3rd Suffolk Regiment will remember it with the greatest pride.
>
> Knowing the interest your Royal Highness takes in past history, I should like to bring to your special notice that the first time this Regiment was under fire was in July, 1667, when the Suffolk Militia, under the Earl of Suffolk, repulsed a landing of the Dutch, some 3,000 strong, at Aldeburgh and Landguard. They defeated the enemy, sustaining a loss of only 12 to themselves.
>
> Since then, the 3rd Suffolk Regiment has contributed officers and men to the Line Battalions, and has helped to win the battle honours that are now enrolled on the new Colours.
>
> I beg to offer my heartfelt thanks to your Royal Highness for giving us our Colours, and especially at a time when I know your Royal Highness has so many important engagements.
>
> I can assure your Royal Highness that it will be the object of every officer, warrant officer, non-commissioned officer and man, to uphold in peace and war the good traditions the Battalion has hitherto

PRESENTATION OF COLOURS TO THE 3RD BATTALION, SUFFOLK REGIMENT.
1912.

borne, and in doing so to show in a practical way our devotion to our Sovereign and our Country."

Her Royal Highness, in replying, said :—

"Colonel Massy Lloyd, Officers, Warrant Officers, Non-commissioned Officers, and Men of the 3rd Suffolk Regiment; I sincerely appreciate the honour you have done me to-day in asking me to present you with your new Colours. The Colours of a Regiment are visible emblems and valued historical records of honours and glories won by the Regiment. They are the sacred trust of the men serving in the Regiment, and an incentive to success. Its good name and fame should always be remembered, even after the men now serving have left the Colours. Their significance should never fade from their memory. As Reservists in this country, or as settlers in the British Dominions over the seas, the pride of the Regiment should make all old soldiers live up to the glories of the past history of those men who have gone before them; a lasting guide to help them through life. This Regiment will, I feel sure, show the same devotion and loyalty to God, King, and Country, as has ever been the tradition in His Majesty's Army, and will keep one word as its motto, 'Loyalty,' which covers all."

The Commanding Officer (Colonel Massy Lloyd) then called upon the Battalion to give three cheers for Princess Louise, and these were given with enthusiasm and effect, the general public also taking part. Her Royal Highness returned to the dais, and the new Colours were received with a General Salute and marched to their place in line, the band playing "God Save the King." The Battalion then marched past, the Princess graciously acknowledging each Company in turn, as also did the Duke of Argyll. There was an advance in Review Order, and a most interesting and historical function came to a conclusion with the Battalion giving the Royal Salute and the playing of the National Anthem by the band. The Battalion, headed by the band, proceeded to the Barracks via Nelson Road, with the new Colours proudly flying. Thus ended a memorable occasion. On leaving the ground the Princess and party drove to the Barracks, where they were entertained by the officers, who, the same night, gave a ball at the Athenæum, which was largely attended.

Two days later (May 25th) the laying up of the old Colours took place in Ickworth Church, where they were conveyed by an escort of 600 non-commissioned officers and men, accompanied by the band and all the officers of the 3rd Battalion; the ceremony was conducted, at a special service, by the Rev. Lord Manners Hervey, Rector of Horringer and Ickworth, in the presence of a large congregation, and the Colours were deposited above the vault containing the remains of the late Marquis of Bristol, who had been Honorary Colonel of the 3rd Battalion Suffolk Regiment. After the National Anthem had been sung, the service concluded.

1913. The battalion proceeded for its annual training on the 19th May, to Shorncliffe, for 27 days, strength 22 officers and 440 of other ranks.

APPENDICES

APPENDIX I.

SUCCESSION OF COLONELS

OF THE

12TH REGIMENT OF FOOT.

The Suffolk Regiment.

APPENDIX II.

BIOGRAPHIES OF COLONELS

OF THE

12TH (SUFFOLK) REGIMENT OF FOOT .

HENRY DUKE OF NORFOLK, K.G.

Appointed 20th June, 1685.

HENRY HOWARD, son of Henry, sixth Duke of Norfolk, sat in the House of Lords by the title of Lord Mowbray in the lifetime of his father, and on the death of Prince Rupert, in 1682, he was nominated Captain-Governor and Constable of Windsor Castle and Warden of the Forest at Windsor, also Lord Lieutenant of the Counties of Berks and Surrey. On the decease of his father, in 1684, he succeeded to the dignity of DUKE OF NORFOLK, and of Earl Marshal of England, and he was also constituted Lord Lieutenant of Norfolk. On the accession of King James II, he was one of the peers who signed the order for His Majesty's proclamation, and on the 22nd July, 1685, was installed a Knight Companion of the most noble Order of the Garter. He took an active part in favour of the King on the breaking out of the rebellion of James, Duke of Monmouth, and interested himself in the raising of a corps of pikemen and musketeers, now the 12th Foot, of which he was appointed colonel, and of which his garrison company at Windsor Castle later formed a part. In a few months after tranquillity was restored, he relinquished the command of the regiment, but continued to attend at court, and witnessed, with painful emotions, the predilection of the King in favour of papacy and arbitrary government. On one occasion His Majesty gave the Duke of Norfolk the Sword of State to carry before him to the Roman Catholic chapel; but on arriving at the door, His Grace stopped, not being willing to enter the chapel, when the King said, "My Lord, your father would have gone further"; to which the Duke replied, "Your Majesty's father was the better man, and he would not have gone so far." [1]

The DUKE OF NORFOLK continued faithful to the interests of the Protestant Religion, and was one of the peers who invited the Prince of Orange to come to England with an army to oppose the proceedings of the court. When the Prince landed, His Grace was in London, and signed the petition to the King for a free Parliament; His Majesty replied, "They should have a Parliament, and such a one as they asked for, when the Prince of Orange had quitted the realm": and commenced his journey, on the same day, to place himself at the head of his army. His Grace set out for his seat in Norfolk, declared for the Prince of Orange, and brought over that, and some of the neighbouring counties, to the Prince's interest. On the accession of the Prince and Princess of Orange to the throne, His Grace was sworn a member of the privy council, and he took an active part in raising a regiment for the King's service, the 22nd Foot, of which he was appointed Colonel, by commission dated the 16th of March, 1689.

On the 20th February, 1695, he was appointed a Commissioner of Greenwich Hospital, and in 1697, Colonel in the Berks, Norwich, Norfolk, Surrey and Southwark Regiments of Militia, and Captain of the first troop of Surrey Horse Militia.

On the 18th January, 1691, he attended William III to Holland. He died on the 2nd April, 1701, and was buried on the 8th, at Arundel, Sussex.[2]

EDWARD EARL OF LITCHFIELD.

Appointed 14th June, 1686.

SIR EDWARD HENRY LEE, of Ditchley, Baronet, was advanced to the peerage by King Charles II, in 1674, by the titles of Baron of Spelsbury, in the County of Bucks, and EARL OF LITCHFIELD. He was appointed one of the Lords of the Bedchamber to King James II, also Custos Rotulorum for the County of Oxford, high steward of the Borough of Woodstock, and

[1] Bishop Burnet. [2] *Dictionary of National Biography.*

Lord Lieutenant of Woodstock Park. In 1686 he succeeded the Duke of Norfolk in the Colonelcy of the regiment, later the 12th Foot, which he continued to command until November, 1688, when, being a staunch supporter of the measures of the court, he was removed to the Colonelcy of the 1st Regiment of Foot Guards, which he only held a few weeks, the Prince of Orange conferring that appointment on the Duke of Grafton. The Earl of Litchfield was not afterwards employed in a military capacity. He died on the 14th of July, 1716.

Robert Lord Hunsdon.

Appointed 30th November, 1688.

Sir Robert Carey, Knight, served in a military capacity in the reign of King Charles II, and succeeded, on the decease of John Earl of Dover without issue, to the dignity of Lord Hunsdon. He was one of the supporters of the measures of King James II, who appointed him Lieut.-Colonel of the old Holland Regiment (now 3rd Foot) in 1685,and in November, 1688, promoted him to the Colonelcy of the 12th Foot, from which he was removed, at the Revolution, by the Prince of Orange. He died in 1692.

Henry Wharton.

Appointed 31st December, 1688.

Henry Wharton, brother of the Marquis of Wharton, served in the Foot Guards in the reign of King Charles II and, in the summer of 1685, when the Duke of Monmouth's rebellion commenced in the west of England, he raised a company of Foot for the service of King James II, which was incorporated in the Duke of Norfolk's Regiment. He proved a very zealous and determined supporter of the interests of the Protestant Religion, and on the 31st of December, 1688, the Prince of Orange promoted him to the Colonelcy of the regiment. He served in Ireland under the Duke of Schomberg, signalized himself at the siege of Carrick-fergus, and evincing, on all occasions, much personal bravery and spirit of enterprise, united with a generous disposition, and a kind regard for the interests of his soldiers, he was beloved by his regiment. He was the composer of the famous political song "Lillibullero," which he sang before King James II at the Playhouse in 1688.[1] He died at Dundalk in October, 1689, and was buried in a vault in Dundalk Church.

Richard Brewer.

Appointed 1st November, 1689.

Richard Brewer raised a company of pikemen and musketeers for Sir Edward Hales' Regiment, now 14th Foot, in the summer of 1685, and served in that corps until the Revolution. He prized the established religion and constitution of his country too highly to permit himself to aid in their destruction, and he espoused the principles of the Revolution with great warmth. On the 31st of December, 1688, he was promoted to the Lieut.-Colonelcy of the 12th Foot, with which corps he served in Ireland, and displayed signal bravery on several occasions, for which he was rewarded with the Colonelcy of the regiment on the 1st of November, 1689. He commanded the 12th Regiment at the Battle of the Boyne in 1690, also in the action at Lanesborough, and was appointed commandant at Mullingar, near which place the troops under his immediate command had several encounters with detachments of the enemy. He continued to serve in Ireland until the deliverance of that country from the power of King James was accomplished, and in 1692 he commanded his regiment in the expedition under the Duke of Leinster. He also served at the head of his regiment in the Netherlands, during the campaign of 1694; in the attack on Fort Kenoque, and the defence of Dixmude in 1695 (on which last-mentioned occasion he opposed the Governor in the resolution to surrender), and in the protection of the maritime towns of Flanders in 1696. After the Peace of Ryswick, he proceeded with his regiment to Ireland; and on the breaking out of the war, in the reign of Queen Anne, he retired from the service.

[1] Dalton's *Army List*, Vol. I, p. 167.

John Livesay.

Appointed 28th September, 1702.

This officer was appointed Lieutenant in the Royal Fusiliers in 1685 ; he served in the army during the wars of King William III, and was promoted Lieut.-Colonel in the 12th, on the 6th January, 1697. He was distinguished for gallantry and a strict attention to duty, and was rewarded by Queen Anne, in September, 1702, with the Colonelcy of the 12th Regiment, which he commanded in the West Indies in 1703, 1704, and 1705. On the 1st of January, 1707, he was promoted to the rank of Brigadier-General, and on the 1st of January, 1710, to that of Major-General. Political events, connected with the removal of the Duke of Marlborough from the command of the army, and the measures pursued by the new ministry of Queen Anne, occasioned Major-General Livesay to retire from the command of the regiment in 1712. He died on the 23rd of February, 1717, and in the north aisle of Puddington Church there is a neat monument to his memory, on which are his arms: Arg. a lion rampant, gules, between three trefoils slipped proper.

Richard Philipps.

Appointed 16th March, 1712.

Richard Philipps, born in 1661, was great-grandson of Sir John Philipps, Bart., of Picton Castle, Pembrokeshire, and entered the army as lieutenant in Lord Morpeth's Regiment on the 23rd February, 1678. He forwarded the Prince of Orange's cause at the Revolution, and was imprisoned by the Mayor of Dartmouth for circulating the Prince's "Declaration" (Historical Records, 40th Regiment, p. 496). Was appointed captain in the Earl of Drogheda's Regiment, and was present at the battle of the Boyne in 1690. The pistols which he then used are in possession of his descendant, Lord St. Davids, by whose courtesy a drawing of them is reproduced at page 35. In January, 1692, he became captain in Major-General Kirk's Regiment. According to Cannon, he served in 1706 with Bretton's Regiment (afterwards disbanded) and proceeded with it to the relief of Barcelona in that year. Served in Flanders and Spain, and was taken prisoner at the battle of Almanza. He subsequently served with his company on board the fleet as marine officer, and was promoted to the Lieut.-Colonelcy of the regiment. Queen Anne rewarded his services, in 1712, with the Colonelcy of the 12th Foot, from which he was transferred in 1717, to the 40th Regiment, then newly formed of independent companies, at Placentia, Annapolis, and other parts of America. In 1719, he was appointed Captain-General and Governor of Nova Scotia, with instructions to form its first separate council. Was promoted to the rank of Brigadier-General in 1735, to that of Major-General in 1739, and Lieut.-General in 1743. In 1750, he was removed to the 38th Foot. He died 24th October, 1754, at the advanced age of 90, and is buried in Westminster Abbey.

Thomas Stanwix.

Appointed 25th August, 1717.

Thomas Stanwix served in the Netherlands, with reputation, under King William III, and afterwards in Holland and Germany under the Duke of Marlborough. In April, 1706, he was commissioned to raise, form, and discipline a regiment of Foot, in Ireland, with which corps he embarked from Cork, in May, 1707, for Portugal, where he served under the Marquis de Montandre, the Marquis de Fronterira, and the Earl of Galway. In 1709 he was at the battle of the Caya, where his regiment highly distinguished itself, and in 1710 he commanded the storming party at the capture of Xeres de los Cavaleras: at the Peace of Utrecht his regiment was disbanded. In 1715, when the partisans of the Pretender sought to elevate him to the throne, Colonel Stanwix was commissioned to raise a regiment of Foot, for the service of King George I., and in July, 1717, he was transferred to the 30th Regment, which he only commanded five weeks, when he was appointed to the 12th Foot, with the rank of Brigadier-General. He died 14th of March, 1725.

Thomas Whetham.

Appointed 22nd March, 1725.

This officer obtained a commission in Sir William Clifton's Regiment, now 15th Foot, on the breaking out of the rebellion of James, Duke of Monmouth, in June, 1685, and served under King William in Ireland and Flanders, where he acquired a reputation for gallantry and zeal. On the 29th of August, 1702, Queen Anne rewarded him with the Colonelcy of the 27th Regiment of Foot, with which corps he served in the West Indies in 1703 and 1704, and was engaged in the unsuccessful attack on the Island of Guadaloupe. In 1707 he was promoted to the rank of Brigadier-General, and in 1710 to that of Major-General; he served in Spain during the latter part of the War of Succession, commanded in Catalonia, 1711–12, and also at the Island of Minorca for a short period; in 1715, he commanded the left wing of the Royal Army at Sheriffmuir, under the Duke of Argyll, during the rebellion of the Earl of Mar. In 1725, he was transferred to the 12th Foot, and in 1727 obtained the rank of Lieut.-General; he was promoted to the rank of General in 1739, and was appointed Governor of Berwick and Holy Island in 1740. He died on the 28th of April, 1741.

Scipio Duroure.

Appointed 12th August, 1741.

Scipio Duroure, the elder son of Francis Duroure, a refugee French officer in Ireland, obtained a commission in the army in December, 1705, prior to which, in 1704, he had served as a volunteer in the ranks of Lieut.-General Varenne's Regiment, and was with it at the battle of Hochstet (Blenheim), and in 1705 at the passages of the Lines at Tirlemont, and the siege of Southlewen, receiving, at the latter, a dangerous wound in the neck. He now had the offer of a " pair of colours " in the Hessian Guards from the King of Sweden, who commanded at the siege.

In 1706, Scipio Duroure was appointed ensign in General Grumbkow's Regiment, and was with it at Ramillies. Went as a volunteer to the Siege of Ostend with General Ivoy (Quarter-Master-General to the Dutch), and sent a daily account, and plan of the approaches to General Grumbkow, to be transmitted to the King at Berlin. After this siege, he attended General Trossel as A.D.C. at the siege of Menin. In 1708, he was with his regiment at the battle of Oudenarde, and attended General Trossel at the Siege of Lisle, after which he rejoined his regiment, and was detached to Honscote, with a battalion of the Margrave Albrecht's, and two squadrons of horse, under Colonel Kale. They were attacked, and taken prisoners, but were exchanged the year following, and took the field. In 1709, Scipio Duroure attended General Grumbkow at the siege of the town and citadel of Tournay, and was with General Trossel at the battle of Malplaquet, where he received two slight wounds in the body.

After the battle, he attended General Trossel at the Siege of Mons. This year, he was made a lieutenant, and on being recommended to the Duke of Marlborough, his Grace encouraged him to come to England, with the promise of a captaincy in the new Levies it was then intended to raise. In 1710, he purchased a company in Lord North and Grey's (10th) Regiment, and served with it at the Sieges of Douay and St. Venant, and in 1711, at the Siege of Bouchain.

In 1722, he purchased his majority in Lord North and Grey's Regiment, and was Brigade Major to the infantry encamped at Salisbury. This year, the King of Prussia did him the honour to offer him a post vacant at that time, if he was disposed to return to that service, viz.:—Deputy-Assistant-Quarter-Master-General, with a lieut.-colonel's commission, and 1500 crowns per annum; but, with all expressions of gratitude and respect, he declined it, though at the time he had lost nearly £5000 in a South Sea transaction, and might have sold his majority for £2000.

In 1725, General Wade appointed him Brigade Major to the forces employed at Inverness for putting into execution the Disarming Act, and he was continued in that post up to 1731. He served many years in the 12th Foot, of which corps he was appointed Lieut.-Colonel on the 25th August, 1734. He also became Captain and Keeper of the Castle of St. Mawes, and was promoted to the Colonelcy of the regiment on the 12th August, 1741.

By a warrant, signed at Ghent, 1st February, 1743, from Field-Marshal the Earl of Stair, K.T. (Commander-in-Chief in the Austrian Netherlands), Colonel Duroure was appointed Adjutant-General to the Forces.

He distinguished himself at the battle of Dettingen in 1743, and behaved with great gallantry at the head of his regiment at the battle of Fontenoy, in 1745, where he was mortally wounded.

He died on the 21st May, 1745, and was buried on the ramparts at Ath; there is a monument to him in the East Cloister at Westminster Abbey. (See Plate 45.)

Colonel Scipio Duroure had had the advantage of serving three campaigns under the Duke of Marlborough, which included 4 battles and 9 sieges.

Henry Skelton.

Appointed 28th May, 1745.

Henry Skelton entered the army in December, 1708, and served two campaigns in the Netherlands. He was many years an officer in the 3rd Foot Guards, was promoted Major of the regiment with the rank of Colonel in the army, in 1739, and in April, 1743, he was advanced to Lieut.-Colonel in the same corps. In August following, King George II rewarded him with the Colonelcy of the 32nd Regiment, which he commanded at Fontenoy; His Majesty also promoted him to the rank of Major-General, transferred him to the 12th Foot in 1745, and advanced him to the rank of Lieut.-General in 1747. He died on the 9th of April, 1757, leaving his ancestral home, Branthwaite Hall, Cumberland, to a former A.D.C., Captain James Jones, of the 3rd Guards, who had saved his life in Flanders.[1]

Robert Napier.

Appointed 22nd April, 1757.

Robert Napier was appointed ensign in the 2nd Foot, on the 9th of May, 1722, and after performing regimental duty a few years, was placed on the staff, and employed in the Quarter-Master-General's Department. In 1745, he was appointed one of the six aides-de-camp to the Duke of Cumberland, at Fontenoy, and, as a reward for his services there, was promoted to the rank of Lieut.-Colonel, and appointed Deputy-Quarter-Master-General; in 1746, he was advanced to the rank of Colonel, and was afterwards appointed Adjutant-General of the Forces. In 1755, King George II. appointed him Colonel of a newly-raised regiment, now 51st Foot; in 1756 he was promoted to the rank of Major-General, and in 1757 was transferred to the 12th Foot. In 1759, he was promoted to the rank of Lieut.-General, and died in November, 1766.

Henry Clinton, K.B.

Appointed 21st November, 1766.

Henry Clinton, grandson of Francis, sixth Earl of Lincoln, served in an independent company of Foot at New York, and in 1751 he was appointed Lieutenant and Captain in the 2nd Foot Guards, from which he was promoted, in 1758, to Captain and Lieut.-Colonel in the 1st Foot Guards. He served in Germany during the Seven Years' War, was promoted to the rank of Colonel in 1762, and in 1766 he obtained the Colonelcy of the 12th Foot. He was promoted, in 1772, to the rank of Major-General. On the commencement of the American War, in 1775, he was sent with reinforcements to Boston, with the local rank of Lieut.-General, and at the Battle of Bunker's Hill he joined the troops engaged with additional forces from Boston during the conflict, and contributed materially to the gaining of the victory. He afterwards proceeded to North Carolina, with the local rank of General; assumed the command of the troops which arrived from Great Britain, and in 1776 he undertook the reduction of Charleston, but was not able to accomplish his object from the want of a sufficient force. He then joined General Sir William Howe, was engaged in the reduction of Long Island,and commanded the leading column of the army at the Battle of Brooklyn. General Clinton also commanded the division which took possession of New York Island, was at White Plains and other engagements, also commanded the troops which took Rhode Island, and was rewarded with the dignity of Knight of the Bath. In 1777 he commanded at New York, and, in order to create a diversion in favour of General Burgoyne's army, he proceeded up the river and captured Forts Clinton and Montgomery. In the following spring he was nominated Commander-in-Chief in North America, and assuming the command of the army at Philadelphia, marched from thence to New York, repulsing the attacks of the enemy during the movement.

[1] Skrine's *Fontenoy*, p. 134.

In the winter of 1778, he was transferred from the 12th Foot to the command of a corps of Royal Highland Emigrants, and in 1779 was appointed Colonel of the 7th, or Queen's Own Light Dragoons.

The departure of the French Fleet from North America enabled General Sir Henry Clinton to fit out an expedition against Charleston, which he captured in 1780, for which he received the thanks of Parliament; and this success was followed by important results in North and South Carolina; but the tide of success did not long flow in favour of the British cause, and some reverses taking place, he was succeeded as Commander-in-Chief in North America by General Carleton. He arrived in England in June, 1782, and afterwards published a vindication of his conduct. The appointment of Governor of Limerick was conferred upon General Sir Henry Clinton; he was also groom of the bedchamber to the Duke of Gloucester, and was many years a Member of Parliament; in 1795, he was appointed Governor of Gibraltar. He died in December of the same year.

William Picton.

Appointed 21*st April*, 1779.

This officer first served in the marines, in which corps he was promoted to the rank of Captain, in March, 1755; and in August, 1756, was appointed Captain of the grenadier company in the 12th Foot. He served at the head of his company, in Germany, during the Seven Years' War, and evinced great gallantry on numerous occasions. In the neighbourhood of Warburg (31st July, 1760) his company commenced the attack on the hill of Ochsendorf in which 24 non-commissioned officers and grenadiers were killed and wounded. Was specially mentioned in Prince Ferdinand's General Orders for the surprise attack on Zierenberg in September, 1760. Was severely wounded near Campen (October 1760), and had a horse shot under him. After the raising of the siege of Wesel, was taken prisoner on account of his wound preventing his being removed. In 1762 he was promoted Major, and in 1765, Lieut.-Colonel of his regiment. He performed all the duties of commanding officer of the 12th Regiment in the United Kingdom, and afterwards at Gibraltar, with reputation to himself and advantage to the service, for thirteen years; in 1778 was appointed Colonel of the 75th Foot, then newly raised, and afterwards disbanded: in the following year he was transferred to the 12th Regiment.

King George III frequently selected individuals of merit on whom he conferred distinguished marks of his Royal approbation, and the promotion of Colonel Picton furnishes an instance of His Majesty's attention to meritorious services, which had not the advantage of Ministerial or Parliamentary patronage. (See page 112.) He joined his regiment at Gibraltar, and distinguished himself in the defence of that fortress, under General Eliott.

In 1782 he was promoted to the rank of Major-General, in 1793 to that of Lieut.-General, and in 1798 to that of General. He died on the 14th October, 1811, at the age of 87.

Sir Charles Hastings, Baronet.

Appointed 15*th October*, 1811.

Charles Hastings, natural son of Francis, tenth Earl of Huntingdon, was appointed Ensign in the 12th Foot in July 1770, and joined the regiment at Gibraltar. In 1776 he was promoted Lieutenant, and was permitted to serve with the 23rd Regiment in America, where he was appointed aide-de-camp to Earl Percy, and afterwards to Sir Henry Clinton. He was at the actions at Pelham Manor and White Plains, and at the capture of Fort Washington; also in the successful expedition against the American magazines at Danbury. He accompanied Sir William Howe to Pennsylvania, was engaged at Brandywine and Germantown, and was twice wounded. In 1780 he was promoted Captain in the 12th Foot, and joined his regiment at Gibraltar, where he had several opportunities of distinguishing himself during the siege of that fortress. He displayed great gallantry whilst covering the retirement of the sortie, made by the garrison in November, 1781, with the grenadier and light companies of his regiment, and was thanked in general orders.[1]

In 1782, he was appointed Major in the 76th; in 1783 he was promoted Lieut.-Colonel in the 72nd, which regiment was disbanded in the same year. He obtained the Lieut-Colonelcy of the 34th Regiment in 1786, and was afterwards transferred to the 61st, and subsequently

[1] *MS. Records, R.U.S.I.*

to the 65th. Was promoted to the rank of Major-General in 1796, and to that of Lieut.-General in 1803. In February, 1806, he was created a Baronet, of Willesley Hall, in the County of Derby; and in November following was appointed Colonel of the 4th Garrison Battalion, from which he was transferred to the 77th Regiment in July, 1811; and in October following, to the 12th Foot. In 1813, promoted to the rank of General. He died on the 8th October, 1823.

Honourable Robert Meade.

Appointed 9th October, 1823.

This officer embarked with the 87th for Holland in 1794. In 1801, he accompanied Sir Ralph Abercromby's army to Egypt and was actively engaged in the several actions of the 8th, 13th, and 21st March, proceeding afterwards to Cairo. In 1804, he was appointed to the staff as a Brigadier-General, and was subsequently sent with four regiments to reinforce the army in Sicily, with which he remained until an expedition was fitted out for a second attack on Alexandria, under General Mackenzie Fraser; and, at the capture of Rosetta, General Meade was severely wounded in the head, with the loss of an eye, and obliged to return to England.

He was soon after sent as Governor and Commander-in-Chief to Madeira, where he remained for four years, and then proceeded to the Cape of Good Hope, where he was quartered until after the proclamation of peace.

He received the Gold Medal from the Grand Seigneur, and the Silver War Medal, with one clasp, for Egypt. He became Lieut.-Colonel, 10th March, 1795; Colonel, 29th April, 1802; Major-General, 25th October, 1809; Lieut.-General, 4th June, 1814; and General, 10th January, 1837.

He died on the 28th July, 1852.

Sir Richard G. Hare Clarges, K.C.B.

Appointed 29th July, 1852.

Entered the Army 6th July, 1796, and served in the following campaigns, viz:—in Egypt, 1801, under Sir Ralph Abercromby; in Hanover, under Lord Cathcart, and in Spain, under Sir John Moore, including the retreat to Corunna. Was on the staff of General Sir T. Graham in the Walcheren Expedition, and, having attained the rank of Lieut.-Colonel in September, 1813, became Assistant-Adjutant-General of a Division in the Duke of Wellington's army in the Peninsula. He received the Gold Medal and one clasp for Nivelle and Nive, and the Silver War Medal with nine clasps for Egypt, Corunna, Busaco, Barrosa, Ciudad Rodrigo, Badajoz, Salamanca, Pyrenees, and St. Sebastian. He became Colonel in 1830, Major-General in November, 1841, and was appointed Colonel of the 73rd Regiment on the 18th May, 1849. Was promoted Lieut.-General in November, 1851, and transferred to the Colonelcy of the 12th on the 29th July, 1852. He died on the 13th April, 1857.

Charles A. F. Bentinck.

Appointed 14th April, 1857.

Joined the Army on the 16th November, 1808, entering the Coldstream Guards, and became Captain and Lieut.-Colonel, 27th May, 1825. Served with two companies of his regiment at the defence of Cadiz and the Isle of Leon, from March 1810 to June 1811; was wounded at the Battle of Barrosa, which prevented him joining the 1st Battalion in Portugal.

He was appointed Adjutant to the 2nd Battalion, and accompanied the 6 companies sent to Holland, under Lord Lynedock, in 1813; was engaged in the successful attack on Merxem, bombardment of Antwerp, and operations against Bergen-op-Zoom.

Was attached to the 2nd Division, as Deputy-Assistant-Adjutant-General, under Sir Henry Clinton, at the Battle of Waterloo, and capture of Paris, for which he received the brevet rank of Major, and on being appointed Assistant-Adjutant-General to the Division, continued to serve in the Army of Occupation until its dissolution at the end of 1818. Was awarded the Waterloo Medal, and the War Medal, with one clasp. He became Colonel, 28th June, 1838, Major-General in 1843, Lieut.-General in 1858, and was appointed Colonel 12th Foot, 14th April, 1857. He died on the 28th October, 1864.

Henry Colvile.

Appointed 29th October, 1864.

Entering the Army on the 29th December, 1813, he was gazetted to the Scots Fusilier Guards, in which he attained the rank of Captain and Lieut.-Colonel on the 6th July, 1830, becoming Colonel in 1844; Major-General, 1854; Lieut.-General, 1860; and was appointed to the Colonelcy of the 12th Foot, four years later. He had seen no war service, and died on the 1st November, 1875.

John Patton.

Appointed 2nd November, 1875.

Gazetted Ensign in the 12th Regiment on the 18th September, 1817, and became Lieut.-Colonel on the 18th August, 1843; Colonel, 20th June, 1854 and Major-General, 20th February, 1859. Was appointed to the Colonelcy of the 47th Foot on the 8th December, 1867, and promoted Lieut-General on the 1st January, 1868. He became a General on the 10th October, 1874, and was transferred a year later to the Colonelcy of his old regiment. General Patton did not see any war service, and died on the 21st February, 1888. (See Plate 32.)

John Maxwell Perceval, C.B.

Appointed 28th February, 1888.

Entered the army on the 21st June, 1833, and became Lieut.-Colonel on the 2nd April 1850. Commanded the Reserve Battalion of the 12th Regiment in the Kaffir War of 1852–3 at the affairs in the Waterkloof and Fish River Bush, for which he received the medal, and was nominated a Companion of the Bath. He became Colonel in 1854; Major-General, 1864; Lieut.-General, 1873; and the following year was appointed to the Colonelcy of the 97th Foot, from which he was transferred to that of the 12th in 1888, having in the meantime attained the rank of General on the 1st October, 1879. He died on the 24th January, 1900. (See Plate 32.)

The Honourable Sir P. R. B. Fielding, K.C.B.

Appointed 25th January, 1900.

Gazetted Ensign 8th August, 1845, and, serving in the Coldstream Guards, attained the rank of Captain and Lieut.-Colonel on the 23rd November, 1855, and that of Colonel ten years later. He served in the Crimean Campaign of 1854–55 as Brigade-Major to the Brigade of Guards, at the Battle of Alma (mentioned in despatches), and on the staff of the 1st Division at the Battles of Balaklava, Inkerman (severely wounded), and Siege of Sevastopol. Received the Medal with four clasps, 5th Class of the Medjidie, and Turkish Medal; became a Knight of the Legion of Honour, and was later nominated a Knight Commander of the Bath. He was promoted Major-General in 1870, Lieut.-General in 1886, General, 1st April, 1891, and died on the 9th January, 1904.

The Honourable Bernard M. Ward, C.B.

Appointed 10th January, 1904.

Obtained his commission in the 47th Foot in January, 1850, and four years later served in the Crimea, taking part in the Battle of Inkerman, and the Siege of Sevastopol. Mentioned in despatches, and received the medal with two clasps, and the Sardinian and Turkish Medals. In January, 1864, he went as Major to the 32nd Foot, which he commanded from June 1869 to March 1876, acting as commandant at Natal in 1870–71, during the regiment's stay at the Cape. Became Colonel in 1874, Major-General in 1885, and Lieutenant-General in the following year. Was appointed Colonel of the Suffolk Regiment on the 10th January 1904, and nominated a Companion of the Bath on the occasion of the Jubilee of the Crimea.

APPENDIX III.

SUCCESSION LIST OF LIEUTENANT-COLONELS.

(Asterisks, thus *, show that these officers did not succeed to the command.)

Date of Appointment.	Names.	Remarks.
June 20, 1685	Thomas Salusbury	Joined James's army at the Revolution.
Dec. 31, 1688	Richard Brewer	Promoted to the Colonelcy, 1st November 1689.
Nov. 1, 1689	William Barnes	
1693	Thomas Dowsett *	Died 5th January, 1697.
Jan. 6, 1697	John Livesay	Promoted to the Colonelcy, 28th September 1702.
Sept. 28, 1702	Isaac Foxley	
Feb. 22, 1703	Richard Franks	
Dec. 18, 1710	Edmund Arwaker	
1711–12	John Ligonier *	Later became a Field-Marshal and Earl Ligonier.
May 28, 1722	James Long	
Nov. 17, 1731	Lewis Ormsby	
Aug. 25, 1734	Scipio Duroure	Promoted to the Colonelcy, 12th August, 1741. Died of wounds received at Fontenoy 21st May, 1745.
Aug. 12, 1741	John Hayes	
March 30, 1742	William Whitmore *	Killed at the battle of Fontenoy.
May 27, 1745	John Cosseley	Appointed Lieutenant-Governor of Chelsea Hospital before 1748.
Oct. 28, 1748	Anthony Meyrac	
Nov. 10, 1755	William Robinson	
May 1, 1761	Lord Charles Brome *	Later became the 1st Marquis Cornwallis.
Aug. 21, 1765	William Picton	Transferred to the Colonelcy, 21st April 1779.
June 6, 1778	Thomas Trigge	
Oct. 22, 1794	H. H. Aston	Killed in a duel.
April 25, 1795	John Perryn	Promoted Lieut.-Colonel in succession to Colonel Trigge.
Jan. 26, 1796	Thos. Grey *	Died in 1797.
Dec. 1, 1796	J. S. Wortley *	
Jan. 18, 1798	James Taylor *	
March 1, 1799	G. W. D. Harcourt	Transferred from 40th Regiment. Took over command in May, 1800.
March 7, 1799	Robert Shaw [1]	Transferred temporarily from 74th, for one year.
Nov. 30, 1809	John Picton (1st Battalion)	
Dec. 25, 1811	Julius Stirke (2nd Battalion)	Posted on its formation.
June 4, 1813	H. F. Cooke	
Dec. 16, 1813	Nicholas Eustace	
May 5, 1815	William Henry Forssteen (1st Battalion)	
Oct. 12, 1815	John Castle (2nd Battalion)	

[1] The transfer not shown in the Army Lists for 1799 or 1800.

Date of Appointment.	Names.	Remarks.
Jan. 11, 1816	Thomas Carnie	
April 20, 1817	Hon. Henry Cecil Lowther	
Sept. 18, 1828	Richard Bayly	
Oct. 8, 1830	Gervas Turbervill	
Aug. 28, 1835	Joseph Jones	
Aug. 18, 1843	John Patton	Promoted as a General Officer to the Colonelcy, 2nd November, 1875.
April 14, 1846	S. F. Glover (2nd Lieut.-Colonel)	Commanded Reserve Battalion.
March 30, 1849	Wm. Bell do.	Reserve Battalion.
Feb. 8, 1850	Rundal Rumley	Exchanged to the 27th Regiment.
April 2, 1850	J. M. Perceval, C.B.	Transferred as a General to the Colonelcy, 28th February, 1888.
March 3, 1854	Edward St. Maur (Res. Battalion)	Exchanged from the 27th Regiment in March, and retired in May the same year.
May 19, 1854	Thomas Brooke do.	Reserve Battalion.
Feb. 22, 1861	H. M. Hamilton (1st Battalion)	
Oct. 22, 1861	E. G. Hibbert (2nd Battalion)	
April 21, 1863	A. E. V. Ponsonby (2nd Battalion)	Died, 16th June, 1868, while serving.
June 17, 1868	Richard Atkinson (2nd Battalion)	Exchanged to the 35th Regiment.
Nov. 17, 1868	Brevet-Colonel J. McN. Walter, C.B. (2nd Battalion)	do. from do.
May 1, 1871	John McKay (2nd Battalion)	Promoted from the ranks. Became a Major-General.
April 1, 1873	E. Foster (1st Battalion)	
April 3, 1878	G. F. Walker (1st Battalion)	
April 10, 1878	F. Bagnell (2nd Battalion)	
July 1, 1881	C. J. C. Sillery (2nd Battalion)	
July 1, 1881	W. C. O'Shaughnessy (2nd Battalion)	
July 1, 1881	W. T. Baker * (1st Battalion)	
April 1, 1882	H. P. Pearson (1st Battalion)	
Sept. 7, 1882	H. M. Lowry * (1st Battalion)	
July 14, 1883	Wm. Keough (2nd Battalion)	
Aug. 27, 1884	R. H. O'Grady Haly (1st Battalion)	
April 1, 1886	J. E. Harris (2nd Battalion)	
May 19, 1886	Arthur Tower (1st Battalion)	Exchanged from the "Sherwood Foresters" and died while serving.
Sept. 7, 1886	C. H. Gardiner (1st Battalion)	Exchanged to the "Sherwood Foresters."
Dec. 9, 1890	J. C. R. Glasgow (1st Battalion)	
April 2, 1892	Owen Williams (2nd Battalion)	
Dec. 9, 1894	R. T. E. Dowse (1st Battalion)	
Feb. 24, 1897	C. R. Townley (2nd Battalion)	
Sept. 19, 1898	A. J. Watson (1st Battalion)	Killed in action at Colesberg, South Africa, 6th January, 1900.
Jan. 7, 1900	G. F. C. Mackenzie (1st Battalion)	
Feb. 24, 1902	A. C. Cubitt (2nd Battalion)	
Jan. 7, 1904	F. Graham (1st Battalion)	
Feb. 24, 1906	V. W. H. Graham (2nd Battalion)	
Jan. 7, 1908	Edward Montagu (1st Battalion)	
Feb. 24, 1910	C. H. C. van Straubenzee (2nd Battalion)	
Jan. 7, 1912	W. B. Wallace (1st Battalion)	

APPENDIX IV.

Succession List of Adjutants.

(The stars mark unavoidable gaps, owing to missing Army Lists.)

Date of Appointment.	Names.	Remarks.
June 20, 1685	Charles Houston	
Nov. — 1687	John Blake	
* *	* * *	* * * *
Oct. 31, 1707	John Nuttall	
Dec. 20, 1709	William Newbury	
—— — 1711	Arthur Gambell	
* *	* * *	* * * *
Jan. 11, 1722	John Cosseley	
* *	* * *	* * * *
July 13, 1743	James Wolfe	
* *	* * *	* * * *
Nov. 19, 1748	Thomas Lawless	This officer shown as Adjutant of the regiment, until August, 1765.
Aug. 21, 1765	Thomas Tutteridge	
May 3, 1771	Thomas Hitchbone	
Feb. 25, 1788	Joseph Moore	
Aug. 14, 1797	William Langford	
Dec. 25, 1800	John Jagger (Ensign)	Gazetted Ensign (probably from the ranks) November 3, 1800.
May 14, 1807	David Hayes	
July 9, 1811	Robert Jenkins	
Feb. 6, 1812	A. P. Walsh	
Dec. 30, 1813	Jonathan Priestley	
June 12, 1817	J. K. Leith	
Sept. 17, 1818	F. Clarke	
Jan. 25, 1831	John Thompson	
Nov. 22, 1836	H. D. Persse	
Nov. 10, 1840	F. G. Hamley	
March 29, 1844	W. E. Crofton	Reserve Battalion
April 14, 1846	Edward Foster	"
Feb. 9, 1849	Thomas Dundas	1st Battalion
Feb. 27, 1852	I. C. Munro	1st Battalion
July 27, 1855	George Gibson (Ensign)	Reserve Battalion
Oct. 1, 1858	F. A. Fitzgerald	1st "
June 15, 1860	J. S. Richardson	1st "
June 30, 1863	H. J. MacDonnell	2nd "
Aug. 21, 1863	G. de L. Lacy	1st "
Sept. 20, 1865	G. L. B. Thomas	1st "
June 9, 1866	W. J. Boyes	2nd "
Dec. 29, 1866	E. C. C. Foster	1st "
Feb. 3, 1869	S. B. Triphook	1st "
Jan. 19, 1873	Charles Hely	2nd "
Sept. 24, 1873	C. R. Townley	1st "
April 3, 1878	C. Kennedy	2nd "
Nov. 10, 1880	J. B. McDonell	1st "

Date of Appointment.	Names.	Remarks.	
March 1, 1884	H. Cautley	2nd Battalion	
Nov. 10, 1885	L. J. Shadwell	1st ,,	
Jan. 9, 1888	V. M. Grantham	2nd ,,	
Jan. 10, 1891	E. P. Prest	1st ,,	
April 1, 1892	W. F. Coleman	2nd ,,	
Jan. 10, 1895	C. A. H. Brett	1st ,,	
Aug. 25, 1895	W. G. Thomson	2nd ,,	
Aug. 25, 1899	E. C. Doughty	2nd ,,	
Jan. 10, 1899	F. A. P. Wilkins	1st ,,	(Killed in action at Colesberg, South Africa, Jan. 6th, 1900)
April 23, 1900	F. T. D. Wilson	1st ,,	
Dec. 30, 1900	S. J. B. Barnardiston, D.S.O.	1st ,,	
Nov. 1, 1903	F. A. White, D.S.O.	2nd ,,	
Dec. 30, 1904	W. N. Nicholson	1st ,,	
Nov. 1, 1906	G. H. Walford	2nd ,,	
Dec. 30, 1907	E. N. Jourdain	1st ,,	
Nov. 1, 1909	G. C. Stubbs	2nd ,,	
Dec. 30, 1910	F. S. Cooper	1st ,,	
Nov. 1, 1912	A. M. Cutbill	2nd ,,	
Dec. 30, 1913	D. V. M. Balders	1st ,,	

APPENDIX V.

Regimental Plate, Pictures, &c.

Officers' Messes are stated to have become general throughout the army in 1795, when Government, beyond providing rooms, and fuel and light for them, made no further concessions, and no furniture was supplied. The Mess was at first purely a voluntary association, which no officer need join unless so disposed, and it will be seen from Colonel Mair's "Reminiscences" (Chapter IX.) that even as late as 1858, the dinner hour (then at 3 p.m. daily) was the only recognised time of the day at which officers assembled at the mess, apparently making their own arrangements for other meals.

1st Battalion.

The Officers' Mess Plate of the 1st Battalion Suffolk Regiment appears (from hall marks on some of the old silver) to date back to 1798. Some of it had, however, been lost on service in Holland, in 1794–95, for which a claim of £50 was disallowed. (See page 182.)

The custom of presenting some article to the Mess, on promotion, or on leaving the regiment, may be said to have become general in the early fifties of the Nineteenth Century, though, on the formation of the Reserve Battalion of the 12th in 1842 (now the 2nd Battalion) it will be found that their presentations date back as far as that year.

In 1894, a centre-piece, height 3 feet, was purchased for the Mess of the 1st Battalion, of the following handsome design:—

> A Corinthian column, with four crossed Colours, rising from a large silver base, representing the Rock of Gibraltar, on the sides of which are chased four battle scenes. A raised border of laurels, with the regimental badge in the centre, adorns the lower part of the base, with the regimental motto on a ribbon beneath, on which are also inscribed the battle honours. The base stands on four shell and scroll feet, with two larger ones in the centre, which are decorated with oak leaves in high relief, and the column is surmounted by a figure of "Victory." (See Plate 42.)

The following is a list of plate of the 1st Battalion:—

Ram's Horn Snuff Mull, 1849.	Presented by Lieutenant T. Dundas.
Oak Tree Centre-piece on glass mounted plateau, 1853.	,, Captain Sir George Bishop, Bart.
Plated Cigar Lighter, 1853.	,, E. C. Bisshopp, Esq.
Coffee Service, 1854.	,, 4 officers.
Small Silver Snuff Box, 1860.	,, C. Littlehales, M.D. "As a memorial of his son."
Silver Pint Mugs (25) of standard pattern, 1863–73.	Given by 25 officers on promotion and retirement, as per inscriptions thereon.
Claret Jugs (2) partly gilt, vase shaped, 1867.	
Claret Jug, 1875.	Presented by 5 officers.
,, ,,	,, 5 other officers.

Challenge Cup for Officers' Polo Ponies (Racing), 1880–1912.

1880.	Won by	Lieutenant C. D. Cave.
1882.	,,	,, Do.
1883.	,,	Captain O. Williams.
1883.	,,	Lieutenant E. A. Kemble.
1884.	,,	,, P. W. Scudamore.
1884.	,,	,, J. M. McDonell.
1908.	,,	,, P. W. Brooks.
1911.	,,	Captain F. S. Cooper.
1912.	,,	Lieutenant L. H. Clark.

Cup. 1881.—Presented by 20 officers (names inscribed).

Silver (Cashmir) Oval Tea Trays (2) 1881.	Presented by 9 officers (names inscribed).
,, Band Programme Frames (2) 1881.	,, 6 ,, ,,
,, Mounted "Aghdan" (Indian) 1883.	,, Lieutenant Fayrer.
,, (Cashmir) Cream Jug, 1885.	,, ,, Glossop.
,, Triple Cigar Cutter, 1885.	,, Captain J. M. Walter.
,, Large Bowl (Lucknow work), 1891.	Made from 36 champagne goblets, which were presented to the mess in 1867.
,, Castle Match Holder, 1892.	Presented by Captain E. Montagu.
,, ,, (8) smaller, 1892.	,, 8 officers.
,, Specimen Vases (4) 1892.	,, 2nd Lieutenant Rice.
Marble Clock, 1892.	,, 8 officers.
Silver Statuette, (Private, 1742), 1893.	,, 8 officers.
,, Castle Match Holder, 1893.	,, Captain L. J. Shadwell.
,, Whist Markers (2), 1893.	,, Captain A. D. Thorne, "King's Own," Lieutenant Marriott, "The Buffs," Lieutenant E. C. Curtis, Bedford Regt.

CENTRE-PIECES.—OFFICERS' MESSES.

1ST BATTALION.

2ND BATTALION.

Silver Glass Shot Container, 1893.	Presented by	3 officers.
,, Centre-piece, 1894	Purchased.	
,, Statuettes (4) of Privates, (1685, 1759, 1799, 1893), 1894.	,,	
,, ,, (2) Equestrian, of Polo Players, 1894.	,,	
,, Tea Pot and Cream Jug, 1894.	Presented by	4 officers.
,, Cream Jug, 1894.	,,	Surgeons G. Mansfield and F. Faichnie, A.M.S.
,, Vase-shaped Sugar Dredger, 1894.		Surgeons G. Mansfield and F. Faichnie, A.M.S.
,, Frame for Hunting Appointments, 1894.	,,	Lieutenant E. C. Doughty.
,, Egg-shaped Muffineers (2), 1894	,,	,, R. J. Cumming.
,, Statuette (Pikeman, 1685), 1895.	,,	9 officers.
,, Inkstand (in fox's mask), 1896.	,,	Colonel J. H. Hornby.
,, Square Ash Trays (2), 1898.	,,	2nd Lieutenant Barnadiston.
,, Mustard Pots (8), 1898.	,,	four 2nd Lieutenants.
,, Pepper Pots (24), 1898.	,,	4 officers (6 each).
,, Cigarette Box, 1898.	,,	Captain L. C. Arbuthnot.
,, Glass Preserve Jars (2), 1898.	,,	Lieutenant A. B. Morgan.
,, Swans (2) on ebonised bases, 1899.	,,	officers, 3rd Bn. Suffolk Regt.
,, Salad Bowl, 1899.	,,	Lieut.-Colonel A. J. Watson.
,, Statuette (Private, 1900, field service kit), 1902.	,,	10 officers.
,, Card Boxes (2), 1902.	,,	Major A. E. Kemble.
,, Statuette of Private, 12th Regt. (1850), 1903.	,,	12 officers.
,, Statuette of Private, 12th Regt. (M. I. 1899–1902), 1903.	,,	11 ,,
,, Statuette of the late Lieutenant S. J. Carey, 1903.	,,	his parents
,, Jug for hot milk, 1903.	,,	4 officers
,, Engagements Frame, 1903.	,,	three 2nd Lieutenants
,, Perpetual Calendar, 1903.	,,	two do.
,, President's Frame, 1904.	,,	2nd Lieutenant C. Jackson.
,, Dining Out Frame, 1904.	,,	Captain W. B. Wallace
,, Sugar Basin, 1906 (to match coffee service)	,,	4 officers.
,, Cup, with lid, 1908.	"In remembrance of Lavinia R. Morris and her brother Major C. I. Morris, both of whom were devoted to the regiment, presented by their old friends, Colonels J. E. Harris and Owen Williams."	
,, Table Bells (2), 1910.	Presented by	4 officers
,, Cayenne Pepper Pots (2), 1910.	,,	Captain Mulligan, R.A.M.C.
,, Cigar Cutters (2), 1912.	,,	Lieutenant F. Moysey.

The following cups and shields for competitions are also in possession of the battalion :—

1879. Silver-gilt Vase (Shooting). Presented by Commander-in-Chief in India (General Sir F. P. Haines, G.C.B.). Won by regimental team.
1880. Cup. Presented by, and won, as above.
1898. Cup. Officers' tug-of-war, Malta. Won by regimental team.
1908. Cup. Officers' Bayonet Team Combat.
1909. Cup. Military Rifle Association. Won by 4 officers competing.
1909. Cup. (Shooting). Malta Command Rifle Meeting. Won by regimental team.
1909. Cup. (Shooting) do. Won by E Company. Presented by 1st Battalion Royal Inniskilling Fusiliers.
1909. Cup. (Bayonet Fighting). Malta Infantry Brigade. Won by E Company.
1909. Cup. (Running). Malta Command. Won by Private Hall. Presented by Lady Curzon-Howe.

1909. Cup. (Hockey). Malta. Won by regimental team. Presented by "The Malta Chronicle."
1910. Cup. Regimental (Shooting). Malta. Won by regimental team.
1910. Cup. (Bayonet fighting, officers). Won by officers' team.
1910. Cup. (Bayonet fighting). Malta Infantry Brigade. Won by "E" Company.
1910. Cup. (Shooting). Won by regimental team. Presented by the King's Royal Rifles.
1911. The "Foster Shield," presented by E. W. D. Foster, Esq., C.M.G., in memory of his father, who commanded the 1st Battalion. Inter-Company Athletic Challenge Shield.
1912. Cup. (Shooting). Regimental Corporals' Challenge. Won by Lance-Corporal Quantrill.
1912. Cup. (Running). Egyptian Command. Won by regimental team.
1912. Cup. (Running). Regimental Challenge. Won by "A" Company.
1912. Cigar Box, Silver Mounted. (Bayonet Fighting) Officers' Competition. Egypt. Won by the regimental team.
1912. The Montagu Shield. Presented by Colonel E. Montagu on completion of term of command. Inter-Company Challenge Shield, for Ceremonial Drill.
1913. Cup. (Shooting). Cairo Command Rifle Meeting. Championship won by Corporal Mothersole, "G" Company.

Other articles of interest in the Officers' Mess are two very handsome old silver entrée dishes, dated 1798; an old silver double snuff-box, hall mark 1821; old silver nutmeg graters, given by Captain Elwes in 1836, and a large silver tooth-pick holder, given by Colonel Patton in 1843. Also, a King's Colour, in a frame, which probably dates about 1770, and bears the family motto of Colonel Scipio Duroure. (See Chapter XIV, "Notes on Colours.") Miss Craigie, who gave it to the regiment, was presented with a silver statuette. Also, a miniature of Lieutenant W. G. Shafto, 12th Regiment, *circa* 1820, presented by 2nd Lieutenants Trelawny and Hotham; a West African war horn, by Lieutenant Hall (1899); two 4·7 shells, from Pretoria, 4th June, 1900; a Boer sword and rifle, from the Transvaal, 1900, and a fragment of shell (relic of the Siege of Gibraltar) given by Lieutenant H. Castle Smith, in 1910.

In 1895, a case of 15 medals was given by 2nd Lieutenant A. S. Peebles, and contains the following:—

Ireland, 1689 to 1691.	Gibraltar, 1782.	New Zealand, 1864–6.
The Boyne, 1690.	„ 1783.	Afghanistan, 1878–80.
Fontenoy, 1745.	Seringapatam, 1799.	Hazara, 1888.
War with France, 1758–9.	South Africa, 1853.	Good Conduct.
Minden (3), 1759.		

The following medals in the Officers' Mess are in separate cases:—

Seringapatam (Silver Gilt).	Presented in 1897, by Major W. R. Lloyd, and Second Lieutenant C. A. White.
Colonel Picton's Gibraltar Medal.	Presented in 1904, by Lieut.-Colonel F. Graham.
Seringapatam (Silver) Medal. Eliott's Gibraltar Medal.	Purchased in 1904.

Also, a Cabinet of 9 Medals, mostly won by 12th men, which comprise:—

Sevastopol, 1854. Crimea, 1855. New Zealand, 1861–66. Long Service and Good Conduct.	All belonged to Sergeant-Instructor of Musketry, John Carroll, 1st Battalion 12th Regiment.
South Africa, 1853.	
New Zealand, 1863–6.	
Distinguished Conduct Medal.	Private R. Longworth, 1st Battalion, May, 1880.
Africa General Service, 1902–4 (Somaliland),	Lance-Corporal T. Maher (the only man of the regiment in the expedition).

Amongst the pictures are many of interest, including all the principal prints that have ever been published of the Great Siege of Gibraltar and the Storming of Seringapatam; some bought and others presented; 17 of the former between 1885 and 1912, and 9 of the latter between 1894 and 1908.

Others have been given and purchased as follows :—

Caricature of Officers, 1st Battalion, 1886.	Drawn by Lieutenants Saulez and Wallace.
His Royal Highness Albert Edward, Prince of Wales, 1888.	Presented by General Sir Martin Dillon.
Battle of the Boyne (2).	Purchased, 1894.
,,	Presented by 2nd Lieutenant F. Hall, 1895.
Plan of the Battle of Dettingen.	,, Captain S. B. Stotherd, 1896.
Minden.	,, ,, ,,
Six Etchings by Paton (Artist's proofs).	Purchased, 1892.
Old uniforms, about 1855 (2).	Presented by Lieutenants Saunders and Turner, 1892.
Wreck of the "Birkenhead" (with key).	Purchased, 1894.
New Zealand (Water Colours) (3).	Presented by Surgeon-General Bartley, 1895.
,, (Picquet, 1st Battalion).	,, ,, ,,
,, (Maori chapel).	,, ,, ,,
Mauritius, Views (8).	,, Major W. F. Scudamore,
Type of 12th Soldier, 1742.	Presented by Lieutenant C. Stanbridge, 1902.
Old Uniforms, 1850 (one series).	,, Sir W. Onslow, Bart., 1902.
Colonel Watson, Suffolk Regiment.	,, Mrs. Watson, 1902.
General Sir Thos. Picton.	Purchased, 1903.
Henry, 7th Duke of Norfolk.	,, 1904.
H.M.S. "Suffolk."	Presented by the Ward Room Officers, 1904.
Type of Officer, 6th Foot, 1799.	Presented by Major Lennock, 1904.
Plan of Fontenoy.	,, Captain W. H. Glossop, 1904.

Portraits of General the Hon. R. Meade; Colonel Thomas Grey (1797); Major Cave; 3 commanding officers, between 1850 and 1904; and Officers, 1st Battalion, before leaving for South Africa, in 1899.

Officers, H.M.S. "Suffolk," 1908–10.	Presented by themselves.
Types of 12th Regiment, 1871–77.	Presented by Lieut.-Colonel E. Montagu, 1909.
,, (6 coloured drawings).	,, ,, ,, ,, 1910.
,, (Privates, in marching order).	Purchased, 1911.
,, (1742, coloured engraving).	,, ,,
,, (Bugler, 1st Battalion).	Presented by Major J. Macdonald, R.A., 1911.
General Sir Henry Clinton.	,, Lieutenant D. Eley, 1911.
Major-General James Wolfe.	,, Captain V. Currey, 1911.
Death of ,,	,, Captain G. Brander, 1911.
The Colours, 12th Regiment.	,, Captain E. Jourdain.
Abbas Hilmi, Khedive of Egypt.	,, himself, 1911.
King Edward VII.	

Frame containing patterns of private's lace (1769–1836) and drummer's lace (1820–71).
General Sir Robert Sale, K.C.B. Purchased, 1913.

2nd Battalion.

In 1878, a centre-piece, height 2 feet 6 inches, was purchased for the Mess of the 2nd Battalion, of beautiful workmanship, and of the following design :—

A vase, supported by a Corinthian column, draped with colours; at the base of column, the regimental badge; the whole, on an ebony plinth decorated with two silver reliefs, representing the Rock of Gibraltar. At one end, a silver statuette of an officer in the uniform of 1799, and at the other, a pikeman of 1670. The centre-piece rests on a large silver-mounted and glass plateau. (See Plate 42.)

The plate of the Officers' Mess dates from the formation of the Reserve Battalion in April, 1842, and is as follows :—

Silver Salver, 1842.	Presented by Captain R. A. Phillips.
" " "	" " H. C. Elwes.
" Vegetable Dish, 1842.	" Sir R. A. Douglas, Bart.
Silver Snuff Box, 1842.	" Lieutenant Marcon.
" Claret Jug, 1863.	" " T. Gunn. "Boys, here's something for you."
" Cigar Lighter (Indian), 1867.	" " J. E. Harris.
Cigar Box, Silver Mounted, 1868.	" Lieut.-Colonel Atkinson.
Silver Goblets (24), 1870–71, with names of donors and dates engraved.	
" Salver, 1872.	Presented by Lieutenant J. Holder.
" Claret Jug, 1874	" 5 officers.
" " "	" 5 other officers.
" Centre-piece, with Silver Mounted and Glass Plateau. 1878.	Purchased.
" Table Bell, 1880.	Presented by 3 officers.
" Castle Cigar Lighter, 1880.	" 4 officers.
" Frame, 1885.	" Captain A. F. Poulton.
" Cup, 1889.	" 20 officers.
" Crumb Scoop (ivory handle), 1890.	" Captain L. H. Bazalgette.
" Mounted Castle Inkstand, 1890.	" Captains Bayly and Withington.
" Jug (Indian), 1892.	" Lieutenant S. B. Stotherd.
" Whist Markers (2), 1894.	" Lieutenant E. C. Doughty.
" do. (2), 1894.	" 3 officers.
" Bowl (Burmese), 1896.	" Colonel Owen Williams.
" Ash Trays (12,) 1896.	" Lieutenant C. Murphy.
" Double Preserve Stand, 1897.	" Captain J. R. Hopkins.
" Sugar Dredgers, 1897.	" Captain L. C. Arbuthnot.
" Bells (2), 1897.	" Captain K. Campbell.
" Castle Cigar Lighter, 1898.	" Lieut.-Colonel C. R. Townley.
" Bugle, 1900.	" Lieutenant and Adjutant E. C. Doughty.
" and Horn Cigar Lighter, 1903.	" Lieutenant Lorriner.
" Dining Out Slate, 1904.	" Major W. B. Wallace.
" Band Programme Stands (2), 1905.	" 4 officers.
" Flower Vases (2), 1906.	" Colonel A. C. Cubitt.
" Cigarette Box, 1906.	" Mrs. Owen Williams.
" Bell, 1907.	" Captain C. H. Turner.
" Mounted Stationery Cabinet, 1908.	" Lieutenant V. F. Currey.
" Cigar Cutters (2), 1909.	" N. A. Bittleston.
" Mounted Blotters (2), 1910.	" Captain Thomas and Lieutenant Stubbs.

The following cups, &c. for competitions are also in possession of the battalion :—

1867. Cup (Shooting). Indian Army, inter-regimental. Won by officers' team.
1888–1906. Silver Bowl (Shooting). Officers' Regimental Rifle Challenge ; 6 names inscribed.
1895. Silver Burmese Pagoda (Shooting). Burma Rifle Meeting. Won by officers' team.
1896. Cup (Shooting). Burma Rifle Meeting. "
1900. Statuette (Polo). Mounted Polo Player, Quetta Polo Tournament. "
1901. Cup (Hockey). Quetta District Challenge. "
1904. Cup (Shooting.) Sind District, inter-regimental Rifle Meeting. "
1905. Cup (Hockey.) Madras Challenge. "
1905. Cup. Southern India Polo Association, Infantry Tournament. Won by officers' team.
1910. Cup (Shooting). Presented by Major-General Lawson, Aldershot Command. Won by regimental team.
1910. Clock (Shooting). Presented by Brigadier-General Simpson. Aldershot Brigade Miniature Rifle Competiton. Won by regimental team.

1911. Cup (Shooting). Berdoe Wilkinson Challenge (replica), Aldershot. Won by regimental team.
,, Cup (Shooting). Running Man Company Competition. Aldershot.
,, Cup (Shooting). Detached Post, ,, ,,
1912. Cup for Judging Distance. 5th Brigade, Aldershot.
,, Cup (Shooting). Henniker Challenge, Young Soldiers' Competition, Aldershot.
,, Cup (Shooting). Army Challenge (replica), Aldershot Command Rifle Meeting.
,, Cup (Shooting). Earl Roberts (replica) ,,
1913. Cup (Shooting). Berdoe Wilkinson Challenge (replica), Aldershot.

Other articles of interest in the Officers' Mess are: a flint-lock musket of the 12th Regiment, taken in a Burmese dacoity, and presented in 1905 by Captain C. H. Turner; Swords from Seringapatam, given in 1906, by Captain Walford and Lieutenant Hepworth; a miniature, on ivory, of General Sir Thomas Picton, G.C.B., given in 1910 by Captain H. Castle Smith, a fragment of a shell (relic of the Siege of Gibraltar), and also the Centre of a Regimental Colour, 12th Regiment, 1805–27, found amongst the effects of Colonel W. H. Forssteen, and presented by Miss Craig.

In 1904, an oak frame containing medals was given by Lieut.-Colonel Cubitt, Major F. Graham, and Lieutenant E. C. Doughty; in 1908, an aneroid barometer was given by Lieutenant G. Brander; in 1909, a locker cabinet by 4 officers, and a letter-box of polished oak by Major W. Naish, 4th Battalion Hampshire Regiment; in 1911, a reference book rack, by 2nd Lieutenants Thomas and Gadd, and a large oak letter-rack by Majors Unwin and C. P. Crooke.

Amongst the pictures are many of interest, including a unique collection comprising every old print worthy of note that has ever been published of the Great Siege of Gibraltar and the Storming of Seringapatam, most of which are presentations; also a photogravure of Queen Victoria and the Prince Consort, presented on the 1st January, 1891, by Her Majesty the Queen, in commemoration of a review of troops of the Aldershot Division in 1859, at which the 2nd Battalion was present; and autograph portraits of their Royal Highnesses, the Duke and Duchess of Connaught, given by themselves in 1909.

Others have been given as follows:—

Gravelotte.	Presented by Captain Watson, 1884.
Her Majesty Queen Victoria.	Purchased.
His Majesty King Edward VII.	,,
Her Majesty Queen Alexandra.	,,
His Royal Highness, the Duke of Cambridge.	,,
"After Neuville."	Presented by Captain J. M. Walter, 1886.
Le Bourget.	,, Major Tillyer Blunt, 1886.
General the Hon. R. Meade.	,, Lieut.-Colonel Harris, 1888.
Battle of Minden.	,, Captain Coleman, 1888.
Halt of the 2nd Suffolk at Mandarah.	,, Lady Butler, 1890.
Mahomet Tewfik.	,, himself, 1890.
Wreck of H.M.S. "Birkenhead."	,, Lieut.-Colonel W. R. Routh, 1896.
Lord Heathfield.	,, Captain Glossop, 1897.
Officers, 1st Battalion, at Dover.	Purchased, 1899.
Colonel Dowse.	Presented by himself, 1899.
Colonel Townley.	,, himself, 1902.
Battle of the Boyne.	,, Major Keates, 1904
Island of Guadeloupe.	,, Major E. Montagu, 1904.

Soldier, 12th Foot, 1742.	Presented by Major E. Montagu, 1904.
Lieut.-General Sir T. Picton, G.C.B.	,, Captain F. A. White, 1904.
H.M.S. "Suffolk."	,, The Ward Room Officers, 1904.
H.M.S. "Resistance," with 1st Battalion on board from Mauritius, 25th February, 1848.	Purchased.
Major-General James Wolfe.	Presented by Lieut.-Colonel V. Graham, 1908.
Map of Suffolk, published in 1662.	,, Lieut. L. Hepworth, 1908.
Henry, 7th Duke of Norfolk.	,, Henry, 14th Duke of Norfolk, 1909.
Colonel Thomas Grey.	,, Lieut.-Colonel E. Montagu, 1909.
Sergeant Thomas Mackenzie.	,, himself, 1909.
Saxton's Map of Suffolk, 1575.	,, 2nd Lieutenants White and George, 3rd Battalion.
Old Uniforms, 12th Foot (3).	,, Lieut.-Colonel Montagu and Captain Wood-Martin, 1911.
,, ,, (3).	,, Captain V. Currey, 1911.
Wreck of H.M.S. "Birkenhead."	,, Captain R. Cockburn, 1911.
Battle of the Boyne. Battle of La Hague.	Bequeathed by the late Captain W. Saunders, 1912.
Sir F. J. Bolton.	Presented by himself.
Hunting caricatures (5).	,, Lieut. E. S. Roberts, R.E., 1891.
Colonel and Mrs. Williams.	,, Mrs. Williams, 1896.
Corporal Smith (late 12th), survivor of the "Birkenhead."	,, himself, 1904.
Colonel A. C. Cubitt.	,, himself, 1905.
Officers, Suffolk Regiment Donkey Race, 1883.	Purchased.

It is worthy of note that the officers of H.M.S. "Suffolk" are permanent honorary members of the officers' messes of the Suffolk Regiment, and also that the officers of the Suffolk Regiment and the 2nd Battn. Royal Irish Rifles (86th) are permanent honorary members of each others' messes.

Sergeants' Mess, 1st Battalion.

The sergeants of the battalion possess a good deal of handsome plate, some of which has been won at shooting, and at various sports, besides presentations. Their collection dates from 1879, since which time they have been always keen to enter for all competitions open to them. In their mess are also some very interesting pictures and portraits, which are much prized and well cared for.

The following is a list of the plate :—

Silver Cup (Shooting), 1879. Challenge Cup. Presented by sergeants.
Silver Cup (Pontoon Race), 1885, India. Presented by Bengal Sappers and Miners' Regatta.
Silver Bowls (2) ,, 1891. Presented by sergeants.
Clock in Case ,, 1895. ,, ,,

Silver Cup (Shooting), 1898. 1st Prize, Malta Rifle Meeting. Won by sergeants.
" " 1903. Replica, Eastern District Skirmishing. Won by G Company.
" " " 1st Prize, Essex County. Won by sergeants.
" " 1904. Replica, Eastern District Championship. Won by Private Hasket.
" " " " " " United Service. Won by regimental team.
" " " " Essex County Mafeking Cup. Won by D Company.
Silver Cup (Shooting), 1907. Presented by Sergeant G. Langer.
" (Billiards) " " Sergeant-Major Crane.
Fragment of Shell (relic of the Siege of Gibraltar, 1910. " Captain Castle Smith.
Silver Cup (Football), 1911. Replica. Presented by 2nd Battalion Malta Militia.
" (Shooting) " Challenge Cup. Presented by Mr. Donovan.

Sergeants' Mess, 2nd Battalion.

The sergeants of the 2nd Battalion also possess some very handsome plate, which dates from 1866, mostly shooting prizes, including challenge cups and shields, besides presentations. When quartered in India in 1865–66, there appears to have been particular scope for their zeal, Colonel Gardiner relating, in his "Annals of the 12th," that, in the former year, the battalion had been very successful in winning individual prizes, offered by the Northern India Rifle Association, and, in 1866, was victorious in the Grand Inter-Regimental Rifle Match, officers and sergeants winning respectively first and second prizes in the contests open to all officers and sergeants of European regiments quartered in Bengal.

They also possess some very interesting pictures and portraits.

The following is a list of the plate :—

Silver Cup (Shooting), 1866–67. Indian Army Cup. Won by sergeants.
Gibraltar Trophy, 1880. Presented by Officers of the Battalion.
Silver Cup (Shooting), 1888. Vanishing Target. Won by Colour-Sergeant C. S. Hatten.
Company Individual Challenge Shield (Shooting), 1890. (Last fired for in 1900). Won by H Company.
Company Field Firing Challenge Shield (Shooting), 1890. (Last fired for in 1900). Won by F Company.
Silver Cup (Shooting), 1892. Secunderabad. Won by sergeants.
" 1893. Presented by Officers of the Battalion.
" " " " "
" 1895. Presented by Lieutenant Roberts, on promotion to commissioned rank.
" (Shooting), 1895. Burma. Won by sergeants.
" " 1896. " "
" " 1899. Quetta Rifle Meeting. "
" " 1900. Presented by Colonel Tulloch, 126th Baluchis.
Silver Salvers (3). 1900. " Officers of the Battalion.
Silver Cup (Shooting), 1901. Individual Challenge. Won by sergeants.
" 1905. Presented by Sergeants, Royal Warwickshire Regiment.
" 1907 " " Royal Scots.
" (Shooting), Stanhope Cup. Aldershot Command Rifle Meeting. Won by sergeants.

APPENDIX VI.

The Rood Screen,

St. Mary's Church, Bury St. Edmunds.

DEDICATION CEREMONY.

A special service was held at St. Mary's Church, Bury St. Edmunds, on Sunday morning, 2nd February, 1913, for the purpose of dedicating To the Glory of Almighty God the new Rood Screen which has been erected by the Officers of the 1st and 2nd Battalions of the (12th) Suffolk Regiment, as a memorial to their comrades who are killed in action or die upon the active list.

The Rood Screen will form a memorial to Officers of the Suffolk Regiment for many years to come, the intention being to incise upon its panels the names of Officers of the Regiment as they are killed in action or die whilst on the active list. This memorial will take the place of the customary tablet which is usually erected by the regiment to each officer when he dies. In order to carry out this plan, the panels are all made so that they can be easily taken out when fresh names have to be added. There are at the present time four names upon the left-hand panels, which read as follows :—

Lieutenant and Quartermaster R. Felstead, died at Cambridge, November 9th, 1902, *aged* 38.
Captain J. R. G. Hopkins, died at Scarborough, September 15th, 1904, *aged* 36.
Captain P. W. Brooks, died at Malta, September 9th, 1908, *aged* 42.
Captain H. G. Chapman, killed in action at Agbedi, West Africa, May 15th, 1910, *aged* 29.

The memorial inscription, which appears in raised letters on the top framework of the panels, is as follows :—

> "This screen was erected in the year 1913 by the Officers of the 1st and 2nd Battalions (12th) the Suffolk Regiment, to the Glory of GOD and to commemorate their comrades killed in Action or dying whilst on the Active List."

The Regimental Coat of Arms is also carved on the north side of the screen.

The "rood loft," which now joins the old rood doors of the chancel, so that it is possible to walk right across the chancel from one side to the other, is four feet broad. The whole screen is twenty feet high, and it is a very handsome piece of work, both in its artistic conception and in the way it has been executed. It is made of Austrian oak, and the carving is of a heavy description. Not only is the whole appearance of the church greatly improved thereby, but the screen seems to act as a sounding board over both pulpit and lectern—as the architects said that they thought it would—and consequently the acoustic qualities of the church are considerably improved.

Whilst this immense advantage is gained on the one hand, the corresponding loss of spiritual power, which is so often experienced in churches where screens are erected, in the obstruction of sight and sound, has in this

ROOD SCREEN ERECTED IN ST. MARY'S CHURCH, BURY ST. EDMUNDS,
by the Officers 1st and 2nd Battalions (12th), The Suffolk Regiment.

case been avoided by the architects following the plan of the original rood screen, as shown in the picture now hanging in the coffee room of the Angel Hotel, Bury St. Edmunds.

The Rood Screen has been erected by Mr. R. Hedley, of St. Mary's Place, Newcastle.

The new oak cresting to the old choir screens has been given by Mr. F. W. King, a local resident. The screen between the Lady Chapel and the chancel—the gift of the "Crusaders" three years ago—has been darkened, so as to harmonise in colour with the older woodwork. The chancel begins to assume its final and perfect form, only one screen being still required to complete it—that between the chancel and the Suffolk Chapel.

The Dedication Service, which was at half-past nine, was essentially a military one, and there was a large attendance of soldiers, both officers and men, who were seated in the nave, besides the Mayor and Corporation of Bury and friends of the officers. The general public occupied seats in the north and south aisles. The Mayor (Alderman A. Mitchell), who wore his robe and chain of office, was accompanied by the Town Clerk, the Deputy Mayor, the Aldermen and Councillors.

APPENDIX VII.

EXTRACT FROM A LETTER FROM THE CAMP AT SERINGAPATAM, DESCRIBING HOW TIPPOO SULTAN HAD BEEN KILLED (PRESUMABLY BY A PRIVATE OF THE LIGHT COMPANY, 12TH REGIMENT), AT THE ASSAULT AND CAPTURE OF SERINGAPATAM, MAY 4TH, 1799.

"I SEND you the following particulars, relative to the conduct of the late Tippoo Sultan, on the 4th May, collected chiefly from the Killadar of Seringapatam, and from accounts given by some of his own servants.

The Sultan went out early in the morning, as was his custom daily, to one of the cavaliers of the outer rampart of the North Face, whence he could observe what was doing on both sides. He remained there till about noon, when he took his usual repast under a pandal. It would appear that he had at that time no suspicion of the assault being so near, for, when it was reported to him that our parallels and approaches were unusually crowded with Europeans, he did not express the least apprehension, or take any other precaution, beyond desiring the messenger to return to the West Face, with orders to Meer Gofhar, and the troops on duty near the breach, to keep a strict guard.

A few minutes afterwards he was informed that Meer Gofhar had been killed by a cannon shot near the breach, which intelligence appeared to agitate him greatly. He immediately ordered the troops that were near him under arms, and his personal servants to load the carbines which they carried for his own use, and hastened along the ramparts towards the breach, accompanied by a select guard, and several of his chiefs, till he met a number of his troops, flying before the van of the Europeans, whom he perceived had already mounted and gained the ramparts. Here he exerted himself to rally the fugitives, and, uniting them with his own guard, encouraged them, by his voice and example, to make a determined stand.

He repeatedly fired on our troops himself, and one of his servants asserted that he saw him bring down several Europeans near the top of the breach. Notwithstanding these exertions, when the front of the European flank companies of the left attack approached the spot where the Sultan stood, he found himself almost entirely deserted, and was forced to retire to the traverses of the North Ramparts. These he defended, one after another, with the bravest of his men and officers, and, assisted by the fire of his people on the inner wall, he several times obliged the front of our troops, who were pushing on with their usual ardour, to make a stand. The loss here would have been much greater on our part, had not the light company and part of the battalion companies of the 12th Regiment, crossing the inner ditch and mounting the ramparts, driven the enemy from them, and taken in reverse those who, with the Sultan, were defending the traverses of the outer ramparts. While any of his troops remained with him, the Sultan continued to dispute the ground, until he approached the passage across the ditch to the gate of the inner fort. Here, he complained of pain and weakness in one of his legs, in which he had received a bad wound when very young, and, ordering his horse to be brought, he mounted; but, seeing the Europeans still advancing on both the ramparts, he made for the gate, followed by his palanquin, and a number of officers, troops, and servants. It was then probably his intention either to have entered and shut the gate, in order to attack the small body of our troops which had got into the inner fort (and, if successful in driving them out, to have attempted to maintain it against us), or to have endeavoured to make his way to the Palace, and there make his last stand; but, as he was crossing to the gate, by the communication from the outer ramparts, he received a musket ball in the right side, nearly as high as the breast; he, however, pressed on, till he was stopped, about half way through the arch of the gateway, by the fire of the 12th light company from within, when he received a second ball close to the other. The horse he rode, being also wounded, sank under him, and his turban fell to the ground. Many of his people fell at the same time, on every side, by musketry, both from within and without the gate. The fallen Sultan was immediately lifted by some of his adherents and placed upon his palanquin, under the arch, and on one side of the gateway, where he lay, or sat, for some minutes, faint and exhausted, till some Europeans entered the gateway.

A servant who has survived, relates that one of the soldiers seized the Sultan's sword-belt, which was very rich, and attempted to pull it off; when the Sultan, who still held his sword in his hand, made a cut at the soldier with all his remaining strength, and wounded him about the knee; on which the latter put his piece to his shoulder and shot the Sultan through the temple, when he instantly expired. Not less than three hundred men were killed, and numbers wounded, under the arch of this gateway, which soon became impassable, excepting over the dead and dying. About dusk, General Baird, in consequence of information he had received at the Palace,

LAST MOMENTS OF THE SULTAN TIPPOO.
(SERINGAPATAM, MAY 4TH, 1799.)

came with lights to the gate, accompanied by the late Killadar of the fort, and others, to search for the body of the Sultan, and, after much labour, it was found, and brought from under a heap of slain to the inside of the gate. The countenance was in no way distorted, but had an expression of stern composure. His turban, jacket, and sword-belt were gone, but the body was recognised by some of his people who were there, to be that of the Sultan, and an officer who was present, with General Baird's leave, took from his right arm the talisman, which contained, sewn up in pieces of fine flowered silk, an amulet of a brittle metallic substance of the colour of silver, and some manuscripts in magic Arabic and Persian characters, the purport of which, had there been any doubt, would have sufficiently ascertained the identity of the Sultan's body.

It was placed on his own palanquin, and by General Baird's orders, conveyed to the court of the Palace, where it remained during the night, furnishing a remarkable instance, to those who are given to reflection, of the uncertainty of human affairs.

He, who had left his palace in the morning, a powerful imperious Sultan, full of ambitious projects, was brought back a lump of clay; his kingdom overthrown, his capital taken, and his palace occupied by the very man (Major-General Baird), who, about fifteen years before, had been, with other victims of his cruelty and tyranny, released from nearly four years of rigid confinement in irons, scarcely three hundred yards from the spot where the corpse of the Sultan now lay.

Thus ended the life and power of Tippoo Sultan. It will require an able pen to delineate a character, apparently so inconsistent, but he who attempts it must not decide hastily."

APPENDIX VIII.

Major Allan's Account of the Interview with the Princes, in the Palace of Seringapatam, and of the Finding of the Body of the late Tippoo Sultan.

A short time after the troops were in possession of the works, Major Beatson and I observed, from the South Ramparts, several persons assembled in the Palace, many of whom, from their dress and appearance, we judged to be of distinction.

I particularly remarked that one person prostrated himself before he sat down, from which circumstance I was led to conclude that Tippoo, with such of his officers as had escaped from the assault, had taken shelter in the Palace. Before any attempt could be made to secure the Palace (when it was thought the enemy, in defence of the Sovereign and his family, would make a serious resistance), it became necessary to refresh the troops, who were greatly exhausted by the heat of the day, and the fatigue which they had already undergone. In the meantime, Major Beatson and I hastened to apprise General Baird of the circumstances, and, on our way, we passed Major Craigie and Captain Whittle with some Battalion Companies of the 12th Regiment.

As soon as we reached General Baird, we proposed to bring these troops to him, to which he assented. On my return, General Baird directed me to proceed to the Palace, with the detachment of the 12th, and part of Major Gibbing's battalion of Sepoys. He directed me to inform the enemy that their lives would be spared, on condition of their immediate surrender, but that the least resistance would prove fatal to any person within the Palace walls; so having fastened a white cloth on a sergeant's pike, I proceeded to the Palace, where I found Major Shee, and part of the 33rd Regiment, drawn up opposite the gate. Several of Tippoo's people were on a balcony, apparently in the greatest consternation.

I informed them I was deputed by the general who commanded the troops in the forts, to offer them their lives, provided they did not make resistance, of which I desired them to give immediate intimation to their Sultan.

In a short time after, the Killadar, another officer of consequence, and a confidential servant, came over the terrace of the front building, and descended by an unfinished part of the wall. They were greatly embarrassed, and appeared inclined to create delays, probably with a view to effecting their escape, as soon as the darkness of the night should afford them an opportunity.

I pointed out the danger of their situation, and the necessity of coming to an immediate determination, pledging myself for their protection, and proposing that they should allow me to go into the Palace, that I might, in person, give these assurances to Tippoo. They were very averse to the proposal, but I positively insisted on returning with them.

I desired Captain Schohey (who spoke the native language with great fluency), and Captain Hastings Frazer, to accompany me. We ascended by the broken wall, and lowered ourselves down on to a terrace, where a large body of armed men were assembled. I explained to them that the flag which I held in my hand was a pledge of security, provided no resistance was made, and the stronger to impress them with this belief, I took off my sword, which I insisted on the Killadar having. The Killadar and many others affirmed that the Princes and the family of Tippoo were in the Palace, but not the Sultan, and they appeared greatly alarmed and averse to any decision. I told them that delay might be attended with fatal consequences, and that I could not answer for the conduct of our troops, by whom they were surrounded, and whose fury was with difficulty restrained.

They then left me, and shortly after I observed people moving hastily backwards and forwards in the interior of the Palace, and, as there were many hundreds of Tippoo's troops within the walls, I began to think our situation rather critical.

I was advised to take back my sword, but such an act on my part might, by exciting their distrust, have kindled a flame, which, in the present temper of the troops, might have been attended with the most dreadful consequences, probably in the massacre of every soul within the Palace walls. The people on the terrace begged me to hold the flag in a conspicuous position, in order to give confidence to those in the Palace, and prevent our own troops from firing the gates. Growing impatient at these delays, I sent another message to the Princes, warning them of their critical situation, and that my time was limited. They answered, they would receive me as soon as a carpet could be spared for that purpose, and soon after that, the Killadar came to conduct me.

I found two of the Princes seated on the carpet, surrounded by a great many attendants; they desired me to sit down, which I did, in front of them. The recollection of Moiza Deen, whom, on a former occasion, I had seen delivered up with his brother hostages to the Marquis Cornwallis, the sad reverse of their fortunes, their fears (which, notwithstanding their struggles to conceal, were but too evident), excited the strongest emotions of compassion in my mind. I took Moiza Deen (to whom the Killadar, etc., principally directed their discourse) by the hand, and endeavoured by every mode in my power to remove his fears, and to persuade him that no violence should be offered to him, or his brother, or to any person in the Palace. I then entreated him, as the only means to save his father's life, whose escape was impracticable, to inform me of the spot where he was concealed. Moiza Deen, after some conversation apart with his attendants, assured me the Padshah (Sultan) was not in the Palace. I requested him to allow the gates to be opened. All were alarmed at this proposal, and the Princes were reluctant to take such a step except by the authority of their father, to whom they desired to send. At length, however, having been promised that I would post a guard of their own Sepoys within, and a party of Europeans on the outside, and having given them the strongest assurance that no person should enter the Palace but by my authority, and that I would return and remain with them until General Baird arrived, I convinced them of the necessity of compliance, and I was happy to observe that the Princes, as well as their attendants, seemed to rely with confidence on the assurances I had given them.

On opening the gate, I found General Baird, and several officers, with a large body of troops assembled. I returned with Lieut.-Colonel Close to the Palace, for the purpose of bringing the Princes to the General. We had some difficulty in conquering the alarms and objections which they raised to quitting the Palace, but they at length permitted us to conduct them to the gate. The indignation of General Baird was justly raised by a report, which had reached him soon after he had sent me to the Palace, that Tippoo had inhumanly murdered all the Europeans who had fallen into his hands during the siege. This was heightened, probably, by a momentary recollection of his own sufferings during more than three years' imprisonment in that Palace. He was, nevertheless, sensibly affected by the sight of the Princes, and his gallantry at the assault was not more conspicuous than the moderation and humanity he displayed on this occasion. He received the Princes with every mark of regard, repeatedly assured them that no violence or insult should be offered them, and he gave them in charge of Lieut.-Colonel Agnew and Captain Marriot, by whom they were conducted to head-quarters in camp, escorted by the Light Company of the 33rd Regiment. As they passed, the troops were ordered to pay them the compliment, and presented arms.

General Baird now determined to search the most retired parts of the Palace, in hopes of finding Tippoo. He ordered the Light Company of the 74th, followed by others, to enter the Palace Yard. Tippoo's troops were immediately disarmed, and we proceeded to make the search through many of the apartments, having entreated the Killadar, if he had any regard for his own life, or that of the Sultan, to inform us where he was concealed.

He put his hand on the hilt of my sword, and in most solemn manner, protested that the Sultan was not in the Palace, but that he had been wounded during the storming, and lay in a gateway in the north face of the fort, whither he offered to conduct us, and if it was found he had deceived us, he said the General might inflict on him what punishment he pleased. General Baird, on hearing the report of the Killadar, proceeded to the gateway, which was covered with many hundreds of the slain. The number of the dead, and the darkness of the place, made it difficult to distinguish one person from another, and the scene was altogether shocking, but, aware of the great political importance of ascertaining, beyond possibility of doubt, the death of Tippoo, the bodies were ordered to be dragged out, and the Killadar and the other two persons were desired to examine them one after another. This, however, appeared endless, and, as it was now becoming dark, a light was procured, and I accompanied the Killadar into the gateway.

During the search, we discovered a wounded person lying under the Sultan's palanquin ; this man was ascertained to be Rajah Cawn, one of Tippoo's most confidential servants, who had attended his master during the whole of the day, and, on being made acquainted with the object of our search, he pointed out the spot where the Sultan had fallen. By a faint glimmering light, it was difficult for the Killadar to recognise the features, but the body being brought out, and satisfactorily proved to be that of the Sultan, was conveyed on a palanquin to the Palace, where it was again recognised by the eunuchs, and the servants of the family.

When Tippoo was brought from under the gateway his eyes were open, and the body was so warm that, for a few moments, Colonel Wellesley and myself were doubtful whether he was not alive. On feeling his pulse and heart, that doubt was removed. He had four wounds, three on the body and one on the temple, the ball having entered a little above the right ear, and lodged in the cheek. His dress consisted of a jacket of fine white linen, loose drawers of flowered chintz, with a crimson cloth of silk and cotton round his waist, and a handsome pouch with a red and green silk belt hung across his shoulder ; his head was uncovered, his turban being lost in the confusion of his fall, and he had an amulet on his arm, but no ornament whatever.

Tippoo was of low stature, corpulent, with high shoulders, and a short thick neck, but his feet and hands remarkably small; his complexion was rather dark, eyes large and prominent, with small arched eyebrows, and his nose aquiline. He had an appearance of dignity, or perhaps of sternness, in his countenance, which distinguished him above the common order of people.

MEMORIALS

WESTMINSTER ABBEY.

APPENDIX IX.

The Late Major Jeremiah O'Keefe.

Elegy on the lamented death of Major Jeremiah O'Keefe of the 12th (late of the Royal) Regiment, who was killed at the attack on the Isle of France (Mauritius), 29th November, 1810.

O'Keefe! accept the tribute of a tear
A brother soldier drops upon thy bier.
In distant climes together we have been,
Together war and pestilence have seen.
Thy spirit, dauntless, saw the frequent grave,
Receive alike the good, the just and brave;
The yellow fiend it saw, with ruthless fang,
Thy bold companions rend; 'twas then a pang
First pierced thy heart, thy gallant nerves unstrung,
And pity in thy feeling bosom wrung,

What tho' remotest India's burning strand,
Thy steps impressed at honour's bright command;
What tho' on Egypt's favoured field thy blood,
When British valour shone, triumphant flowed,
Again, to assert thy injured country's cause,
To chasten France, and vindicate her laws,
Disdaining to indulge ignoble ease,
Fame, tho' acquired, the occasion bid thee seize,
Bid thee seek India's fervid shores again.
And reap new laurels on his sultry plain,
Alas! the laurel that should bind thy brow,
Provokes this sigh,–and prompts this tear to flow.

Cold, cold is now the hand I oft have pressed,
And cold the heart that warmed thy generous breast,
That manly form which every eye admired,
That countenance each noble passion fired,
That pulse, so sensitive to others' pain,
That sorrow vibrated through every vein,
That voice, which grief or gladness could convey,
Insatiate Death has claimed an early prey,
Not modest, unassuming worth like thine,
The tyrant's fatal summons could decline,
But why repine, why mourn thy fall, my friend?
A life so prized, deserved thy glorious end.

APPENDIX X.

Memoir of Private John Dorman, 12th Regiment, who is reported to have fought at Dettingen, Fontenoy, Minden, and Culloden, and who lived to the age of 110 years.

John Dorman, or Diermott, was born at Hoigh, or the Bullock House, in the Parish of Clonlee, and County of Donegal, on August 24th, 1709; he was baptised by the Rev. Dunwith, rector of that parish, who then lived at Lifford, on the spot where the gaol has since been erected.

His father, after whom he was called John, was a labourer, and lived to the age of 111 years. His mother's name was Margaret Sharkey, who lived to be nearly 113 years old.

These circumstances, combined with his own great age, seem to favour the opinion of those who think longevity is hereditary; he was, however, the youngest of twelve children, of the rest of whom only one female lived to any great age. His grandfather, Bryan Diermott, of Temple Douglas, near Letter Kenny, lived to be a very old man, and had considerable property in that neighbourhood, which he forfeited to the Crown in the rebellion of 1641. The wife of this Bryan Diermott was Giles McGennis, of a respectable family, and cousin of a Major Stafford, a gentleman of property in the County of Donegal at that time. His father (John Dorman's) was brought up to be a Roman Catholic priest; but, as the term is, he was spoilt in the making, for he fell in love with Margaret Sharkey and married her. By this step, he displeased his family, and was obliged to earn his bread as a day labourer, until he got into the service of Dr. Nicholas Forster, Bishop of Raphoe, whom he served for many years in the capacity of land steward. In the year 1721, Bishop Forster confirmed this John Dorman, then twelve years old, and the boy was sent to school, to John Campbell of Clonlee, where some of his relations lived. Here he was taught to read, but neglected to learn to write, which afterwards proved a heavy loss to him, as his inability to keep accounts prevented his rising in the world, as he might otherwise have done, from the opportunities that occurred to him. After he arrived at the age of manhood, he joined in his father's labours, and remained at home till the year 1736, when he resolved to try his fortune in France, where he had an uncle, on his father's side, a Captain of Lord Clare's Regiment, in the Irish Brigade. With this view he traversed the coast of Ireland, from Donaghadee to Dingle and back, without being able to procure a passage, an embargo having at that time been laid on all the Irish ports, in consequence of the apprehensions of a rupture with Spain. Still determined to push his way, he passed from Donaghadee to Port Patrick, and thence to Dumfries, where his money failed him, in consequence of which he gave up his intention of proceeding to France, and enlisted in the 12th Regiment of Foot, then commanded by Colonel Duroure. The officer who enlisted him was Captain Conyngham, of Crauford, in the County of Donegal. This gentleman behaved very kindly to him in his distress at Dumfries, and offered him a guinea, to take him home, if he should not wish to enlist. But Dorman was ashamed to return, as he had left home contrary to his parents' wishes, and he found a friend in Captain Conyngham, whose servant he became, and found his situation very comfortable. With Captain Conyngham he lived eight years, and was in his service when that gentleman died of pleurisy at Limerick, and was buried in the churchyard of St. Mary's, in the year 1744.

Soon after Dorman enlisted, his Regiment was ordered on Foreign Service, and he passed with his master through London to Holland, landing at the port of Helvoetsluys, whence he proceeded with the Regiment to Amsterdam, after he had spent the winter of 1736 in quarters at Bergen-op-Zoom.

The British Army, on the Continent at this time, was commanded by John Dalrymple, Earl of Stair. He (Dorman) remained with the army on the Continent till the beginning of the year 1739, when a draft was made up of seven men from each company in every regiment, to form a body of Marines. With these Dorman volunteered, and went with Admiral Vernon's squadron of seven ships to the coast of Spain. Here he was engaged at the taking of Porto Bello, and St. Serengo, and the bombardment of Carthagena. At the same time, a strong armament was sent to the West Indies, under the command of Lord Cathcart, to curb the insolence of the Spaniards. The frost this year was extremely severe, even in Spain. He returned to Plymouth under the command of General Hobson, and shortly rejoined the 12th Regiment, then quartered at a village within eight stones (24 miles) of Fontenoy.

For six years he remained either in Hanover or its neighbourhood, and was frequently engaged in skirmishes, and out on guards. His health was firm; he was seldom indisposed, except after drinking excessively of "foozle," a liquor somewhat like our whiskey. He was, however, in general, a temperate man, and all his life an early riser. In the year 1743, he was engaged with his Regiment at the Battle of Dettingen. The order of this battle was directed by George II, who commanded the army in person. The King, advancing to the front of the line, gave fresh spirits to the soldiers. The British troops fired too soon this day, upon the marching up of the enemy, on which the French Black Musqueteers, detaching themselves from their lines, and galloping between the Allied Foot, were all cut to pieces. The firing now became general, when the presence of His Britannic Majesty (who was in posts of the greatest danger, and behaved with the noblest intrepidity), decided the fate of the day. Marshal Noailles showed great bravery in this battle. The Duke of Cumberland, being in the hottest of the engagement, was wounded in the calf of the leg. After losing the flower of the French Army, hewn down in every direction by British valour, Marshal Noailles ordered a retreat. In this battle, the French lost 6000 men and a multitude of officers, and the English, 2500. Had the enemy been properly pursued, before they recovered themselves from their first confusion, in all probability they would have sustained a total overthrow. The Earl of Stair proposed that a body of cavalry should be detached on this service, but his advice was overruled. The English Generals Clayton and Murray were killed in this battle, and the Earl of Albemarle, General Huske, and several other officers of distinction, wounded. The Battle of Dettingen was fought on the 27th June, 1743. The hostile armies remained after the battle on each side of the river Maine, till the 12th July, when the French General, receiving intelligence that Prince Charles of Lorraine had approached the Neckar, suddenly retired, and re-passed the Rhine, between Worms and Oppenheim. On the 27th August, the Allied Army passed the Rhine at Mentz, and the King of England fixed his Head-Quarters in the Episcopal Palace of Worms. Here the forces lay encamped till the latter end of September, when they advanced to Spire, where they were joined by 20,000 Dutch auxiliaries from the Netherlands. In the month of October, the King of Great Britain returned to Hanover, and the Army separated. The troops in British pay marched back to the Netherlands, and the rest took their route to their respective countries.

On the 30th April, 1745, John Dorman was present and engaged at the Battle of Fontenoy. The King of France had resolved on conquering the Netherlands, and assembled a prodigious army for that purpose, under the auspices of Marshal Saxe. The King and the Dauphin, on arrival at the camp, near Tournay, in the latter end of April, laid siege to that strong town. The Dutch garrison there, consisting of 8000 men, commanded by old Baron Dorth, made a vigorous defence. The allies were resolved to prevent the loss of the city by a battle. Their army was much inferior to that of the French. The Duke of Cumberland took command, having the Earl of Stair second in command under him. The Duke, having made the proper dispositions, began his march towards the evening. At 2 A.M., a brisk cannonade ensued, and about 9 o'clock, both armies were engaged, the village of Antoine being on the French right, a wood on their left, and the town of Fontenoy before them. The French had very great advantages in their position. Notwithstanding this, the British Infantry pressed forward, bore down all opposition, and for nearly an hour was victorious. Dorman, with the 12th Regiment, was in the hottest part of this action, and received a flesh wound in the right shoulder. So closely were the two armies engaged, that the muskets of each clashed against those of their respective opponents. Marshal Saxe was, at this time, sick with the disorder of which he afterwards died. He visited all parts in a litter, and saw, notwithstanding all appearances to the contrary, that the day was his own. One circumstance occurred, which subsequent historians were unwilling to record of the Duke of Cumberland, which was, that in the midst of the battle he resigned the command to the Earl of Stair, and Dorman alleges that if he had not done so, the whole army would have been cut to pieces. The gallantry of his Royal Highness, however, could not be doubted, and it shone forth as conspicuously in this battle, as in that of Dettingen. The English Column having driven the French beyond their lines, advanced so far as to pass the several columns of the enemy, which had opened, and made an avenue for them, and closed behind them as they passed on. The French Artillery then began to fire upon them, and though they continued a long time unshaken, yet, being wholly unsupported by the other wing, and exposed back and front, flank and rear, to a dreadful fire, which did great execution, the British were obliged to make the necessary dispositions for a retreat, which was effected in tolerable order, at 3 P.M. This was one of the most bloody battles that had been fought for a century. The Allies left upon the field nearly 12,000 slain, and the French bought the victory with nearly an equal number. Among the many British Officers killed in this battle were Lieut.-General Campbell, and Major-General Ponsonby. The latter commanded a squadron, consisting of the Scots Greys, Ligonier's Horse, and Inniskilling Dragoons. He fell in the rear of General Sperkin's Brigade, and not far from the spot on which John Dorman was then engaged.

Although the attack on the French Army at Fontenoy was generally adjudged rash and precipitate the British and Hanoverian troops fought with such courage and perseverance that, if they had been properly supported by the Dutch forces, and their flanks covered by a sufficient body of cavalry, the French, in all probability, would have been obliged to abandon the siege of Tournay, which, after a gallant resistance, surrendered to them on the 21st June. After dismantling Tournay and surprising Ghent, the French army invested Ostend, which, though defended by an English garrison, and open to the sea, was, after a short siege, surrendered on the 14th August. Dendermond, and afterwards Newport and Weth, underwent the same fate, while the Allied Army lay entrenched, beyond the Canal of Antwerp, and the King of France, having subdued the greater part of the Austrian Netherlands, returned to Paris, which he entered in triumph.

Flushed with his successes on the Continent, and resolved, if possible, to humble the pride of England, the King of France furnished the Young Pretender with a supply of money and arms, and sent him to Scotland, for the purpose of recovering the Crown which his family had forfeited. On the 14th July, 1745, he sailed in a small frigate from the port of St. Lazare, accompanied by the Marquis of Tullibardine, Sir Thomas Sheridan, and a few other Irish and Scotch adventurers. Off Belleisle, he was joined by the "Elizabeth," a French ship of war, mounted with 60 guns, as his convoy. Their design was to sail round Ireland, and land in the western part of Scotland, but falling in with the "Lion," an English ship of the line, a very obstinate and bloody action ensued. The "Elizabeth" was so disabled, that she could not proceed on the voyage, and with difficulty reached the harbour of Brest, but the "Lion" was shattered to such a degree, that she floated like a wreck upon the water. The Pretender, in the frigate, continued his course to the Western Isles of Scotland, and, landing on the coast of Lochaber on the 27th July, brought with him seven officers, and arms for 2000 men. In a short time, he found himself at the head of 1500 men, and he invited others to join him, by manifestoes scattered through the Highlands. Sir John Cope was then sent to oppose his progress. A requisition was made of 6000 auxiliaries, and six British regiments, which had remained with the Duke of Cumberland in Flanders, after the Battle of Fontenoy.

With the 12th Foot, Dorman returned, and was engaged at the Battle of Culloden,[1] where, in an hour, the rebels were totally routed, and the field covered with upwards of 1500 of their wounded and slain. Smollett says, in his continuation of Hume's History of England, that Lord Balmerino was conveyed, with the Lords Kilmarnock, Cromarty and Macleod, by sea to London, to be tried for joining in this rebellion. John Dorman said this was a mistake of the historian's, and alleged that he

[1] If this was the case, he must have been detached from the regiment, as it is shown at page 87 that the 12th missed being present at that engagement.

himself was one of the guard that accompanied them by land the whole way, under the command of Captain Eyre, who was afterwards made Governor of Galway for his services at that time. Dorman also said, that a person, by the name of M'Keuzir, the nephew of Lord Balmerino, and an officer in Sir John Bruce's Battalion, conceiving that his uncle had been used harshly by Captain Eyre, followed him to Galway, and remained there for a considerable time, endeavouring to get a shot at him, but was obliged to leave without effecting his purpose.

After the rebellion in Scotland had been suppressed, Dorman marched with the 12th Regiment to Portsmouth, and embarked for Flanders, where the veteran remained with the Duke of Cumberland's army until the peace which was concluded at Aix-la-Chapelle, on the 7th October, 1747. Immediately after the conclusion of peace, Monsieur Dupleix, who commanded for the French in the East Indies, began, by his intrigues, to sow the seed of dissension among the natives, that he might be the better able to accomplish certain designs which he had formed. His headquarters were at Pondicherry, whence he supplied the deposed Nabob of Arcot, Sundar Sabel, with 2000 sepoys, 60 Kaffirs, and 420 French soldiers, which enabled Sabel to defeat his rival Anaword Khan, whom they killed in battle. He then re-possessed himself of the Government of Arcot, and, according to a previous stipulation, ceded to the French the town of Vellore, in the neighbourhood of Pondicherry, with its dependencies, consisting of 45 villages.

Mahomed Ali Khan, son of the deceased Nawab (Anaword Khan), fled to Terricherapall, and solicited the assistance of the English, who gave him a reinforcement of money, men and arms, under the conduct of Major Lawrence, a brave and experienced officer. Thus commenced the celebrated War in India, which terminated with the reduction of the province of Arcot, after the army of Sundar Sabel had been completely routed, and its unfortunate commander put to death by the Nawab of Tanjour, an ally of the English Company, who cut off his head with his own hand in order to prevent any disputes about the manner of disposing of him. Among the reinforcements sent out to the East Indies, on this occasion, was a draft from the 12th Regiment, and with it, John Dorman. The vessel in which he sailed stopped for water at the Cape of Good Hope, and arrived at her destination in eleven months, which was then reckoned a good passage. Eight ships sailed under the same convoy, and in each of them about 100 soldiers. He was at the taking of Madras and Pondicherry, and received no wound at either place; he remained with the British army for three years, between Madras and Pondicherry, and the climate agreed with him, but neither he, nor any of the privates, could acquire any share of the wealth which was there accumulated by the officers of this army. Their food was chiefly rice, and they drank arrack with water. Those who perspired profusely, as Dorman did, enjoyed good health, but those who did not were sickly, and many of them died. Their uniform consisted of nankeen coats and trousers. The intemperate use of fresh arrack caused much sickness and several deaths in the European armies. The successes of Colonel (afterwards Lord) Clive, at this time in India, were almost incredibly great, and laid the foundation of the present amazing extent of riches and territory which the English possess in the East Indies.

In the year 1752, Dorman returned to the King's service, and to Europe, with about £30 prize money, which he had shared in the taking of Pondicherry. The vessel in which he returned to England, stopped for water at St. Helena. On his being discharged at Charing Cross, from the service of the company, Adjutant-General Napier, who then commanded the 12th Regiment, enquired if there were any men there, who had belonged to it. Dorman replied that he was one, and immediately re-enlisted in it, upon which the General gave him five guineas, which, with the £30 he had brought to Europe, he dissipated in a very short time, reserving only what bore his expenses to Aberdeen, where the Regiment was quartered.

In the month of May, 1756, the King of England declared war against the French, on account of the infringements and encroachments made by them upon the British Territories in America after the Peace of Aix-la-Chapelle; and, in the ensuing month, the French King, in his turn, declared war against his Britannic Majesty, in terms of uncommon asperity. On the 20th September, Prince Ferdinand of Brunswick, after driving the French out of Hanover (with an army which his Britannic Majesty had authorised him to raise in that country), took possession of Leipsic, for the purpose of forwarding the King of Prussia's designs upon Poland. A sharp war ensued, in which the King of Poland was deprived of his Electorial Dominions, and his troops, arms, artillery and ammunition. In the latter end of the year, the Hanoverian auxiliaries were transported from England to their own country, which was, at this time, in great danger of invasion. After various successes and reverses, the French, on the 24th July, 1757, laid part of Hanover under contribution, which led to the action of Hoslenbeck, in which the Allied Army, under the command of the Duke of Cumberland, were obliged to retreat. After which the French took possession of the whole of the Electorate of Hanover, and also of Hesse Cassel.

Dorman was on the Continent during the whole of this war, and engaged in different battles and skirmishes, the particulars of none of which he remembers, except those of Minden, where he was severely wounded in the left hand. In this memorable engagement, one of the most glorious in the English annals, Prince Ferdinand of Brunswick, with about 7000 English troops, defeated 80,000 of the French regular troops in fair battle. In the middle of this battle, Lord George Sackville behaved so extremely badly, that, when it was over, Prince Ferdinand took his sword and sash from him, and ordered him to retire from the army, as he had no occasion for his services. His command was given to the Marquis of Granby, who had highly distinguished himself in the battle, and the unfortunate Lord George was afterwards tried, and forfeited his commission. His crime was, not bringing up in due time a body of cavalry which he commanded. So great was the indignation against him at the moment, that six Regiments nearly mutinied in the field, because he was not immediately shot.

Dorman was carried out of the field on a waggon, and brought with other wounded men to a military hospital at Bremen, on the Weser. On his recovery, he was discharged, with a pension of £7 18*s*. a year, which he forfeited eleven years afterwards, by refusing to remove to a depot in England, from Strabane, where he was carrying on, with some success, the trade of a baker, and where he died, after a short illness, on January 13th, 1819, in his 110th year.

APPENDIX XI.

Regimental Music.

Marching past in slow time having for some years been practically abolished, there has been some difficulty in ascertaining what slow marches were adopted by the 1st and 2nd Battalions respectively, owing to their having been lost sight of.

The evidence available shows that the slow march of the 1st Battalion is known merely by the name of "The 1st Suffolk Regimental Slow March."

That of the 2nd Battalion (raised in 1858) was "The Druids' Chorus" from the opera "Norma."

Apparently, the only quickstep known to have been in use by the 1st Battalion is "Speed the Plough," and, that of the Reserve Battalion, prior to 1858 (played on their fifes and drums, as they had no band), was "The Duchess."

On the formation of the 2nd Battalion, their German bandmaster (Herr Richs) introduced, as a quickstep, a German March (name forgotten), which was disliked.

This continued until Colonel Ponsonby assumed command in 1863, when he changed it to "Milanollo," the quickstep of the Coldstream Guards, which was followed for periods by "The Men of Harlech," "The Dashing White Sergeant," and "The White Cockade."

At Colonel Ponsonsby's death in 1868, "The Duchess" was introduced as a march past by Major Espinasse, and continued in use until about 1898, when it was changed to "Speed the Plough," in order that the 1st and 2nd Battalions should be alike.

On the date of closing this History (1913) the 1st, 2nd, 3rd, 4th and 5th Battalions, Suffolk Regiment, all go by to "Speed the Plough."

Officers' special Mess Calls, on guest nights, in the 12th, originated in the 2nd Battalion, at the Curragh, in 1863, when they were sounded on bugles with chromatic attachments. Before 1881, it became customary for the band to play them, and they were introduced into the 1st Battalion in 1909.

The Officers' 2nd Mess Call of the 1st Battalion is "The Roast Beef of Old England," its 1st Mess Call, and both those of the 2nd Battalion, being shown in the pages which follow.

The Company Bugle Calls of the 1st and 2nd Battalions were in use prior to the formation, in October, 1913, of the double-company system, and were also adopted by the 3rd (Special Reserve) Battalion.

The Battalions now use the calls of A, B, C and D Companies for the new four-company organisation.

APPENDIX XII.

OLD COLOURS.

1ST BATTALION. 12TH REGIMENT.

1827–1849.

THE following correspondence took place with reference to the laying up of the old Colours (presented in 1827) in the church of St. Mary Le Tower, Ipswich, after the presentation of new ones at Weedon, on the 14th July, 1849.

Lieut.-Colonel Patton's suggestion as to their disposal, contained in the following letter, was considered most appropriate :—

Weedon, September, 1849.

Sir,

You are perhaps aware that the 12th Regiment has been connected with the Eastern Division of the County of Suffolk, from which it has been styled "The East Suffolk Regiment," ever since the year 1783, at a time when they were engaged in the glorious defence of Gibraltar; and that it is an old and distinguished regiment. The old Colours of the regiment, which have been recently replaced by new, are now in my custody; and I have been desired by General The Hon. R. Meade, our Colonel, to communicate with you as Incumbent of the principal Church in Ipswich with a view to their being suspended therein. It will be very gratifying to the Corps which I have the honour to command, to have the Colours which they have honourably borne for a period of 22 years in every vicissitude of climate, preserved in a Church associated with the early recollections of many who have served, as well as those now serving, in the Regiment. Should you see no objection, and give your sanction to their being handed down and preserved in the manner pointed out, I shall be most happy to hear from you, and receive your direction as to their being forwarded.

I have the honour to be, Sir.
Your most obedient, humble servant,
John Patton, Lieut.-Colonel,
Commanding 12th Regiment

The Rev. W. Nassau St. Leger.
Incumbent, St. Mary Le Tower.

The proposal was received in the spirit which prompted it, as the subjoined paragraph extracted from Mr. St. Leger's communication will testify :—

St. Mary Le Tower, Ipswich.

Dear Sir,

I shall have great pleasure in receiving the old Colours of your distinguished Regiment, and shall assign them a suitable place of distinction in the Venerable Mother Church of Ipswich. I shall direct the Churchwardens, on the reception of the Colours in the Church, to register them in the Parochial Books as the future property of the Parish. . . .

In another letter, dated 8th September, 1849, the arrival and admission of the old Colours into the Chancel of St. Mary Le Tower are thus described :—

Dear Sir,

The Colours arrived here safely last Wednesday, and after reposing during the interval in the Barracks, were this morning escorted with all the honours to St. Mary Le Tower by the Heroes of Aliwal. The Band of the 16th Lancers happened to be here, and added much to the ceremony ; Major Gavin commanded. In addition to the Lancers we had the Staff of the East Suffolk Militia, with Captain Beecham, an old 45th man. I had also a sergeant and two privates of the 12th. It will be gratifying to you to know that the Colours were brought into the Church up to the Altar by two Officers—the Queen's Colour was borne by Lieutenant Barton, and the Regimental by the Grandson of Sir Walter Scott. The Colours are—for the present—north and south of the Altar.

Yours faithfully,
W. N. St. Leger.

Lieut.-Colonel Patton,
12th Regiment.

Major Gavin, 16th Lancers, forwarded the following official notification of the circumstances to the Regiment :—

Ipswich Barracks, September 8th, 1849.

Sir,
I beg to inform you that we this day placed the old Colours of the 12th Regt. in the Tower Church of Ipswich, paying them all due honours.

I have the honour to be, Sir,
Your most obedient servant,
S. Gavin, Major,
16th Lancers.

Lieut.-Colonel Patton, 12th Foot.

Thus far the wishes of the Regiment were fulfilled. The Old Colours however were not allowed to take their rest undisturbed. A petty jealousy of the disinterested zeal evinced by Mr. St. Leger, and increased unhappily by that prevailing spirit of contempt for the sacred character of these "Insignia of Honour and Loyalty" engendered an opposition which tried to disguise itself beneath a sentiment of piety. This opposition was sufficiently powerful to effect their removal from the Church, as will be seen by the following correspondence, which begins with a letter from a Churchwarden, Mr. Silburn, to Mr. St. Leger, intimating that the Colours would not be allowed to remain in the Church. This was soon followed by an order from the Archdeacon of Suffolk for them to be removed. The following are the letters on the subject :—

September 13th, 1849.

Dear Sir,
Mr. Rabett and myself called on Mr. Stewart to-day. He says he has no doubt about the Archdeacon's objecting to the Colours being put up in the Church ; under which circumstances we will, if you like, write to the Archdeacon on the subject, or, as they were given to the Town, suggest to the Regiment that they be committed to the care of the Corporation in the Council Chamber.

I am, dear Sir, yours &c.,
C. Silburn.

To the Rev. W. N. St. Leger.

September 14th, 1849.

Dear Sir,
I cannot see the propriety of applying to the Archdeacon in this matter ; it would appear ungracious to the Regiment and rather unfriendly towards the Archdeacon himself. The practice of the Church in all ages down to the present days as to receiving Regimental Colours with all honour, is so clear that the "objection" of the Archdeacon, if expressed, would by our interference be brought out in contrast to the sentiment and example of the whole Catholic Church. I am, of course, only supposing that the Archdeacon has any "objection." If he has, we can only recognise it as the "objection" of an individual Priest, which would certainly

prevent the suspension of Colours in his own Parish Church; but as Archdeacon, his "objection," if any, is of no weight, because it is against the plain practice of the Church. On these grounds, I must, for my part, decline any reference to the Archdeacon.

One word more. You are in error in supposing that these Colours were presented to the Town. THEY WERE NOT. In the entire correspondence, there never was any allusion to either Mayor or Corporation. They were presented with true Soldierly feeling to the Mother Church of the Town. They are consecrated Banners. They were received from the Church, and now that their work is over, with the blessing which still lingers upon them, they are returned to the Church again.

I am, Your Faithful Servant,
W. N. St. Leger.

To Mr. Silburn, Churchwarden,
St. Mary Le Tower.

Here the matter rested until the decision of the Venerable The Archdeacon Ormerod, which was given on the 3rd October. On the same day the Colours were conveyed to the house of the Incumbent, Mr. St. Leger, who wrote that evening the following letter to General Meade :—

Ipswich, October 3rd, 1849.

Sir,

I feel very great regret in reporting to you the decision of the Archdeacon of Suffolk, relative to the suspension of the Colours of the 12th Regiment. I beg at once and briefly to say that it is the Archdeacon's desire that these honoured Colours should be removed from the Church and forwarded to Chelsea Hospital.

As I do not feel comfortable in the idea of the Colours being in the Church without authority, I shall remove them to-night to await your orders. It has been suggested that the Colours should be presented to the Corporation of Ipswich. I have rejected the proposal with a feeling akin to contempt. I cannot imagine a greater degradation than for them to be handed over to the municipal authorities of an unimportant county town, especially as the presiding Magistrate in this place is, as I have been informed, a member of the Peace Society. It is most satisfactory to myself that I have had the opportunity of paying honour to the Colours of your Regiment, and very painful it is that the indignity which has been offered has proceeded from those who, as Suffolk Men, ought to have felt highly honoured in having amongst them the banner of the distinguished Regiment of their own County.

I am happy to remember the fine soldierly feeling displayed by Major Gavin and the officers and men of the 16th Lancers, in escorting the Colours to the Church, and it is a consoling reflection to me, as I am sure it must be to every Irishman in that celebrated Regiment, that in our own beloved Ireland, the humiliating occurrence, so much to be deplored, could not have taken place.

I have the honour to be,
Your faithful servant.
W. N. St. Leger.

To General the Honourable Meade,
Bryanston Square.

The following letter from General the Hon. R. Meade is in reply to the above :—

Grimston Park, October 5th, 1849.

My Dear Sir,

In the first place, I must offer you my very sincere thanks for the trouble you have taken for the fulfilment of the great wish of myself and every Officer in the 12th Regiment, to have the remains of their late Colours placed in your Church. I have now to thank you for your letter of the 3rd instant, relating to the interference and attempted indignity of your superior officer to the Insignia of Honour and Loyalty. From whatever motive it proceeded I know not, and I would not wish to assign one, but I cannot believe that he ever visited the edifices of Westminster Abbey and St. Paul's Cathedral, or he would have learned that the insignia of honour and military glory are there Consecrated and Hallowed. Indeed, Sir, your ready and anxious wish to obtain due honour for our Colours will ever be remembered by myself and every officer of the 12th Regiment, and I am sorry that any embarrassment should have been incurred by you in fulfilment of our wishes.

I have requested Colonel Patton to communicate with you, and relieve you from further anxiety.

Believe me ever, my Dear Sir,
With very great esteem,
Your very Obedient Humble Servant,
Robert Meade, General.

To The Rev. W. N. St. Leger.

The reverse which Mr. St. Leger had received from the Archdeacon did not abate his ardour in a cause which he considered collateral with that of the British Army. This led him to appeal in the following letter to the Right Hon. the Earl of Stradbroke, Lord Lieutenant of the County, entreating him to "vindicate the honour of the County" :—

Ipswich, October 6th 1849.

My Lord,

I presume your Lordship has seen by this day's journal the order of the Archdeacon of Suffolk for the removal of the old Colours of the 12th Regiment from my Church.

These venerable relics, after having been forwarded with true soldierly feeling by the East Suffolk Regiment, to repose in the sacred custody of the Metropolitan Church of the County, are, I grieve to say it, ordered to be removed by an authority which, until the Bishop speaks, I felt bound to obey.

In justice to the Archdeacon, I am happy to believe that whatever may be his own private feelings, he would not have interfered were it not for the objection of a Churchwarden; who, I am led to think, has been greatly fortified in his unbecoming officiousness by the sympathy of our very unsoldierly M.P., Mr. Cobbold.

I cannot but lament the decision of the Archdeacon, opposed as it is to the plain practice of the Church. However, my Lord, I am unwilling to let the Colours be removed from the town without appealing to your Lordship, as Lord Lieutenant of Suffolk, to vindicate the honour of the County, from any imputation which this unlooked-for, and most unhappy interference, so much to be deplored, might possibly cast upon it.

In justice to the venerable Colours of a distinguished Corps, to these once proud banners no longer able to defend themselves, I look to your Lordship for sympathy and protection.

I have the honour to be, my Lord,
Your Lordship's faithful servant,
W. N. St. Leger.

To the Right Hon. The Earl of Stradbroke.

Henham, October 7th, 1849.

Dear Sir,

I much regret that any step should be taken by the Archdeacon of Ipswich, hostile to the feelings of the parishioners of St. Mary Le Tower in removing the old Colours of the 12th Regiment of Foot, from the Church. No doubt it is common in Churches and many places of worship in England, and on the Continent of Europe, to have these insignia of military glory placed within their walls, a record of victories gained in defence of our liberties. But the right to remove them may be given to the high authorities of the Church, and the Lord Lieutenant has no power to resist their orders in any matter touching the administration of Church regulations.

If, however, a strong feeling were to be expressed by the inhabitants of St. Mary's against the removal of these relics, perhaps the Archdeacon would reconsider the question after consulting with the Bishop of the Diocese.

I am, dear Sir, yours faithfully,
Stradbroke.

To the Rev. W. N. St. Leger.

Mr. St. Leger also wrote the following letter to the Member of Parliament on the same day :—

Ipswich, October 7th, 1849.

Sir,

I am encouraged by the assurance of the full sympathy that I shall receive from you to request your powerful and honourable protection for the Colours of the 12th Regiment, which, in obedience to the Archdeacon, I have removed from my Church.

The proud relics appeal to every honourable feeling of my heart, and I received them with all loyalty and reverence; with loyalty, because one is the flag of England; with reverence, because the principle which influenced the Regiment to place their Colours in my Church was full of religious sentiment, and love of home; feelings which, while they actuate the British Army, will make that Army the pride as well as it has been the conqueror of the world.

I am sure you must have read the account of our enthusiastic reception of those Colours. Thank God, they were paid all honour, escorted to the Church by the heroes of Aliwal, both Colours were borne by Officers in testimony of respect to the whole Army of the Line.

By the interference of a Parish Officer, I have been obliged to remove those venerable relics from the Holy Place which I fondly hoped would have been their peaceful sanctuary. I am anxious to do all justice to the Archdeacon.

He seemed to sympathise with me in feelings of deep respect, but his inclination (having been appealed to) was so strong for sending the Colours to Chelsea Hospital, that he gave the order to remove them from the Church.

I now write to you, Sir, and I also write to Lord Stradbroke by this post, to assure you that I cannot suffer those venerable Standards to leave Ipswich without affording those of the loftiest station in the County an opportunity to vindicate the military pride of Suffolk.

I have the honour to be, Sir,
Your faithful servant,
W. N. St. Leger.

To Lieut.-General Sir Ed. Kerrison, Bart., M.P.,
Oakley Park.

Sir Ed. Kerrison's answer to Mr. St. Leger's letter of 6th October :—

Oakley Park, October 19th, 1849.

Dear Sir,

I much regret with you that the authorities of Ipswich should not, as has been common in Great Britain, have received the Colours of the East Suffolk Regiment of Foot with the honour and respect due to that gallant Corps. The Holy desire of finding for their Colours a resting-place in the Sanctuary of their Church and Home, might have been met with more grateful feelings from their County men. However, I hope with Lord Stradbroke that this mistake may be reconsidered by those in power.

I regret it is out of my humble influence to render you the aid you desire. With a soldier's thanks for the feelings you have throughout evinced, that England not only expects, but honours those who do their duty.

I am, etc., etc.,
(signed) Ed. Kerrison.

Mr. St. Leger had conveyed to the Regiment the intimation of what had happened in a letter which he wrote to Lieut.-Colonel Patton, which is given below.

Ipswich, October 3rd, 1849.

My Dear Sir,

I am very sorry, and were I a Suffolk man, I should be very much ashamed, to tell you that after our enthusiastic reception of your Colours, the proud relics have been ordered out of the Church by the Archdeacon of Suffolk. In common justice to the Archdeacon, I am happy to say he expressed perfect sympathy with me in offering all honour to the Colours, but his feeling was so strong for sending them to Chelsea Hospital, as the fitting and appropriate place for their reception, that he decided upon removing them from the Church.

I think, in confidence, it is just as well. You perhaps are aware that Churchwardens are sometimes very officious personages, so that were the Colours left here, it is not impossible that they might have been taken down by some bustling functionary.

I am, however, thankful that we have had an opportunity of paying honour to your venerable Standards, and most gratifying was it to witness the fine soldierly feeling of Major Gavin and Officers of the 16th Lancers on the occasion.

I write to General Meade by this post, and I shall keep the Colours in my own house, until I receive instructions from him, or from you. It has been suggested to hand those relics over to the Corporation of Ipswich. I could not entertain the proposal. The Corporation, you may easily imagine, consists of persons who, most probably, possess little sympathy for the emblems of glorious war. Moreover, their Magistrate, I understand, is a member of the Peace Society.

In conclusion, as the unforeseen objection has appeared, we shall consult our own dignity best by removing the Colours from a place which it yet remains to be proved is worthy of them.

I am, my dear sir,
Yours faithfully,
W. N. St. Leger.

To Lieut.-Colonel Patton,
Commanding 12th Regt., Weedon.

When this indignity became known to the officers and men of the Battalion, they were justly excited by a strong emotion of mingled indignation and regret, that their religious and chivalrous intention should have been so unworthily and unwisely frustrated. Lieut.-Colonel Patton expresses this in the following letter to Mr. St. Leger :—

Weedon, October 5th, 1849.

My Dear Sir,

As you may suppose, I have not been a little astounded at the result of our well intended and well executed transfer of the old Colours to the Church of St. Mary Le Tower; it is a most untoward affair, and I cannot but feel deeply mortified that both the Archdeacon and Churchwarden should feel otherwise than highly flattered by having the consecrated banner of a distinguished Corps of their County in their keeping.

To you, my dear sir, our most grateful thanks are due, and I beg you will hold the tattered emblems in your keeping until you shall hear from General the Hon. Meade. They cannot be in better hands. Pray on no account encourage the idea that they can ever grace the hall of a Corporation. With every respect for the Corporation body, I do not see that they can have anything to do with these Colours. Certainly not with my consent. May I request, when you have the opportunity, that you will reiterate my obligations to Major Gavin and the officers of the 16th Lancers, and I shall look forward to making, at some future day, your personal acquaintance, to render thanks for your ready (but as it turned out unlucky) co-operation, in what we both looked upon as fitting and appropriate.

Yours very faithfully,
J. Patton, Lieut-Colonel.

To the Rev. W. N. St. Leger.

It remained therefore to select another asylum. Lieut.-Colonel Patton communicated immediately with the Governors of Chelsea Hospital. Here his wishes were again defeated; only one case of Regimental Colours being received into this institution had occurred. This case originated in a mistake, as appeared from an official letter wherein it was stated that Trophies of War were the only authorised admissions.

Meanwhile the Old Colours were conveyed from the house of the Incumbent to the residence of General Meade, upon which occasion Mr. St. Leger dispatched this letter to Lieut.-Colonel Patton's address with the request that if any record of the affair be kept it might be inserted :—

St. Mary Le Tower, Ipswich, November 5th, 1849.

My Dear Sir,

I fear you are sick as myself of St. Mary Le Tower, Ipswich. Nevertheless, I must do my duty to your venerable Colours—the Colours which thousands of fine fellows have saluted in their time, and which thousands, yourself among the number, would have shed their heart's blood to defend. My last duty to your proud Banners is to inform you that they leave my house to-night to go by the mail train to London, according to your direction. I send a Sergeant and two Privates of your own Regiment who are recruiting here to escort them to the train. I am extremely gratified that my feeble but heartfelt defence of the Colours and (to say truth) of the military profession, shall be preserved in the Records of the 12th Regt. It may be a stimulant hereafter to my beloved infant boys, whom with God's blessing I intend for a military life, to find that my feelings have been confirmed as those feelings are by the testimony of the whole Catholic Church towards your honourable and highminded profession. If I am enthusiastic in this matter, it is a fault on the right side in modern times, and may be accounted for, not more by the associations of early life, than by the regard I bear to my dearest and best loved friends, who are all in the service; chief among these is that upright soldier, Colonel Chatterton, on whose behalf I wrote those letters, which I think I sent you some time ago with reference to the Cork Election. In conclusion, my dear Colonel, let me assure you that the 12th Regiment shall be ever remembered by me with lively interest, and that in my memory, whatever promotion you may yet acquire, and I trust it will be all you may wish for, the gallant Corps shall still live under the command of that Lieut.-Colonel, whose correspondence with me throughout has expressed the feelings of a British soldier, and a well-bred gentleman.

I am, My dear Sir,
Yours faithfully,
W. N. St. Leger.

Lieut.-Colonel Patton,
12th Regiment.

The ultimate fate of the Old Colours is confirmed by the subjoined extract from a letter from General The Hon. R. Meade to Lieut.-Colonel Patton :—

After all their vicissitudes the Old Colours of the 12th Regiment have arrived here, and now decorate my Hall, and during my life shall remain free from further molestation.

A tribute of respect, which might perpetuate the gratitude of the officers and men of the 1st Battalion was unanimously subscribed to, and a handsome piece of plate presented to the Rev. Gentleman bore this inscription :—

Presented by General The Honourable R. Meade, Lieut.-Colonel Patton, the Officers, Non-commissioned Officers and Men 1st Battalion, 12th, or East Suffolk Regiment, to The Reverend William Nassau St. Leger, A.B., Incumbent of St. Mary Le Tower, Ipswich, in grateful appreciation of his chivalrous and reverent reception of their Old Colours into the Chancel of his Church, and noble protest against the violation of their consecrated repose. Weedon, December 15th, 1849.

The piece of plate was an ornamented inkstand. The base, which rests upon four scroll-formed feet of alternately bright and frosted silver, rises in the centre supporting a beautiful model of the Rock and Castle of Gibraltar between two silver vases. At each end, and reposing against these vases, are the Colours of the Regiment, most chastely executed in frosted silver. On the Colours are *Minden*, *Gibraltar*, *Seringapatam*, *India*. Around the summit of the Rock is a scroll with the inscription *Montis Insignia Calpe*. The Castle is executed in frosted silver, and on each of its doors is raised in high and bright relief, the Key, being part of the Badge of the Regiment. The covers of the vases are ornamented with the Crest, and on the reverse of the base are the family Arms of the Reverend Gentleman, whilst round the Salver runs a wreath composed of oak leaves and acorns. The Plate was accompanied by the following letter :—

Weedon, January 3rd, 1850.

My dear Sir,

I do not remember any event throughout the period of my service more gratifying than that afforded by my present pleasurable duty, in transmitting to you the accompanying enduring expression of our grateful feelings, for the manner in which you identified yourself with the Corps under my command on a recent occasion. Your generous sense of the spirit which prompted us to place our Old Colours in the Mother Church of the Chief Town in that Division of the County wherein the Regiment was raised, will not be forgotten by the officers and men now serving, and I confidently indulge the hope that our successors will esteem the simple record of this affair as a most interesting episode in the history of the Old Twelfth. It is difficult to comprehend how a proposal so complimentary to the town of Ipswich, so reverent, and in such fine accord with the genius of our Military Institutions, could fail in receiving the respectful acquiescence, if it did not awaken the enthusiastic sympathy, of the Authority who decided upon its rejection. Was it a fair return to men who by a firm, patient, and loyal attitude have so frequently contributed to maintain public order and safety, to repel their unaffected homage in giving back to the sainted care of the Church the Banners she had consecrated? *Hic pietatis honos?* I am rejoiced to hear that the 23rd Fusiliers met with a happy and honourable accomplishment of a similar wish, and to conclude a letter longer than I purposed, let me again and again thank you for the chivalrous regard you testified in our behalf, which reminds me that I am requested to intimate on the part of the Officers of the Battalion their desire that you be considered *in perpetuum* an Honorary Member of their Mess, where you may always expect their warmest welcome.

Believe me ever,
Faithfully yours,
J. Patton, Lieut.-Colonel.

The Rev. W. N. St. Leger,
Ipswich.

To the above the following reply was transmitted :—

To Lieut.-Colonel J. Patton, Officers, Non-commissioned Officers and Men of the 12th Regiment.

My Brethren,

With deep and unaffected gratitude I receive your splendid testimonial, beautiful and splendid as a costly work of English skill, but chiefly valuable to me for the generous English sentiment which it records. Had I expected any reward when in obedience to the Church I received your consecrated Colours with reverence, and dutifully defended their sacred character when assailed by an ignorant or affected piety, your noble present would have far outgone my largest expectations. Your manly sympathy, my dear Colonel, was to me sufficient encouragement under the difficulties that are past; as the warm-hearted approval which your high-principled and soldierly letter conveys would have been a happy consolation after our defeat on the occasion which you so generously remember. I felt it a high honour to myself, as well as my duty as a Priest, to identify myself with your Corps, not because of the proud position which your Regiment holds in the noble ranks of the noblest Army in the World, but because of that firm, patient and loyal attitude which has so frequently contributed to maintain public order and safety; safety even for those whose base ingratitude you can pity and forgive. Not because your Regiment was the early school of the illustrious Picton, where he strengthened the valour and learned those duties which were at last consecrated by his life's blood in the field of Waterloo by a death which is enshrouded in imperishable fame. Not because of the honours which adorned your venerable Banners, honours which they had unfurled in many a clime, and under many a sky. *Minden, Gibraltar, Seringapatam,* honours as renowned as the hand of Heraldry has ever emblazoned—not for these associations, enough as they are to kindle apathy into enthusiasm, yet not for these did I consider it an honour and a duty to identify myself with your Corps, but because of the patriotic love of Old England and of Home, because of the pious reverence for religion and the Corps—because of the unaffected homage which gave back to the Church of God the sacred Banners which that Church had consecrated. May such holy feeling ever influence the British soldier. May such holy feeling never again be outraged! It will not, I trust, be out of place here to revert for a moment to circumstances which will be dwelt upon with satisfaction by every soldier in the Regiment, and which make some amends for the indignity which an official accident committed; but which the Catholic Church would condemn. Soldiers! the Old Colours of your Regiment which left your loyal ranks with that same unsullied honour which had ever rested upon them from the hour of their consecration, were received in this place with the highest pomp of military ceremonial; recording India amongst your scenes of service, your Colours were escorted to the Mother Church of your native county by Heroes of Sobraon and Aliwal. To the Holy Altar of that Church they were borne by Officers of high distinction, whose chivalrous enthusiasm coveted the honour of carrying the Sacred Banners to the peaceful Sanctuary. The Queen's Colour was borne by an Officer whose decorations told of many a hard-fought field in India. The Regimental Colour was borne by one of illustrious birth, the grandson of Sir Walter Scott, the descendant of the High Priest of Human Nature, and the great Master of the human heart; of him who has bequeathed to us this spirit-stirring character of the British Soldier,—

> "For the rights of fair England his broadsword he draws,
> Her King is his Leader, her Church is his Cause,
> His watchward is honour, his pay is renown,
> God strikes with the gallant, that strikes for the Crown."

Thus were those standards for which you would have shed your dearest blood received, thus were they honoured in the hour of their burial. Reverting, almost of necessity, to those past occurrences, I must notice that epithet, "Emblems of Blood," which at the time was unhappily applied to all Military Colours and which was effectually alleged as a reason why your Colours should not be permitted to remain in my Church. I notice this, not to excite an indignant feeling, but to refute the inconsiderate and injudicious assertion. It is not perhaps worth an argument, but I would briefly remark, that there is no reason against receiving Colours into a Church which will not tell more powerfully against bestowing upon them, in the first instance, a consecrated character, and if Colours be inadmissible, much more should be all accoutrements and weapons of war rejected; and if these be forbidden, *a fortiori* should all soldiers themselves who are sworn, and are ever ready, to use these weapons, be excluded from the Sacred Edifice and from Divine Service. That is, in other and plainer terms, all soldiers should be excommunicated from the Church of God; such is the wicked absurdity of this necessary conclusion. But my Soldier Brethren, as one perhaps as well acquainted with sacred subjects as the authority alluded to, I will take leave to assert, that the false and offensive terms "Emblems of Blood" and "Blood Stained," are no more applicable to Regimental Colours, or to yourselves, than they are to Judicial Robes, or to the venerable Judges of our Country. Soldiers, as well as all Judicial Authorities, are the Servants of Law and Order, are alike Ministers of Justice. The Judge takes his seat on the Bench to administer Justice, to enforce the Law; and if the Law demand the life of the Criminal, to sentence the guilty one to death. Are the robes of the Judge, therefore, emblems of blood, and is he blood-stained and unfit to appear within the sacred walls of the Church? Yet it is that Judge's sentence which sends the convict to the scaffold. In like manner our Army is distributed throughout all our possessions at Home and Abroad, in the protection and défence of the Laws of England. Every soldier is a sworn defender of the Crown, a Minister of Law and Order, and upholder of the peace; the Colours under which he stands, and whose honour he would maintain at the sacrifice of his life, are emblems of religion, for they bear the sign of the cross; they are emblems of Law, Justice and Order. They are not flags of Buccaneers and Murderers; they are not the Standards of a Cromwellian host of renegades and hypocrites; they are Holy Emblems, they are the Standards of Law, of Order and of Peace. Be it ever remembered that Military Colours are not blood-stained emblems, unless the robes of the Judge are blood-stained vestments. In thanking you, my Brethren, I could not express my feelings more briefly. I know you will forgive me if I have been too tedious, you will extend a kind indulgence to the gratitude which your own generous gift has created. Remember, my Brethren, I speak as a Priest, and I speak it parting. Your Colours are Consecrated Banners, they are Sacred Standards, you march under the all-powerful sign of the Cross. May He who died upon the Cross for us sinners, ever defend and protect you through life, through death, and through the awful trial of the Day of Judgment. May His grace, mercy and peace be ever with us, and at the last may he grant to us all, his great salvation. Thus, my dear Brethren, I bid you farewell, as a friend, with kindly words of respect and gratitude; as a Priest, with words of prayer and benediction.

Believe me, my Soldier Brethren, to be,

Ever you sincere and faithful friend and Brother,

Wm. N. St. Leger,

Incumbent, St. Mary Le Tower,

Ipswich,

January 8th, 1850.

THE SUFFOLK REGIMENTAL COTTAGE HOMES

BURY ST. EDMUNDS.

These Homes, which were erected by subscription, and are situated nearly opposite the barrack gates at Bury St. Edmunds, were established for aged or disabled pensioners of the Suffolk Regiment, with their wives and children (if married).

The occupants are admitted on the recommendation of the Officers Commanding 1st and 2nd Battalions, and the Depot, Suffolk Regiment, each being able to appoint a proxy to vote on his behalf.

The building is a very neat construction, and bears the regimental crest carved in relief between the two cottages. A flower garden is provided in front, and a large railed-in portion of ground is allotted at the back for a kitchen-garden.

The Cottage Homes have no connection with the Old Comrades Association of the regiment, formed in 1908.

Lieut.-Colonel S. E. Massy Lloyd has kindly supplied this interesting photograph.

H. I. Jarman] [Photographer.

THE REGIMENTAL COTTAGE HOMES.

BURY ST. EDMUNDS.

INDEX.

E

F

G

H

I

J

K

L

M

R

* Exclusive of R.A., R.E., Cavalry, Foot Guards and Royal Marines.

Numbered Regiments.

Territorial Regiments.

* Exclusive of R.A., R.E., Cavalry, Foot Guards and Royal Marines.

* Exclusive of R.A., R.E., Cavalry, Foot Guards and Royal Marines.

XII
GIBRALTAR 1779-83
MONTIS INSIGNIA CALPE

www.ingramcontent.com/pod-product-compliance
Ingram Content Group UK Ltd.
Pitfield, Milton Keynes, MK11 3LW, UK
UKHW050146280726
14058UKWH00007B/865

9 781843 4211